Soil Biology

Series Editor: Ajit Varma

3

Preface

The inspiration and concept for preparing this volume were conceived while both of us were sipping a cup of Columbian coffee in the Goethe Gallery, Jena (Germany). The idea developed that soils are dynamic biological systems and certainly not a static substrate that supports the life of microbes, plants and animals. Microorganisms play a vital role in creating this universe and maintaining life in it. Dealing with the interrelationships of organisms and the relationships between organisms and their environment was formerly more or less confined to a small group of specialists within the scientific community. This has changed in the current scenario. Growing environmental problems have created a public awareness of the ecological disturbances and dangers related to excessive industrial expansion and the way of life in "disposable societies". As a consequence of the perceived importance of ecology, research in this field has developed rapidly. As one of the three environmental media besides water and air, soils have now become a central concern for a broad range of scientists.

In the golden era of microbiology, the study of soil organisms soon became an area of interest to a large number of early bacteriologists, and the pioneering investigations of Winogradsky, Omeliansky, and Beijerinck still stand as major contributions to our knowledge of the bacterial population. At the same time, it became apparent to soil scientists that the surface crust of the earth is not merely a static physiochemical matrix upon which green plants grow, but also a biological system in a continuous dynamic equilibrium. In the realm of pure science, information on the ecology, function, and biochemistry of microflora has grown considerably so that a clear picture of soil biology is beginning to emerge.

The innumerable developments in recent years make a complete review impossible within the scope of a single volume. Some of the more detailed points have been omitted for brevity, yet, where conflicts still exist, contrasting viewpoints are presented. Time may change these views, but it is in the very nature of science to be in a continual state of flux and for the errors of one generation to be amended by the next. Soil microbiology is not a pure discipline. Its parentage may be traced through bacteriology, mycology, and soil science; biochemistry and plant pathology have also made their mark, especially in recent years.

In the framework of agriculture, the microflora is of significance for man's ability to feed himself. For the microbial inhabitant, the soil functions as a unique ecosystem to which the organism must adapt and from which it must obtain sustenance. However, in the final analysis, microbiologists can find definitive answers as to how these processes are brought about only through biochemical inquiries.

We have attempted to bring together the major aspects of rhizosphere research and principles of rhizosphere ecology for the benefit of developing young scientists and technologies, as well as for the established professional researcher and teacher. A prime objective and hope is that this volume might generate ideas that will bring forth new approaches and methodology leading to further advances in our understanding of rhizosphere interactions and their implications for agriculture and forestry. Nevertheless, even if the rhizosphere is the compartment from which plants acquire their water and nutrients and a hot spot of microbial and animal activity, this compartment can only be understood in the context of whole soil functioning, from soil genesis to the nutrient cycling, and including the exchanges with water and atmosphere. These aspects therefore occupy a large part of the volume.

References are of great value not only to the research worker, but to the advanced student as well. The blind acceptance of secondary sources when primary material is readily accessible is not the hallmark of the serious student. Where available, reviews are included in the reference lists of each chapter so that the finer points of each topic may be sought out. Pertinent original citations are likewise included since these permit the student and researcher alike to examine the original source, observe the techniques utilized, and draw their own conclusions. However, a mere literature review is not intended, since much good work has not been cited. We have deliberately drawn upon some old research information, largely for the benefit of advanced students and young scientists, to show where research has come from and where it may be going. In doing this, we believe we have revealed many gaps in our knowledge which are yet to be filled. Emphasis is given to the more recent papers, but certain classical works are also included, particularly where the studies have been of such a nature as to define a unique approach. It may also be of value to students majoring in other fields, such as soil science, geology, hydrology, plant ecology, zooecology, phytopathology, agronomy, forestry, or the environmental, crop sciences, natural science management, agricultural engineering, biological sciences, animal sciences, and life sciences.

For meaningful contributions to be made in the future, the need for refined technology and a multidisciplinary pooling of expertise by soil microbiologists, phytopathologists, soil physicists and chemists, plant physiologists, and zoologists should be clearly evident.

The presentation is essentially arranged into six main parts. The Introduction or Part I outlines the definition of soils and dynamics to the microbial diversity. Part II deals with varied functions of the microorganisms and soil genesis. In Part III, we highlight the biogeochemical processes. The biotic interactions in terms of plant/microorganisms involving symbiosis are given in Part IV. Functions of microbes in specific soil compartments are discussed in Part V. Modern tools and techniques to understand soil biology are elucidated in Part VI.

It is hoped that the groundwork will be laid herein for a fuller enquiry on the part of the readers. If this goal is achieved even in part, the volume will have served its purpose.

Molecular microbiological studies have focused our attention in recent times on the characterization of known as well as unknown microbial species implicated in soil transformation and plant growth.

While assuming sole responsibility for any omissions or errors in the book, we are grateful to all those unselfish individuals who have contributed to the chapters. Finally, we would like to ask the reader to make allowances for our lack of linguistic proficiency considering that English is not our mother tongue.

We are grateful to the many people who helped bring this volume to light. We wish to thank Dr. Dieter Czeschlik and Dr. Jutta Lindenborn, Springer-Verlag, Heidelberg for generous assistance and patience in finalizing the volume. Finally, specific thanks go to our family, immediate, and extended, not forgetting the memory of those who passed away, for their support or their incentive in putting everything together. Ajit Varma in particular is very thankful to Dr. Ashok K. Chanhan, founder president, An Institute of Ritnand Balved Education Fondation (Amity), New Delhi, for the kind support and constant encouragement received. Special thanks are due to my Ph. D. students Ms. Rina Kamari and Mr. Ram Prasad for compiling the subject index.

Leipzig, Germany
New Delhi, India
June 2004

François Buscot
Ajit Varma

Contents

Part I Introduction

Part IV Biotic Interactions Involving Soil Microorganisms

Part V Function of Microbes in Specific Soil Compartments

Contributors

Azcón, R.
Departamento de Microbiología del Suelo y Sistemas Simbióticos, Estación Experimental del Zaidín, Prof. Albareda 1, 18008 Granada, Spain

Azcón-Aguilar, C.
Departamento de Microbiología del Suelo y Sistemas Simbióticos, Estación Experimental del Zaidín, Prof. Albareda 1, 18008 Granada, Spain

Miguel Barea, J.
Departamento de Microbiología del Suelo y Sistemas Simbióticos, Estación Experimental del Zaidín, Prof. Albareda 1, 18008 Granada, Spain, e-mail: josemiguel.barea@eez.csic.es, Tel: +34-958-181600, Fax: +34-958-129600

Bianciotto, V.
Istituto per la Protezione delle Piante del C.N.R and Dipartimento di Biologia Vegetale dell'Università, Viale Mattioli 25, 10125 Torino, Italy, e-mail: v.bianciotto@ipp.cnr.it, Tel: +39–11–6502927, Fax: +39-11-55839

Binet, F.
Station Biologique de Paimpont, France

Bonfante, P.
Istituto per la Protezione delle Piante del C.N.R and Dipartimento di Biologia Vegetale dell'Università, Viale Mattioli 25, 10125 Torino, Italy

Bonkowski, M.
Technische Universität Darmstadt, Fachbereich Biologie, Schnittspahnstr. 3, 64287 Darmstadt, Germany

Büdel, B.
University of Kaiserslautern, Department of Biology/Botany, P.O. Box 3049, 67653 Kaiserslautern, Germany, e-mail: buedel@rhrk.uni-kl.de

Buscot, F.
University of Leipzig, Institute of Botany, Department of Terrestrial Ecology, Johannisallee 21–23, 04103 Leipzig, Germany, e-mail: buscot@uni-leipzig.de, Tel: +49-341-9738581, Fax: +49-341-9738599

Cappellazzo, G.
Istituto per la Protezione delle Piante del C.N.R and Dipartimento di Biologia Vegetale dell'Università, Viale Mattioli 25, 10125 Torino, Italy

Chotte, J.-L
Laboratoire d'Ecologie Microbienne des Sols, UR Ibis R083, Centre ISRA-IRD Bel Air, BP 1386 Dakar, Sénégal, e-mail: Jean-Luc.Chotte@ird.sn, Tel: +221-849-3308, Fax: +221-832-1675

Deubel, A.
Martin-Luther University Halle-Wittenberg, Institute of Soil Science and Plant Nutrition, Adam-Kuckhoff-Str. 17b, 06108 Halle, Germany

Dilly, O.
Institut für Bodenkunde, Universität Hamburg, Allende-Platz 2, 20146 Hamburg, Germany, e-mail: o.dilly@ifb.uni-hamburg.de, Tel: +49-40-428382010, Fax: +49-40-428382024

Diouf, M.
University of Paris 06, Lab Ecol Sols Trop, UMR 137, IRD, 93143 Bondy, France

Gadd, G.M.
Division of Environmental and Applied Biology, Biological Sciences Institute, School of Life Sciences, University of Dundee, Dundee, DD1 4HN, Scotland, UK, e-mail: g.m.gadd@dundee.ac.uk, Tel: +44-1382-344765, Fax: +44-1382-348216

Garg, A.P.
Ch. Charan Singh University, Meerut, Uttar Pradesh, India

Germon, J.C.
Microbiologie et Geochimie des Sols, INRA-University of Burgundy, 17 rue Sully BP 86510, 21065 Dijon Cedex, France

Gerzabek, M.H.
ARC Seibersdorf Research, Division of Environmental and Life Sciences, 2444 Seibersdorf, Austria

Giang, P.H.
School of Life Science, Jawaharlal Nehru University, New Delhi 110067, India, current address: International Centre for Genetic Engineering & Biotechnology (UNO, Triesta, Italy) New Delhi, India

Giri, B.
School of Life Science, Jawaharlal Nehru University, New Delhi 110067, India

Girlanda, M.
Istituto per la Protezione delle Piante del C.N.R and Dipartimento di Biologia Vegetale dell'Università, Viale Mattioli 25, 10125 Torino, Italy, e-mail: mariangela.girlanda@unito.it, Tel: +39-11-6502927, Fax: +39-11-55839

Gorbushina, A.A.
AG Geomikrobiologie, ICBM, Carl von Ossietzky Universität, Postfach 2503, 26111 Oldenburg, Germany, e-mail: a.gorbushina@uni-oldenburg.de, Tel: +49-441-7983393, Fax: +49-441-7983384

Hobbie, E.A.
Complex Systems Research Center, University of New Hampshire, Durham, New Hampshire 03824, USA, e-mail: Erik.Hobbie@unh.edu, Tel: +1-603-8623581; Fax: +1-603-8620188

Kandeler, E.
Institute of Soil Science, University of Hohenheim, 70599 Stuttgart, Germany, e-mail: kandeler@uni-hohenheim.de

Kersante, A.
Station Biologique de Paimpont, France

Krumbein, W.E.
AG Geomikrobiologie, ICBM, Carl von Ossietzky Universität, Postfach 2503, 26111 Oldenburg, Germany

Kumari, R.
School of Life Science, Jawaharlal Nehru University, New Delhi 110067, India, current address: Ch. Charan Singh University, Meerut, Uttar Pradesh, India

Lavelle, P.
University of Paris 06, Lab Ecol Sols Trop, UMR 137, IRD, 93143 Bondy, France, e-mail: Patrick.Lavelle@bondy.ird.fr

Lazzari, A.
Istituto per la Protezione delle Piante del C.N.R and Dipartimento di Biologia Vegetale dell'Università, Viale Mattioli 25, 10125 Torino, Italy

Merbach, W.
Martin-Luther University Halle-Wittenberg, Institute of Soil Science and Plant Nutrition, Adam-Kuckhoff-Str. 17b, 06108 Halle, Germany, e-mail: merbach@landw.uni-halle.de, Tel: +49-345-5522421, Fax: +49-345-5527113

Oelmueller, R.
Institutes fur Allgemeine Botanik, Dornburger Str 159, 07743 Jena, Germany

Perotto, S.
Istituto per la Protezione delle Piante del C.N.R and Dipartimento di Biologia Vegetale dell'Università, Viale Mattioli 25, 10125 Torino, Italy

Philippot, L.
Microbiologie et Geochimie des Sols, INRA-University of Burgundy, 17 rue Sully BP 86510, 21065 Dijon Cedex, France

Prasad, R.
Ch. Charan University, Meerut, Uttar Pradesh, India

Rouland, C.
University of Paris 06, Lab Ecol Sols Trop, UMR 137, IRD, 93143 Bondy, France

Ruess, L.
Technische Universität Darmstadt, Fachbereich Biologie, Schnittspahnstr. 3, 64287 Darmstadt, Germany

Sachdev, M.
School of Life Science, Jawaharlal Nehru University, New Delhi 110067, India

Scheu, S.
Technische Universität Darmstadt, Fachbereich Biologie, Schnittspahnstr. 3, 64287 Darmstadt, Germany, e-mail: scheu@bio.tu-darmstadt.de, Tel: +49-6151-165521, Fax: +49-6151-166111

Stemmer, M.
Institute of Soil Research, University of Agricultural Sciences, 1180 Vienna, Austria

Tebbe, C.C.
Institut für Agrarökologie, Bundesforschungsanstalt für Landwirtschaft (FAL), Bundesallee 50, 38116 Braunschweig, Germany, (e-mail: christoph.tebbe@fal.de

Varma, A.
School of Life Science, Jawaharlal Nehru University, New Delhi 110067, India, e-mail: ajitvarma73@hotmail.com, Tel: +91-26704511, Fax: +91-26187338/26198234, current address: Amity Institute of Herbal & Microbial Studies, Sector 125, New Super Express Highway, Noida, India, Tel: 95120-2432400, Fax: 95120-2432200

Part I
Introduction

1 What Are Soils?

François Buscot[1]

1 Introduction

Soil is often defined as the earth surface layer exploited by roots. This kind of definition is not the most appropriate to introduce a volume on soil microorganisms as these are also found in soil compartments not colonized by roots. Making this point, Paul and Clark (1996) gave the examples of microbial denitrification, observed much below the rooting depth, and of the numerous bacteria and fungi that colonize small pores and micro-aggregates not accessible to roots or even root hairs.

Another definition (see, for example, Wild 1993) refers to soil genesis by mentioning the intervening factors, i. e., the parent material, the relief, the climate, the organisms involved and time. This approach is more appropriate as it emphasizes that soil edification is a biologically driven process. It also inherently points out the complexity and especially the heterogeneity of the resulting medium. In order to mark the diversity that results from combining the interaction of very diverse and complex organism communities on different types of rock material under variable climatic and topographic conditions and over a time scale (the unit of which may vary from decades to thousands or even millions of years), many soil scientists avoid using the term "soil", but prefer to speak of "soils".

The line that results from genesis-based definitions of soils and from the complex and heterogeneous soil functions that such definitions underline has deeply inspired the structure of this volume. If we follow this line, we have to consider at least five questions when addressing the fate of soil microorganisms:

- What are the functions of microorganisms in soil genesis?
- What are the roles played by microorganisms in the energy and matter fluxes and in their transformation within functioning soils?

[1]University of Leipzig, Institute of Botany, Department of Terrestrial Ecology, Johannisallee 21–23, 04103 Leipzig, Germany, e-mail: buscot@uni-leipzig.de, Tel: +49-341-9738581, Fax: +49-341-9738599

Soil Biology, Volume 3
Microorganisms in Soils: Roles in Genesis and Functions
(ed. by F. Buscot and A. Varma)

- As the soil genesis and functioning involve complex and tightly integrated bioceonoses, in which kind of biotic interactions do the soil microorganisms participate?
- What is the function of microorganisms in specific domains of soils that are highly influenced by biotic or abiotic factors?
- Finally, considering that soils are difficult media to work on, especially for microbiologists, which approaches can be used by soil microbiologists, taking the wide structural and functional diversity of soil microbes into account, but avoiding going too far into details that do not provide explanations for emergent properties and processes characteristic of soils?

In the introductory chapter, we will summarize some basic traits of soil genesis and functioning and try to indicate at which stages the processes that are detailed in the different parts and chapters of the book are involved. This first chapter will not replace general soil science books (see, for example, Brady 1990), but aims to be a guideline providing an integration of the matter detailed in the book and pay the correct tribute to the role played by microorganisms for soil genesis and functions.

2 Soil Genesis

2.1 Rock Weathering or Decay

At the beginning of each soil formation, for example, after a volcanic eruption or a glacier or water retreat, the initial mother substrate, in general, displays a reduced capacity to immediately carry an abundant plant and animal biocoenosis. Up to this stage, however, microorganisms such as bacteria, algae and their associations with fungi in biofilms of lichens belong to the early colonizers (see Chap. 2). If the basic substrate is loose, the microbial community, constituting biological crusts, will provide stabilization and avoid erosion. Such crusts form also at the surface of some developed soils. They are analyzed in detail in Chap. 15.

Moreover, if the mother substrate consists of a hard rocky material such as granite or limestone, the initial process of soil formation consists of weathering. Both basic mechanisms of weathering, i. e., the substrate fractionation and its gradual chemical transformation, are bound together. Fractionation enhances the contact surface between substrate and environment, which, in turn, increases chemical reactivity and transformation rate. Each time a monolith is divided into 1000 fractions, the ratio between its surface and its volume increases by an order of magnitude of at least 10. Such a surface increase factor may appear low. However, one has to keep in mind that the smallest particles resulting from weathering are clay minerals that have an equivalent diameter less than 2 μm. At this stage of fractionation, the contact surface is tremendous. According to the mineral type, 1 g of clay has a surface varying from 93 to 800 m^2 (Gisi et al. 1997). As clay particles are negatively loaded, they display a considerable potential for binding and exchanging cations that may be crucial nutrients, but also toxic substances such as heavy metals (see Chap. 16).

In Chap. 3, Gorbushina and Krumbein point out two important traits of weathering and its consequences. The first trait is that even if the basic mechanisms of weathering are of a physical and chemical nature, they are largely biologically driven, with a predominant role of the microorganisms, especially at the initial stage. To take this biological component into account, the authors propose that "wear down" might be a more appropriate term than weathering. This view is in agreement with van Breemen et al. (2000), who has recently used the term "rock eating fungi". The second trait concerns the exchange surface that results from substrate fractionation. The idea here is that not the classical surface of the terrestrial ecosystem, but its fractal surface, i. e., the real contact interface of soils with water and air, should be considered. This radically inverts our current view that the exchange surface between the atmosphere and oceans is higher than the one with the terrestrial component. As microorganisms represent the widest biota fraction in soils, both in terms of biomass and number of organisms, and as they are tightly associated with this tremendous fractal surface, they play a key role in biogeochemical cycles including those of climate-relevant gases. The importance of this role is even reinforced when considering the carbon flux driven by so called chemolithotroph bacteria that mobilize important quantities of CO_2 from the atmosphere during their attack on rocks. As the time scale at which such processes operate is very high, they are to be considered as a driving force of geomorphological processes. The consequences of wear down are therefore much broader than the sole contribution to soil edification. It influences the geology and the climate (see Chap. 3).

2.2 Importance of Soil Texture

In the soil matrix, the solid phase of soils, the skeleton and the fine earth correspond to particle fractions with an equivalent diameter higher and lower than 2 mm, respectively. The fine earth is split into three particle size fractions, the sand fraction with an equivalent diameter of 50 or 63–2000 µm, the silt fraction (2–50 or 63 µm) and the already mentioned clay fraction (< 2 µm). The proportion of each fraction of the fine earth defines the soil texture which is a crucial property as it determines at two levels the volume available for the two other soil phases, the gaseous (soil–air) and the aqueous ones (soil–water or soil solution). Sandy soils not only have a higher total available volume for water and air, they warrant a better water percolation and evaporation, resulting in rapid shifts of soil moisture versus soil aeration. In contrast, clay soils have numerous capillary pores that retain water and lower aeration and water circulation. These properties are determinant for microorganisms themselves, as they influence the balance between oxidative and reductive biological processes which drive biogeochemical cycles in soils (see Chaps. 7–9). Soil texture and soil pore size are also important as they rule the distribution of soil organisms. The classification of organisms used by soil biologists refers to the dimension of soil particles and soil pores (Fig. 1).

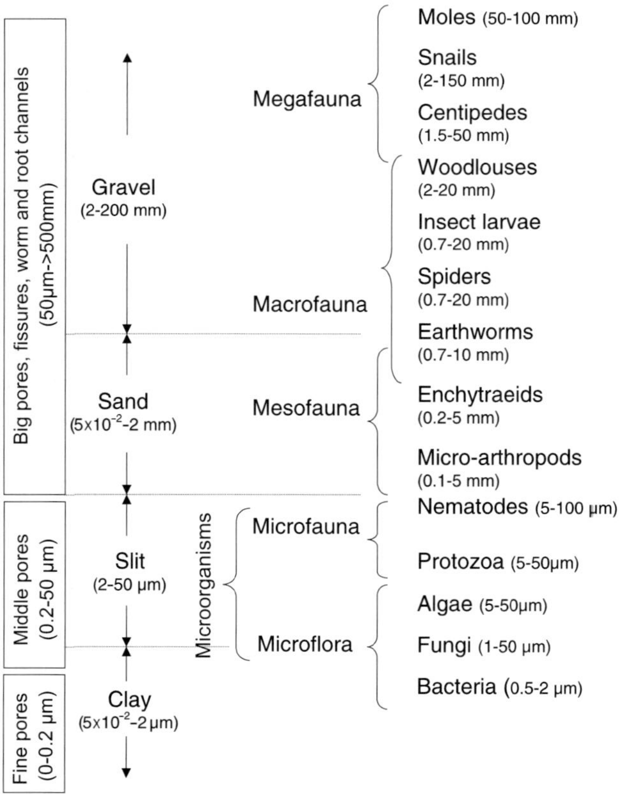

Fig. 1. Classification of soil biota in relation to size of pores and particle in soils used in soil biology. (Adapted from Gisi et al. 1997)

2.3
Input of Organic Matter into Soils and Aggregation

After the wear down or weathering of the initial substrate that results in producing a matrix with a huge reactive surface and determines complex niches for the soil organisms, the second important event in soil genesis is the input and transformation of organic matter. This organic matter originates from the organisms that establish themselves on and within soils. Here, two mechanisms are balanced that are detailed in Chap. 4. On the one hand, organic matter may be mineralized, resulting in production of CO_2, phosphate, sulfate, nitrate, etc. that may be remobilized by the soil biota, undergo further microbial transformation through oxidation or reduction, or eventually leave soils as gas or by transfer into the groundwater (for details, see Chaps. 6–9). On the other hand, it may only be partially decomposed into more or less complex organic radicals that polymerize and form humus, a very stable complex component of soils. Both the mineralization and humification processes are driven by soil bacteria and fungi and involve a broad set of enzymes with either narrow substrate specificity like cellulases or, on the contrary, the ability to attack a diversity of substrates. This is the case for oxidative exo-enzymes such as laccases (Luis et al. 2004). Although humus rarely represents more than several percent of the soil matrix, it confers many properties to soils. Due to its colloidal structure, it increases the soil water holding capacity and participates in the formation of aggregates with soil minerals, contributing to building what is called the soil structure. Like clay, humus particles are loaded negatively and bind and exchange cations. Therefore, this fraction influences soil fertility, especially in many soils of the tropics, where the clay fraction is less stable than in temperate region.

In addition to an indirect role in aggregation and soil structure via their contribution to humification, soil microorganisms also act directly. Bacteria and fungi exude colloidal polysaccharides that can glue soil particles. Soil fungi, for example, produce glomalin, which has been demonstrated to represent a high proportion of soil organic matter promoting aggregate formation (Rillig et al. 2002). The mechanical role of microorganisms is also considerable, given their biomass of 40–200 $g\,m^{-2}$ and their hyphal structure (Dighton and Kooistra 1993; Thorn 1997) that contributes to anchoring soil components to each other. To pay tribute to this phenomenon, Chap. 5 is devoted to the role of microorganisms in soil aggregation.

2.4 Migration Processes

A third basic phenomenon involved in soil formation is the differential migration of components, which result in forming complex horizons the chemical composition of which may be very different to both that of minerals delivered by the mother rock substrate and the organic components originating from the biota. Microorganisms are not involved directly in the migration processes themselves as these result from a mass flow of particles suspended in the soil solution under the driving force of gravitation and from diffusion of ions along concentration gradients. However, microbes play a role in the decay and solubilization processes delivering the fine particles or ions that migrate. By synthesizing enzymes and organic acids, they also directly influence the oxidation level and the solubility of components that determine their mobility across or their precipitation within soil horizons. Another indirect role of microorganisms is their association to soil animals often called the “soil engineers” that are actively involved in bioturbation. This aspect is treated in Chap. 12.

3 Biogeochemical Processes in Soils

As soils are one of the three environmental media besides air and water, they are involved in biogeochemical processes not only during their genesis, but also once they are formed. At the genesis stage, the input of carbon and energy and the bioavailable forms of nitrogen and phosphorus is crucial.

3.1 Energy and Carbon

As for all ecological niches, soils need energy for their biota. Most of this energy comes indirectly from the sun via the primary producers. These deliver energy-rich organic compounds into soils either in the form of litter or by direct exudation through the roots. As enlightened in Chaps. 6 and 7, microbes are highly involved in energy and carbon cycling in soils. According to Smith and Read (1987), soil fungi forming mycorrhizal symbioses with plant roots actively contribute to directly derive 10–20% of the photoassimilation by plants into soils (see also Chaps. 11 and 14).

The fate of microbial energy in soils is especially complex and integrated. Soils are characterized by a high variety of energy-carrying substrates and of redox and pH conditions. Consequently, a diversity of energy-related metabolic processes in which more or less specialized microorganism groups are involved coexist in each soil, where they are heterogeneously distributed in space and undergo succession in time. Finally, various compounds with biogeochemical and climatic relevance are released from soils into the atmosphere. The most important are carbon dioxide, methane, diverse nitrous oxides such as N_2O, NO and NO_2, often cited under the term NO_x. Information on the involved microorganisms and their functions is given in Chaps. 6–8.

In Chap. 6 we learn that, globally considered, life in soils is very energy limited. Mean rates of two to four divisions per year for bacteria reflect that energy economy of soil microorganisms differs highly from that known from classical microbiology acting in a Petri dish or bioreactors. In this context, two points should be considered. First, the biotic interactions discussed in Part D are often found in soil compartments with particular energetic contexts. This is the case for the microorganisms associated with the rhizosphere (Chaps. 11 and 14) or with members of the micro- and mesofauna, but also for specific groups such as chemolithotrophic bacteria that gain their energy autotrophically from mineral substrates without photosynthesis (Bartosch et al. 2002). Second, there are methodological consequences, as studying the functions of soil microorganisms *in vitro* may lead to misinterpretations. Development of *in situ* investigation methods such as the molecular approaches reported in Chap. 17 is, therefore, of special importance. These methods are extremely sensible and precise, but the disadvantage of the scale at which they operate is that their results are difficult to integrate in up-scaling procedures. Therefore, analytical techniques that allow the measurement of microbially driven integrated functions, in addition to detailed molecular analyses, are of special importance. Besides classical soil biological methods to measure sum parameters (see, for example, Ritz et al. 1994 or Shinner and Sonnleitner 1996), techniques based on specific fractionating of natural isotopes have recently appeared to be an important tool for elucidating functions of soil microorganisms in biogeochemical cycles and integrating them at higher scales. These techniques are also important to detect through which microbial pathways elements are transformed in soils. Hobbie et al. (2001), for example, showed that using specific fractionation rates of natural N isotopes allows us to determine whether nitrogen is mobilized by plants via mycorrhizal fungi or directly from the soil. In fungi, analyses of C isotope fractionation allow us to determine whether they gain their carbon as saprophytes or as symbionts. Chapter 18 (Part VI) is devoted to soil ecological investigations using the isotope fractionation technique.

3.2 Nitrogen and Phosphorus

Nitrogen and phosphorus, two essential biological elements, are good examples to illustrate how diverse and specific the entering and cycling of nutrients during formation and function of soils may be.

Molecular nitrogen (N_2) constitutes about 79% of the atmosphere. As geological substrates are poor in nitrogen, they cannot deliver sufficient amounts of this key element to meet the biological demand in soils. In fact, almost the total soil nitrogen pool originates from the atmosphere and has been originally fixed by soil bacteria alone or in symbiosis with plant roots or fungi. Biological nitrogen fixation consists in a reduction of N_2 to NH_4^+ (ammonium). This biological reduction is impressive in that to create an identical reaction chemically, the industrial process developed by Haber-Bosch in the nineteenth century requires temperatures of 500 °C and a pressure of about 200 bars, which represents a tremendous amount of energy. The biological N_2 fixation is also highly energy consuming as it requires 28 ATP or 675 kJ to produce two NH_4^+. To fix 1 g N, nodulating plants have to provide 12 g glucose to their symbiotic bacterial partners.

Once nitrogen has entered the food chain, it is transformed into organic compounds such as amino acids, nucleotides or adenosine and also structures molecules like chitin and lignin which are difficult to break down. When they return to soils as litter, such organic nitrogen forms are partly recycled by soil microorganisms in the course of the ammonification and nitrification processes detailed in Chap. 8. The rest is more or less durably immobilized in soil humus. A fraction of the recycled nitrogen can leave soils as nitrates due to out-washing into the ground water or as gaseous NOx after denitrification by soil bacteria operating in reductive soil compartments. A part of these nitrogen losses is balanced by *de novo* fixation of nitrogen from the atmosphere and, in the mean time, by fertilization. Soils are especially sensible to nitrate outwashing, because this anion is very mobile and its negative charge does not allow its adsorption onto the negatively loaded clay or humus particles. Apart from nitrate outwashing, all steps of the global nitrogen cycling are driven by microorganisms, most of which live in soils. This complex cycle is detailed in Chap. 8.

In soils, the primary phosphorus sources are the orthophosphates released by weathering of mother rock compounds like apatite. After its release, phosphate can be absorbed by organisms and included in organic compounds, bound on oxide–hydroxides surfaces or precipitated as salts. Without input through fertilization, the P concentration in the soil solution is in the order of 10^{-6} mol l^{-1} which is much too low to meet the biological demand of plants and soil microorganisms for this essential element. In addition, phosphorus has a low mobility, so that depletion zones occur around absorbing biological structures such as roots (see Chap. 11). Soil microorganisms are involved at all levels of the management with this basic P limitation in soils. According to the classical scheme proposed by Stewart and Sharpley (1987), four pools contribute to compensate the phosphorus uptake from the soil solution: the stable and labile organic soil-P and the stable and labile mineral soil-P. Directly or indirectly, microorganisms are involved in the fluxes between these P pools. They are directly involved in the mineralization of organic compounds, and at least indirectly in the mobilization from both mineral pools. This complex cycle is presented in detail in Chap. 9. The general phosphorus limitation also has consequences for biotic interactions in soils. With few exceptions, plants alone are not able to gain enough P from soils, but do this in association with mycorrhizal fungi (Read and Perez-Moreno 2003). Traits of this symbiosis, which was probably essential for plants to colonize terrestrial habitats (Redecker et al. 2000), are explained further in Chap. 11. Of special importance is that the different biological pathways in the N and P cycles are not independent. As elucidated in Chap. 10, there are numerous reciprocal facilitative interactions between bacteria-fixing N_2 and mycorrhizal fungi involved in the P mobilization from the soils.

4 Biotic Interactions Involving Soil Microorganisms

4.1 Competition Versus Facilitation

Biotic interactions involving soil microorganisms are discussed in Part IV as well as in Part II with regard to soil genesis and in Part III with regard to biogeochemical processes. Here, some additional aspects should, however, be underlined. Within classical ecological concepts that were largely inspired from observations or experimentation on plants and animals, competition processes play an essential role (see, for example, Begon et al. 1996). Mutualistic to neutral symbioses in the initial sense of de Bary were, for a long time, difficult concepts to handle for ecologists (see, for example, Boucher 1985). On the one hand, numerous pathogenic microbes inhabit soils (e.g., Renker et al. 2003), and there is great competition between soil microorganisms for the sparse nutrient and energy resources. On the other hand, soil microorganisms are also involved in diverse biotic interaction processes, some of which modify emergent soil properties. The facts presented in Chaps. 10–12 (Part IV) and Chaps. 13–16 (Part V) demonstrate how important such interactions are for the key processes of soil formation and function, and for the adaptation of higher organisms to life in the soil, which constitutes the sum of heterogeneous and complex microcompartments.

4.2 The Example of Mycorrhizas

According to Douglas (1994), an important trait of symbioses taken in a broad sense is that they may provide metabolic capabilities that the partners do not possess intrinsically. This consideration applies especially within the soil compartment. A good illustration is given by Read (1993) in his hypothetical interpretation of how different types of mycorrhiza mediate the adaptation of more or less diverse plant biomes to the limiting resources in their respective soil contexts. In forest soils of boreal and temperate regions, nitrogen is not only limiting, but also distributed heterogeneously and under a wide range of forms. In this context, the occurrence of a relatively species-poor tree vegetation can only be explained by the fact that a diversity of ectomycorrhizal fungi, i.e., several thousands of species of the asco- and basidiomycetes can mobilize the variable forms of nitrogen, before homogenizing, centralizing and finally distribut-

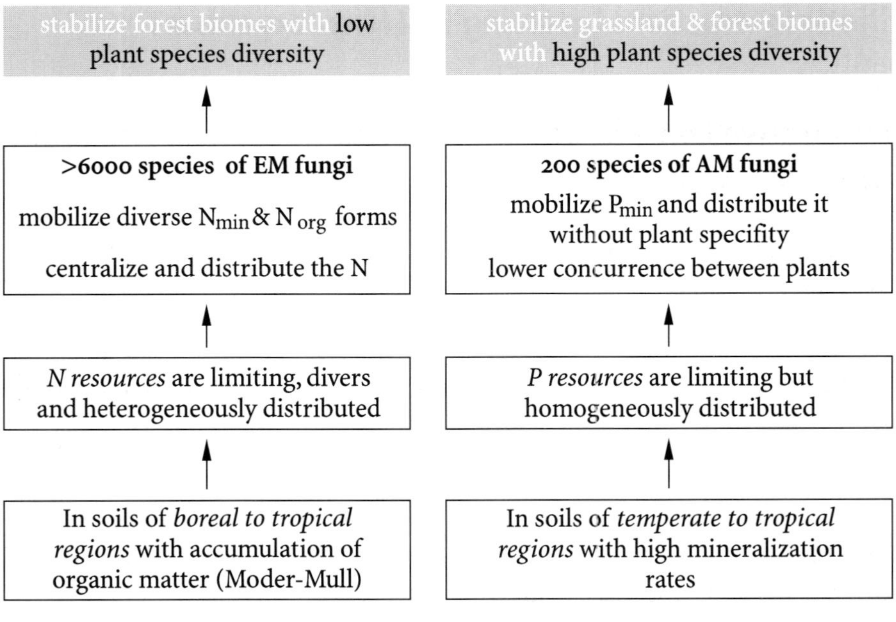

Fig. 2. Method by which ecto- (*EM*) and arbuscular mycorrhizas (*AM*) provide a link allowing adaptation of different types of plant biomes to their respective soil context and influence the diversity of their constituting plant species. (After Read 1993)

ing it to the trees via a network of mycelium and rhizomorphs. Without this network mediating adaptation to the heterogeneous medium, such soils would be colonized by a diversity of small plants specialized for each micro-niche, but not as it is the case by trees sometimes building almost mono-specific forests over thousands of square kilometers. For species-rich plant communities such as grasslands or the tropical forests that grow on well-mineralized soils, the limiting factor is phosphorus. Here, a few hundred species of arbuscular mycorrhizal fungi (AMF), compatible with a broad spectrum of plant partners, mediate the mobilization and equal distribution of this element, lowering the competition between plants and thus promoting high plant diversity. Without this mediating agent, vegetation with few species highly competitive for the P mobilization under the respective climatic and geologic conditions would normally be expected to establish (Fig. 2). Most recent works have questioned the species concept (Clapp et al. 2001) and the lack of host plant specificity (van der Heijden et al. 1998; Vandenkoornhuyse et al. 2002, 2003) of arbuscular mycorrhizal fungi. Ongoing investigations in the tropics also describe abundant presence of AMF in organic forest soils (Kottke et al. 2004), so that the scheme proposed by Read might have to be nuanced (i.e., that EM come rather in organic and AM rather than in mineral soils or horizons), but there is no doubt that mycorrhizas mediate adaptation of the plant community to the different soil contexts.

5 Integrative Considerations on Functions of Microorganisms in Specific Soil Compartments

The importance of biotic interactions in soils and the high involvement of microorganisms both justify the interest in analyzing the behavior of integrated soil microbial communities as a whole. This can be approached by investigating the influence of man-made or natural disruptions in such communities, or by trying to integrate single facts in comprehensive models. Both approaches are addressed in Part V.

5.1 Release of Transgenic Organisms as a Tool to Trace Effects of Ecological Disruptions on Soil Microorganisms

The release of transgenic plants or microorganisms into soils questions both their ability to establish themselves permanently and to interact with the indigenous soil biota. Apart from problems related to the biological hazard, such a release constitutes an experimental system of choice for investigating the ecological disruption that the import of allochthonous organisms into soils represents and for elucidating some mechanisms ruling the function of microorganisms in soil compartments. In the first release experiment of transgenic forest trees in Europe, the analysis of mycorrhizal patterns in trees was part of the accompanying biological monitoring program. Although the transformation modified the plant hormonal balance, a factor known to have influence on mycorrhizal associations (Herrmann et al. 2004), it had no direct impact on the mycorrhizal pattern in the field, at least at the initial stage (Kaldorf et al. 2002). However, the analysis revealed a clone-specific, reduced compatibility for one particular mycorrhizal fungal partner in one of the transgenic tree lines and, moreover, allowed characterizing patterns of site colonization by mycorrhizal fungi in young tree plantations (Kaldorf et al. 2003). Chapter 13 reviews this aspect in more depth by investigating the impact that released transgenic bacteria may have on rhizospheric fungi.

5.2 Soil Pollution by Heavy Metals as a More Complex Disruption

Heavy metal pollution in soils constitutes a disruption of ecological equilibrium with a higher level of complexity, and it is therefore worth investigating how soil microorganisms deal with this kind of pollution. Soils naturally contain not only a broad diversity of metallic elements; in addition, each metal may be present at variable concentrations and under different chemical species. While some metals have no biological relevance, others belong to essential trace elements that, however, become toxic over a certain concentration level. As soil metals are often in an ionized form, they react with the negatively loaded soil particles so that not only their concentration, but also their bioavailability is relevant. The resulting situation is that soil biota must permanently regulate activities not only to make available and take up the necessary concentration of essential metals, but also to exclude or detoxify detrimental forms or concentrations. In particular, the soil microorganisms must display great physiological adaptation. Taking the space and time variability of soils into account, the selection pressure resulting from the metal status in soils probably constituted a motor for the adaptation of physiological pathways in soil microorganisms and for their evolution. This is one example of the soil medium complexity that may explain why the biodiversity of soil microorganisms is so high. The fate of adaptation of soil microorganisms in heavy metal-polluted soils is discussed in more depth in Chap. 16.

5.3
Understanding Complex Functional Domains in Soil Habitats

As shown by Bonkowski et al. (2000), the rhizosphere and invertebrates constitute functional domains of soils that are inhabited by a diversity of microorganisms (see also Chap. 12). Biological crusts formed at the interface between soil surface and atmosphere offer a further example of a functional domain with complex and integrated microbial communities (see here Chap. 15). Such domains are characterized by multilevel interactions that cannot be explained by reductionist approaches. Chapter 14 proposes a conceptional model to understand the biological functions in one of such domains. According to this model, a hierarchy of factors at different space and time scales rules the general microbial activities in soils. The microbial soil community constitutes a dormant metabolic potential that needs input of energy and substrate sources to be activated. Bacteria in particular cannot move toward such sources, therefore, the focus of their activities is limited to energy- and nutrient-rich functional domains such as the rhizosphere or the domains provided by animals of the meso- and macrofauna acting as soil engineers. Plant roots and soil animals in turn have developed several strategies such as predation or mutualism to adapt to this microflora.

The elucidation of the role of specific microorganism groups in soil functional domains is not only important to understand basic traits of soil functions. The resulting knowledge can be used for practical issues. For example, efforts to restore man-disturbed ecosystems increasingly consider the importance of target groups of the soil microflora acting in functional domains such as mycorrhizas and the possibility of manipulating their species composition to influence the reestablishment of a sustainable plant cover (Renker et al. 2004).

6
Conclusion or Back to Biodiversity of Soil Microbes

The goal of this first introductory chapter was to present a general view of what soils are, i. e., a complexity of media that form slowly from mineral substrates under the involvement of organisms and the influence of climate. With respect to their participation in the soil biomass, microorganisms play a crucial role, e. g., at the microscopic level of soil micro-aggregates to global cycling of elements such as C and N. This role involves a complex network of biotic interactions between the microorganisms themselves and the other members of the edaphon (plants and animals). Functional groups and soil functional domains have to be considered in studies aiming to understand the complex ecological role of microbes in the edification and function of soils and have historically influenced the specific approach of soil microbiology as a scientific discipline. The second chapter of this introductory section reviews the diversity of soil microorganisms with regard to classical systematics and the differentiation into functional groups or inhabitants of functional domains in soils.

Acknowledgements. The financial support by the German Science Foundation (Deutsche Forschungsgemeinschaft) in different research projects dealing with soil biological aspects (especially grant numbers Bu 941–1/1–3 and Bu 941–2/1–2) is gratefully acknowledged. I am indebted to my friend Professor Ajit Varma (Jawaharlal Nehru University, New Delhi, India) for his collaboration in this book project. Finally, I would like to dedicate this chapter to my mentor, Prof. Dr. Jean Charles Munch (GSF München, Germany), who not only provided guidance during my habilitation, but also aroused my interest in soil biology.

References

Bartosch S, Hartwig C, Spieck E, Bock E (2002) Immunological detection of nitrospira-like bacteria in various soils. Microb Ecol 43:26–33

Begon ME, Harper JL, Townsned CR (1996) Ecology, 3rd edn. Blackwell, Oxford

Bonkowski M, Cheng W, Griffiths BS, Alphei J, Scheu S (2000). Microbial-faunal interactions in the rhizosphere and effects on plant growth. Eur J Soil Biol 36:135–147

Boucher DH (1985) The idea of mutualism, past and present. In: Boucher HD (ed) The biology of mutualism. Croom Helm, London, pp 1–28

Brady NC (1990) The nature and properties of soil, 10th edn. Macmillan, New York

Clapp JP, Rodriguez A, Dodd JC (2001) Inter- and intra-isolate rRNA large subunit variation in *Glomus coronatum* spores. New Phytol 149:539–554

Dighton J, Kooistra M (1993) Measurement of proliferation and biomass of fungal hyphae and roots. Geoderma 56:317–330

Douglas AE (1994) Symbiotic interactions. Oxford Univ Press, Oxford

Gisi U, Schenker R, Schulin R, Stadelmann F, Sticher H (1997) Bodenökologie, 2nd edn. Thieme, Stuttgart

Herrmann S, Oelmüller R, Buscot F (2004) Manipulation of the onset of ectomycorrhiza formation by indole-3-acetic acid, activated charcoal or relative humidity in the association between oak microcuttings and *Pilderma croceum*: influence on plant development and photosynthesis. J Plant Physiol 161:509–517

Hobbie EA, Weber NS, Trappe JM (2001) Mycorrhizal vs saprotrophic status of fungi: the isotopic evidence. New Phytol 150:601–610

Kaldorf M, Fladung M, Muhs H-J, Buscot F (2002) Mycorrhizal colonization of transgenic aspen in a field trial. Planta 214:653–660

Kaldorf M, Renker C, Fladung M, Buscot F (2003) Characterization and spatial distribution of ectomycorrhizas colonizing aspen clones released in an experimental field. Mycorrhiza 14:295–306

Kottke I, Beck A, Oberwinkler F, Homeier J, Neill D (2004) Arbuscular endomycorrhizas are dominant in the organic soil of a neotropical montane forest. J Trop Ecol 20:125–129

Luis P, Walther G, Kellner H, Martin F, Buscot F (2004) Diversity of laccase genes from basidomycetes in a forest soil by. Soil Biol Biochem 36:1025–136
Paul EA, Clark FE (1996) Soil microbiology and biochemistry, 2nd edn. Academic Press, San Diego
Read DJ (1993) Mycorrhiza in plant communities. In: Ingram DS, William PH, Tommerup IC (eds) Advance in plant pathology. Academic Press, London, pp 1–31
Read DJ, Perez-Moreno J (2003) Mycorrhizas and nutrient cycling in ecosystems – a journey toward relevance? New Phytol 157:475–492
Redecker D, Kodner R, Graham LE (2000) Glomelean fungi from the Ordovician. Science 289:1920–1921
Renker C, Alphei J, Buscot F (2003) Soil nematodes associated with the mammal pathogenic fungal genus *Malassezia* (Basidiomycota: Ustilaginomycetes) in central European forests. Biol Fertil Soils 37:70–72
Renker C, Zobel M, Öpik M, Allen MF, Allen EB, Buscot F (2004) Structure, dynamics and restoration of plant communities: do arbuscular mycorrhizae matter? In: Temperton W, Hobbs R, Nuttle T, Halle S (eds) Assembly rules and restoration ecology. Bridging the gap between theory and practice. Island Press, Washington, DC, pp 189–229
Rillig MC, Wright SF, Eviner VT (2002) The role of arbuscular mycorrhizal fungi and glomalin in soil aggregation: comparing effects of five plant species. Plant Soil 238:325–333
Ritz K, Dighton J, Giller KE (1994) Beyond the biomass. Wiley, Chichester, 275 pp
Shinner F, Sonnleitner R (1996) Mikrobiologie und Bodenenzymatik: Grundlagen, Klima, Vegetation und Bodentyp. Springer, Berlin Heidelberg New York
Smith SE, Read DJ (1997) Mycorrhizal symbiosis, 2nd edn. Academic Press, San Diego
Stewart JWB, Sharpley AN (1987) Controls on dynamics of soil and fertilizer phosphorus and sulfur. SSSA Spec Publ 19:101–121
Thorn G (1997) Modern soil microbiology. In: Van Elsaß JD, Trevors JT, Wellington EMH (eds) Modern soil microbiology. Dekker, New York, pp 63–128
Van Breemen N, Lundström US, Jongmans AG (2000) Do plants drive podzolization via rock-eating mycorrhizal fungi? Geoderma 94:163–171
Van der Heijden MG. A, Klironomonos JN, Ursic M, Moutoglis P, Streitwolf-Engel R, Boller T, Wiemken A, Sanders IR (1998) Mycorrhizal fungal diversity determines plant biodiversity, ecosystem variability and productivity. Nature 396:69–72
Vandenkoornhuyse P, Husband R, Daniell TJ, Watson IJ, Duck JM, Fitter AH, Young JPW (2002) Arbuscular mycorrhizal community composition associated with two plant species in a grassland ecosystem. Mol Ecol 11:1555–1564
Vandenkoornhuyse P, Ridgway KP, Watson IJ, Fitter AH, Young JPW (2003) Co-existing grass species have distinctive arbuscular mycorrhizal communities. Mol Ecol 12:3085–3095
Wild A (1993) Soils and environment. Cambridge Univ Press, Cambridge

2 Microbial Diversity in Soils

Bhoopander Giri[1], Pham Huong Giang[2], Rina Kumari[3],
Ram Prasad[3], Ajit Varma[1]

1 Introduction

Soil microbiology emerged as a distinct branch of soil science in 1838 after the French agricultural chemist and farmer, Boussingault, showed that legumes could obtain nitrogen from air when grown in soil which was not heated. Fifty years later, a Dutch scientist, Beijerinck, isolated bacteria from nodules of legume roots. Since then, a number of investigations have been conducted in the area of soil microbiology. However, scientists are still investigating soil microbial diversity.

Soil is the outer covering of the earth, which consists of loosely arranged layers of materials composed of inorganic and organic compounds in different stages of organization (Tate 1995; Kapoor et al. 2002). It is a natural medium in which microbes live, multiply and die. Microbial diversity in the soil is a critical environmental topic that concerns people from all walks of life. Interest in microbial diversity has grown rapidly in the scientific community (Wilson 1988; Franklin 1993; Benizri et al. 2002). Increasing attention is being drawn to microorganisms because the fertility of soil depends not only on its chemical composition, but also on the qualitative and quantitative nature of microorganisms inhabiting it. Maintenance of viable, diverse populations and functioning microbial communities in the soil is essential for sustainable agriculture (Beare et al. 1995; Benizri et al. 2002). Soil contains a wide range of microorganisms descried as a 'black box' (Paul and Clark 1989).

Microorganisms are generally divided into five major taxonomic categories: algae, bacteria, fungi, protists and viruses (Prescott et al. 1996; Hurst 2002). In soil, they are closely associated with soil particles, mainly clay–organic matter complexes (Foster 1988). Often, microbes can be found as single cells or as microcolonies embedded in a matrix of polysaccha-

[1]School of Life Science, Jawaharlal Nehru University, New Delhi 110067, India and Amity Institute of Herbal and Microbial Studies, Sector 125, New Super Express Highway, Noida, Tel: 95120-2432400, Fax: 95120-2432200, e-mail: ajitvarma@aihmr.amity.edu
[2]International Centre for Genetic Engineering and Biotechnology (UNO, Triesta, Italy) New Delhi, India
[3]Ch. Charan Singh University, Meerut, Uttar Pradesh, India

Soil Biology, Volume 3
Microorganisms in Soils: Roles in Genesis and Functions
(ed. by F. Buscot and A. Varma)

rides (Smiles 1988; Wood 1989). Their activity and interaction with other microbes and larger organisms and with soil particles depend largely on conditions at the microhabitat level that may differ among microhabitats even over very small distances (Wieland et al. 2001). The microhabitats for soil microorganisms include the interior as well as exterior surfaces of soil aggregates for varying sizes and compositions. Soil can therefore be regarded as highly heterogeneous with respect to the distribution of soil matter and organisms (Beare et al. 1995).

2 Origin of Microbial Diversity

The diversity of microorganisms has a much longer evolutionary history than plants or animals and thus has had more time to evolve into diverse forms. Microorganisms have been exposed to and have survived cataclysmic conditions unknown by higher animals and plants. Plants and animals are relative newcomers and have only had to prove their adaptive capacity for several hundred million years, a fairly short period in evolutionary time. During this time, conditions on the earth's surface were conducive to the survival of plants and animals. Certainly, there have been many examples of species extinction, however, by and large, the temperature has remained fairly stable, there have been few collisions with really large meteors, volcanic activity has been moderate, and the oceans have remained homogeneous and oxygenated.

Microorganisms have proved their ability to face challenges unimaginable to us today. Moreover, microorganisms did not simply occupy various niches offered by earth. Through their chemical activities, they transformed the earth and its atmosphere in a number of ways. Some of these changes actually contributed to making the earth habitable for the plants and animals that appeared much later.

The earth is about 4.5 billion years old. Scientists estimate that the first living creature appeared about 4 billion years ago, shortly after the earth's surface had cooled enough to allow liquid water to form (Fig. 1). These creatures were most similar to modern-day prokaryotes – bacteria and archaea. Because some microorganisms living on earth today are capable of growing in boiling water, life could clearly have begun while the earth's surface was still very hot. Moreover, the sun was only about two-thirds as bright as it is today, therefore, the earth's surface would have become habitable faster than if the sun had been brighter.

Life during the high-impact period would not have been easy. Some impacts were powerful enough to vaporize oceans, creating clouds of steam that would have sterilized the earth's surface. These events may not have

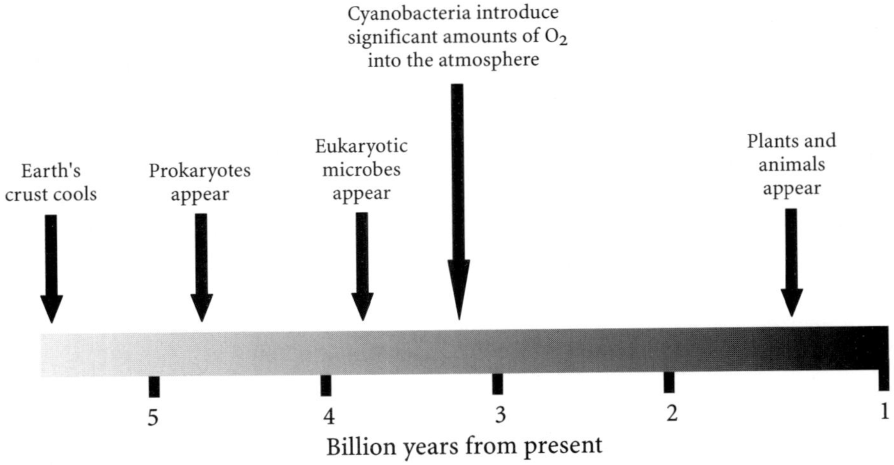

Fig. 1. Approximate timing of major events in the history of life on earth. (Salyers and Whitt 2001)

completely obliterated emerging life forms. Microorganisms could have survived this period deep underground. Some may have had the capacity of modern microorganisms to produce tough survival forms called spores. Although direct exposure to steam would have killed them, some spores could have survived under slightly cooler conditions that would still have been hot enough to kill an actively growing microorganism.

Some microorganisms may actually have been able to live on the earth's surface. One bacterium, *Deinococcus radiodurans*, can survive doses of radiation 3000 times greater than the lethal dose for humans. Most organisms, however, probably developed in the subsurface of landmasses or beneath the ocean surface where they were protected to some degree from UV radiation.

2.1 Oxygen Revolution

The revolutionary development occurred between 2.5 and 2 billion years ago, changing the earth and its atmosphere completely. Oxygen began to appear in significant amounts in the earth's atmosphere as a result of a microbial metabolic process called oxygenic photosynthesis. Although many compounds such as water contained bound oxygen, there had been no oxygen in the atmosphere. Oxygen photosynthesis differed from earlier forms of photosynthesis, in that it splits water and released oxygen. The bacteria responsive to this new type of photosynthesis are called cyanobacteria. The first appearance of oxygen left a tangible geological record: banded iron formations in rock. Iron in the earth's crust combined with oxygen to form black iron oxides, producing dark bands. Cyanobacteria also left a fossil record. Some cyanobacteria accumulated to form large mounds called stromatolites. Geologists have found fossilized stromatolites dating back 3 billion years and microfossils of individual cyanobacteria cells that date to 3.5 billion years ago. Cyanobacteria brought the oxygen level of the earth's atmosphere up to about 10% of today's level, high enough to create conditions that favored the evolution of oxygen-utilizing organisms.

2.2 Origin of the First Eukaryotes

Because many eukaryotes are oxygen-dependent, scientists had theorized that protozoa first appeared about 2 billion years ago. However, there are modern protozoa that live in anoxic environments, so protozoa could have emerged before the appearance of oxygen in the atmosphere. It is estimated that the time of appearance of the first protozoa dates back to about 3 billion years ago. Algae presumably appeared after cyanobacteria because their chloroplasts were derived from cyanobacteria. They probably evolved within the last 2 billion years. The fungi appeared only comparatively recently, during the last several hundred million years. It is thought that terrestrial fungi might have co-evolved with plants because they are closely associated with them. Fungi are often thought to be exclusively terrestrial. However, they are also reported in marine and other locations far from land (Salyers and Whitt 2001).

3 Types of Soil Microorganisms

Microscope studies led to the recognition of a profoundly important dichotomy among the various groups of organisms with respect to their internal architecture of the cell; two radically different kinds of organisms co-exist in the contemporary living world. The more complex cells constitute eukaryotes (organisms with a true nucleus), which include algae, fungi and protists (Fig. 2). Evolutionary studies revealed a great diversity of eukaryotic organisms as compared to prokaryotic microorganisms (Fig. 3). The less complex cell constitutes prokaryotes, comprising two microbial groups: the eubacteria (including cyanobacteria, the group once known as blue-green algae) and the archaebacteria, a heterogeneous group of microorganisms with prokaryotic structure. These organisms show characteristic features and play some beneficial roles to mankind (Table 1). Considering the cell structure and function as criteria, there are three groups of cellular organisms: eukaryotes, eubacteria, and the archaebacte-

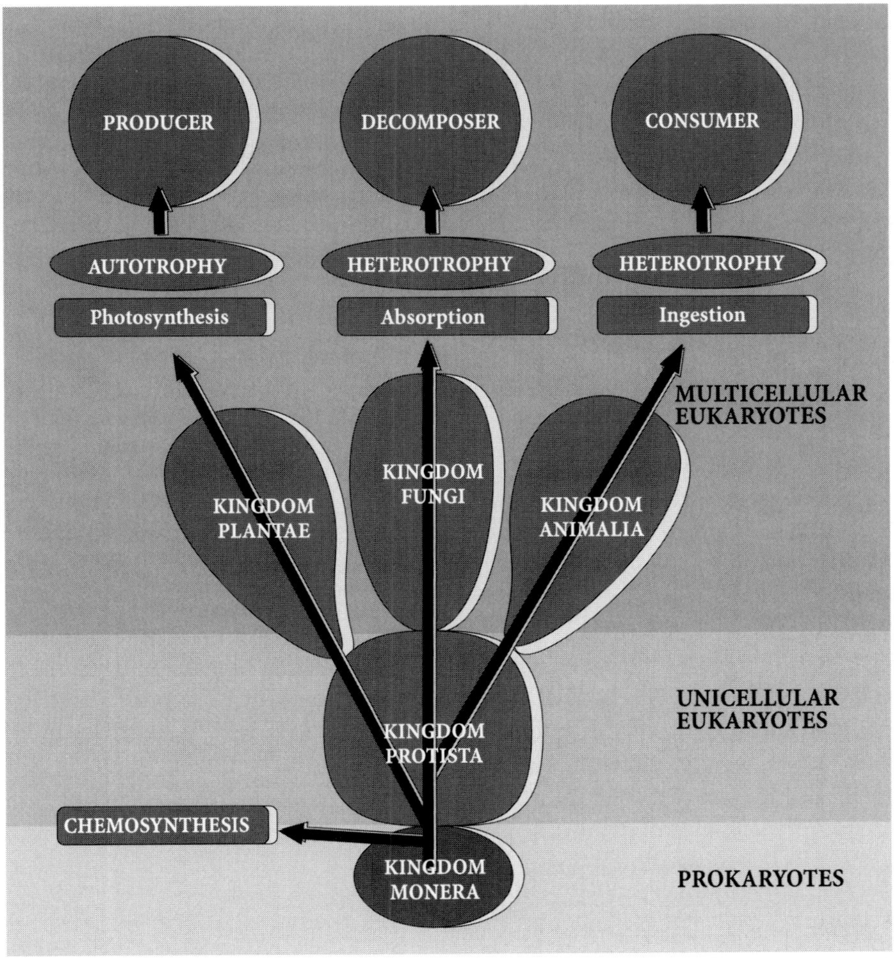

Fig. 2. The five-kingdom system showing diversity of organisms (http://www.npc.edu/Bio105/media_htm/M1_L7.01.htm)

ria. The eukaryotes can be subdivided into three further groups: the plants, animals and fungi. The eubacteria can be subdivided into purple, green, gram-positive and gram-negative eubacteria on the basis of the cell wall. On the basis of their nutritional requirements, prokaryotes have been categorized as Photoautotrophs, Photoheterotrophs, Chemolithoautotrophs, And Chemolithohetetrpophs (Table 2). Bacteria have also been classified as oxybionts and anoxybionts on the basis of their oxygen metabolism. Prokaryote diversity, however, is not only restricted to relationships to molecular oxygen or to their ability to utilize radiant energy to capture energy. Optimal diversity also depends on soil pH, temperatures (cold, ambient, hot), inorganic salts, etc. (Herman et al. 1993; Hurst 2002).

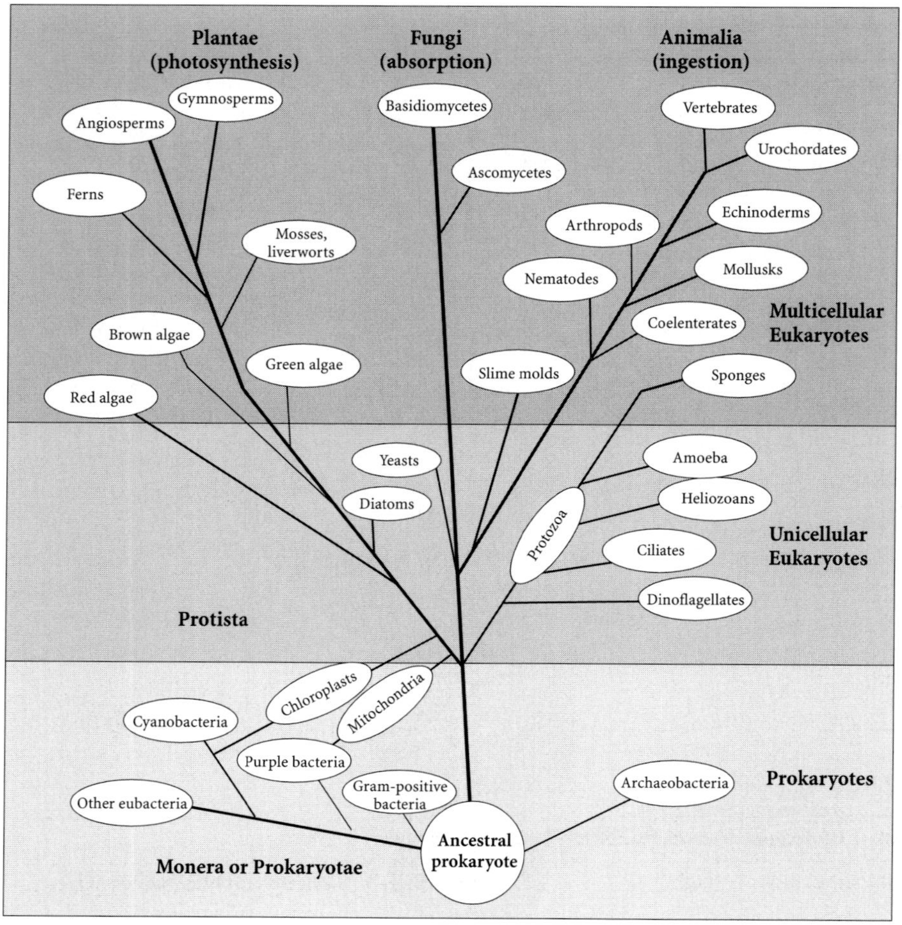

Fig. 3. A simplified diagram showing the diversity of organisms. (Modified after Prescott et al. 1996)

3.1
Eubacteria

Eubacteria are prokaryotic microorganisms. They are recognized as the most dominant group of microorganisms among the various kinds of soil (Table 3; Liesack and Stackebrandt 1992; Visscher et al. 1992; Borneman et al. 1996). They are present in all types of soil, but their population decreases as the depth of soil increases (Duineveld et al. 2001; Wieland et al. 2001). In general, horizon A (soil with organic matter) of a soil profile consists of more microorganisms than horizon B (silicate clay minerals plus organic

Table 1. Comparison of the main types of microorganisms

Microorganisms	Characteristics	Beneficial roles
Prokaryotes		
Bacteria	Rigid cell wall, divided by binary fission, some capable of photosynthesis	Recycle biomass, control atmospheric composition, component of phytoplankton and soil microbial populations
Archaea	Rigid cell wall, unusual membrane structure, photosynthetic membrane, lack chlorophyll	Produce and consume low molecular weight compounds, aid bacteria in recycling dead biomass, some are extremophiles
Eukaryotes		
Fungi	Rigid cell wall, single-cell form (yeast), reproducing by budding, multicellular form (hyphae, mycelium), no photosynthetic members	Recycling biomass, stimulate plant growth
Algae	Rigid cell wall, photosynthetic	Important component of phytoplankton

Table 2. Nutritional aspects of microbial diversity

Nutritional type	Energy source	Carbqon source	Examples
Photoautotroph	Light	Carbon dioxide (CO_2)	Photosynthetic bacteria (green sulfur and purple sulfur bacteria), cyanobacteria, extreme halophiles
Photoheterotroph	Light	Organic compounds	Purple non-sulfur and green non-sulfur bacteria
Chemolitho-autotroph	Inorganic compounds	Carbon dioxide (CO_2)	*Nitrosomonas*, *Nitrobacter*
Chemolitho-heterotroph	Organic compounds	Organic compounds	Most bacteria, fungi, and all animals

matter) and C (weathered parent material; Bruns and Slatar 1982; Subba Rao 1997).

Bacteria live in soil as cocci (sphere, 0.5 µm), bacilli (rod, 0.5–0.3 µm) or spiral (Fig. 4). The bacilli are common in soil, whereas spirilli are very rare in natural environments (Baudoin et al. 2001, 2002). Bacteria have been

Table 3. Microbial diversity of major groups in soils. (Modified after Hawksworth 1991a)

Major groups	Number of microbial species			Species in culture	
	Described species	Estimated species	Total species (%)	Number	Total estimated species (%)
Bacteria	3,000	30,000	10	2,300	7.0
Fungi	69,000	1,500,000	5	11,500	0.8
Algae	40,000	60,000	67	1,600	2.5

classified into two broad categories, the autochthonous and the zymogenous organisms. The autochthonous or indigenous populations are more uniform and constant in soil, since their nutrition is derived from native soil organic or mineral matter (*Arthrobacteria* and *Nocardia*; Herman et al. 1993). The zymogenous bacteria require the input of an external substrate, and their activity in soils is variable. They often produce resting propagules (*Pseudomonas* and *Bacillus*). When specific substrates are added to soil, the number of zymogenous bacteria increases and gradually declines when the added substrate is exhausted (cellulose decomposers, nitrogen utilizing bacteria, *Nitrosomonas*, *Nitrobacter*).

Ten orders are included in the class Schizomycetes. Of these, three orders, *Pseudomonas*, *Eubacteria* and *Actinomycetes*, contain the species of bacteria which are predominantly reported in the soil (Gaskins et al. 1984; Benson 1988; Paul and Clark 1989; Liesack and Stackebrandt 1992; Benizri et al. 2001). The most common bacteria belong to the genera *Pseudomonas*, *Arthrobacter*, *Clostridium*, *Achromobacter*, *Bacillus*, *Micrococcus*, *Flavobacterium*, *Corynebacterium*, *Sarcina*, *Azosprillium*, and *Mycobacteria* (Loper et al. 1985; Bruck 1987; Lynch 1987a, b). *Escherichia* is encountered rarely in soils except as a contaminant from sewage, whereas *Aerobacter* is frequently encountered and is probably a normal inhabitant of certain soils (Subba Rao 1997). Another group of bacteria common in soil is the Myxobacteria belonging to the genera *Myxococcus*, *Chondrococcus*, *Archangium*, *Polyangium*, *Cytophaga* and *Sporocytophaga*. The latter two genera are cellulolytic and, hence, are dominant in cellulose-rich environments (Slater 1988; Benizri et al. 2001).

Bacteria can withstand extreme climates, although temperature and moisture influence their population (Woese 1987; Benizri et al. 2002). In Arctic zones where the temperature is below freezing point, bacteria can thrive as luxuriantly as they do in arid desert soils, where temperatures are very high (Moreno et al. 1986). Such bacteria form spores possessing a tough outer covering, facilitating the survival of bacteria in all adverse environments. Survival by spore formation under extreme conditions should

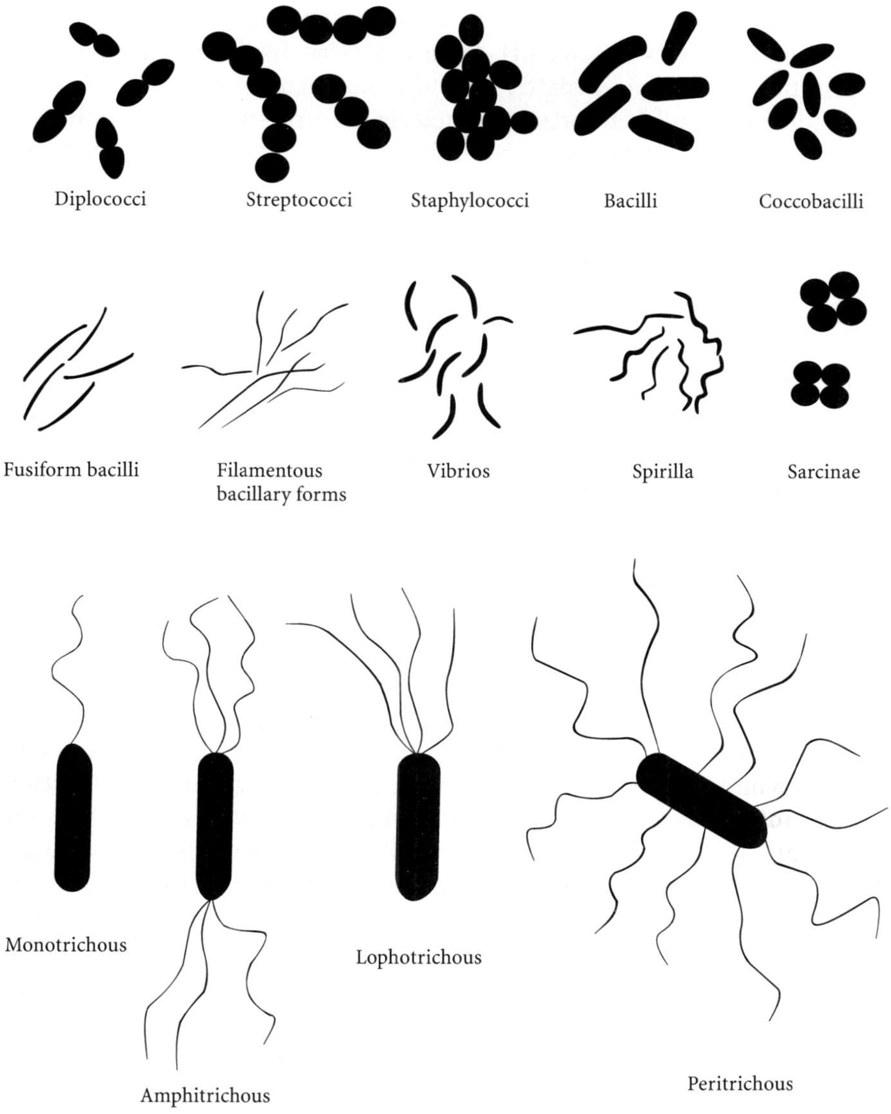

Fig. 4. Diversity of major forms of soil bacteria

be differentiated from tolerance to different temperature ranges, which is one of the factors determining the population of bacteria in soil (Burr and Caesar 1984).

Based on the temperature tolerance, bacteria are grouped as mesophyllous (15–45 °C), psychrophilous (below 20 °C) and thermophilous (45–65 °C; Subba Rao 1997). However, mesophyllous bacteria constitute the bulk of soil bacteria (Barber and Lynch 1997). Other factors affecting bacterial population in soil are pH, farm practices, fertilizers and pesticide applications and organic matter amendments (Tate 1987).

Autotrophic and heterotrophic bacteria are present in a wide range of soils (Tate 1995). Autotrophic bacteria (purple and green bacteria) synthesize their own organic matter from CO_2 or inorganic carbon sources, whereas heterotrophic bacteria depend on pre-formed organic matter for their nutrition and energy support. Photoautotrophs derive their energy from sunlight that they catch and transform into chemical energy through the bacteriochlorophyll pigment. Chemoautotrophs oxidize inorganic materials to derive energy and at the same time, they gain carbon from CO_2 (Tate 1995). There is a group of bacteria known as obligate chemoautotrophs. Within this group, *Nitrobacter* utilizes nitrite and *Nitrosomonas* ammonium, while *Thiobacillus* converts inorganic sulfur compounds to sulfate and *Ferrobacillus* converts ferrous ions to ferric ions (Alexander and Clark 1965; Baudoin et al. 2002).

The cyanobacteria are a structurally diverse assembly of gram-negative eubacteria characterized by their ability to perform oxygenic photosynthesis. They are considered true prokaryotic microorganisms (Stanier et al. 1986). They have characteristics common to bacteria and algae and are therefore often named "blue-green algae". Cyanobacteria contain a pigment known as phycocyanin, in addition to chlorophyll, which gives a special blue-green color to these organisms. The dominant cyanobacteria belong to the genera *Chrococcus*, *Aphanocapsa*, *Lyngbya*, *Oscillatoria*, *Phormidium*, *Microcoleus*, *Cylindrospermum*, *Anabaena*, *Nostoc*, *Scytonema*, and *Fischerella* (Subba Rao 1997; Benizri et al. 2002). Some cyanobacteria also possess heterocysts, which are implicated in nitrogen fixation. The rice fields are a good habitat for the development of certain cyanobacteria where they fix atmospheric nitrogen (Prescott et al. 1996).

Actinomycetes are soil microorganisms with sufficient distinctive features to delimit them into a distinct group within the prokaryotes. Actinomycetes are clubbed with further bacteria in the class of the Schizomycetes, but confined to the order Actinomycetales. They bear certain similarities to Fungi Imperfecti in the branching of the aerial mycelium, which profusely sporulates, and in the formation of distinct clumps or pellets in liquid cultures (Benson 1988).

The number of actinomycetes increases in the presence of decomposing organic matter. They are intolerant to acidity and their numbers decline below pH 5.0. The most conducive range of pH is between 6.5 and 8.0. Waterlogging of soil is unfavorable for the growth of actinomycetes, whereas desert soils of arid and semi-arid zones sustain sizeable populations, probably due to the resistance of spores to desiccation. The percentage of actinomycetes in the total microbial populations increases with the depth of soil. Actinomycetes can be isolated in sufficient number even from the C horizon (weathered parent material) of soil profiles. The commonest genera of actinomycetes are *Streptomyces* (nearly 70%). In contrast, *Nocardia* and *Micromonspora* and in particular, *Actinomyces*, *Actinoplanes* and *Streptosporangium* are only encountered occasionally (Prescott et al. 1996; Subba Rao 1997). Temperatures between 25 and 30 °C are conducive for the growth of actinomycetes although thermophilic cultures growing at 55 and 65 °C are common in compost heaps where they are numerically extensive and belong mostly to the genera *Thermoactinomyces* and *Streptomyces.*

3.2 Archaebacteria

Archaebacteria is a group of primitive prokaryotes, which were the earliest organisms to have appeared on the earth. Therefore, they are called the ancient bacteria. They even live in extreme hostile environments, like salt

Table 4. Diversity of archaebacteria

Archaebacteria	Characteristics
Methanogens *Methanococcus, Methanosprillum*	Generate methane when they oxide hydrogen gas as an energy source, using CO_2 as a terminal electron acceptor
Extreme halophiles *Halobacterium, Halorubrum, Natrinobacterium, Natronococcus*	Found near salt lakes, soda lakes, and brines. They produce pigments and can be seen as pink blooms in concentrated saltwater ponds
Methane-generating thermophiles *Methanothermus*	Found near hydrothermal vents; can grow at temperatures near 100 °C
Sulfur- and sulfate-reducing hyperthermophiles *Thermococcus, Archaeoglobus, Thermoproteus, Pyrodictium, Pyrolobus*	Obligate anaerobes that use sulfur or sulfate as a terminal electron acceptor, generating hydrogen sulfide. *Thermococcus*, and *Archaeoglobus* oxidize organic compounds as an energy source; *Thermoproteus*, *Pyrodictium*, and *Pyrolobus* oxidize H_2 as an energy source
Sulfer oxidizers *Sulfolobus*	Oxidize sulfur as a source of energy, using O_2 as a terminal electron acceptor to generate sulfuric acid
Thermophilic extreme acidophiles *Thermophilus, Picrophilus*	Grow only in extremely hot, acid environments

pans, salt marshes, hot sulfur springs, etc. Archaebacteria is a heterogeneous group that is phylogenetically very distant from the eubacteria and possesses very distinct characteristics (Table 4). They are characterized by the absence of peptidoglycan in their wall. Instead, their wall contains proteins and non-cellulosic polysaccharides. Their cell membrane contains branched chain lipids that enable them to bear extreme temperatures and pHs. Their rRNA nucleotides are quite different from those of other organisms (DeLong and Pace 2001; Huber et al. 2002).

Archaebacteria comprise two subgroups which are respectively obligate and facultative anoxybiont. Obligate anoxybionts live exclusively in the absence of oxygen and are killed in the presence of O_2. They comprise the methanogen and halophile species. Facultative anoxybionts are found in the presence of oxygen, but can live under anaerobic conditions. They are represented by thermoacidophiles (Tate 1995; Barns et al. 1996; Kyrpides and Olsen 1999).

3.2.1 Methanogens

Methanogens are strict anoxybionts occurring in marshy areas and characterized by their habit of producing CH_4 (methanogenesis) from CO_2 or fumaric acid. Methanogens are ubiquitous in highly reducing habitats. Some of them live as a symbiont in the rumen or first chamber of the stomach of ruminant animals. The most common species among methanogens are *Methanobacterium*, *Methanobrevibacter*, *Methanococcus*, *Methanospirillum*, and *Methanosarcina*. Methanogenesis is now attributed to more than 50 species of bacteria (Jones 1991). Their growth and survival depend directly on the activities of associated microflora, which enhance methanogenesis through the release of C substrates and the maintenance of reducing conditions (Tate 1995; Prescott et al. 1996).

3.2.2 Halophiles

Highly saline environments harbor large populations of a small and distinctive group of halophiles (*Halococcus* and *Halobacterium*). These archaebacteria live in extremely strong brine or salt solutions, salt beds and salt marshes. Some halophiles occur in deep sea volcanic vents at 100 °C, a temperature at which water remains liquid because of extreme hydrostatic pressures. In strong light, halophiles develop a purple pigmented membrane, which can absorb solar radiations. The absorbed light is utilized in the synthesis of ATP. These archaebacteria are unique because they carry out their metabolic processes directly by the ATP produced by their pigmented membrane. They cannot convert CO_2 to sugar as in photosynthesis. Halophiles growing in salt beds give an offensive smell and undesirable pigmentation to the salt (Beare et al. 1995; Barns et al. 1996).

3.2.3 Thermoacidophiles

The thermoacidophiles occur in high temperature environments like hot sulfur springs, where temperature may be as high as 80 °C and pH as low as 2. These archaebacteria are chemoautotrophic and obtain energy and carbon by oxidizing sulfur under consumption of CO_2. Under aerobic conditions they oxidize sulfur to sulfuric acid. Some archaebacteria can also reduce sulfur to hydrogen sulfide in the absence of oxygen (Stanier et al. 1986; Tate 1995; Prescott et al. 1996).

3.3 Fungi

Fungi dominate all types of soils and represent the greatest diversity among soil microorganisms (Table 1). Fungi possess filamentous mycelium composed of individual hyphae. The hyphae may be uni-, bi- or multinucleate and nonseptate or septate (Hawksworth 1991b). All the environmental factors that influence the distribution of bacteria also apply in fungal flora of soils. However, the quality and quantity of organic matter have a direct bearing on fungal numbers in soils since fungi are heterotrophic organisms. Fungi are dominant in acid soils because an acidic environment is not conducive to the existence of either bacteria or actinomycetes, resulting in the monopoly of fungi for utilization of organic substrates (Bolton et al. 1993). They are also present in neutral or alkaline soils and some can tolerate a pH over 9.0. Arable soils contain abundant fungi since they are strictly aerobic and an excess of soil moisture decreases their numbers. Fungi exhibit a selective preference for various soil depths. Species common in lower depths are rarely found on the surface. This specific distribution is ruled by the availability of organic matter and by the ratio between oxygen and carbon dioxide in the soil atmosphere at various depths. Farm practices including crop rotation and fertilizer or pesticide applications influence the nature and dominance of fungal species (Hawksworth 1991a,b).

Fungi are classified into Phycomycetes, Ascomycetes, Basidiomycetes and Fungi imperfecti (Table 5; Alexander 1977). Many fungi, which are commonly isolated from soils, come under the class Fungi Imperfecti by virtue of the fact that they produce abundant asexual spores, but lack sex-

Table 5. Major groups of soil fungi

Group and representative members	Distinguishing characteristics	Asexual reproduction	Sexual reproduction
Zygomycetes *Rhizopus stolonifer* (black bread mold)	Multicellular, coenocytic mycelia	Asexual spores develop in sporangia on the tips of aerial hyphae	Sexual spores known as zygospores can remain dominant in adverse environment
Basidiomycetes *Agaricus campestris* (meadow mush-room), *Cryptococcus neoformans*	Multicellular, un-inucleated mycelia. group includes mushrooms, smuts, rusts that affect the food supply	Commonly absent	Produce basidiospores that are born on club-shaped structures at the tips of the hyphae
Ascomycetes *Neurospora*, *Saccharomyces cerevisiae* (baker's yeast)	Unicellular and multicellular with septate hyphae	Common by budding, conidio-phores	Involves the formation of an ascus on specialized hyphae
Deuteromycetes (Fungi Imperfecti) *Penicillium*, *Aspergillus*	A number of these are human pathogens	Budding	Absent or unknown

ual stages (Lynch 1987a, b). Members of this class are distinguished by their septate mycelium and a structure called conidiophore from which conidia or spores are continuously produced. The other three classes of fungi have both sexual and asexual means of reproduction. Phycomycetes members possess nonseptate and unicellular mycelia and produce an undefined number of specialized spore cells called sporangia. In Ascomycetes, the sporangium produces a species-specific number of meiotic spores (often four or eight) and different types of active or passive spore extrusion mechanisms are encountered. A higher specialization degree of the sporangium, the basidia, is reached in Basidiomycetes. Here, the number of produced meiotic spores (generally four) is constant. These result either from fragmentation of the basidia or from their budding in so-called ballistospores. The most important vegetative trait of soil fungi is their producing a mycelium capable of polarized growth toward adequate substrate sources. Fungi and especially members of the Asco- and Basidimycetes are able to degrade very complex organic compounds such as cellulose or lignin, but many of them also live as root symbionts (mycorrhizas) and obtain simple sugars from their plant partners (Lynch and Hobbie 1988).

The following genera of fungi are most commonly encountered in soils (Fig. 4): *Acrostalagmus*, *Aspergillus*, *Botrytis*, *Cephalosporium*, *Gliocladium*, *Monilia*, *Penicillium*, *Scopulariopsis*, *Spicaria*, *Trichoderma*, *Trichothecium*, *Verticillium*, *Alternaria*, *Cladosporium*, *Pillularia*, *Cylindrocarpon* and *Fusarium*, *Absidia*, *Cunninghamella*, *Mortierella*, *Mucor*, *Rhizopus*, *Zygorynchus*, *Pythium*, *Chaetomium*, and *Rhizoctonia* (Newman 1985; Hawksworth 1991a; Subba Rao 1997). Many yeasts belonging to true Ascomycetes such as *Saccharomyces* and those belonging to Fungi Imperfecti such as *Candida* have been isolated from soils. However, their number in soil is relatively low. Filamentous fungi in soil degrade organic matter and help in soil aggregation. Certain fungi like *Alternaria*, *Aspergillus*, *Cladosporium*, *Dematium*, *Gliocladium*, *Helminthosporium*, *Humicola*, and *Metarhizium* produce substances similar to humic substance in soil and, hence, may be important in the maintenance of soil organic matter (Hawksworth 1991b).

3.4 Algae

Soil algae are ubiquitous in nature where moisture and sunlight are available. The algae, which are dominant in soils, are members of the class Chlorophyceae. Diatoms have also been found in soils. These microorganisms are visible to the unaided eye in the form of green scum on the surface of soils, whereas some algae are microscopic. In the soil, algae are not as plentiful as fungi (Table 3; Metting 1988). They may be unicellular (*Chlamydomonas*) or filamentous (*Spirogyra*, *Ulothrix*). Algae are photoautotrophic organisms by virtue of the presence of chlorophyll in their cells. They use CO_2 from the atmosphere and produce O_2. Algae have been found below the surface of the soil and beyond the reach of sunlight. However, their number here is low compared to that of algae inhabiting the surface of soil (Metting 1988; Subba Rao 1997). Some of the common green algae occurring in most soils belong to the genera *Chlorella*, *Chlamydomonas*, *Chlrococcum*, *Oedogonium*, *Chlorochytrium*, and *Protosiphone* (Metting 1988; Lynch 1990).

4 Microbial Diversity and Biological Spheres

Factors such as resource availability, microclimatic conditions, soil solution chemistry and soil structure can significantly influence the size, composition and distribution of soil biotic communities (Wolters 1991; Baudoin et al. 2001, 2002). Soils can be viewed as being composed of a number of biologically relevant spheres of influence that define much of their spatial and temporal heterogeneity. Examples of these spheres include the detritusphere, the drilosphere, the porosphere, the aggregatusphere and the rhizosphere. Although not mutually exclusive, each sphere has fairly distinct properties that regulate the interactions among organisms and the biogeochemical processes that they mediate (Beare et al. 1995).

4.1 The Detritusphere

The detritusphere corresponds to the zone of recognizable plant and animal detritus undergoing decay. Numerous studies have shown that the structure of decomposer communities is influenced by the chemical composition of plant detritus (Swift et al. 1979; Kjoller and Struwe 1982). In many cases, distinct communities of soil organisms, such as fungi (Wicklow et al. 1974) can be ascribed to ecosystems of similar vegetation cover. Diversity in microfungal communities often correlates well with the variance in the composition of the plant community (Christensen 1989), and can be related to the patchy distribution of resources. Disruptions to the soil ecosystem such as overgrazing, cultivation and fertilizer applications tend to reduce microhabitat heterogeneity and the diversity of corresponding microfungal communities (Gochenauer 1981; Boddy et al. 1988; Christensen 1989). Furthermore, microhabitat patches may create a mosaic of aerobic and anaerobic microsites that promote the activities of N_2-fixing and -denitrifying microorganisms in the detritusphere (Lynch and Harper 1985; Lynch 1990). Patterns of microbial colonization are influenced by nutrient fluxes in litter (Beare et al. 1992). Nutrient release from rapidly decaying litter fractions stimulates decomposition of adjacent recalcitrant litter (Seastedt 1984), while others suggest that inhibitory compounds such as phenolics and tannins may lower the decomposition of litter mixtures. Recent studies by Blair et al. (1990) provide support for these hypotheses, showing that interaction between litter types can alter decomposer communities and rates of nutrient release from single species litter.

4.2 The Drilosphere

The zone of earthworm influence, including maiden litter and soil volume descending along the burrow walls, is often referred to as the "drilosphere" (Hamilton and Dindal 1983; Lavelle et al. 1989). Drilosphere soils are enriched in N, P and humified organic matter in comparison to the surrounding soils. They are also estimated to contain a high percentage of the whole soil N_2-fixing and -denitrifying bacteria (Wolters 1991). However, the nature of these influences differs between the earthworm species, in accordance to their ecological classification. Shaw and Pawluk (1986) observed that deep burrowing anecic earthworms had effects on the soil fabric that were localized in the drilosphere. However, wherever endogeic species were also present, their activities tended to homogenize the surface soil horizons. Clearly, these interactions can greatly affect the heterogeneity of organisms and processes in soils.

4.3 The Porosphere

Soil structure can be defined as the arrangement of solids and voids in soils, covering a range of sizes from nanometers to centimeters (Oades 1993). The influence of soil biota spans the full range of sizes, affecting the pore size distribution through biopore development and the formation and disruption of soil aggregates. This milieu, termed the "porosphere" (Vannier 1987), is occupied by organisms the smallest of which range from bacteria, protozoa, nematodes to fungi. Larger soil biota such as plant roots, earthworms and other members of the macrofauna create smooth, cylindrically shaped macropores. These biopores extend considerable distances in the soil and change soil structure. Ants and termites form mounds and have patchy effects on soil structure (Lobry de Bruyn and Conacher 1990). However, mounds are also sites of nutrient enrichment due to subsoil nutrients brought to the surface and the storage of plant detritus in their galleries. In this zone several mycorrhizal fungi have been reported (Friese and Allen 1993). Evans and Miller (1988) demonstrated that macropores are the sites of concentrated mycorrhizal inocula. They found increasing rates of mycorrhizal infection and phosphorus availability related to increased plant growth.

4.4 The Aggregatusphere

Soil organisms have many wide-ranging effects on aggregation that can influence the physical, chemical and biological properties of soils (Lee and Foster 1991). Aggregates are comprised of a number of components, ranging from clay microstructures and fine particulate organic matter to microaggregates (50–250 µm), made up of these primary particles and macroaggregates (> 250 µm diameter), themselves composed of microaggregates (Oades and Waters 1991). The aggregatusphere encompasses all these constituents, and defines a complex of constraints for the exchange of biota, solutes and gases, the properties of which depend on the scale at which it is viewed. The contribution of soil microorganisms to aggregation is most apparent in soils of lower clay content and low shrink–swell capacities, where the abiotic effects of wet–dry and freeze–thaw cycles are reduced (Oades 1993).

Microorganisms are the primary agents of aggregate stabilization. Both fungi and bacteria contribute to stabilization of soil aggregates through deposition of extracellular polysaccharides and formation of degraded, aromatic humic materials that form clay–polyvalent metal–organic matter complexes. Though not as persistent, fungi also contribute to aggregate stabilization through hyphal anchoring of particles. The influence of fungi and bacteria on aggregate stabilization varies widely among species and depends considerably on the nature of the available substrates (Aspiras et al. 1971) and on the products of rhizodepositions (Reid and Goss 1981). Furthermore, the type of land-use management can influence both the composition of microbial communities and their contribution to aggregate stabilization (Beare et al. 1994).

4.5 The Rhizosphere

The zone of primary root influence can be termed the "rhizosphere". It is a temporally and spatially variable environment where the products of rhizodeposition stimulate microbial activity and populations, thereby altering the balance between N mineralization and immobilization (Fig. 5; Clarholm 1985; Coleman et al. 1988). The biomass of soil microflora is usually greater in the rhizosphere than in root-free soil (Bowen and Rovira 1991). Some studies show that fungal species diversity is lower while the morphological diversity of bacteria and Actinomycetes is higher in the rhizoplane, the root surface, as compared to soil outside this zone (Bowen and Rovira 1991). The extent of these effects depends on the characteristics of root growth, including their production, turnover and architecture.

Root architecture influences and is influenced by the physical, chemical and biological properties of soils. Root system development responds strongly to soil fertility. The proportion of total plant production allocated below ground and the architecture of the root system (root length, branching frequency and mycorrhizal development) depend greatly on the distribution and availability of nutrients in soils (Fitter 1985). Increases in fine root proliferation, slower root turnover and greater allocation of plant C to mycorrhizal associates tend to occur when nutrients are low or patchily distributed. Depending on their source, root exudates can inhibit

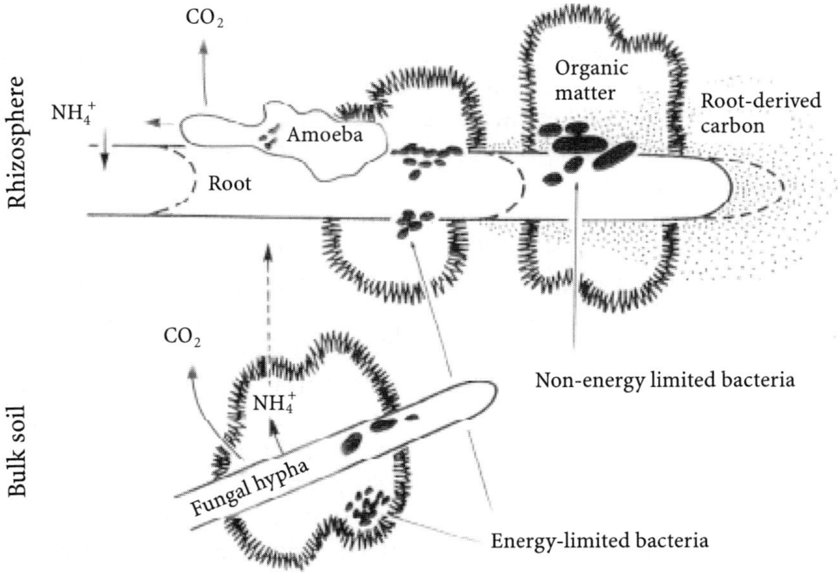

Fig. 5. Model of proposed interactions in the rhizosphere and in the bulk soil. Note the excessive production of root exudates towards the distal end of the root, intermingled with fungal hyphae

the growth of phytopathogenic microorganisms and alter the composition of the rhizosphere community. Though not well studied, mycorrhizal symbionts also influence characteristics of root exudates that shape the composition and activity of the rhizosphere community (Meyer and Linderman 1986).

5
Microbial Diversity and Chemical Transformation

The importance of microbial diversity for biogeochemical transformations can be viewed most directly through the specific chemical transformations that organisms perform. Their effects on biogeochemical transformations occur through both direct and indirect means. In this section, we review the direct effect of microorganisms (particularly bacteria, fungi) on biogeochemical transformations in soils (Beare et al. 1995).

5.1 Nitrogen Transformation

Nitrogen availability is a key factor regulating the biological productivity of many ecosystems (Herbert 1999; Capone 2000). Soil microorganisms have long been recognized as important agents affecting N pools through various transformations. The assimilation into the organic form and subsequent release of inorganic N, as performed by a broad array of prokaryotic and eukaryotic organisms, comprise the inner core of the N cycle in nature (Alexander 1977; Paul and Clark 1989). However, it is the uniquely bacterial processes of N_2 fixation, nitrification, and denitrification that define the broader cycle and can affect directly the availability and form of N within particular ecosystems (Postgate 1987). In the nitrogen cycle, many bacteria (eubacteria and archaea) are involved in ammonification, but other N transformations are carried out by taxonomically narrow groups of microorganisms. Chemoautotrophic nitrification is accomplished by relatively few obligate aerobic soil bacteria (ammonium oxidizers and nitrite oxidizers) which oxidize NH_3 to NO_2 (*Nitrosomonas*, *Nitrococcus*) and NO_2 to NO_3 (*Nitrobacter*; Kaplan 1983). Heterotrophic nitrification is also known in several bacteria (*Arthrobacter*) and Actinomycetes, but probably accounts for relatively low levels of NO_3 production. Other steps in the N cycle, such as dissimilatory NO_3 and NO_2 reduction (*Mycobacterium*, *Clostridium*) and denitrification (*Pseudomonas*, *Bacillus*, *Thiobacillus*), are carried out by a few, widely distributed genera (Payne 1981). Asymbiotic N_2 fixation is carried out by aerobic (*Azotobacter*, *Beijerinckia*), microaerophilic (*Clostridium*) organotrophic bacteria as well as by free-living cyanobacteria that are sometimes abundant in soils. Symbiotic N_2 fixation is best known for bacterial (*Rhizobium*, *Bradyrhizobium*) associations with legumes, but also concerns some plant genera of nonleguminous angiosperms (*Alnus*, *Casuarina*, *Ceanothus*, *Myrica*) associated to specific Actinomycetes such as *Frankia*.

Fungi are major components of the soil biomass (Hawksworth 1991a,b) and are of considerable importance in regulating ecosystem processes (Dighton and Boddy 1989; Cromack and Cadwell 1992; Wainwright 1992). Though often grouped according to their specific enzymatic capabilities, most fungi have broad versatility in their chemoheterotrophic metabolisms. Despite this versatility and their prominent role in plant litter decomposition (Kjoller and Struwe 1982; Cromack and Cadwell 1992), many fungi maintain more specialized mechanisms for obtaining energy and nutrients (Wainwright 1992).

The important role of many fungi, including ectotrophic mycorrhizal species (Wainwright 1992; Lakhanpal 2000), in the ammonification of organic N is well established, but their contribution in other areas of the N cycle has received little attention. Nitrification has long been known for *Aspergillus flavus*, but the broader range of fungal involvement has only recently been described (Killham 1987). Though autotrophic nitrification by bacteria is often assumed to dominate, the heterotrophic activities of fungi may account for a significant proportion of the nitrification in acid forest soils (Schimel et al. 1984). The extent of fungal nitrification in other soil systems remains poorly known. In contrast, several genera of fungi are known to play a role in nitrate reduction (*Fusarium*, *Acremonium* and *Aspergillus* spp.) though few studies have demonstrated significant levels of complete denitrification in fungi.

5.2 Phosphorus Transformation

Phosphorus is considered to be a major growth-limiting nutrient and unlike the case for nitrogen, there is no large atmospheric source that can be made biologically available (Ezawa et al. 2002). It is essential in both cellular energetics (ATP) and cellular structures (DNA, RNA, and phospholipids). Therefore, phosphate-dissolving soil microorganisms play a profound role (Schachtman et al. 1998). The role of bacteria in the P cycle appears somewhat less specialized. Although there is no microbially mediated gaseous flux of P, however, *Pseudomonas* and *Bacillus* are involved in the solubilization of inorganic phosphorus. Although bacteria have been used in the growth of plants, fungi seem to be better agents in the dissolution of phosphates (Barea 2000; Barea et al. 2002; Chalot et al. 2002). Phosphate-dissolving bacteria are known to reduce the pH of the substrate by secretion of a number of organic acids such as formic, acetic, propionic, lactic, glycolic, fumaric and succinic acids. As a group, soil bacteria are important to the short-term immobilization of P and the mineralization of organic phosphorus (Subba Rao 1997).

Somewhat more specialized groups of bacteria are involved in the transformation of metals in soils. Examples of these transformations include the reduction (*Bacillus*) and precipitation (Chladobacteriaceae) of iron as well as the chemolithotrophic oxidation of Fe^{2+} under acid conditions (*Thiobacillus ferroxidans*; Table 6). Some free-living fungi (*Aspergillus* and *Penicillium*) also excrete organic acids and Fe siderophores that solubilize insoluble forms of phosphate and contribute to the weathering of soil minerals (Mehta et al. 1979; Sollins et al. 1981).

Several enzymes are involved in the decomposition of the organic phosphorus compounds (Jennings 1995). Those enzymes that hydrolyze P-esters are commonly called phosphatases. The function of the phosphatase is to break down organic phosphates and polyphosphates, thus releasing or-

Table 6. Metabolism of chemolithoautrotrophs

Common name of organism	Source of energy	Oxidation reaction (energy yielding)	Important features of group	Common genera in group
Hydrogen bacteria	H_2 gas	$H_2 + ½O_2 \rightarrow H_2O$	Can also use simple organic compounds for energy	*Hydrogenomonas*
Sulfur bacteria	H_2S	$H_2S + ½O_2 \rightarrow H_2O + S$ $S + 1½O_2 + H_2O \rightarrow H_2SO_4$	Some organisms of this group can live at a pH of less than 1	*Thiobacillus*, *Beggiatoa*, *Thiothrix*
Iron bacteria (nonphotosynthetic)	Reduced iron (Fe^{2+})	$2Fe^{2+} + ½O_2 + H_2O \rightarrow 2Fe^{3+} + 2OH^-$	Iron oxide present in the sheaths of these bacteria	*Sphaerotilus*, *Gallionella*
Nitrifying bacteria	NH_3	$NH_3 + 1½O_2 \rightarrow HNO_3 + H_2O$	Important in nitrogen cycle	*Nitrosomonas*
	HNO_2	$HNO_2 + 1½O_2 \rightarrow HNO_3$	Important in nitrogen cycle	*Nitrobacter*

thophosphate (Tabatabai 1982). In soils, there are two groups of phosphatases, the phosphoric monoester hydrolases and the phosphoric diester hydrolases. In the first group are enzymes such as phytase, nucleotidase and sugar phosphatases, while the second group contains the nuclease and phospholipases. More generally, these enzymes are divided into two groups named after their optimal pH activity. In soils, phosphatases generally exhibit three pH optima (5.0, 7.0, and 9.5), consequently representing acid, neutral and alkaline phosphatases, respectively. Apart from influencing the substrate, changes in the proton concentration and thus in the pH strongly influence the enzymes by altering their ionization state and solubility. Phosphatases are the most stable around their pH optimum and are irreversibly denatured at extreme pH values (Tabatabai 1982). Some fungi exhibit their highest acid phosphatase activity at acidic pH values. These fungi also display some activity at natural pH (Tarafdar and Rao 1996; Pant and Warmen 2000). Alkaline phosphatase from mycorrhizal fungi also showed some activity at natural pH (Bae and Borton 1989). Very little is known about the origin and production of natural phosphatases. Nannipieri et al. (1996) showed that part of the neutral phosphatase activity in soil could be correlated to the microbial biomass.

5.3 Sulfur Transformation

Sulfur is an important element from both a biochemical and geochemical point of view. It constitutes approximately 1% of the dry mass of organisms in which it has many structural and enzymatic functions. Sulfur also acts as a significant electron donor and acceptor in numerous bacterial metabolic pathways (Prescott et al. 1996; Hurst 2002). Sulfur can be found in a range of valence states from the highly reduced sulfide to the most oxidized form in sulfate. Microbial S transformations are closely linked with the carbon cycle in which S reduction coupled with organic matter utilization is a major mineralization pathway in anoxic habitats, while S oxidations can occur aerobically and anaerobically, whereby the concerned bacteria can be auto- and/or phototrophic (Jorgensen 1982, 1994).

Microorganisms of the S cycle are extremely diverse. They can be either oxybiont or anoxybiont. The anoxybiont sulfate-reducing bacteria (SRB), which are unique physiologically and genetically, are represented by several genera (Devereux and Stahl 1993). Sulfate-reducing bacteria are capable of utilizing iron and manganese as electron acceptors (Lovley and Phillips 1994). Oxygen-reduction has been demonstrated, but O_2-dependent growth has not been confirmed (van Niel and Gottschal 1998). Chemolithotrophic sulfur oxidation is mediated aerobically by colorless sulfur bacteria, some purple sulfur bacteria and SRB (Table 6). Anaerobically, nitrate respiring chemolithotrophs oxidize sulfide, and both oxygenic and anoxygenic phototrophic bacteria use sulfide as an electron donor for photosynthesis (Prescott et al. 1996). Sulfate-reducing bacteria may diminish the availability of sulfur for plant nutrition and thus influence agricultural production. *Desulfovibrio desulfuricans* is a species belonging to this class of bacteria (Hurst 2002). Bacteria capable of oxidizing inorganic sulfur compounds vary morphologically from nonfilamentous (*Thiobacillus*) to filamentous forms (*Beggiatoa, Thiothrix and Thioploca*). Among these bacteria, *Thiobacillus* deserves special mention as it produces sulfuric acid when elemental sulfur is added to soil with the result that the soil pH may fall as low as 2.0 after prolonged incubation with the bacterium. Several fungi and Actinomycetes have also been reported to be sulfur oxidizers (*Aspergillus, Penicillum, Microsporeum*). *Thiobacilli* can also be used in the manufacture of 'Biosuper', a form of organic fertilizer once favored in Australia. In Biosuper, a mixture of rock phosphate and gypsum is inoculated with *Thiobacillus thiooxidans*. Sulfuric acid produced in the mixture dissolves the phosphate, thus enhancing the phosphorus nutrition of plants (Widdel and Hansen 1991).

5.4 Iron Transformation

Certain bacteria oxidize ferrous iron to the ferric state, which precipitates as ferric hydroxide around cells (Table 6; Quastel 1995). These bacteria, commonly known as iron bacteria, are usually nonfilamentous and spherical or rod-shaped (*Gallionella*, *Siderophacus*, *Siderocapsa*, *Siderophaera*, *Ferribacterium*, *Naumanniela*, *Ochrobium*, *Sideromanas*, *Sideronema*, *Ferrobacillus*, *Siderobacter*, and *Siderococcus*). Filamentous forms resembling algae are also encountered (*Leptothrix*, *Sphaerotilus*, *Toxothrix*, *Crenothrix*, and *Colnothrix*). In addition to these bacteria, certain algae belonging to Cyanophyceae, also transform ferrous salts to the ferric state and deposit the precipitation around their filaments. The ferric hydroxide deposits give a brown or rust-red color to these organisms.

The iron bacteria can be grouped into: (1) obligate chemoautotrophs, capable of utilizing energy released in the process of ferric hydroxide formation (*Gallionella ferruginea*, *Thiobacillus ferroxidans*, and *Ferrobacillus ferroxidans*), (2) facultative chemoautotrophs, utilizing energy derived in the process of ferric hydroxide formation or alternatively from organic matter (*Leptothrix ochraceae*) and (3) heterotrophs represented by most other iron bacteria which do not derive energy from iron oxidation, but depend upon organic matter for their nutrition.

6 Microbial Diversity and Biotic Interactions

Due to their vast diversity, large populations and long evolutionary history, microorganisms have contributed greatly to the rich and complex interactions among soil organisms (Barea 2000; Barea et al. 2002). These interactions range from highly specific symbioses to diffused mutualisms.

Mycorrhizal symbioses are among the best-known examples of plant–microbe interactions and play a key role in regulating plant productivity and nutrient cycling (Barea et al. 1998, 2002; Berreck and Haselwandter 2001). Mycorrhizal fungi are found in 75–80% of all vascular plant species. Although these associations are often assumed to have weak specificity, it has been shown that many are highly specific, emphasizing the importance of diversity to ecosystem functioning. The root–microbe interactions are the key to understanding ecosystem function, and places mycorrhizas in perspective with the many other complex interactions taking place in the rhizosphere.

Mycorrhizal fungi interact with a wide range of other microorganisms in the rhizosphere (Bowen and Rovira 1999). These interactions may be stimulatory or inhibitory; some may be competitive, while others may be mutualistic. Mycorrhizal fungi are found in the endorhizosphere, in the rhizosphere and in the bulk soil. In all these zones, they interact with the soil microbiota. The internal mycelium interacts mainly with the host root and other microorganisms inhabiting the area. The external mycelium interacts with many organisms, including bacteria, fungi, protozoa, nematodes, arthropods and large animals. Some interactions may be mutualistic while others may be difficult to define. Some bacteria such as fluorescent pseudomonas may proliferate in the hyposphere of mycorrhizal fungi (Lynch 1990). Competitive interactions between the mycorrhizal fungus and bacteria and other fungi have been observed, and there may be allelochemical interactions similar to antibiosis which can, however, be either stimulatory or suppressive (Srivastava et al. 1996; Bansal et al. 2000).

Formation of arbuscular mycorrhiza (AM) fungi changes plant physiology and certain nutritional and physical properties of the rhizosphere soil (Giri et al. 2001). This, in turn, affects colonization patterns of this region by soil microorganisms by the so-called mycorrhizosphere effect (for details, see Chap. 11). AM fungi thus interact with natural and introduced microorganisms in the mycorrhizosphere, hence affecting soil properties and quality (Gryndler 2000). The interactions of plant growth-promoting rhizobacteria (PGPR) and AM fungi have great importance in plant health and soil fertility (Azcon-Aguilar and Barea 1996). Conversely, soil organisms are known to affect AM formation and functioning (Barea et al. 2002). The microbial population in the rhizosphere can either interfere with or benefit the establishment of AM fungi (Vosatka and Gryndler 1999). Deleterious rhizosphere bacteria (Nehl et al. 1996) and mycoparasitic relationships (Jeffries 1997) have been found to interfere with AM development, while many microorganisms can stimulate AM formation and/or functioning (Gryndler 2000; Barea et al. 2002).

The microbial interactions in the mycorrhizosphere may involve a variety of bacteria and fungi with specific functional capabilities that may influence plant growth. This may include microbes such as strict or facultative anaerobes, extracellular chitinase producers, phosphate solubilizers, siderophores, antibiotic, hormone producers, and plant growth promoters (Linderman 1988; Barea 1997; Mukerji et al. 1997).

Recently, Varma and his colleagues have discovered a new root endophyte designated *Piriformospora indica*, belonging to the Hymenomycetes (Basidiomycota; Fig. 6; Verma et al. 1998; Varma et al. 1999; Koch et al. 2004; Pham et al. 2004a). *P. indica* hyphae colonize the root and show inter- and intracellular structures (vesicles and hyphal coils). The fungus grows on a wide range of synthetic simple and complex media (Pham et al. 2004b). The temperature range of the fungal growth is 20–35 °C; the optimum temperature and pH being 30 °C and 5.8, respectively. This new

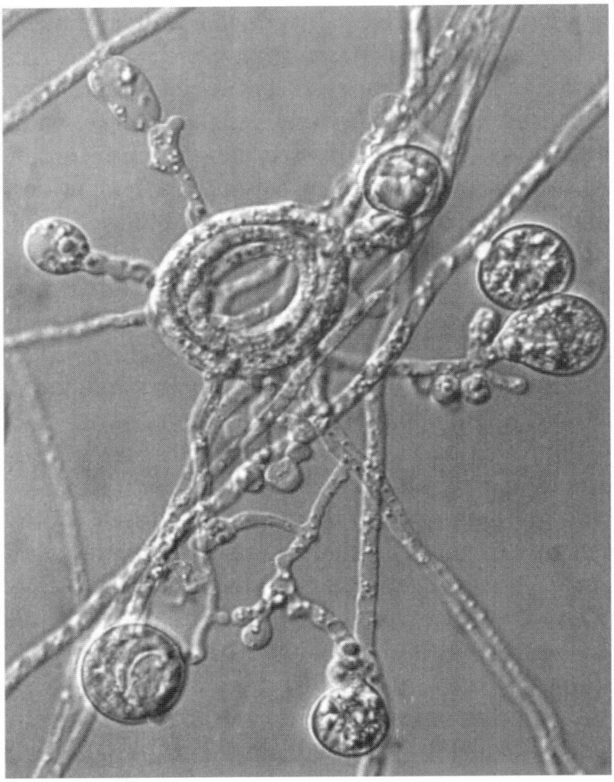

Fig. 6. *Piriformospora indica*: typical growth and differentiation on solidified nutrient medium. Note the pear-shaped spores and hypha coils. (Courtesy G Kost, Marburg, Germany)

fungus shows interactions with a wide range of soil microbiota. *P. indica* interacts with rhizobacteria, including *Pseudomonas florescence*, *Azotobacter chrococcum*, *Pseudomonas putrida*, *Bacillus subtilis*, *Azospirillum*, and *Bradyrhizobium* (Pham et al. 2004a,b).

On MMN media, a green alga *Chlamydomonas rienhardtii* and the *P. indica* showed a positive interaction. Both microorganisms grew well in perfect harmony. On the Kaefer medium, *P. indica* and a symbiotic fungus *Sebacina vermifera* grew normally without inhibiting each other. The most interesting part was after 7 days at the intersection of two colonies, when hyphae turned highly intertwined, thickened and produced a large number of chlamydospores (Singh et al. 2003).

Several commonly occurring soil fungi were tested for the interaction with *P. indica*. The results were highly diverse (Varma et al. 2001; Pham et al. 2004a). The growth of *Aspergillus sydowi*, *Rhizopus stolonifer*, and

Aspergillus niger was completely blocked by *P. indica*. *Cunninghemella echinulata* was partially blocked by *P. indica*, whereas *Rhizopus oryzae*, *Aspergillus flavus*, and *Aspergillus* sp. had completely blocked the growth of *P. indica*. Results indicate that *P. indica* interacts with a diverse group of soil fungi and its interaction varied from the negative to positive association (Kumari et al. 2003; Pham et al. 2004a).

P. indica showed a profound effect on disease control when challenged with a virulent root and seed pathogen *Gaeumannomyces graminis*. *P. indica* completely blocked growth of this pathogen. It indicates that *P. indica* acted as a potential agent for biological control of root diseases, however, the chemical nature of the inhibitory factor is still unknown.

Geosiphon pyriforme, a coenocytic soil fungus, lives in endocytobiotic association with a cyanobacterium, *Nostoc punctiforme* (Schüßler 2002). The symbiotic nature of the system was first recognized by von Wettstein (1915), who described it as a symbiosis between a heterotrophic siphonal chlorophycean alga and *Nostoc*. The fungus lives together with the cyanobacterium on the surface and in the upper layer of wet soils poor in inorganic nutrients, particularly in phosphate (Schüßler and Kluge 2001; Kluge et al. 2002). When a fungal hypha comes into contact with free-living *Nostoc* cells, the latter are incorporated by the fungus at the hyphal tip, which thereafter swells and forms a unicellular "bladder", about 1–2 mm in size and appearing on the soil surface (Fig. 7). Inside this bladder, the cyanobacteria are physiologically active and dividing. Due to the physiological activities of the endosymbiont, the consortium is capable of C- and N-autotrophic life. *Geosiphon* can be considered as a primitive endocytobiotic system, because the photobiont can be experimentally separated and cultured without the fungal partner, which is obligate symbiont. It has been suggested that *Geosiphon* could provide an important model system for another symbiosis, the arbuscular mycorrhiza. It bears a great potential for the study of many fundamental mechanisms and evolutionary questions concerning arbuscular mycorrhizas (Kluge et al. 1997)

Geosiphon pyriforme representing a symbiotic association between a glomalean fungus and a photoautotrophic prokaryotic alga could reflect an ancestral partnership. Thus, it is very plausible to assume that in the beginning of terrestrial plant life, other associations between glomalean fungi and photoautotrophic organisms also existed (Redecker et al. 2000).

Mollenhauer et al. (1996) studied the development of the symbiotic association *Geosiphon pyriforme*. Initially, the cells of the cyanobacterium *Nostoc punctiforme* live freely together with the future fungal partner in and on the soil. There, the partners come into contact, but a successful interaction of the fungus with *Nostoc* to form the symbiosis depends on the appropriate developmental stage of the cyanobacterium (Schüßler and Kluge 2001).

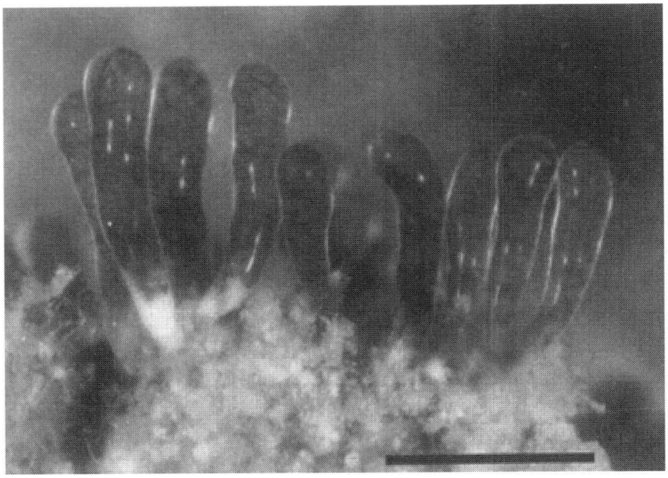

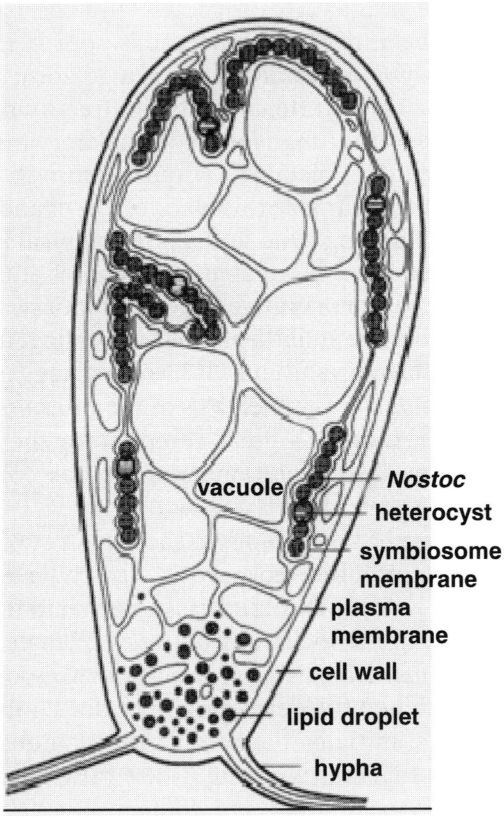

Fig. 7. *Above* Bladder of *Geosiphon pyriformae* on a natural substrate (Schüßler et al. 2001). *Below* Overview showing schematic drawing of bladder compartmentation. (Schüßler and Kluge 2001)

7 Conclusion

Microbiology is the study of microorganisms which exist in natural or artificial environments. The origin of scientific research in this field rests in the observation of Antony van Leeuwenhoek that was published in 1677 as "animalcula" or the "little animals", which lived and replicated in water. During the intervening centuries, the expansion of our knowledge has been based on increasingly detailed observations and experimentation, in which we have been aided by advancements in microscopy and the development of biochemical and mathematical tools. We have discovered that microorganisms cover our planet, living even in the fumaroles of surface volcanoes and in the sedimentary rocks within dry valleys. Microbes can be found as deep down as several kilometers, both in glacial ice sheets and in bedrock. At deep-ocean thermal vents, where the temperature of the water can reach several hundred degrees above its normal boiling point, the extremely high barometric pressure keeps water in its liquid state and microbial life bounds. The microorganisms chemically interact with their physical environment, and their most notable effect has been the creation of an oxidizing atmosphere on this planet. By way of these chemical interactions, microbes remain crucial to the biogeochemical cycling which supports the continuance of life on our planet, producing the elements that represent the basic ingredients of life such as carbon, hydrogen, nitrogen, oxygen, phosphorus and sulfur.

During the last few decades, we have begun learning how to harness microbial biosynthetic and degradative activities. This harnessing, including the intentional manipulation of microbial activities, constitutes the basis of microbial biotechnology, whereby we direct the activity of microorganisms within both natural and artificial environments for a variety of purposes. As one example, we utilize microorganisms as tools to help us achieve goals such as the production of materials which are beneficial to our existence, including numerous antibiotics, vitamins, and fuels such as biogas and ethanol. Microorganisms also are used as tools to help us intentionally degrade both natural and anthropogenic materials in wastewater digesters, compost, landfills, natural terrestrial environments, and natural or artificial aquatic environments. Sometimes we use microorganisms as tools to achieve agricultural goals such as protecting plants from insect damage. Furthermore, microbial processes, such as using microorganisms to leach metals from ores and to enhance the recovery of petroleum from wells, have been used as a means of minimizing the application of hazardous chemicals in geochemical recovery operations. Just as we sometimes use our knowledge of beneficial microbial processes to optimize their usefulness, at other times we try to prevent natural microbial activities such as those which contribute to corrosion and decay of objects exposed to the environment.

Presently, we still use a microbial classification scheme which is very traditional and divides the microorganisms into five major taxonomic groups. Four of these are considered to be cellular, meaning that they possess cell membranes. These four are the algae, bacteria, fungi, and protozoans. The fifth group, the viruses, is acellular. Biochemically based phylogeny studies constantly provide us with suggestions for revising such groupings. The most recent suggestions divide the older "bacteria" group into two domains, the *Bacteria* and *Archaea*, while assigning the algae, fungi, and protozoa to be part of the domain *Eucarya* (Pennisi 1999). The viruses and some of their biological relatives, which previously were never included within any kingdom, could fit into the proposed domain *Akamara* (Hurst 2002). The most important aspect is our understanding that within ecosystems these groups of microorganisms naturally organize among themselves as they go about their interactions both with one another and with the macroorganisms on this planet. These interactions occur and can be studied on many levels: spatially, biochemically, and even genetically.

A rough estimate indicates that 10, 5 and 67% of soil bacteria, fungi and algae, respectively, have been described. Out of this, only 7, 0.8 and 2.5% of bacteria, fungi and algae, respectively, have been axenically cultured. This makes it difficult to ascribe their phylogenetic taxonomic position and biotechnological recognition. The number of species that compose the functional groups or the transformation power of one group is more important to earth. We do not know the importance of one species inside dynamic biological systems, what one species represents within the biological dynamic and especially, what importance can one species have in nutrient cycling? These questions could lead us to conclude that we need to review our vision of the soil microcosm, extend our understanding of the biological processes and interactions that occur in the soil–plant system. Functional aspects are more important than biodiversity in natural ecosystems. Functional groups which take part are: carbon, phosphorus, nitrogen and sulfur biogeochemical cycles.

Acknowledgements. The authors are grateful to the Department of Biotechnology, Science and Technology, Council of Scientific and Industrial Research, and University Grant Commission for partial financial support.

References

Alexander M (1977) Introduction to soil microbiology, 2nd edn. Academic Press, New York

Alexander M, Clark FE (1965) Nitrifying bacteria. In: Black CA (ed) Methods of soil analysis, part 2. Chemical and microbiological properties. American Society of Agronomy, Madison, Wisconsin, USA, pp 1477–1483

Aspiras RB, Allen ON, Harris RF, Chester G (1971) The role of microorganisms in the stabilization of soil aggregates. Soil Biol Biochem 3:347–353

Azcon-Aguilar C, Barea JM (1996) Arbuscular mycorrhizas and biological control of soil-borne plant pathogens: an overview of the mechanisms involved. Mycorrhiza 6:457–464

Bae KS, Barton LL (1989) Alkaline phosphates and other hydrolyases produced by *Cenococcum graniforme*, an ectomycorrhizal fungus. Appl Environ Microbiol 55:2511–2516

Bansal M, Chamola BP, Sarwar N, Mukerji KG (2000) Mycorrhizosphere: interaction between rhizosphere microflora and VAM fungi. In: Mukerji KG, Chamola BP, Singh J (eds) Mycorrhizal biology. Kluwer Academic Press/Plenum, New York, pp 143–152

Barber DA, Lynch JM (1997) Microbial growth in the rhizosphere. Soil Biol Biochem 9:305–308

Barea JM (1997) Mycorrhiza/bacteria interactions on plant growth promotion. In: Ogoshi A, Kobayashi L, Homma Y, Kodama F, Kondon N, Akino S (eds) Plant growth-promoting rhizobacteria, present status and future prospects. OECD, Paris, pp 150–158

Barea JM (2000) Rhizosphere and mycorrhiza of field crops. In: Touant A (ed) Biological resource management: connecting science and policy. OECD, INRA Editions and Springer, Berlin Heidelberg, New York, pp 110–125

Barea JM, Andrade G, Bianciotto V, Dowling D, Lohrke S, Bonfante P, O'Gara F, Azcon-Aguilar C (1998) Impact on arbuscular mycorrhiza formation of *Pseudomonas strains* used as inoculants for the biocontrol of soil-borne plant fungal pathogens. Appl Environ Microbiol 64:2304–2307

Barea JM, Toro M, Orozco MO, Campos E, Azcon R (2002) The application of isotopic (^{32}P and ^{15}N) dilution techniques to evaluate the interactive effect of phosphate solublizing rhizobacteria, mycorrhizal fungi and *Rhizobium* to improve the agronomic efficiency of rock phosphate for legume crops. Nutri Cycl Agroecosyst 65:35–42

Barns SM, Delwiche CF, Palmer JD, Pace NR (1996) Perspectives on archaeal diversity, thermophily and monophily from environmental rRNA sequences. Proc Natl Acad Sci USA 93:9188–9193

Baudoin E, Benizri E, Guckert A (2001) Metabolic structure of bacterial communities from distinct maize rhizosphere compartments. Eur J Soil Biol 37:85–93

Baudoin E, Benizri E, Guckert A (2002) Impact of growth stages on bacterial community structure along maize roots by metabolic and genetic fingerprinting. Appl Soil Ecol 19:135–145

Beare MH, Parmelee RW, Hendrix PF, Cheng W, Coleman DC, Crossley DA Jr (1992) Microbial and faunal interactions and effects on litter nitrogen and decomposition in agroecosystems. Ecol Microorg 62:569–591

Beare MH, Cabrera ML, Hendrix PF, Coleman CD (1994) Aggregate-protected and unprotected pools of organic matter in conventional and no-tillage soils. Soil Sci Soc Am J 57:392–399

Beare MH, Coleman DC, Crossley DA Jr, Hendrix PF, Odum EP (1995) A hierarchical approach to evaluating the significance of soil biodiversity to biogeochemical cycling. Plant Soil 170:5–22

Benizri E, Baudin E, Guckert A (2001) Root colonization by plant growth promoting Rhizobacteria. Biocont Sci Technol 5(11):557–574

Benizri E, Dedourge O, Di Battista-Leboeuf C, Nguyen CS, Piutti, Guckert A (2002) Effect of maize rhizodeposits on soil microbial community structure. Appl Soil Ecol 21:261–265

Benson DR (1988) The genus *Frankia*: actinimycetes symbionts of plants. Microb Sci 5:9–12

Berreck M, Haselwandter K (2001) Effect of the arbuscular mycorrhizal symbiosis upon uptake of caesium and other cations by plants. Mycorrhiza 10:275–280

Blair JM, Parmelee RW, Beare MH (1990) Decay rates, nitrogen fluxes and decomposer communities of single-and mixed species foliar litter. Ecology 7:1976–1985

Boddy L, Walting R, Lycon AJE (eds) (1988) Fungi and ecological disturbance. Proc R Soc Edinb 94:1–188

Bolton H Jr, Fredrikson JK, Elliot LE (1993) Microbiology of the rhizosphere. In: Metting FB Jr (ed) Soil microbial ecology. Dekker, New York, pp 27–63

Borneman J, Skroach PW, O'Sullivan EW, Palus JA, Rumjanek NG, Jansen JL, Nienhuis J, Triplett EW (1996) Molecular microbial diversity of an agricultural soil in Wisconsin. Appl Environ Microbiol 62:1935–1943

Bowen GD, Rovira AD (1991) The rhizosphere, the hidden half. In: Waisel Y, Eshel A, Kafkafi U (eds) Plant roots: the hidden half. Dekker, New York, pp 641–669

Bowen GD, Rovira AD (1999) The rhizosphere and its management to improve plant growth. Adv Agron 66:1–10

Bruck TD (1987) The study of microorganisms in situ: progress and problems. In: Fletcher M, Gray TRG, Jones JG (eds) Ecology of microbial communities. SGM symposium 41. Cambridge Univ Press, Cambridge, pp 1–17

Bruns RG, Slatar JH (1982) Experimental microbial ecology. Blackwell, Oxford, 683 pp

Burr JJ, Caesar A (1984) Beneficial plant bacteria. CRC Crit Rev Plant Sci 21:1–20

Capone DG (2000) The marine nitrogen cycle. In: Kirchman D (ed) Microbial ecology of the ocean. Wiley-Liss, New York, pp 455–493

Chalot M, Javelle A, Blaudez D, Lambilliote R, Cooke R, Sentenac H, Wipf D, Botton B (2002) An uptake on nutrient transport processes in ectomycorrhizas. Plant Soil 244:165–175

Christensen M (1989) A view of fungal ecology. Mycologia 81:1–19

Clarholm M (1985) Possible roles of roots, bacteria, protozoa and fungi in supplying nitrogen to plants. In: Fitter AH, Atkinson D, Read DJ, Usher MB (eds) Ecological interactions in soil. Blackwell, Oxford, pp 297–317

Coleman DC, Crossley DA Jr, Beare MH, Hendrix PF (1988) Interactions of organisms at root/soil and litter/soil interfaces in terrestrial ecosystems. Agric Ecosyst Environ 24:117–134

Cromack K, Caldwell BA (1992) The role of fungi in litter decomposition and nutrient cycling. In: Carroll GC, Wicklow DT (eds) The fungal community, its organization and role in the ecosystem. Dekker, New York, pp 601–618

DeLong EF, Pace NR (2001) Environmental diversity of bacteria and archaea. Syst Biol 50:470–478

Devereux R, Stahl DA (1993) Phylogeny of sulfate-reducing bacteria and a perspective for analysing their natural communities. In: Odom JM, Singleton R Jr (eds) Sulfate-reducing bacteria: contemporary perspectives. Springer, Berlin Heidelberg New York, pp 131–160

Dighton J, Boddy L (1989) Role of fungi in nitrogen, phosphorous and sulphur cycling in temperate forest ecosystems. In: Boddy L, Marchent R, Read DJ (eds) Nitrogen, phosphorus and sulphur utilization by fungi. Cambridge Univ Press, Cambridge, pp 269–298

Duineveld BM, Kowalchuk GA, Keijzer A, van Elsas JD, van Veen JA (2001) Analysis of bacterial communities in the rhizosphere of *Chrysanthemum* via denaturing gradient gel electrophoresis of PCR-amplified 16S rRNA as well as DNA fragment coding for16S rRNA. Appl Environ Microbiol 67:172–178

Evans DG, Miller MH (1988) Vesicular-arbuscular mycorrhizas and the soil-disturbance-induced reduction of nutrient absorption in maize. New Phytol 110:67–74
Ezawa T, Smith SE, Smith FA (2002) P metabolism and transport in AM fungi. Plant Soil 244:221–230
Fitter AH (1985) Functional significance of root morphology and root system architecture. In: Fitter AH, Atkinson D, Read DJ, Usher MB (eds) Ecological interactions in soil. Blackwell, Oxford, pp 87–106
Foster RC (1988) Microenvironment of soil microorganisms. Biol Fertil Soils 6:189–203
Franklin JF (1993) Preserving biodiversity: species, ecosystems, or landscapes? Ecol Appl 3:200–205
Friese CF, Allen MF (1993) The interaction of harvester ants and vesicular arbuscular mycorrhizal fungi in a patchy semi-arid environment: the effects of mound structure on fungal dispersion and establishment. Funct Ecol 7:13–20
Gaskins MH, Albrecht SL, Hubell DH (1984) Rhizosphere bacteria and their use to increase plant productivity: a review. Agric Ecosyst Environ 12:99–116
Giri B, Kapoor R, Mukerji KG (2001) VAM/VA mycorrhizal technology in establishment of plants under salinity stress conditions. In: Mukerji KG, Manoharachari C, Chamola BP (eds) Techniques in mycorrhizal studies. Kluwer, Dordrecht, pp 51–85
Gochenauer SE (1981) Responses of soil fungal communities to disturbance. In: Wicklow DT, Carroll GC (eds) The fungal community: its organization and role in the ecosystem. Dekker, New York, pp 459–479
Gryndler M (2000) Interactions of arbuscular mycorrhizal fungi with other soil microorganisms. In: Kapulink Y, Douds DD Jr (eds) Arbuscular mycorrhizas: physiology and function. Kluwer, Dordrecht, pp 239–262
Hamilton WE, Dindal DL (1983) The vermisphere concept: earthworm activity and sewage sludge. Biocycle 24:54–55
Hawksworth DL (1991a) The biodiversity of microorganisms and invertibrates: its role in sustainable agriculture. CAB International/Redwood Press, Melksham, UK, 302 pp
Hawksworth DL (1991b) The fungal dimension of diversity: magnitude, significance, and conservation. Mycol Res 95:641–655
Herbert R (1999) Nitrogen cycling in coastal marine ecosystems. FEMS Microbial Rev 23:563–590
Herman RP, Provencio KR, Torrez RJ, Seager GM (1993) Effect of water and nitrogen additions on free-living nitrogen fixer populations in desert grass root zones. Appl Environ Microbiol 59:3021–3026
Huber H, Hohn MJ, Rachel R, Fuchs T, Wimmer VC, Stetter KO (2002) A new phylum of archaea represented by a nanosized hyperthermophilic symbiont. Nature 417:63–67
Hurst CJ (2002) An introduction to viral taxonomy and the proposal of Akamara, a potential domain for the genomic acellular agents. In: Hurst CJ (ed) Viral ecology. Academic Press, San Diego, pp 41–62
Jeffries P (1997) Mycoparasitism. In: Wicklow DT, Sodertom BE (eds) Environmental and microbial relationship. The Mycota IV. Springer, Berlin Heidelberg New York, pp 95–113
Jennings DH (1995) The physiology of fungal nutrition. Cambridge Univ Press, Cambridge
Jones JW (1991) Diversity and physiology of methanogens. In: Rogers JE, Whitman WB (eds) Microbial production and consumption of greenhouse gases: methane, nitrogen oxides, and halomethanes. American Society of Microbiology, Washington, DC, pp 39–35
Jorgensen BB (1982) Ecology of the bacteria of the sulfur cycle with special reference to anoxic-oxic interface environments. Philos Trans R Soc Lond 298:543–561
Jorgensen BB (1994) Sulfate reduction and thiosulfate transformations in a cyanobacterial mat during a diel oxygen cycle. FEMS Microbiol Ecol 13:303–312

Kaplan WA (1983) Nitrification. In: Carpenter EJ, Capone DG (eds) Nitrogen in the marine environment. Academic Press, New York, pp 139–190
Kapoor R, Giri B, Mukerji KG (2002) Soil factors in relation to distribution and occurrence of vesicular arbuscular mycorrhiza. In: Mukerji KG, Manoharachari C, Chamola BP (eds) Techniques in mycorrhizal studies. Kluwer, Dordrecht, pp 51–85
Killham K (1987) Heterotrophic nitrification. In: Prosser JI (ed) Nitrification. Society of General Microbiology, Spec Public IRL Press, Oxford, pp 117–126
Kjoller A, Struwe S (1982) Microfungi in ecosystems: fungal occurrence and activity in litter and soil. Oikos 39:391–422
Kluge M, Gehrig H, Mollenhauer D, Schnepf E, Schubler A (1997) News on *Geosiphon pyriforme*, an endocytobiotic consortium of a fungus with a cyanobacterium. In: Schenk HEA, Herrmann R, Jeon KW, Muller NE, Schwemmler W (eds) Eukaryotism and symbiosis. Springer, Berlin Heidelberg New York, pp 469–476
Kluge M, Mollenhauer D, Wolf E, Schüßler A (2002) The *Nostoc* – *Geosiphon* endocytobiosis. In: Rai AN, Bergman B, Rasmussen U (eds) Cyanobacteria in symbiosis. Kluwer, Dordrecht, pp 19–30
Koch B, Kaldorf M, Rexer KH, Kost G, Varma A (2004) Patterns of interaction between *Populus* esch5 and *Piriformospora indica*: a transition from mutualism to antagonism. Plant Biol (in press)
Kumari R, Yadav HK, Bhoon YK, Varma A (2003) Colonization of Cruciferous plants by *Piriformospora indica*. Curr Sci 85:1672–1674
Kyrpides NC, Olsen GJ (1999) Archaeal and bacterial hyperthermophiles: horizontal gene exchange or common ancestry? Trends Genet 15:298–299
Lakhanpal TN (2000) Ectomycorrhiza–an overview. In: Mukerji KG, Chamola BP, Singh J (eds) Mycorrhizal biology. Kluwer/Plenum, New York, pp 101–118
Lavelle P, Barois I, Martin A, Zaidi Z, Schaefer R (1989) Management of earthworm populations in agro-ecosystems: a possible way to maintain soil quality? In: Clarholm M, Bergstrom I (eds) Ecology of arable land. Kluwer, Dordrecht, pp 109–122
Lee KE, Foster RC (1991) Soil fauna and soil structure. Aust J Soil Res 29:745–775
Liesack W, Stackebrandt E (1992) Occurrence of novel groups of the domain bacteria as revealed by analysis of genetic material isolated from an Australian terrestrial environment. J Bacteriol 174:5072–5078
Linderman RG (1988) Mycorrhizal infection with the rhizosphere microflora: the mycorrhizosphere effect. Phytopathology 78:366–371
Lobry de Bruyn LA, Conacher AJ (1990) The role of termites and ants in soil modification: a review. Aust J Soil Res 28:55–93
Loper JE, Haack C, Schroth MN (1985) Population dynamics of soil Pseudomonads in the rhizosphere of potato (*Solanum tuberosum* L.). Appl Environ Microbiol 49:416–422
Lovley DR, Phillips (1994) Novel processes for anaerobic sulfate production from elemental sulfate by sulfur-reducing bacteria. Appl Environ Microbiol 60:2394–2399
Lynch JM (1987a) Microbial interactions in the rhizosphere. Soil Microorg 30:33–41
Lynch JM (1987b) Soil biology – accomplishments and potential. Soil Sci Soc Am J 51:1409–1412
Lynch JM (1990) The rhizosphere. Wiley, New York
Lynch JM, Harper SHT (1985) The microbial upgrading of straw for agricultural use. Philos Trans R Soc Lond 310:221–226
Lynch JM, Hobbie JB (1988) Microorganisms in action: concepts and application in microbial ecology. Blackwell, Oxford, 363 pp
Meyer JR, Linderman RG (1986) Selective influences on population of rhizosphere or rhizoplane bacteria and actinomycetes by mycorrhizas formed by *Glomus fasciculatum*. Soil Biol Biochem 18:191–196

Mehta AP, Torma AE, Murr LE (1979) Effect of environmental parameters on the efficiency of biodegradation of basalt rock by fungi. Biotechnol Bioeng 21:875–885

Metting B (1988) Micro-algae in agriculture. In: Borowitzka MA, Borowitzka LA (eds) Micro-algal biotechnology. Cambridge Univ Press, Cambridge, pp 288–304

Mollenhauer D, Mollenhauer R, Kluge M (1996) Studies on initiation and development of the partner association in *Geosiphon pyreforme* (Kuiz.) v. Wettstein, a unique endocytobiotic system of a fungus (Glomales) and the cyanobacterium *Nostoc punctiforme* (Kuiz.). Hariot Protoplasma 139:3–9

Moreno J, Gonsalez Loper J, Vela GR (1986) Survival of *Azotobacter* spp. in dry soils. Appl Environ Microbiol 51:123–125

Mukerji KG, Mandeep K, Varma A (1997) Mycorrhizosphere microorganisms: screening and evaluation. In: Varma A (ed) Mycorrhiza manual. Springer, Berlin Heidelberg New York, pp 85–98

Nannipieri P, Sastre I, Landi L, Lobo MC, Pietramellara G (1996) Determination of extracellular neutral phosphomonoesterase activity in soil. Soil Biol Biochem 28:107–112

Nehl DB, Allen SJ, Brown JF (1996) Deleterious rhizosphere bacteria: an integrating prospective. Appl Soil Ecol 5:1–20

Newman EI (1985) The Rhizosphere: carbon sources and microbial populations. In: Fitter AH, Atkinson D, Read DJ, Usher MB (eds) Ecological interactions in soil, plants, microbes and animals. Blackwell, Oxford, pp 107–121

Oades JM (1993) The role of biology in the formation, stabilization and degradation of soil structure. Geoderma 56:377–400

Oades JM, Waters AG (1991) Aggregate hierarchy in soils. Aust J Soil Res 29:815–828

Pant HK, Warman PR (2000) Enzyme hydrolysis of soil organic phosphorus by immobilized phosphatases. Biol Fertil Soils 30:306–311

Paul EA, Clark FE (1989) Soil microbiology and biochemistry. Academic Press, San Diego

Payne JW (1981) Denitrification. Wiley, New York

Pennisi E (1999) Is it time to uproot the tree of life? Science 284:1305–1307

Pham GH, Singh A, Malla R, Kumari R, Prasad R, Sachdev M, Luis P, Kaldorf M, Tatjana P, Harrmann S, Hehl S, Declerck S, Buscot F, Oelmuller R, Rexer KH, Kost G, Varma A (2004a) Interaction of *P. indica* with other microorganisms and plants. In: Varma A, Abbott L, Werner D, Hampp R (eds) Plant surface microbiology. Springer, Berlin Heidelberg New York pp 237–265

Pham GH, Kumari R, Singh A, Sachdev M, Prasad R, Kaldorf M, Buscot F, Oelmuller R, Tatjana P, Weiss M, Hampp R, Varma A (2004b) Axenic cultures of *Piriformospora indica*. In: Varma A, Abbott L, Werner D, Hampp R (eds) Plant surface microbiology. Springer, Berlin Heidelberg New York, pp 593–616

Postgate JR (1987) Nitrogen fixation, 2nd edn. Arnold, London

Prescott LM, Harley JP, Klein DA (1996) The diversity of the microbial world. In: Prescott LM, Harley JP, Klein DA (eds) Microbiology. WCB Publishers, Dubuque, Iowa

Quastel JH (1995) Soil metabolism. Proc R Soc 143:159–179

Redecker D, Morton JB, Bruns TD (2000) Ancestral lineages of arbuscular mycorrhizal fungi (Glomales). Mol Phylogenet Evol 14:276–284

Reid JB, Goss JM (1981) Effects of living roots of different plant species on the aggregate stability of two arable soils. J Soil Sci 52:521–541

Salyers AA, Whitt DD (2001) Diversity and history of microorganisms. In: Salyers AA, Whitt DD (eds) Microbiology: diversity, diseases and the environment. Fitzgerald Science Press, Bethesda, Maryland, pp 19–32

Schachtman DP, Reid RJ, Ayling SM (1998) Phosphorus uptake by plants: from soil to cell. Plant Physiol 116:447–453

Schimel JP, Firestone MK, Killham K (1984) Identification of heterotrophic nitrification in a Sieerrs forest soil. Appl Environ Microbiol 48:802–806
Schüßler A (2002) Molecular phylogeny, taxonomy, and evolution of *Geosiphon pyriforme* and arbuscular mycorrhizal fungi. Plant Soil 244:75–83
Schüßler A, Kluge M (2001) *Geosiphon pyriforme*, an endocytosymbiosis between fungus and cyanobacteria, and its meaning as a model system for arbuscular mycorrhizal research. In: Hock B (ed) The Mycota IX. Springer, Berlin Heidelberg New York, pp 151–161
Schüßler A, Wolf E, Kluge M (2001) *Geosiphon pyriforme* and *Nostoc punctiforme*: a unique symbiosis with implications for mycorrhizal research. ISS Symb Int 1:4–5
Seastedt TR (1984) The role of microarthropods in decomposition and mineralization processes. Annu Rev Entomol 29:25–46
Shaw C, Pawluk S (1986) The development of soil structure by *Octolasion tyrtaeum*, *Aporrectodea turgida* and *Lumbricus terrestris* in parent materials belonging to different textural classes. Pedobiologia 29:327–339
Singh An, Singh A, Kumari M, Rai MK, Varma A (2003) Biotechnology importance of *Piriformospora indica* – a novel symbiotic mycorrhiza-like fungus: an overview. Ind J Biotech 2:65–75
Slater JH (1988) Microbial population and community dynamics. In: Lunch JM, Hobbie JB (eds) Microorganisms in action: concepts and application in microbial ecology. Blackwell, Oxford, pp 51–74
Smiles DE (1988) Aspects of the physical environment of soil organisms. Biol Fertil Soils 6:204–215
Sollins P, Cromack K Jr, Li CY, Fogel R (1981) Role of low-molecular weight organic acids in the inorganic nutrition of fungi and higher plants. In: Carroll GC, Wicklow DT (eds) The fungal community, its organization and role in ecosystem. Dekker, New York
Srivastava D, Kapoor R, Srivastava SK, Mukerji KG (1996) Vesicular arbuscular mycorrhiza: an overview. In: Mukerji KG (ed) Concepts in mycorrhizal research. Kluwer, Dordrecht, pp 1–39
Stanier RY, Ingraham JL, Wheelis ML, Painter PR (1986) The microbial world. Prentice-Hall, Englewood Cliffs
Subba Roa NS (1997) Soil microbiology. IBH Publ, Oxford
Swift MJ, Heal OW, Anderson JM (1979) Decomposition in terrestrial ecosystems, studies in ecology, vol 5. Blackwell, Oxford, UK
Tabatabai MA (1982) Soil enzymes. In: Page AL, Miller Rh, Keeney DR Methods of soil analysis, part 2. Chemical and microbiological properties – Agronomy monograph, No 9, 2nd edn. Wisconsin, pp 903–947
Tate RL II (1987) Soil organic matter: biological and ecological effects. Wiley, New York, 291 pp
Tate RL III (1995) Soil microbiology. Wiley, New York
Tarafdar JC, Rao AV (1996) Contribution of *Aspergillus* strains to acquisition of phosphorus by wheat (*Triticum aestivum* L.) and chick pea (*Cicer arietinum* Linn.) grown in a loamy sand soil. Appl Soil Ecol 3:109–114
Van Niel EWJ, Gottschal JC (1998) Oxygen consumption by *Desulfovibrio* strains with and without polyglucose. Appl Environ Microbiol 64:1034–1039
Vannier G (1987) The porosphere as an ecological medium emphasized in Professor Gilarov's work on soil animal adaptations. Biol Fertil Soil 3:39–44
Varma A, Verma S, Sudha Sahay N, Britta B, Franken P (1999) *Piriformospora indica* – a cultivable plant growth promoting root endophyte with similarities to arbuscular mycorrhizal fungi. Appl Environ Microbiol 65:2741–2744

Varma A, Singh A, Sudha Sahay N, Sharma J, Roy A, Kumari M, Rana D, Thakran S, Deka D, Bharti K, Franken P, Hurek T, Blechert O, Rexer K-H, Kost G, Hahn A, Hock B, Maier W, Walter M, Strack D, Kranner I (2001) *Piriformospora indica*: A cultivable mycorrhiza-like endosymbiotic fungus. In: Hock B (ed) The Mycota IX. Springer, Berlin Heidelberg New York, pp 123–150

Verma S, Varma A, Rexer K-H, Hassel A, Kost G, Sarbhoy A, Bisen P, Buetehorn P, Franken P (1998) *Piriformospora indica* gen. nov., a new root-colonizing fungus. Mycologia 90:895–909

Visscher PT, Vandenede FP, Schaub BEM, van Gemerden H (1992) Competition between anoxygenic phototrophic bacteria and colorless sulfur bacteria in a microbial mat. FEMS Microbial Ecol 101:51–58

Von Wettstein F (1915) *Geosiphon* Fr. v. Wettst., eine neue, interessante siphone. Österr Bot Z 65:145–156

Vosatka M, Gryndler M (1999) Treatment with culture fractions from *Pseudomonas putida* modifies the development of *Glomus fistulosum* mycorrhiza and the response of potato and maize plants to inoculation. Appl Soil Ecol 11:245–251

Wainwright M (1992) The impact of fungi on environmental biogeochemistry. In: Carroll GC, Wicklow DT (eds) The fungal community, its organization and role in the ecosystem. Dekker, New York, pp 601–618

Wicklow MC, Billen WB, Denison WC (1974) Comparison of soil microfungi in 40 year-old stands of pure alder, pure conifer, and alder-conifer mixtures. Soil Biol Biochem 6:73–78

Widdel F, Hansen TA (1991) The dissimilatory sulfate and sulfur-reducing bacteria. In: Balows A, Truper HG, Dworkin M, Harder W, Schleifer KH (eds) The prokaryotes, 2nd edn. Springer, Berlin Heidelberg New York, pp 583–634

Wieland G, Neumann R, Backhaus H (2001) Variation of microbial communities in soil, rhizosphere, and rhizosphere in response to crop species, soil type, and crop development. Appl Environ Microbiol 67:5849–5854

Wilson EO (1988) Biodiversity. National Academy Press, Washington, DC

Woese CR (1987) Bacterial evolution. Microbiol Rev 51:221–271

Wolters V (1991) Soil invertebrates – effects on nutrient turnover and soil structure: a review. Z Pflanzenern Bodenkd 154: 389–402

Wood M (1989) Soil biology. Chapman and Hall, London

Part II
Microorganisms and Soil Genesis

Biological input into rock wear down must be regarded as a process with global consequences. Physical and chemical processes at the scale of single rock samples are not of general interest. Kant (1747, 1754, 1755, 1756a, b) may have been the first author to clearly suggest that geomorphogenesis includes biogeomorphogenesis (Viles 1984), and that a constant often nonevolutionary dialogue goes on between living and inert matter (Krumbein and Gorbushina 1996; Vernadsky 1997). Herder (1784), Hutton (1795), Krumbein (1983), Tetsuro (1935, 1992) and many others later verified the original assumptions of Kant (1756a). The intimate relation between the biological production of reduced carbon compounds and biological precipitation of huge amounts of highly oxidized carbon in the form of calcium carbonate over more than 3 billion years leads to the theory that plate tectonics (i. e., uplifting and shifting of large continents and mountain ranges) and rock and mountain wear down are processes driven by sun-powered biological reactions on a global biogeochemical and even geophysiological scale (Anderson 1984; Krumbein and Schellnhuber 1992; Krumbein 1996). Biogenic (mostly microbiologically driven) enhancement of rock wear down has a strong influence on the overall evolution of the Earth system and is able to extend the life span of the biosphere significantly by factors up to 1.2 Gyr (von Bloh et al. 2003). Weather and its long-term average – climate – is a consequence of biologically driven rock wear down and not the cause of "weathering" (Schwartzman and Volk 1991b; Paterson 1993; Krumbein 1996).

Biological attack on bare mineral, rock and mountain surfaces does not necessarily lead to soil formation. Soil formation is a multiple organized process, which is not an undeviating consequence of rock wear down. Both soil formation and soil stabilization involve biotic processes with different functions on a local and global biogeochemical scale. Carbon compounds rich in energy (organic matter) and those very poor in energy (limestone) are almost always channeled through the multiple functions of carbon dioxide, the anhydride of carbonic acid. Therefore, neither rock formation nor rock wear down can be seen clearly without referring to the carbon dioxide cycle and actual problems associated with it (Golubic et al. 1979; Krumbein and Swart 1983; Krumbein and Schellnhuber 1990; von Bloh et al. 2003).

2 Carbon Dioxide and Rock Wear Down

Vernadsky (1924, 1929, 1930, 1944, 1997) and Lovelock (1979) claimed that the chemical composition of the atmosphere is biologically controlled. The total amount of carbon in the atmosphere is less than 1% of that represented in living matter, and practically each atom of carbon present in living matter passed many times through the atmosphere and back to the living organisms during its life time (Vernadsky 1929, 1930, 1944; Krumbein 1990). It was further claimed that in its present composition the atmosphere is kept more or less stable against external changes and impacts by the force of living matter (e. g., Berner 1992). Anderson (1984), Krumbein (1983, 1996), and Krumbein and Schellnhuber (1990, 1992) expanded this view to the status of the Earth's crust and its relations to the Earth's mantle in a global geophysiological view. This view of a mainly microbially driven energy and material pumping system has been recently supported by complex calculations of dynamic equilibriums on a global scale by climate impact research (von Bloh et al. 2003). The authors suggest on the basis of reactivity calculations that biological erosion, in fact, stabilizes the carbon dioxide/carbonate equilibrium via increased or reduced carbon dioxide exchange with rocks and soils (von Bloh et al. 2003).

In 1840, Liebig had already realized that life is the driving force of the global carbon cycle. However, Vernadsky (1924) and Lovelock (1979) seem to have first pointed to the fact that the atmosphere of a living Earth is deviating from the theoretical atmosphere of a planet Earth without life in the same position in the solar system. Thus, the increase and decrease in carbon dioxide in the atmosphere must have occurred from the beginning of life on Earth (Arrhenius 1896; Callendar 1938, 1939). The limits of periods of shifting concentrations larger than the seasonal fluctuations of about 2% annually are unknown in the geological past, although long-term fluctuations in this order of magnitude can be seen from gas analyses in ice cores taken near the Polar region (Barnola et al. 1987; Walker 1993). The work of Milankovitch (1920) suggested that deviations of climate and thus of the terrestrial carbon dioxide equilibrium may be related to changes in the sun activity and its impact on Earth's atmosphere (Milankovitch cycles). Whether equilibrium between changes in solar energy emission, changes in its trapping and storage in the Earth's system and geophysiological temperature effects in the crust is the main driving force stabilizing life conditions is still debated (Krumbein and Schellnhuber 1990, 1992).

Since 1958 the carbon dioxide content of the atmosphere has been monitored. It was Arrhenius (1896) who first pointed out that carbon dioxide concentration may change surface temperatures on Earth. Buffon (1749) was already convinced that major human impacts influence the climate. Interactions between man and climate and its bearing on numerous themes of life and society have long been a topic in philosophical discussions and science theory (Herder 1784; Kant 1747, 1754, 1756a; Tetsuro 1992; Gumilev 1990). Many statements since Pliny the Elder peak in the assumption that mankind is culpable of changing climate and environment. In contrast, the much stronger (biogeomorphogenetic) power of organisms (including microorganisms) was first expressed by authors like Paracelsus (see Krumbein et al. 2003b) and Kant (see Krumbein 1996).

Callendar (1938) suggested that production of carbon dioxide by fossil fuel combustion, limestone burning and anthropogenic forest fires influences temperature and climate on the basis of calculated data. The exact transfer and balances, however, became known when the so-called Keeling curve was brought to our attention (Callendar 1939, 1958). Fossil fuel burning can be very clearly calculated from commercial statistics. From this and from rough estimates of the increase of deforestation for agricultural use (artificial versus natural forest fires), a theoretical increase of carbon dioxide in the atmosphere can be calculated. The so-called greenhouse effect is a composite feature that is caused by many gases escaping into the atmosphere. Carbon dioxide, which makes up about 50 and 75%, seems to stem from fossil fuel burning. In general, the increase in rain acidity is practically 99% due to carbon dioxide and less than 1% by $SO_2$and NO_x. Therefore, the indirect increase in rock degradation rates through acidity of rain, runoff water and atmosphere must be attributed mainly to the effects of carbonic acid and perhaps also additionally to the heavy load of air-borne anthropogenic organic compounds such as hydrocarbons, sugars, and fatty acids, which may all be transformed by rock-dwelling microorganisms into aggressive organic acids.

All major groups of microorganisms, i. e., chemolithotrophic, chemoorganotrophic and phototrophic bacteria, algae, plants, fungi and protozoa, have been reported to exist in such subaerial environments and to contribute to rock wear down. Phototrophic microorganisms like lichens, algae and cyanobacteria were the first to receive attention. Figure 1 gives an example of extensive biopitting by desert lichen under relatively dry and hostile conditions. The idea that heterotrophic life forms rather than autotrophic ones are the most enduring and important rock dwellers came up in the last century (Krumbein 1966; Staley et al. 1982; Gorbushina and Krumbein 1999, 2000). Initially, it was thought that the biological attack on rocks and minerals is driven either by photosynthesis of algae and lichen (Sollas 1880; Bachmann 1890), or by chemolithotrophic processes through nitric or sulfuric acid-producing bacteria (Müntz 1890; Isacenko 1936). However, Bachmann (1916) and Paine et al. (1933) had already hinted at the chemical attack by heterotrophic fungi and bacteria on rock surfaces. Fungi have been reported in a wide range of rock types including limestone, soapstone, marble, granite, sandstone, andesite, basalt, gneiss, dolerite, amphibolite and quartz, even in the most extreme environments, e. g., hot and cold deserts (Staley et al. 1982; Sterflinger 2000). Even in basaltic rocks in Iceland, the increase in porosity flaking and etching in rocks has been attributed to fungal presence (Etienne and Dupont 2002). The most obvious subaerial biofilm is the epilithic and endolithic lichen

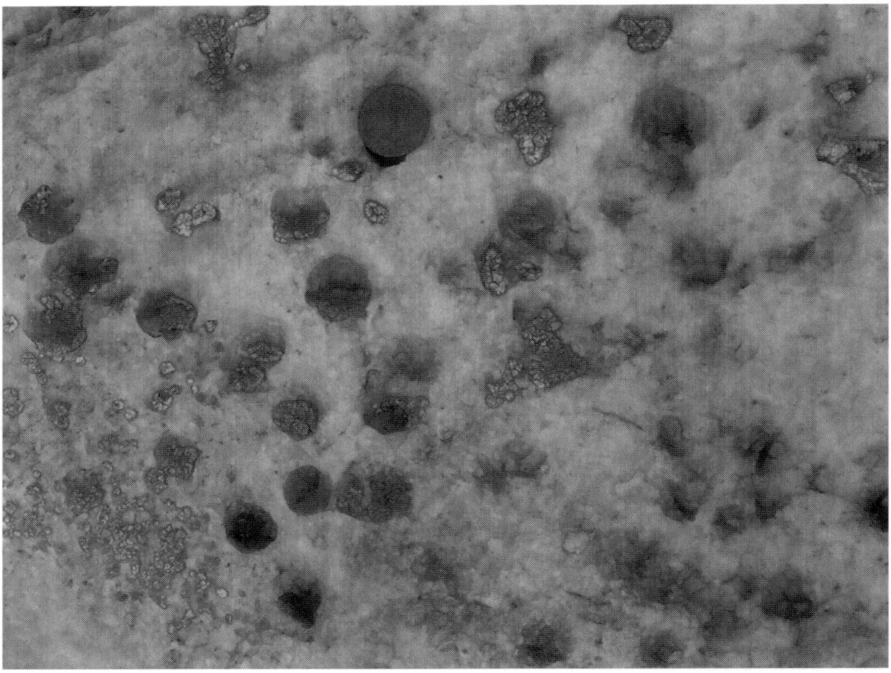

Fig. 1. Dolomitic marble from the vicinity of the Gobabeb Desert research station (Namib Desert, Namibia) showing prominent biopitting pattern. Such wear-down phenomena can be observed on many subaerial rock surfaces as a result of mechanical and chemical attack by epilithic and endolithic lichens (Gehrmann et al. 1988) and by fungi (Sterflinger and Krumbein 1997) with pits of a smaller scale

crust that covers surfaces of rocks (Fig. 1). It thrives and metabolizes in all regions of the world and is only out-competed by specialized fungal, actinomycetal and bacterial biofilms. Practically invisible to the naked eye is the thin spiderweb-like growth of fungi and actinomycetes that occurs on bare rock even under the harshest environmental conditions (Figs. 2–4). First attempts and experiments to explain the potential of mechanical/physical attack of rocks and minerals by microorganisms were made by Krumbein (1969) and Dornieden et al. (1997, 2000), whereby these functions are well established for macroorganisms (basidiomycetes, plant roots, trees). The most astonishing findings of biodeterioration research in the past decade are that the number of chemoorganotrophic microorganisms (including heterotrophic free-living bacteria and fungi) that settle on and within stones is mainly related to the amount of organic energy-rich compounds contained in the surrounding atmosphere. Organic pollution in large cities drastically increases the numbers of heterotrophic organisms attacking rock and thus their deteriorative activity (Krumbein 1969;

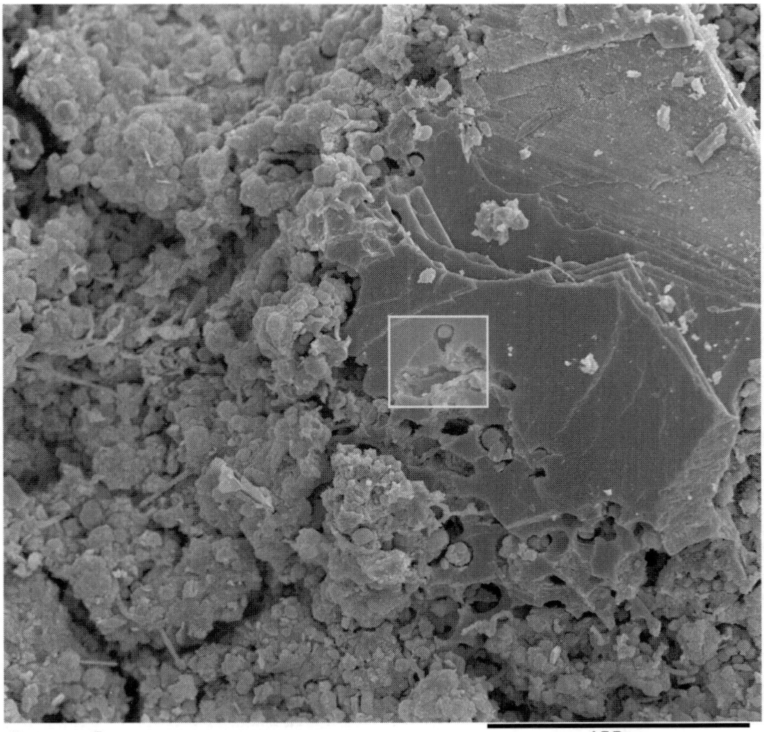

Fig. 2. SEM-micrograph of etching and mechanical decohesion activity by a complex microbial community on marble from a quarry in Carrara (Italy). The multi-compound subaerial biofilm is composed of heterotroph free-living fungi (filamentous and yeast-like), microscopic unicellular algae and accompanying bacteria. Macroscopically, the marble surface has a slightly gray surface appearance created by numerous tiny microcolonies disguising the original shiny white of the marble. The biological growth and wear down partially follow the cleavage zones between the marble grains. However, some hyphae and yeast-like cells of the fungi penetrate through the crystals without following any weakness of the mineral structure (see Fig. 3), thus drastically increasing the contact surface between the biofilm and the rock material

Warscheid et al. 1991). Even in remote places such as the world's large deserts, heterotrophic fungi settle on rocks without any support from autotrophic algae or symbiotic algal partners within lichens (Staley et al. 1982; Gorbushina and Krumbein 1999, 2000). As a matter of fact, free-living fungi are the most enduring organisms under extremely changeable desert conditions with rainfall below 180 mm/year (Perry 1979). They are widely abundant cosmopolite invaders of air-exposed rock surfaces and prevail in all soil formation processes as well.

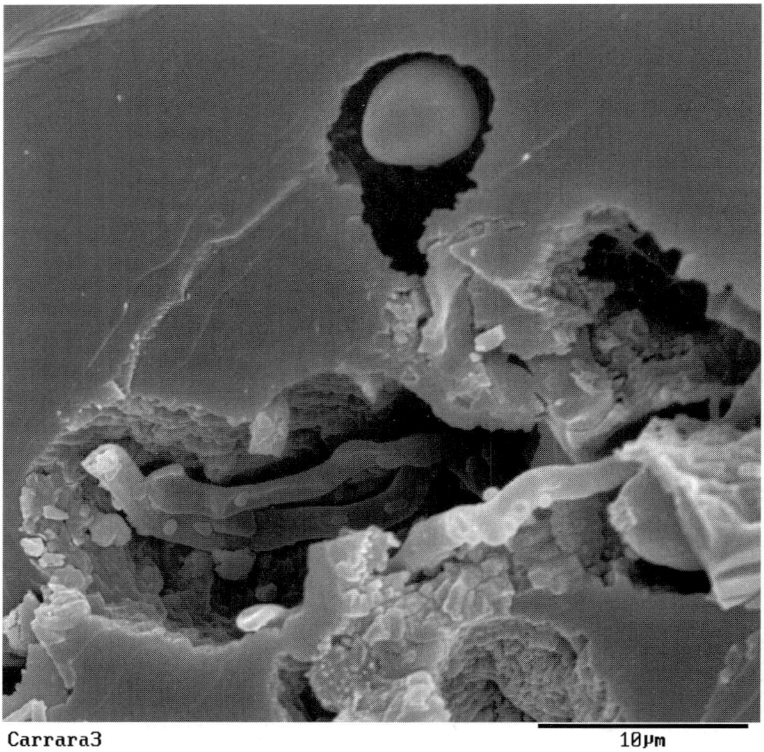

Fig. 3. Magnification of the area marked in Fig. 2. Fungal hyphae with accompanying bacteria and algal cells are visible in the pits and channels of the marble crystals. The congruence of the shape of the microorganisms and the new rock surface microrelief witness a direct physical or chemical biological impact. The etched surface of the channel interior suggests a chemical mechanism of etching by organic acid and acid polysaccharides present in the abundant extracellular material of the biofilm

3
Balance of Carbon Dioxide Sources and Sinks

An important fact to keep in mind is that only 45–50% of the theoretical increase in carbon dioxide and other greenhouse gases is measured over the years in the Keeling curve. In addition, the Keeling curve is very straight and does not reflect major energy crises and changes in the behavior of the industrial society, while the annual change in photosynthesis and respiration between summer and winter is absolutely clear and constant, and even the differences in photosynthesis and respiration in the annual cycle between the southern and the northern hemisphere are detectable (Boden et al. 1990). Some physicists actually claim that even the slow-down effects of deciduous trees transporting water up in summer and down in winter

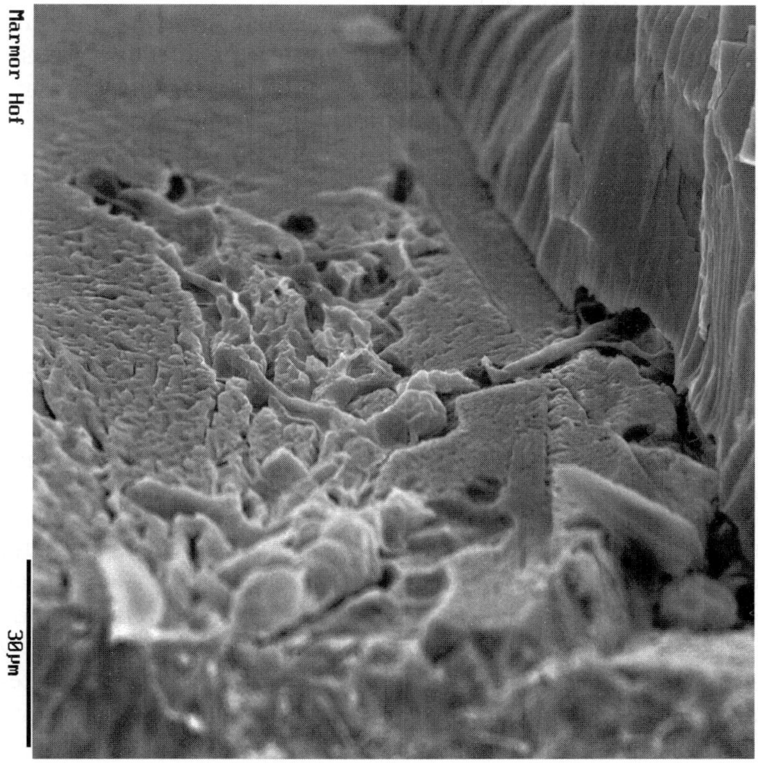

Fig. 4. Biocorrosion on recent grain detachment on a pre-weathered and biofilm-covered Carrara marble sample from Sans-Souci (Potsdam) exposed in a field experiment at the geomicrobiology laboratory (ICBM, Oldenburg University, Germany). The fungal hyphae penetrate deeper into the rock following weak cleavage lines between the crystals, which bring about a mechanical separation of marble grains. The absence of the covering grain reveals an etching pattern created on the underlying cleavage surface of the marble grain by the penetrating and branching fungal hyphae. Once in firm contact with the marble surface, and in a place protected from high solar irradiation, hyphae interact with the mineral by chemical (CO_2, organic acids, acid EPS) and physical means and hereby create exact imprints of living and former (lysed) cellular structures

in the northern hemisphere forests change the rotational pattern of Earth itself (Krumbein and Schellnhuber 1992). Thus, it can be assumed that the major part of gases from fossil fuel burning must be stored immediately elsewhere and not in the atmosphere, or that with perhaps a lag of a few years a re-storage mechanism operating in a yearly cycle and involving fluxes equal to the ones of the input into the atmosphere must operate. The increase in photosynthesis and respiration in land biota cannot be the buffering factor of the fossil fuel burning and cannot serve as a reservoir. The input of carbon dioxide into the atmosphere, therefore, must be balanced by some unknown or not yet clearly analyzed trapping factor(s) and mechanisms in the continental or oceanic surfaces must account for this. Such trapping factors and sinks discussed so far in the scientific literature include:

1. Diffusion into shallow and deep ocean, followed by eddy mixing involving the main marine currents and the up-welling or sinking of the water body.
2. Increase in oceanic and terrestrial photosynthetic binding of carbon dioxide.
3. Increase in oceanic and/or terrestrial calcium carbonate and humus (kerogen) production and trapping in terrestrial or oceanic sinks.
4. Increase in terrestrial and/or oceanic weathering (physical and chemical) of carbonate and silicate rocks under binding of CO_2.

The last assumption strictly relates to this chapter on mechanisms and functions of rock wear down. The recent calculations of von Bloh et al. (2003) make it most probable that the scenario of increased carbon dioxide uptake by the solid rock system and increased biological weathering hereby is a most effective balancing mechanism of excessive carbon dioxide release into the atmosphere. This in turn seems to be related to the amount of organic carbon species available in the atmosphere to fuel microbial metabolism on and in rocks.

4
Rock Wear Down as a Potential Carbon Dioxide Sink

Life processes, in general, accumulate and disseminate energy and matter at speeds between 1000 and 100,000 times higher than abiotic chemical and physical processes. As references, we may quote Vernadsky (1924, 1929, 1930, 1997), Schwartzman and Volk (1990) and Krumbein and Schellnhuber (1992). Furthermore, we know that the continents and oceans as they are today have not always been in the same relation with regard to size and shape. However, with the exception of a few authors (e.g., Krumbein 1983; Anderson 1984; Krumbein and Schellnhuber 1990, 1992), we have not come to the right conclusions from this. Anderson (1984) stated that plate tectonics may perhaps only exist because life creates limestones on Earth which may explain some of the mantle–crust relations. Krumbein (1983, 1990) and Krumbein and Schellnhuber (1992) claim that back-coupling on a million-year scale is as reasonable for a living planetary system as it is on an annual scale for a living human system or a minute scale for a bacterial system. They suggested that crust dynamics, including sediment filling of oceanic basins and biologically accelerated wear down of mountain ranges, should have a deep impact on the atmospheric cycles and on the climate as well (Paterson 1993).

The concepts of subaerial biofilms covering practically all rock surfaces with poikilophilic and poikilotroph heterotrophic microorganisms living on a minimum amount of water and going through the most extreme changes in nutritional and environmental conditions (Gorbushina and Krumbein 2000) have largely changed our view of biological wear down of rock materials under all climatic conditions (Krumbein 1972; Dornieden et al. 1997, 2000; Gorbushina and Krumbein 1999, 2000; Saiz-Jimenez 1999; Gonzalez-del Valle et al. 2003; Gorbushina et al. 2003a; Krumbein et al. 2003b). Furthermore, these rock-dwelling microorganisms and subaerial biofilms (Gorbushina and Krumbein 2000) have been shown by modern molecular ecology techniques to be able to endure 10,000 times longer periods of life suspension than subaquatic biofilms (Aardema et al. 1983; Krumbein et al. 2003a). Our understanding of diversity in such environments has been largely enlightened since the development of PCR-based techniques like denaturing gradient gel electrophoresis (DGGE; e. g., Gorbushina et al. 2003a). Aardema et al. (1983) were among the first authors to realize the survival potential, not only of microorganisms, but also of their genetic information in natural environments. This, together with new insights into the preponderance of terrestrial surface areas or geomorphologies over aquatic ones, makes the air-exposed rock environment one of the most important issues in modern ecological research.

5 The Fractal Dimension of Biological Rock Wear Down

It is anchored in our minds from school lessons that the Earth consists of 70% aquatic and 30% terrestrial reactive surface area. This is most probably wrong for life processes. Fractal physics or nonlinear dynamics ask for views different from the Ptolemaic or Mercator projections we learned at school and keep in our minds. Fractal science and calculations have recently invested a lot of energy into so-called conformal maps of the globe. Actually, data of the whole planet are not good enough to create relief maps and calculate areas which would be suitable for making global climate calculations. However, it is worth making an attempt on a small scale to clarify our thoughts about rock wear down and climate. Vernadsky (1997), and before that, D'Arcy Thompson (1917, 1961), spent a lot of thought and energy on the optimal topographical relief and on the consequences of geographical projections. Another morphogenetically oriented mind in global ecology was the zooecologist Beklemishev (Levit and Krumbein 2001). Beklemishev regarded the biosphere as a "relational functional harmony" or "a weakly individualized morphoprocess". All these attempts have one and the same in mind: the morphogenetic potential of Earth as a living system which was so enigmatically introduced by Kant when he stated that the deviations of Earth from the spherical shape need further attention (Krumbein 1993, 1996). Exactly this is being done today in "global biogeomorphogenesis" research as an alternative to "global biogeochemistry".

The fractal analysis of rocks and terrestrial surfaces forces a different view upon us. According to fractal physics and calculus, the reactive surface of the ocean shows a relatively low fractal dimension, while the continental surface – be it a soil system, a rock system, a corn field, forests, or the Alps and the Himalayas – shows an incredibly high reactive surface. Calculations of geomorphologies (Wong and Howard 1986; Dietler and Zhang 1992), of pore space of different rock systems (Hansen and Skjeltorp 1988; Krohn 1988) and in a general and global frame (Turcotte 1989) teach us to understand that the sensitive surface morphology of the terrestrial environment is much larger than that of the oceanic surface.

If we go to the level where microbial influences on chemical and physical weathering are perceived, we may assume surface areas interacting with atmospheric carbon dioxide and humidity that are about 1 million or even 10 million times larger than the reactive oceanic interface and its neuston and plankton productivity near the oceanic surface.

Gehrmann et al. (1988) and Krumbein et al. (2003b) have derived that the microbial physical and chemical attack on rock surfaces acts at scales between 20 and 20,000 μm^2 with depth extensions as steep as 10,000 μm and more. From this, we estimate that the terrestrial interactive topography or biogeomorphogenetic surface is by several if not many magnitudes higher than that of the aquatic environment. Exact fractal dimensions are difficult to calculate at present, but we can assume that the area of some parts of the calcareous Alps or the Mediterranean and Namibian limestone hills, in which biogenic weathering prevails, has a reactive surface of a minimum of 10^8 km^2 instead of the topographically derived 10^4 km^2 in a simple Mercator projection.

Under the assumption that the surface area $< 25 \times 25$ cm can be described by a homogeneous fractal surface dimension, Dietler and Zhang (1992) derived the following total spatial surface approach. If a landscape or a smaller section of topography is a self-affine fractal, then it is expected that the so-called mean deviation W (height of the Earth's surface) circumscribes the deviation from the horizontal length scale L in fractal terms such that

$$W \sim L^{\chi} \tag{3.1}$$

in which χ (< 1) is the so-called roughening exponent, which is estimated by the origin of erosional processes. The topographical notion can be derived from the following approach: imagine that the Swiss mountains are partially immersed in water and that a coastline forms in the dimension of the British coastline (Mandelbrot 1967) as seen by a bacterium swimming around sand grains and looking for the chemotactically most promising food source. According to mathematical models, the roughening exponent is related to the fractal dimension D_h of horizontal transects by

$$D_\mathrm{h} = 2 - \chi \tag{3.2}$$

which corresponds to the above-mentioned selected case of an imaginary coastline if the surface was partially immersed in water. The area of the mountainous hypermicroscopic surface described above can also be a fractal with the dimension D_s, and related to the previous equation by

$$D_\mathrm{s} = 3 - \chi\,. \tag{3.3}$$

In this case, the roughening exponent is equivalent to the Hurst exponent H, which will be used for strictly self-affine scaling data. This we can assume to be extended to the micrometer scale at least theoretically, although the satellite surface roughening data have a maximal resolution of 25 cm. If we agree that on the scale of 10 mm to 10 µm the morphology of the rock environment is shaped by lichen, fungi, algae and bacteria (Figs. 1–4), we may also assume that the reactive wear down surface is really biogenic (Gorbushina et al. 2001). Furthermore, it has to be noted again that heterotrophic fungi may play the most important role in these processes on bare rocks as well as under soil cover (Jongmans et al. 1997; Sterflinger and Krumbein 1997; Gadd 1999).

Mathematically, this biogenic roughening component in its equivalence with the Hurst exponent H will be defined by

$$\langle |V(\chi + L) - V(\chi)| \rangle \sim L^\mathrm{H} \tag{3.4}$$

where $V(\chi)$ is a generic variable and L will be data between 10 and 10,000 µm (instead of the usual 250 m). From these fractal considerations, we deduce that the reactive terrestrial surface area for microbial wear down systems in soils and rocks for exchange processes with atmospheric carbon dioxide and other (also organic nutrient- and energy-rich) compounds and gases could be by an order of magnitude of at least 10,000 times larger than that of the oceans.

However, most of the models in the global change literature operate with a dimension of 70:30 of reactive oceanic surface versus reactive continental surface.

On the basis of these general remarks, we now approach the question of the potentially missing or neglected sink in terms of chemical reactions that are usually biologically enhanced at surface conditions by factors between 1 and 1000 (see Krumbein 1988; Schwartzman and Volk 1990). We will consider the trapping capacity of rock and soil internal surfaces for carbon dioxide chemically and biologically.

6
Calcium Carbonate and Silicate Wear Down, Dissolution and Precipitation With Special Reference to Biological Rock Degradation

It is assumed that rock weathering is a potential sink of carbon dioxide (Schwartzman and Volk 1990, 1991a,b; Walker 1993). The mechanisms of the distribution of rocks and reactive compartments of the Earth in terms of weathering and wear down as well as the creation of new reservoirs of rocks have been known since Greek, Hebrew and Roman times (references can be found in the Holy Script, see Krumbein and Jens 1981; Herodotus, and Lucretius). Keller (1957) reviewed the chemical equations and the possible biological factors. Krumbein and Dyer (1985), Leyval and Berthelin (1991), Krumbein and Urzì (1993) and Saiz-Jimenez (1999) have pointed to biological processes as the main accelerating factors in rock wear down.

Li (1972) listed the global amounts and dimensions of the distribution of silicate, carbonate and oxide rocks and their transformations in geochemical budget calculations.

There are only a few rock-forming minerals and rock types which are submitted to constant wear down. Most of the dissolved or detached materials will immediately be submitted to transport and physical and chemical sedimentation forming new rocks. Only a small fraction of rock material will be placed and fixed for extended periods of time in soils.

The following equations imply that large amounts of carbon dioxide are bound by biologically operated mineral and rock wear down processes. Carbon dioxide liberated in these processes does not generally return to the atmosphere. It is normally transported into rivers and oceans or precipitated as fresh water or found in shallow marine carbonate sediments and carbonate-bearing rocks. Two general weathering equations would be needed to explain the process. Before the simplified model is given, it must be pointed out that many elements and minerals are involved, the major groups being depicted and characterized in Tables 1–3.

A typical generalized wear-down and reaction equation for silicates starts with a magmatite of a general composition (3.5)

$$100\,\text{kg Magmatite} + \left\{\begin{array}{c} 56\,HCl \\ 219\,CO_2 \\ 20\,H_2S \\ 160\,H_2O \\ 16\,H_2 \end{array}\right. \sum + \left\{\begin{array}{c} 4300\,H_2O \\ 7\,N_2 \end{array}\right.$$

$$\rightarrow \left\{\begin{array}{c} 595\,SiO_2 \\ 47\,Fe_2O_3 \\ 5{,}5\,FeS_2 \\ 7\,CaSO_4 \\ 153\,CaCO_3 \\ 22\,MgCO_3 \\ 18\,NaCl \\ \sum_{\text{clay minerals}} \end{array}\right. + \left\{\begin{array}{c} Na^+ \\ 4\,Mg^{2+} \\ 38\,Cl^- \\ 2\,SO_4^{2-} \\ 4300\,H_2O \end{array}\right. + 44(CH_2O) + \left\{\begin{array}{c} 17\,O_2 \\ 7\,N_2 \end{array}\right. \quad (3.5)$$

Special wear-down reactions, which always involve the binding of carbon dioxide either in the soluble form (carbonic acid or bicarbonate) or as the freshly precipitated form (calcite or aragonite) into the effluent of mountains and rock or soil surfaces and into the shallow sea are as follows:

$$\begin{array}{l} \text{Olivine } Mg_2SiO_4 + 4CO_2 + 4H_2O \Rightarrow 2Mg^2 + 4HCO_3^- + H_4SiO_4 \\ 2Mg^+ + 4HCO_3^- \Rightarrow 2MgCO_3 + 2CO^2 + 2H_2O \end{array} \quad (3.6)$$

$$\begin{array}{l} \text{Feldspar } CaAl_2Si_2O_8 + 2CO_2 + 3H_2O \\ \quad \Rightarrow Al_2(OH)4Si_2O_5 + Ca^{2+} + 2HCO_3^- \end{array} \quad (3.7)$$

$$\text{Carbonates } CaCO_3 + CO_2 + H_2O \Rightarrow Ca^{2+} + 2HCO_3^- \quad (3.8)$$

It is generally agreed that the dissolution of carbonates via biogenic carbon dioxide, on average, binds two molecules of carbon dioxide, while in the case of carbonate precipitation only one molecule returns to the atmosphere (or to the aquatic system). Thus theoretically, each mole of calcium carbonate dissolved eliminates 1 mol of carbon dioxide from the atmosphere and transfers it to other reservoirs. The same holds true for silicate rocks, at least for composite silicates such as olivine/forsterite, feldspars/feldspatites, biotite/muscovite and many clay minerals.

Schoeller (1955) was one of the first authors to point out the speed of degradation of silicates under the influence of biologically enriched carbonic acid in subterranean waters. Moreover, the dissolution of quartz is possible under the same conditions in relatively alkaline environments. Recently, Schwartzman and Volk (1990, 1991a,b) have constructed a scenario of rock degradation as a global carbon sink. Biotic enhancements of the wear down of rocks with factors between 1 and 1000 are calculated in this way. Factors around 10,000 may even be possible in view of most of the modern literature which indicates that biokarst as the interaction of biogenic carbon dioxide with rock is slower by a factor of about 10 than direct biological attack and transfer. On the other hand, modern views of biological buildup and erosion of carbonate rocks, expressed as positive and negative or productive and destructive biokarst (Wang et al. 1993; Gorbushina et al. 1996), indicate a more accentuated influence of biota on fast solubilization and precipitation of calcium carbonate. Thus, the biotic wear-down equations hold true possibly for all rock and mineral types.

$$\frac{V}{V_0} = \frac{B_{\text{biotic}}}{B_0} \left(\frac{P_{\text{CO}_2}}{P^0_{\text{CO}_2}} \right)^{\alpha} e^{\beta(\Delta T)} e^{\gamma(\Delta T)} \frac{A}{A_0} . \tag{3.9}$$

In this equation, data for carbon dioxide outgassing (V), abiotic and biotically enhanced wear down (B_0, B_{biotic}), the related abiotic and biotic soil carbon dioxide pressures (P_{CO_2}), global temperature deviations from normal (T) and a classical (nonfractal) land area term are used. If we apply the fractal dimensions calculated above, factors of 1000 and more may be added to the biological wear-down potential through an increase in reactive surface area. Even under classical assumptions, it is concluded that biotic denudation must be a fast and efficient carbon dioxide trap. Biological denudation, however, would come to a standstill when no tectonic uplift occurs. Thus, Precambrian versus modern uplift rates can be considered just as suggested by Krumbein (1969, 1983), Anderson (1984) and Krumbein and Schellnhuber (1990, 1992). Uplift rates are measurable today and have occurred in the past. Biogenic denudation is more than 10,000 times faster than nonbiological physical or chemical wear down in the presence of distilled water under actual atmospheric conditions. Further biological denudation fluxes will turn out to be considerably larger when the true fractal reactive volume is used instead of the classical Mercator projection surface area (A and A_0) for ancient and present-day surfaces. Anderson (1984) and Krumbein and Schellnhuber (1992) discuss the probability that the main reason for tectonic uplifting is plate tectonics. Plate tectonics in turn are apparently closely tied to geophysiologically controlled ratios between organic and inorganic carbon reservoirs within the crust and upper mantle.

Biogeomorphogenetic or rock wear-down processes are kept in dynamic equilibrium by life processes on Earth. If we consider further that none of the sinks considered so far (e.g., terrestrial or aquatic photosynthesis increase, increased downward transport of fecal pellets and carbonate particles, diffusion processes, etc.) have been shown to be valid sinks, it seems very probable that both biogenic rock degradation and biologically initiated "biogeomorphogenesis" are the main factors that may counterbalance the overproduction of carbon dioxide by fossil fuel burning on the needed time scales and speed because the reactive reservoir of rocks is large enough to adsorb the carbon dioxide passing annually into the atmosphere by a considerable increase in biotic weathering or rock wear down. The latter is probably driven by a heterotrophic flora fed by organic compounds supplied mainly by the atmosphere (Krumbein and Gorbushina 1996; Gonzales-del Valle et al. 2003; Gorbushina et al. 2001, 2003a). This flora is creating its own niche for life on and within rock materials. They create cavities and dwelling places not unlike those created by any animal searching for a safe place to live and care for its offspring. Figures 2–4 illustrate the enormous potential of rock carving and cavity production of some heterotrophic and phototrophic organisms.

Natural fossil fuel burning as well as natural biological oxidation of hydrocarbons has been documented for many periods of Earth's history. Thunderstorms and volcanism have not only regularly set forests on fire, but also peat, lignite, coal, oil deposits and oil shale. The best-studied example is an oil shale between Jerusalem and Jericho, which was set on fire in the Tertiary and produced temperatures high enough to initiate a "natural" Portland cement oven with its typical minerals from the carbonate/shale content of the rock in question, which contained "only" about 2–5% petroleum. The biological oxidation of hydrocarbons is witnessed by petroleum seeps at 2000–3000 m water depth in the Atlantic Ocean of South America, where a dense microbial population creates an organic matter oxidation oasis comparable to the chemolithotrophic black smoker environments of the Deep Sea. Certainly, the anthropogenic dimension of fossil fuel burning and cement industry may be more severe than these accidental forest fires by lightning or the coal and petroleum deposits seeping out and being set on fire in the geological past. These processes, however, must have occurred at a very high frequency. Overproduction of organic matter under optimal conditions for photosynthesis and simultaneous overproduction of oxygen in the atmosphere may have occurred many times in the geological record. This, in turn, caused even wet leaves to burn when natural fires occurred. Furthermore, it cannot be underestimated that human fossil fuel burning affects less than 0.1% of the total organic matter embedded in the sedimentary rocks. In the geological past and today, 99.5–99.9% of the reduced organic matter rising to the Earth's surface was and is oxidized in a slower way by natural biological or physical processes than described before. The total mass oxidized at these natural rates is considerably larger. Therefore, it seems that fossil fuel burning is only a minor factor in the total release of carbon dioxide from fossil-reduced carbon deposits. The presently observed disequilibrium of carbon dioxide levels in the atmosphere may thus have very little to do with anthropogenic activities. It may be just a response pattern to other geophysiological constraints and reactions that have happened in the past and will happen in the future even without the influence of humankind. This is witnessed also by the carbon dioxide curve from ice cores. One of the most popular explanations of the deviations from average in the geological pre-human past is the sun cycle theory of Milankovitch (1920). Another theory is the shifting equilibrium of sun spot activities in complex time and intensity relations between 10 and 14 years. A further good hypothesis is that Earth's temperatures and global carbon dioxide concentrations in the atmosphere may correlate with changes in the direction and speed of the deep oceanic water circulation, including the model that low latitude oceans are a source and high latitude oceans rather a sink for atmospheric carbon dioxide. Degens (1989) hints that the oceanic water movements over 160,000 years correlate well with the study of ice cores and not with any industrial input. Major volcanic events (e.g., Cracatao), however, are well reflected in ice core data.

Since the annual flux of carbon produced by humankind is less than 2% of the total fluxes, it is, however, impossible to verify these models at present. A last hypothesis is that of oscillating population dynamics on a global scale. Ecosystems are in a shifting equilibrium from high to low biodiversity and high to low individual numbers of a single or several species with the effect of higher and lower productivities. This may hold true also for the global life support system of Earth. Thus, a biologically highly diversified Earth would rather equilibrate carbon dioxide and oxygen levels while a low general biodiversity with associated extremely low or (more probable and frequent) extremely high numbers of individuals would yield disequilibria of the gas pressures and fluxes.

Therefore, the question of a sink for additional anthropogenic carbon dioxide in the atmosphere may be an obsolete one because it relates to only 100 years of oxidation of less than 0.1% of the organic matter that is and was constantly oxidized throughout the history of the Earth, when it reached the surface conditions. On the other hand, many mechanisms have functioned before and continue to function at present, which may bring about the same or even more drastic increases and decreases in atmospheric carbon dioxide. The most promising phenomenon may be increasing and decreasing rock wear down through the trapping of carbon dioxide brought about by a mainly heterotrophic rock dwelling microflora. In the past, acid rains, atmospheric influences and other physical-chemical factors were thought to be responsible for rock weathering. Biological rock wear down was later attributed mainly to photo-autotrophic (lichen) and chemo-autotrophic (Thiobacilli, nitrifying bacteria) organisms. Nowadays, more and more information is collected and solid data are produced which relate the biological impact on rock wear down to the activities of surface dwelling poikilophilic heterotrophic organisms such as chemo-organotrophic bacteria, actinobacteria and fungi (Krumbein and Gorbushina 1996; Krumbein 1998; Saiz-Jimenez 1999; Gorbushina and Krumbein 2000; Gonzalez-del Valle et al. 2003). A few examples of the enormous destructive power of fungi are given in Figs. 1–4.

As to the dimension of the sink for that fraction of organic matter which is presently altered by anthropogenic factors and is recycled at accelerated speed, it may be said that it will vary in its dimension and in the transfer rates and speed with varying carbon dioxide concentrations in the atmosphere, surface and phreatic water bodies. From this biotic weathering and cooling effect, some authors (Schwartzman and Volk 1991b), also derive a primordial Earth scenario in which the first steps of microbial soil formation through biofilms and microbial mats may have helped to cool down surface temperatures of Earth from an expected 55–60 °C to the present 20 °C. Their thought that the initiation of biological rock wear down and subsequent Precambrian soil formation belong to the earliest biological impacts on the stabilization of life on Earth is well conceived (Krumbein et al. 2003a).

7 Conclusions

As a general conclusion to our considerations, we may state that:

1. Biologically driven rock wear down under the binding of carbon dioxide may represent the missing carbon dioxide sink in global carbon flux considerations.
2. Fractal calculus of reactive surface areas makes the terrestrial surface the optimal sink of this kind by far exceeding all other potential sinks and certainly also the present-day dimensions of additional human carbon dioxide production. The reactivity is on the level of microbial physiology, i. e., faster than the physiological speed of macroorganisms including humankind.
3. Oceanic crust regions may bind volcanic carbon dioxide by halmyrolysis (submarine weathering) in the same way.
4. Other potential sinks such as an increase in photosynthetic rates, diffusion storage or others are devaluated by a large number of models and measurements, although some of the reservoir sizes and fluxes can only be modeled, but not exactly measured.
5. Rock formation and wear down of rocks are to a certain extent under rate-limiting biological control since the onset of life on Earth.
6. Reasonable arguments imply that biotic rock wear down and soil formation were already an important cooling mechanism in the Precambrian.

7. Tectonic uplift and downward trends in the Earth's crust are biologically controlled as separate sets of assumptions and data claims (Anderson 1984; Krumbein and Schellnhuber 1992).
8. Production and oxidation of organic matter have occurred for about 3.5 billion years, while the period and amount that concerns human impact are 200 years and embrace less than 1% of the transfer of organic matter (or limestone) into carbon dioxide.
9. Biological impact, especially that of microorganisms (bacteria, fungi, lichens), is the most important control factor of the amount and speed of rock wear down.
10. Heterotrophic rather than photo- or chemolitho-autotrophic organisms are the most active components in this process.

Acknowledgements. The authors acknowledge the support of EU grant BIODAM (EVK4-CT-2002-00098) and DFG grants Go897/2-1 and Kr333/30-1.

References

Aardema BW, Lorenz M, Krumbein WE (1983) Protection of sediment adsorbed transforming DNA against enzymatic inactivation. Appl Environ Microbiol 46:417–420

Anderson DL (1984) The Earth as a planet: paradigms and paradoxes. Science 223:347–354

Arrhenius SA (1896) On the influence of carbonic acid in the air upon the temperature of the ground. Philos Mag 41:237–276

Bachmann E (1890) Die Beziehung der Kalkflechten zu ihrem Substrat. Ber Dtsch Bot Ges 8:141–145

Bachmann E (1916) Ein kalklösender Pilz. Ber Dtsch Bot Ges 34:528–539

.Banfield JF, Barker WW, Welch SA, Taunton, A. (1999) Biological impact on mineral dissolution: Application of the lichen model to understanding mineral weathering in the rhizosphere. Proc Nat Acad Sci USA 96:3403–3411

Barnola JM, Raynaud D, Korotkevich YS, Lorius C (1987) Vostok ice core provides a 160,000-year record of atmospheric CO_2. Nature 329:408–414

Berner RA (1992) Weathering, plants, and long-term carbon cycle. Geochim Cosmochim Acta 56:3225–3231

Boden TA, Kanciruk P, Farrel MP (1990) Trends '90 – a compendium of data on global change. Oak Ridge, Tennessee, 257 pp

Buffon GL (1749–1804) Histoire naturelle, générale et particulière, 44 volumes. Paris Imprimerie royale, after the revolution, Plassan, Paris

Callendar GS (1938) The artificial production of carbon dioxide and its influence on temperature. Q J R Metereol Soc 64:223–240

Callendar GS (1939) The composition of the atmosphere through the ages. Metereol Mag 74:33–39

Callendar GS (1958) On the measurement of carbon dioxide in the atmosphere. Tellus 10:241–248

D'Arcy Thompson W (1917, 1961) On growth and form. Cambridge Univ Press, Cambridge, 346 pp

Degens E (1989) Perspectives on biogeochemistry. Springer, Berlin Heidelberg New York, 423 pp

Dietler G, Zhang YC (1992) Fractal aspects of the Swiss landscape. Physica A 191:213–219

Dornieden T, Gorbushina AA, Krumbein WE (1997) Änderungen der physikalischen Eigenschaften von Marmor durch Pilzbewuchs. Int J Restor Build Monu 3:441–456

Dornieden T, Gorbushina AA, Krumbein WE (2000) Biodecay of mural paintings and stone monuments as a space/time related ecological situation – an evaluation of a series of studies. Int Biodeter Biodegrad 46:261–270

Etienne S, Dupont J (2002) Fungal weathering of basaltic rocks in a cold oceanic environment (Iceland): comparison between experimental and field observations. Earth Surface Proc Landforms 27:737–748

Gadd GM (1999) Fungal production of citric and oxalic acid. In: Poole RK (ed) Advances in microbial physiology. Academic Press, New York, pp 48–92

Gehrmann C, Krumbein WE, Petersen K (1988) Lichen weathering activities on mineral and rock surfaces. Stud Geobotan 8:33–45

Golubic S, Krumbein WE, Schneider J (1979) The carbon cycle. In: Trudinger PA, Swaine DJ (eds) Biogeochemical cycling of mineral-forming elements. Elsevier, Amsterdam, pp 29–45

Gonzalez-del Valle M, Dorronsoro C, Irastorza A, Duenas M, Velasco S, Ibarburu I, Saiz-Jimenez C (2003) Microbial communities in black crusts. In: Saiz-Jimenez C (ed) Molecular biology and cultural heritage. Balkema, Lisse, pp 219–223

Gorbushina AA, Krumbein WE (1999) The poikilotrophic micro-organism and its environment – microbial strategies of establishment, growth and survival. In: Seckbach J (ed) Enigmatic micro-organisms and life in extreme environments. Kluwer, Dordrecht, pp 175–185

Gorbushina AA, Krumbein WE (2000) Subaerial biofilms and their effects on soil and rock. In: Riding RE, Awramik SM (eds) Microbial sediments. Springer, Berlin Heidelberg New York, pp 161–170

Gorbushina AA, Krumbein WE, Vlasov D (1996) Biocarst cycles on monument surfaces. In: Pancella R (ed) Preservation and restoration of cultural heritage. Proceedings of the 1995 LPC Congress. EPFL, Lausanne, pp 319–332

Gorbushina AA, Boettcher M, Brumsack HJ, Krumbein WE, Vendrell-Saz M (2001) Biogenic forsterite and opal as a product of biodeterioration and lichen stromatolite formation in table mountain systems (tepuis) of Venezuela. Geomicrobiol J 18:117–132

Gorbushina AA, Diakumaku E, Müller L, Krumbein WE (2003a) Biocide treatment of rock and mural paintings. Problems of application, molecular techniques of control and environmental hazards. In: Saiz-Jimenez C (ed) Molecular biology and cultural heritage. Swets and Zeitlinger, Lisse, pp 61–72

Gorbushina AA, Whitehead K, Dornieden T, Niesse A, Schulte A, Hedges J (2003b) Black fungal colonies as units of survival: hyphal mycosporines synthesized by rock dwelling microcolonial fungi. Can J Bot 81:131–138

Gorbushina AA, Brehm U, Mottershead D (2004) The role of microorganisms and biofilms in quartz weathering. Paleoecol Paleoclimatol Paleooceanograph (in press)

Gumilev L (1990) Ethnogenesis and the biosphere. Progress Publ, Moskwa, 382 pp

Hansen JP, Skjeltorp AT (1988) Fractal pore space and rock permeability implications. Physical Rev B 38:2635–2638

Herder JG (1784) Ideen zur Philosophie der Geschichte der Menschheit. Hempel, Berlin

Hutton J (1795) Theory of the earth. Edinburgh, I. 620 pp, II. 567 pp

Isacenko B (1936) Sur la corrosion du béton. Dokl Akad Nauk SSSR 2:288–289

Jongmans AG, van Breemen N, Lundström U, van Hees PAW, Finlay RD, Srinivasan M, Unestam T, Giesler R, Melkerud PA, Olsson M (1997) Rock eating fungi. Nature 389:682–683

Kant I (1747) Gedanken von der wahren Schätzung der lebendigen Kräfte und Beurtheilung der Beweise, deren sich Herr von Leibniz und andere Mechaniker in der Streitsache bedient haben, nebst einigen vorhergehenden Betrachtungen, welche die Kraft der Körper überhaupt betreffen. In: Königl. Preuss Akad d Wiss (Hrsg) Kants gesammelte Schriften, Bd 1. Reimer, Berlin, 1910, pp 1–181

Kant I (1754) Untersuchung der Frage, ob die Erde in ihrer Umdrehung um die Achse, wodurch sie die Abwechslung des Tages und der Nacht hervorbringt, einige Veränderung seit den ersten Zeiten ihres Ursprungs erlitten habe und woraus man sich ihrer versichern könne, welche von der königl. Akad d Wissenschaften zu Berlin zum Preise für das laufende Jahr aufgegeben worden. In: Königl Preuss Akad d Wiss (Hrsg) Kants gesammelte Schriften, Bd 1. Reimer, Berlin, 1910, pp 183–213

Kant I (1755) Allgemeine Naturgeschichte und Theorie des Himmels. oder der Versuch von der Verfassung und dem mechanischen Ursprunge des ganzen Weltgebäudes nach Newtonischen Grundsätzen abgehandelt. Kants vorkritische Schriften, 1968. Suhrkamp, Frankfurt/Main, pp 227–400

Kant I (1756a) Geschichte und Naturbeschreibung der merkwürdigsten Vorfälle des Erdbebens, welches am Ende des 1755sten Jahres einen großen Theil der Erde erschüttert hat. In: Königl Preuss Akad d Wiss (Hrsg) Kants gesammelte Schriften, Bd 1. Reimer, Berlin, 1910, pp 431–461

Kant I (1756b) Von den Ursachen der Erderschütterungen bei Gelegenheit des Unglücks, welches die westlichen Länder von Europa gegen das Ende des vorigen Jahres betroffen hat. In: Königl Preuss Akad d Wiss (Hrsg) Kants gesammelte Schriften, Bd 1. Reimer, Berlin, 1910, pp 417–427

Keller WD (1957) The principles of chemical weathering. Lucas Brothers, Columbia, Missouri, 111 pp

Krohn CE (1988) Fractal measurements of sandstones, shales, and carbonates. J Geophys Res 93B4:3297–3305

Krumbein WE (1966) Zur Frage der Gesteinsverwitterung (Über geochemische und mikrobiologische Bereiche der exogenen Dynamik) PhD Thesis, Würzburg, 103 pp

Krumbein WE (1969) Über den Einfluß der Mikroflora auf die exogene Dynamik (Verwitterung und Krustenbildung). Geol Rundsch 58:333–363

Krumbein WE (1972) Rôle des microorganismes dans la genèse, la diagenèse et la dégradation des roches en place. Rev Ecol Biol Sol 9:283–319

Krumbein WE (1983) Introduction. In: Krumbein WE (ed) Microbial geochemistry. Blackwell, Oxford, pp 1–4

Krumbein WE (1988) Biotransfer in monuments – a sociobiological study. Durabil Building Mater 5:359–382

Krumbein WE (1990) Der Atem Cäsars. Mitt Geol Paläontol Inst Hamburg 69:267–301

Krumbein WE (1993) Microbial biogeomorphogenesis – an appraisal of Immanuel Kant. In: Guerrero R, Pedros-Alio C (eds) Trends in microbial ecology. Spanish Soc Microbiol, Barcelona, pp 483–488

Krumbein WE (1996) Geophysiology and parahistology of the interactions of organisms with the environment. Mar Ecol 17:1–21

Krumbein WE (1998) Mikrobenbefall und Steinzerstörung: autotroph oder heterotroph? chemisch oder physikalisch? Strategien der Verhinderung und Behebung – eine Bilanz. In: Snethlage R (ed) Denkmalpflege und Naturwissenschaft. Natursteinkonservierung II. Fraunhofer IRB, Stuttgart, pp 173–205

Krumbein WE, Jens K (1981) Biogenic rock varnishes of the Negev Desert (Israel) an ecological study of iron and manganese transformation by cyanobacteria and fungi. Oecologia 50:25–38

Krumbein WE, Dyer B (1985) This planet is alive – weathering and biology, a multi-faceted problem. In: Drever JI (ed) The chemistry of weathering, vol 149. Reidel, Dordrecht, pp 143–160
Krumbein WE, Schellnhuber HJ (1990) Geophysiology of carbonates as a function of bioplanets. In: Ittekott VI, Kempe S, Michaelis W, Spitzy A (eds) Facets of modern biogeochemistry, chap 2. Springer, Berlin Heidelberg New York, pp 5–22
Krumbein WE, Schellnhuber HJ (1992) Geophysiology of mineral deposits – a model for a biological driving force of global changes through Earth history. Terra Nova 4:351–362
Krumbein WE, Swart P (1983) The microbial carbon cycle. In: Krumbein WE (ed) Microbial geochemistry. Blackwell, Oxford, pp 5–62
Krumbein WE, Urzì CE (1993) Biodeterioration processes of monuments as a part of (man-made?) global climate change. In: Thiel MJ (ed) Conservation of stone and other materials. E & FN Spon, Chapman and Hall, London, pp 558–564
Krumbein WE, Gorbushina AA (1996) Organic pollution and rock decay. In: Pancella R (ed) Preservation and restoration of cultural heritage. Proceedings of the 1995 LPC Congress. EPFL, Lausanne, pp 277–284
Krumbein WE, Brehm U, Gerdes G, Gorbushina AA, Levit G, Palinska KA (2003a) Biofilm, Biodictyon, Biomat – Microbialites, oolites, stromatolites – geophysiology, global mechanism, parahistology. In: Krumbein WE, Paterson DM, Zavarzin GA (eds) Fossil and recent biofilms, a natural history of life on earth. Kluwer, Dordrecht, 495 pp
Krumbein WE, Paterson DM, Zavarzin GA (eds) (2003b) Fossil and recent biofilms, a natural history of life on earth. Kluwer, Dordrecht, 495 pp
Levit G, Krumbein WE (2001) Eine vergessene Seite der Ökologiegeschichte: die Biosphäre als Morphoprozess in der Theorie von V. N. Beklemishev (1890–1962). Ber Gesch Theor Ökol 7:199–214
Leyval C, Berthelin J (1991) Weathering of mica by roots and rhizospheric microorganisms of pine. Soil Sci Soc Am J 55:1009–1016
Li YH (1972) Geochemical mass balance among lithosphere, hydrosphere, and atmosphere. Am J Sci 272:119–137
Lovelock JE (1979) Gaia – a new look at life on earth. Oxford Univ Press, Oxford, 157 pp
Mandelbrot BB (1967) How long is the coast of Britain? Statistical self-similarity and fractal dimension. Science 156:636–638
Milankovitch M (1920) Théorie mathématique des phénomènes thermiques produits par la radiation solaire. Gauthier-Villars, Paris, 339 pp
Mottershead D, Gorbushina AA, Lucas G, Wright J (2003) The influence of marine salts, aspect and microbes in the weathering of sandstone in two historic structures. Building Environ 38:1193–1204
Müntz A (1890) Sur la decomposition des roches et la formation de la terre arable. CR Acad Sci 110:1370–1372
Paine SG, Lingood FV, Schimmer F, Thrupp TC (1933) The relation of micro-organisms to the decay of building stones. Philos Trans R Soc B 222:97–127
Perry RS (1979) Chemistry and structure of desert varnish. Ms. Sc. Thesis, Univ Washington, Seattle
Paterson D (1993) Did Tibet cool the world? New Sci 1880:29–33
Saiz-Jimenez C (1999) Biogeochemistry of weathering processes in monuments. Geomicrobiol J 16:27–37
Schoeller H (1955) Géochimie des eaux souterraines. Rev Inst Fr Pétrol 1955:181–213
Schwartzman DW, Volk T (1990) Biotic enhancement of weathering and the habitability of earth. Nature 340:457–460

Schwartzman DW, Volk T (1991a) Biotic enhancement of weathering and surface temperatures on earth since the origin of life. Palaeoecology (Global and Planetary Change Section) 90:357–371
Schwartzman DW, Volk T (1991b) When soil cooled the world. New Sci 51:33–36
Sollas WJ (1880) On the action of a lichen on a limestone. Report, British Association for the Advancement of Science, 586 pp
Staley JT, Palmer FE, Adams JB (1982) Microcolonial fungi: common inhabitants on desert rocks? Science 215:1093–1095
Sterflinger K (2000) Fungi as geological agents. Geomicrobiol J 17:97–124
Sterflinger K, Krumbein WE (1997) Dematiaceous fungi as a major agent for biopitting on Mediterranean marbles and limestones. Geomicrobiol J 14:219–230
Tetsuro W (Japanese 1935, German 1992) Fudo – Wind und Erde. Der Zusammenhang zwischen Klima und Kultur. Wiss Buchgesellschaft, Darmstadt, 216 pp
Turcotte DL (1989) Fractals in geology and geophysics. Pageoph 131:171–196
Vernadsky VI (1924) La géochimie. Alcan, Paris
Vernadsky VI (1929) La biosphère. Alcan, Paris, 232 pp
Vernadsky VI (1930) Geochemie in ausgewählten Kapiteln. Akad Verlagsgesellschaft, Leipzig, 370 pp
Vernadsky VI (1944) Problems of biogeochemistry II. Trans Connect Acad Arts Sci 35:483–517
Vernadsky VI (1997) The biosphere. Springer, Berlin Heidelberg New York, 192 pp
Viles HA (1984) Biokarst. Review and prospect. Prog Phys Geogr 8:532–542
Von Bloh W, Franck S, Bounama C, Schellnhuber H-J (2003) Biogenic enhancement of weathering and the stability of the ecosphere. Geomic J 20(5):501–512
Walker JCG (1993) Biogeochemical cycles of carbon on a hierarchy of time scales. In: Oremland RS (ed) Biogeochemistry of global change. Chapman and Hall, New York, pp 3–28
Wallander H, Wickman T (1999) Biotite and microcline as potassium sources in ectomycorrhizal and non-mycorrhizal *Pinus sylvestris* seedlings. Mycorrhiza 9(1):25–32
Wang FX, Junfa CJ, Jing JLHJW (1993) Biocarst. Geological Publication House, Beijing, 130 pp
Warscheid T, Oelting M, Krumbein WE (1991) Physico-chemical aspects of biodeterioration processes on rocks with special regard to organic pollutants. Int Biodeter 28:37–48
Wong PZ, Howard J (1986) Surface roughening and the fractal nature of rocks. Phys Rev Lett 57:637–641

4 Humification and Mineralization in Soils

Georg Guggenberger[1]

1 Definitions and Introduction

Organic matter is the second most important constituent in soils next to the mineral phase. Traditionally, organic matter is subdivided into nonhumic substances and humic substances. The former encompass all nonaltered or weakly altered plant materials that are still morphologically identifiable and are composed of defined biomolecules. In contrast, humic substances represent strongly altered organic materials which do not show macroscopically identifiable structures. The process that leads to formation of humic substances is called humification. As can be seen in Table 1, the definition of this process is quite vague compared to other processes related to organic

Table 1. Definitions of key processes associated with organic matter cycling in soils

Term	Definition
Litter deposition	Input of plant residues onto (surface litter) or into (root litter) soils
Transformation	Transfer of organic carbon within a distinct chemical structure to another chemical structure caused by enzymatic attack or chemical reactions
Decomposition	Breakdown of organic macromolecules into smaller organic molecules and inorganic constituents of organic matter; this process is usually mediated by micro-organisms and includes depolymerization and oxidation reactions
Mineralization	Microbial conversion into inorganic constituents of organic matter
Humification	The process whereby the carbon of organic residues is transformed and converted to humic substances through biochemical and abiotic processes
Partitioning	Splitting of organic matter into different pathways of decomposition
Assimilation	Incorporation of organic carbon in microbial biomass; the product is resynthesized organic matter
Stabilization	Decrease in the potential of organic matter to be mineralized due to biological, chemical, or physical processes

[1]Martin Luther University Halle-Wittenberg, Institute of Soil Science and Plant Nutrition, Weidenplan 14, 06108 Halle, Germany, e-mail: guggenberger@landw.uni-halle.de

Soil Biology, Volume 3
Microorganisms in Soils: Roles in Genesis and Functions
(ed. by F. Buscot and A. Varma)

matter cycling in soil. This is due to the fact that, in contrast to, e.g., mineralization and decomposition, there is often no well-defined product of the metabolic activities of soil microorganisms or the catalytic activities of soil minerals associated with the formation of humic substances (Zech and Kögel-Knabner 1994). On the one hand, this can be assigned to the heterogeneity of organic substances formed by humification processes in terms of their source, chemical and physical composition, and variability within time and space. On the other hand, humic substances often resist analytical identification of their individual compounds. Nevertheless, the advent of modern analytical techniques such as chemolytic techniques coupled with gas chromatography-mass spectrometry (GC-MS), analytical pyrolysis, and nuclear magnetic resonance spectroscopy (NMR) provided important insight into the humification processes (Hedges et al. 2000). Since the last decade, substance-specific analysis of the ^{13}C content of a sample ($\delta^{13}C$) using gas chromatography-combustion-isotope ratio mass spectrometry (GC-C-IRMS) and pyrolysis-IRMS allows a more detailed investigation of the carbon flow through soils and a coupling of processes to rates at which they are taking place (Gleixner et al. 2001).

The objective of this chapter is first to summarize current knowledge on the origin and composition of the organic residues entering soils. Based on this information, decomposition processes associated with the different types of organic materials and the humification processes related to these are discussed. Because a wealth of information about decomposition and humification processes can be found in numerable text books, here primarily current concepts are discussed. Following the tradition in soil organic matter research, the terms humification and humic substances are used throughout this chapter. However, as will be elucidated in this text, there may be some doubt upon the usefulness of these terms.

2 Soil Organic Matter Resources

In most soils, plant litter materials provide the primary resources for organic matter formation in soil. Exceptions are soils of extreme ecosystems such as dry and cold deserts, where lichens composed of photoautotroph cyanobacteria (green algae) and fungi can form soils crusts (see Chap. 15) and thus represent primary carbon resources for those soils. Also, microorganisms in soils capable of nonphotoautotroph carbon fixation must be considered as primary resources. However, their contribution to total carbon input into soils is probably minor. In addition to the amount of plant litter, its composition and its properties are essential controlling factors for organic matter formation and humification processes in soils

(Kögel-Knabner 2002). Predictors for plant litter decomposition include its content in nitrogen, cellulose, hemicellulose, lignin, and tannins. During decomposition processes, primary resources are partly mineralized and partly assimilated by microorganisms. After death of the microorganisms, the resynthesized organic substances enter the soil organic matter pool as secondary resources. According to Haider (1992), this represents an important parent material for humus formation. Feces and residues of the soil fauna belong to the secondary resources as well. The soil fauna plays an essential role in controlling litter decomposition in soil (Wolters 2000). However, as the relative input of faunal carbon to the soil organic matter pool is low, the part of parent material for humification attributable to animals is much smaller than that of plant and microbial residues.

2.1 Plant Compounds

Plant tissues include intracellular storage materials and structural components that occur in membranes, extracellular or as cell wall constituents (Kögel-Knabner 2002). The storage materials of plants are more easily degradable than the structural components and thus are important carbon and energy sources for microorganisms. Therefore, decomposition of plant materials usually starts within the cell, leaving cell wall materials intact for a longer time (Olah et al. 1978). This variable accessibility to microbial attack can be explained by the different chemical and physical composition of storage and structural materials.

Intracellular and storage materials are primarily composed of proteins, starch, fructans, chlorophyll and other pigments. Proteins represent the most abundant substance group in plant cells and are usually composed of 20 basic, neutral, or acidic amino acids. They can be decomposed by a multitude of microorganisms and are considered to be very bioavailable (Martin and Haider 1986). Starch is an important storage polysaccharide in vascular plants, and consists of two different polymers of glucose, amylose and amylopectin. As polysaccharides are part of the energy metabolism, starch can be rapidly decomposed by aerobic and anaerobic microorganisms. Another intracellular component is chlorophyll which is present in all photosynthetically active cells. Decomposition of chlorophyll already starts during leaf senescence. Products such as the carotenoids and anthocyanins are degradable, however, their fate and relevance in soils are largely unknown (Kögel-Knabner 2002).

The major structural component of cell walls is cellulose. Although 85% of the cellulose has a fibrillar structure with crystalline properties, it is easily degradable by many microorganisms including fungi and eubacteria, in particular under aerobic conditions. Cellulose fibers are surrounded by noncellulosic polysaccharides, often summarized as hemicelluloses. They are formed by pentose, hexose, hexuronic acids and desoxyhexose monomers, are branched and have a lower degree of polymerization than cellulose. Hemicelluloses are often encrusted with lignin, which is more resistant to metabolic breakdown. This encrustation protects also the hemicellulose from rapid decomposition (Paul and Clark 1996). Lignin stabilizes plant tissues during growth and is specific to vascular plants. It consists of three different alcohols from the C_6–C_3 pool, coumaryl, coniferyl, and sinapyl alcohol. These monomers are embedded in a three-dimensional network by polymerization of free radicals, which contains a multitude of ether and C–C bondings. The involvement of free radicals in the formation of lignin and the resulting disordered molecular structure of lignin is an important reason why only highly specialized organisms (i.e., white rot fungi) are capable of completely breaking down this biomolecule. Lignin decomposers have lignin peroxidase, manganese peroxidase, and laccase that catalyze biological oxidation of lignin by free oxygen radicals. Therefore, lignin is not degraded under anaerobic conditions and accumulate in peats (Kirk and Farrel 1987). Under aerobic conditions, lignin is mineralized completely, but the carbon is not used by microorganisms for metabolic reactions (Haider 1992). Like lignin tannins are also produced via the C_6–C_3 pool. They are polyhydroxyaromatic acids, especially gallic or ellagic acid, are part of the chemical defense system of plants and are considered to be resistant to microbial decomposition. However, little is known in detail about the metabolism and turnover rate of tannins in plant residues (Harborne 1997).

Lipids are a heterogeneous group of substances in plant and microbial tissues and comprise long-chain molecules such as fatty acids or alcohols and branched terpenes including glycerides and related compounds. Waxes, in particular cutin and suberin, are polymerized and cross-linked structures of hydroxy fatty acids containing an even number of carbons in the range from C_{16} to C_{26}. Cutin is found on the outer surface of plant tissue (cuticle) while suberin is a cell wall component of cork cells. Some plants contain also cutan and suberans, nonhydrolyzable biomacromolecules which consist of polymethylene chains in addition to the hydrolyzable polyester material of cutin and suberin, leading to a highly cross-linked structure (Tegelaar et al. 1989).

2.2
Microbial Compounds

Microorganisms mineralize not only all primary resources to inorganic end products, but utilize most plant-derived materials to resynthesize their own biomass. Microbial cell wall materials are a more important source of soil organic matter than the cytoplasm, since organic constituents of the latter are rapidly degraded. While fungi use chitin, glucan, or cellulose to form their cell walls, bacteria use more complex materials, such as glycolipids, peptidoglycans (Fig. 1), proteoglycans, and glycoproteins (Gleixner et al. 2001). Glycolipids are composed of carbohydrates and lipids, whereas pep-

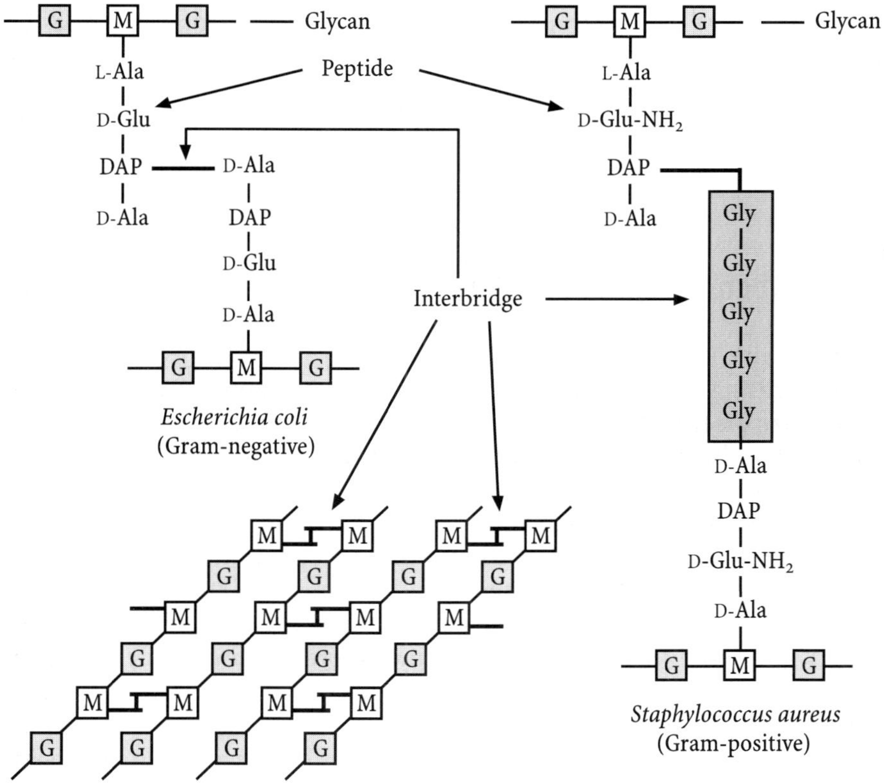

Fig. 1. Schematic representation of the structure of bacterial peptidoglycan (after Madigan et al. 1997). The two images illustrate the linkages of structural units within peptidoglycans of species of Gram-negative and Gram-positive bacteria. The network to the *lower left* shows how these units are assembled into a peptidoglycan sheet (peptide cross-links in *bold*) that is relatively resistant to biodegradation. *G N*-acetylglucosamine, *M N*-acetylmuramic acid, *DAP* meso-diaminopimelic acid, *Ala* alanine, *Gly* glycine, *Glu* glutamic acid. (Hedges et al. 2000)

tidoglycans, proteoglycans, and glycoprotein consist of amino acid polymers and nitrogen-containing sugars (amino sugars) or chitin. The latter three are high molecular weight compounds with a rigid, gel-like structure. Little is known about their fate in soils. However, large concentrations of easily degradable amino sugars released upon acid hydrolysis of soil suggests that these amino sugars are probably well stabilized within the highly cross-linked peptidoglycan after death of the microorganisms (Guggenberger et al. 1999). An additional compound in many fungi, and also some bacteria, are melanin pigments. Melanins consist of a polymeric core of phenolic, indolic, quinone, hydro-quinone and semi-quinone monomers and are non-hydrolyzable (Butler and Day 1998). They are black to brown-colored and are assumed to represent precursors of humic substances in soil (e.g., Saiz-Jimenez 1996). The detailed structure of melanins and rates and pathways of their decomposition are poorly understood. Outside the microbial cell wall extracellular polmeric substances similar to the cell wall components often produce a "diffusion space" that anchors exoenzymes (Gleixner et al. 2001). However, the contribution of these compounds to the pool of microbially derived organic matter in soil is not known.

2.3 Black Carbon

Another type of carbon source in soils is the organic remains of fires, such as charcoal and soot. Due to the black color of such substances, they are referred to as black carbon. Charcoal is the highly condensed, aromatic solid residue of the biomass burned and often still holds the morphological properties of the burned biomass. In contrast, soot is generated de novo in the gas phase of a fire (Kuhlbusch and Crutzen 1995). Soot consists of multilayers of highly condensed aromatic structures that are randomly oriented or well ordered, forming three-dimensional "onion-type" structures (Fig. 2). Fires occur in almost all ecosystems due to lightning or anthropogenic activities, and soot is distributed in the atmosphere worldwide. Hence, black carbon is a ubiquitous carbon species in soils (Goldberg 1985). Estimated global black carbon production is 0.04–0.6 Gt from vegetation fires and 0.007–0.024 Gt from fossil fuel combustion (Kuhlbusch and Crutzen 1995). Quantification of black carbon is difficult because its chemical structure and the physical size of it are extremely heterogeneous. However, in fire ecosystems, black carbon can account for up to 60% of total soil organic carbon (Skjemstad et al. 1996). In soils close to coal processing industries, the contribution of black carbon can even be up to 80%. Hence, black carbon must be considered as an important source of soil organic matter in many ecosystems.

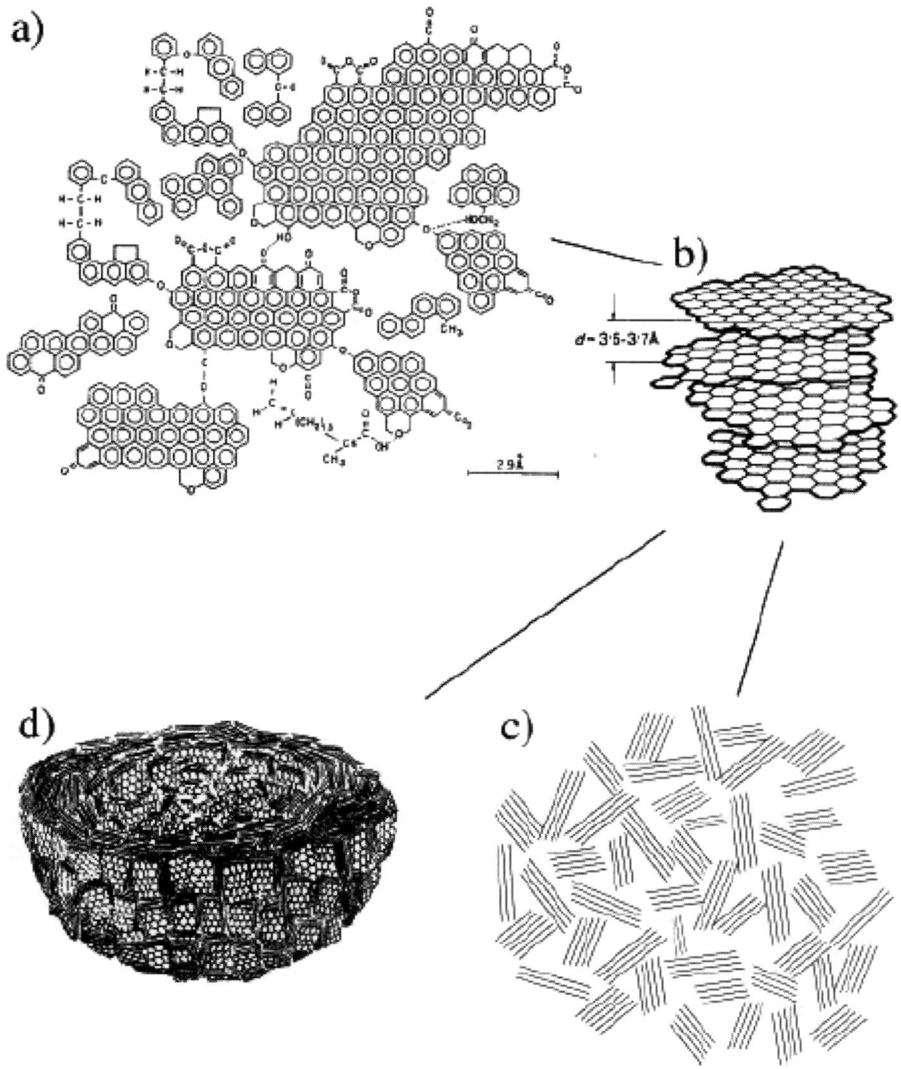

Fig. 2. Soot structure as **a** produced in the laboratory, **b** forming basic structural units of three to four layers, **c** randomly oriented basic structural units shown as a two-dimensional schematic diagram, **d** onion-type particle with several condensation seeds. (Gleixner et al. 2001)

3 Mineralization and Humification Pathways

Fresh litter is very different from older, decomposed litter and organic matter associated with the mineral soil from a chemical point of view. These chemical modifications are a result of primarily biotic decomposition and humification processes occurring in the soil. Vice versa, the different chemical composition influences the rate-regulating factors of organic matter decomposition and the size, composition, and activity of the microbial community.

3.1 Factors Affecting Decomposition and Mineralization

Decomposition of organic residues is the progressive dismantling of organic materials, ultimately into inorganic components, which will then be defined as mineralization (Gregorich and Janzen 2000). Decomposition is mediated mainly by soil microorganisms which derive energy and nutrients from the process. The net effect is the release of carbon and nutrients back into biological circulation.

Decomposition of organic matter is strongly influenced by environmental factors such as water potential, oxygen supply, temperature, nutrient supply and pH, and the quality of organic resource. Mostly, these factors are coupled and cannot be separated from the acting decomposer organisms, the soil fauna and finally the microorganisms (Fig. 3). The three groups of factors: environment, resource quality, and decomposer organism, not only determine the rate of the process, but also the final products of decomposition. The substrates for decomposition and mineralization include a wide range of materials, forming a continuum from fresh primary and

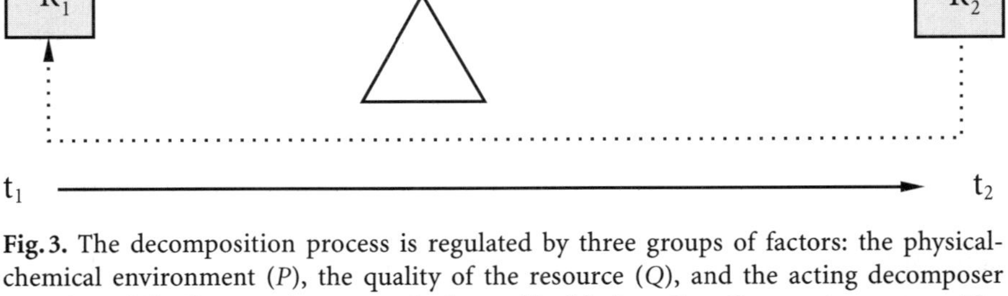

Fig. 3. The decomposition process is regulated by three groups of factors: the physical-chemical environment (*P*), the quality of the resource (*Q*), and the acting decomposer organisms (*O*). The organic resource is changed in this from R_1 to R_2 over time, t_1 to t_2. The changed resource R_2 enters the decomposition cascade where it is further decomposed to $R_3 \ldots R_x$ and redistributed through comminution, catabolism and leaching. (Modified from Gregorich and Janzen 2001)

secondary residues to very stable, humified organic matter. For simplicity, decomposition can be subdivided into primary decomposition, involving the breakdown of fresh litter, and secondary decomposition, involving the progressive breakdown of humified organic matter (Gregorich and Janzen 2000).

Of the environmental factors the water potential affects decomposition because soil microorganisms depend on water for survival and mobility. Optimum soil water potentials for decomposition of organic residues are between − 0.01 and − 0.05 MPa, however, fungi tend to tolerate lower water potentials of −4 to −10 MPa (Sommers et al. 1981). At water potentials higher than about − 0.01 MPa, decomposition is slowed by a low oxygen supply. This situation is typical in peatlands where under anaerobic conditions the metabolism of microorganisms shifts to less energy-efficient fermentation or to nitrate and sulfate reduction and methane production. Under these conditions, breakdown of aromatic and phenolic substances such as lignin, which requires oxygen, is not possible and these compounds accumulate (Haider 1992). The second important abiotic driving variable is the temperature. High temperatures in tropical regions lead to a rapid decomposition of organic matter, whereas in boreal regions cold winters and hot and dry summers restrict biomass production and decay. In temperate regions with a well balanced water supply, the decomposition rate of organic residues is high, unless it is limited by the nutrient supply of the soil or litter quality, i. e., conifer litter.

The substrate quality, i. e., the composition of the different primary and secondary resources as described above, is an intrinsic property that affects decomposition. Destabilization of molecules occurs when the activation energy needed for bond breaking is available. As a rough estimate, double and triple bonds are most stable, followed by homopolar C–C and C–H bonds, whereas heteropolar C–O and C–N bonds are least stable (Gleixner et al. 2001). In biological systems, the activation energy is lowered by specific enzymes that catalyze the breakdown of molecules, and often whole sets of enzymes are involved in decomposition of biomolecules in soils. A critical factor in enzymatic degradation is that most microorganisms are unable to transport molecules larger than about 600 Da through their cell walls. Hence, they must be cleaved to smaller molecules outside the cell by exoenzymes. Further, the enzymes need a specific molecular environment. The substrate must fit exactly into the active center of the enzymes. In polymeric biomolecules, access of the enzymes to their specific reaction sites of the substrate is often impeded. These interactions are illustrated in a classical experiment of Haider and Martin (1975) where mineralization of ^{14}C-labeled monomeric and polymeric substances are studied (Fig. 4). Carbohydrates and proteins are degraded rapidly, being partly mineralized and partly used for microbial biomass formation. Monomeric lignin con-

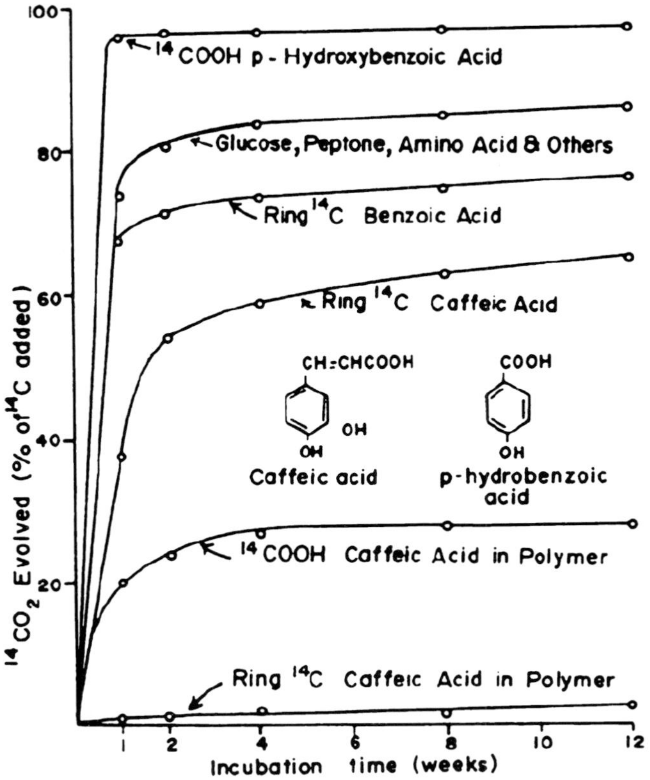

Fig. 4. Decomposition of ^{14}C-labeled lignin moieties. (Haider and Martin 1975)

stituents, even if located in the aromatic ring structure, are also mineralized quite rapidly. In contrast, the stability of lignin constituents is much higher when located in the polymer. The polymeric and disordered structure of lignin prevents rapid enzymatic degradation of this biomolecule. Further, depolymerization of lignin is a co-metabolic process and labile organic carbon species are needed as a co-metabolic substrate, because there is no energy gain during lignin decomposition (Haider 1992). However, recent studies using ^{13}C NMR spectroscopy and analytical pyrolysis have shown that lignin is altered relatively rapidly in soils and does not appear to be stabilized in the long term (Kögel-Knabner 1993; Gleixner et al. 2002). In forest soils, intact lignin structures decrease considerably with increasing soil depth (Kögel-Knabner 1993), which matches the highest potential and diversity of laccase genes in the upper horizons quite well (Luis et al. 2004).

Aliphatic biomacromolecules such as suberans and cutans are also resistant to oxidative and enzymatic attack in soil (Tegelaar et al. 1989), due to reduced mobility of the molecules and blocking of reactive sites (e.g., beta position). For black carbon, the resistance against biodegradation is due to its strongly condensed and disordered molecular structure and the high degree of internal cross-linkages (Almendros and Dorado 1999). Only enzymatically catalyzed reactions forming small radicals from highly polymeric lignins, biomacromolecular aliphatics, and black carbon are able to degrade these substances. These products are further mineralized, but they are also believed to serve as basic units for abiotic spontaneous condensation reactions. The latter process is considered to represent one pathway in the formation of humic substances (Hedges 1988).

3.2 Humification Processes

Soil humic substances and the associated processes of humification have been a subject of intense research in soil science for more than two centuries (Archard 1786). Despite these tremendous efforts, formation of humic substances is probably still the most vague field in soil science and subject to many myths. This problem is mainly due to the long-time inaccessibility of humic substances by analytical methods. Still nowadays, the structural elucidation of humic substances is a challenging task. Nevertheless, the identification of changes in the chemical and physical form of the organic compounds from primary and secondary resources to humic substances is an important prerequisite in the identification of humification pathways. In the following, this will be critically elucidated.

3.2.1 Definition and Composition of Humic Substances

Classically, soil humus is composed of humic substances and nonhumic biomolecules of plant and microbial origin after dissolved organic carbon (DOC) and macroorganic matter has been removed by water extraction and particle size and density fractionation (Oades 1989). Humic substances and nonhumic biomolecules cannot be separated by chemical extraction, and both types of organic materials can be fractionated into fulvic acid, humic acid and the residual humin, based on their solubility in aqueous alkaline and acid solutions. There is criticism that the separation of organic matter into these humic fractions may involve some source of bias (Baldock and Nelson 2000), including (1) the questionable ability of the alkaline extractable organic matter to represent the composition of the nonextracted organic fractions, (2) differences in the chemical characteristics in extracted organic molecules with those of the same materials existing in soils in an adsorbed state, and (3) the formation of artifacts during the extraction procedure. The largest drawback of this concept is that the solubility-based organic matter fractions cannot be related to microbial turnover rates in contrast to fractions that are physically isolated based on size or density characteristics (Christensen 1996). Nevertheless, particular modern spectroscopic techniques to study the chemical structure of humic substances, such as ^{13}C NMR spectroscopy and analytical pyrolysis, provided important insight into the composition of humic substances. It is important, however, that the analytical limitations of these, as in any other degradative and nondegradative technique to study the structure of organic substances, are considered in a proper interpretation of the results (summarized in Guggenberger 2002).

In recent decades, a respectable number of review articles (e.g., Oades 1989; Kögel-Knabner 1993; Preston 1996) and textbooks (Aiken et al. 1985; Hayes et al. 1989; Stevenson 1994) used information from these different analytical techniques to present detailed information on the chemical and physical structure of humic substances. Baldock and Nelson (2000) summarized the results of the chemical composition of humic substances as follows: (1) aromatic rings are a significant component and multiple substitutions with carboxyl, hydroxyl, carbonyl and alkyl groups exist; (2) significant quantities of C1–C20 alkyl C chains either unsubstituted or substituted with O-containing functional groups are present; (3) aromatic and alkyl groups are bound together principally at random by C–C bonds and ether linkages to form the backbone structure of humic substances; (4) simple and polymeric proteinaceous and carbohydrate groups are associated with this backbone. These numerous substances being different in chemical composition, size, and type of bonding suggest that a variety of different mechanisms may be involved in their genesis, which will be discussed next.

3.2.2 Formation Pathways of Humic Substances

One of the proposed pathways of humic substance formation is the spontaneous "heteropolycondensation" reaction between small reactive intermediates released during enzymatic breakdown of biomacromolecules (Hedges 1988). These processes are extracellular and may be catalyzed by microbial exoenzymes. Such humification theories are based on the observation that many simple organic molecules (e.g., amino acids, phenols and sugars) abiotically condense to produce extremely complex assemblages of molecules that exhibit the brown color and many of the physicochemical properties of soil organic matter (Hedges et al. 2000). These condensations include Maillard (or "browning") reactions between carbohydrates and amino acids that form dark, often aromatic, melanoidins (Benzing-Purdie et al. 1983). In soils, such compounds have been identified primarily within foodstuffs of archaeological sites (Evershed et al. 1997). In the polyphenol theory, monomeric phenolic species are produced by enzymatic degradation of lignin. They are capable of forming a quinone structure in the presence of oxygen or polyphenoloxidase enzymes, which spontaneously polymerize with each other or with amines or ammonia to produce polymeric compounds (Baldock and Nelson 2000). Phenols and quinones can also be formed from carbohydrates (Hedges 1978). The heteropolycondensation theory includes also photo-oxidation of polyunsaturated lipids. One argument that is often raised in favor of the heteropolycondensation pathway is the strong resemblance of artificial humic substances produced from well-defined precursors in the laboratory with soil humic substances. However, for the heteropolycondensation pathway, usually an environment of high temperature, high pH, and/or UV radiation is needed that is rarely found in soil.

An important argument against abiotic or enzymatically catalyzed heteropolycondensation comes from ^{15}N NMR spectroscopy. In soils and modern sediments, organic nitrogen is primarily in the amide form (Knicker et al. 1996; Knicker and Hatcher 2001), whereas in ancient sediments aromatic heterocycles are typically found (Derenne et al. 1998), which are probably the product of advanced abiotic condensation (melanoidins). In the ancient sediments, this abiotic formation of aromatic nitrogen can be explained by diagenetic processes. However, in soils, the formation of amides is difficult to explain by abiotic processes. The predominance of amides points strongly toward a remnant biochemical component of the humic substances. Hedges et al. (2000) reported another argument against abiotic heteropolycondensation: extensive in situ formation of new chemical compounds is seldom evident from either NMR (Fig. 5) or molecular-level analyses. It can thus be concluded that during organic matter degradation, relatively resistant biochemicals are selectively concentrated into humic substances implying the concept of selective preservation.

For many aerobic soils, application of NMR and molecular-level analysis showed that the humic substances contain relatively little aromatic carbon and the aromatic carbon present shows little resemblance to lignin (Kögel-Knabner 2000). Hence, it appears that lignin is decomposed rather rapidly under aerobic conditions and is not selectively enriched in humic substances. In contrast, in most soils, the humic materials are rich in highly aliphatic compounds. Possible sources of such aliphatic compounds can be hydrolysis-resistant biomacromolecules in vascular plants, e.g., cutans and suberans (Largeau et al. 1986; Tegellar et al. 1989). Similarly, highly resistant biomolecules have been also identified in microbial and algal cell walls (Derenne et al. 1991; DeLong et al. 1998). Such components may be stabilized against decomposition by spontaneous oxidative cross-linking.

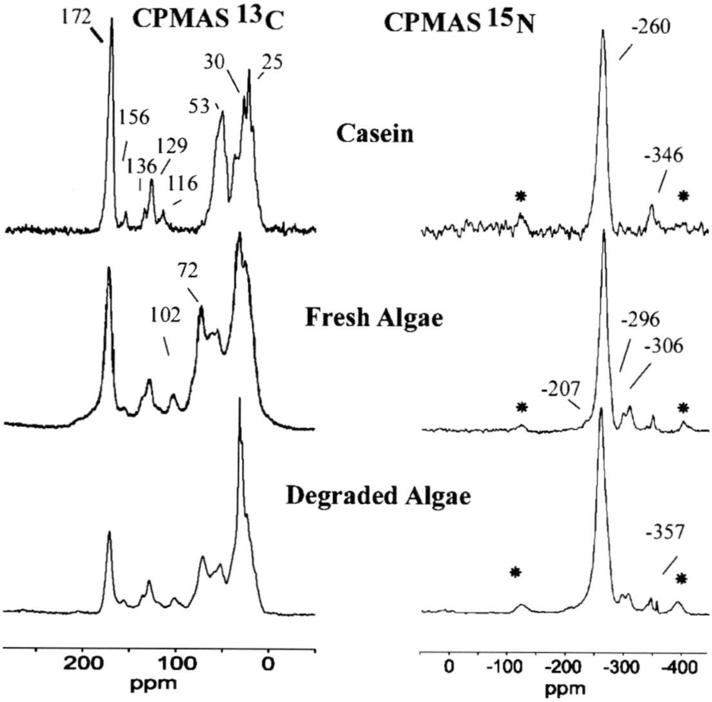

Fig. 5. Solid state CP-MAS ^{13}C and ^{15}N NMR spectra of casein (a milk protein), fresh algae and biodegraded algae. Within the ^{13}C spectra, the chemical shift regions and corresponding major carbon types are: 0–45 ppm (alkyl), 45–60 ppm (*N*-alkyl and methoxyl), 60–95 (*O*-alkyl), 95–115 (di-*O*-alkyl), 115–145 (aromatic), 145–160 (*O*-aromatic) and 160–210 (carboxyl/carbonyl). In the ^{13}C spectra, the predominance of protein in fresh algae is evident, as is the loss of nonalkyl carbon during algal degradation. For the ^{15}N spectra, amide nitrogen (−260 ppm) constitutes the main resonances of all samples. *Asterisks* indicate spinning side bands of larger central resonance. (Knicker 2000)

Such cross-linking and repackaging, however, do not necessarily produce large changes in the bulk chemical composition of the precursor molecule. Even if such resistant biomacromolecules occur only in trace amounts in living organisms, they can be concentrated by degradation to become major components of humic substances in soil.

Besides aliphatic moieties, humic substances are often rich in polysaccharides and nitrogenous organic compounds (Kögel-Knabner 2000; Baldock and Nelson 2000). Hence, humic compounds usually have a C:N ratio of 10 which is much narrower than that of the litter material. As mentioned above, soil organic nitrogen is almost exclusively in the amide form. While it is obvious why resistant aliphatic biomacromolecules are major constituents in humic soil materials, this is not clear for the potentially easily

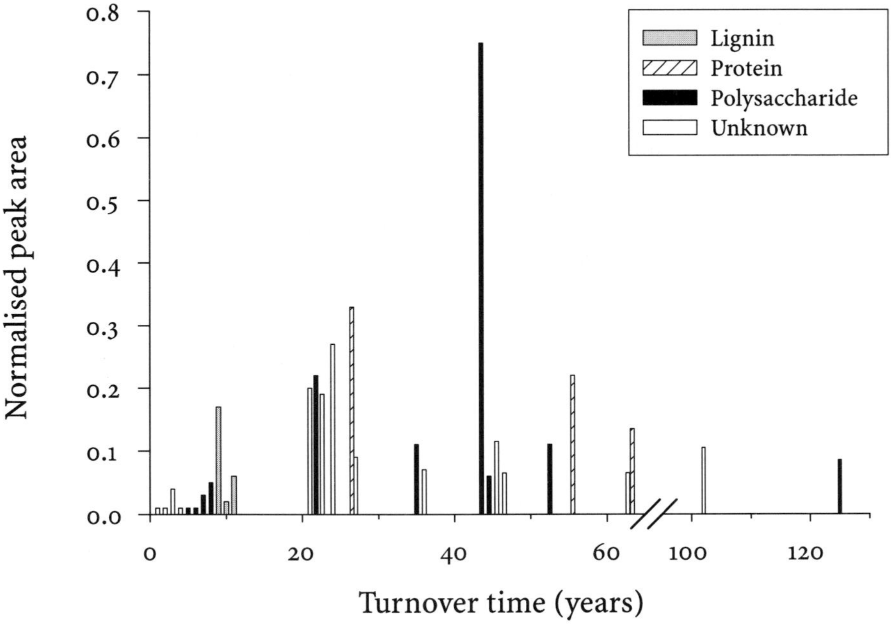

Fig. 6. Peak areas and mean residence time of pyrolysis products from maize leaves, wheat and maize soil. Pyrolysis products are related to their probable precursors. Protein fraction contains pyrolysis products from proteins, amino acid moieties and chitin. Pyrolysis products, only present in maize leaves, are supposed to be degraded within a few years, i. e., all peaks with a turnover under 20 years. In contrast to proteins and polysaccharides, lignin appears to be rapidly degraded in soil. (Gleixner et al. 2002)

degradable polysaccharides and peptides and proteins. However, recently the determination of molecular turnover rates using pyrolysis-isotope ratio mass spectrometry of C3/C4 sequence soils indicated some of the longest turnover rates for pyrolysis products derived from proteins and polysaccharides (Gleixner et al. 1999, 2002; Fig. 6).

One key process that leads to the enrichment of proteinaceous compounds in soil humic substances may be the reduced accessibility of the organic substrate by enzymes. Since most enzymes must be compatible with water, processes that remove water from the immediate surroundings of the macromolecules will slow down their degradation. Folding of large molecules and aggregation of small molecules (Piccolo 2002) can both lead to internal hydrophobic microenvironments whose isolation is strongly favored by the polarity and highly ordered "structure" of the surrounding water network (Hedges et al. 2000). Hydrophobicity, along with hydrogen bonding ion pairing, are strong contributing factors to the complex three-dimensional folding patterns of proteins. Hence, it appears that physical

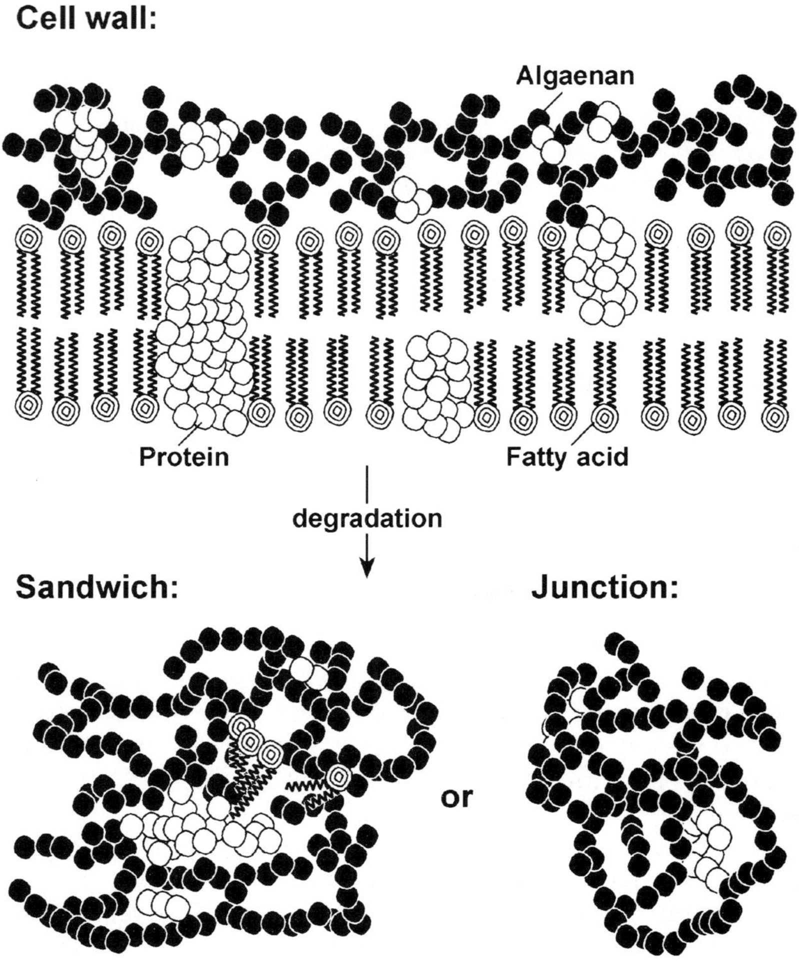

Fig. 7. Schematic illustration of the structure of an algaenan-containing algal cell wall (*above*) and proteinaceous organic matter encapsulated within degraded algal cell wall material (*below*). By encapsulation intrinsically labile proteins and peptides might be physically protected from biodegradation. (Knicker and Hatcher 2001)

availability to attack, rather than intrinsic bonding energies, determines the biological reaction rates and, hence, the processes of selective preservation in humic substances formation. Another example for this is the observation that intrinsically reactive biochemicals such as proteinaceous organic matter appear to be encapsulated in some type of protective matrix (Knicker and Hatcher 1997, 2001; Fig. 7).

In order to react with enzymes, chemical subunits of organic molecules need to be accessible, but they must also be in a very specific alignment.

Relatively subtle chemical or conformational changes leading to minor modifications in low-energy bonding can cause profound changes in enzyme function and stability (Stotzky et al. 1996). Therefore, when portions of proteins are spread over a particle surface they can be resistant to reaction, even though they are physically accessible to enzymes (Nagata and Kirchman 1996). In fact, this process is considered to be primarily responsible for the stabilizing effect of sorption processes on organic substances (Khanna et al. 1998). However, such conformational changes that largely affect the biochemical reactivity of an organic molecule may not be detectable by most bulk and molecular-level analyses.

In addition to the restricted accessibility of enzymes to moderately altered biomolecules, another process may be responsible for large proportions of proteins and polysaccharides in soil humic substances. Microorganisms secrete organic substances, such as enzymes and other biopolymers like polysaccharides and proteins. Baldock et al. (1990) showed by incubating a soil with uniformly labeled ^{13}C-glucose that utilization of glucose by soil microorganisms results predominantly in the synthesis of *O*- and *N*-alkyl C (i. e., polysaccharides and proteins) besides alkyl C. According to molecular isotopic analysis, pyrolysis products derived from microbial polysaccharides and proteins have relatively high mean residence times of several decades (Gleixner et al. 2002). This does not necessarily mean that the distinct molecules are stabilized for this time. Large concentrations of microbial polysaccharides and proteins within the humic substances may be also due to a microbial recycling (Gleixner et al. 2002). Hence, a carbon atom may reside long in soil, but be repeatedly incorporated into polysaccharides or proteins with a more rapid turnover. It could be only the apparent residence time that is long for some polysaccharides and proteins. Such a microbial loop of metabolization, mineralization, and remetabolization may be considered as an important process in stabilization of soil carbon and reducing the carbon losses to the atmosphere. However, this important process is not covered within the current concepts of humification. Another component that is not included in current humification concepts is black carbon.

3.2.3
The Fate of Black Carbon in Soil

Black carbon derived from pyrogenic processes is found in 65-million-year-old marine sediments (Herring 1985) and was considered for a time to be almost inert in the soil environment (Swift 2001). However, recent carbon and oxygen isotopic studies suggest that black carbon degradation in soils can take less than a century (Bird et al. 1999). Hence, it is not reasonable to assume that black carbon is completely inert, i. e., nonreactive. Photochemical breakdown of black carbon was already identified by Ogren and Charlson (1983). Hamer et al. (2004) showed that artificially charred plant materials could be microbially mineralized (about 0.8% during 60 days) in the presence of an easily available carbon source, thus indicating a co-metabolic pathway. According to Glaser et al. (2000), surfaces of black carbon molecules in soil are highly oxidized and able to react with mineral surfaces. Möller et al. (2000) identified benzenehexacarboxylic (mellitic) acid as a highly altered breakdown product of black carbon in a tropical soil. Since this substance cannot be produced from a biogenic precursor (Glaser et al. 1998), it must be the product of charred plant material. Obviously oxidative degradation of black carbon is taking place, leaving smaller and highly reactive aromatic products behind. Based on the facts that lignin is degraded quite rapidly in aerobic soils and that black carbon can be oxidatively altered, Haumaier and Zech (1995) concluded that biooxidative degradation of black carbon is a possible source of highly aromatic humic acids. Since black carbon is almost ubiquitous in soils, it can be generally asked if this elemental carbon source is the major precursor of highly aromatic substances found in humic fractions. However, this needs to be investigated in the future.

4 Conclusions

It appears that the definitions of humification and humic substances depend on the analytical capabilities to elucidate the structure of these biomolecules. Current concepts of humification largely reject random heteropolycondensation reaction as an important pathway in the formation of humic substances (see also Hedges et al. 2000). With the aid of new analytical instrumentation, a new picture emerges showing that the processes of biodegradation and humification cannot be separated, and that relatively resistant organic molecules are selectively concentrated into humic substances. In particular, aliphatic moieties within the humic fractions show strong resemblance to their mother substances. Often subtle changes in the chemical composition and/or conformational changes are decisive whether a biomolecule is degraded rapidly or preserved against decomposition over a long time. Such conformational changes are likely to occur upon sorption to mineral phases in soil. While such processes of selective preservation received considerable attention in the last two decades, other major points that need to be addressed in the future are:

1. The contribution of rapid resynthesis of microbial biomass (microbial loop) to polysaccharide and proteinaceous moieties in the humic

fractions is unclear. The use of stable isotope techniques may help to understand pathways and efficiency of this process as well as to distinguish between the turnover of a carbon atom in soil from that of a distinct molecule.

2. Black carbon appears to be a major source of humic substances at least in some soils. The degradation reactions are poorly understood, and there is even huge analytical uncertainty in distinguishing unaltered black carbon from its decomposition products showing a different degree of biooxidative alteration.
3. Humic substances isolated from soils of different pedogenesis differ strongly in their chemical structural composition (Kögel-Knabner 2000). Probably, the humification pathways are the same in the different soil environments, but the proportions of the individual pathways may differ. Yet, there is little knowledge about the controlling factors.

Nevertheless, a picture emerges that humification can be increasingly identified as a multitude of specific biochemical and biophysical processes that alter the structural composition of the organic matter. Other important processes that influence the fate of organic matter in soil are not discussed here and include faunal communition, turbation and intestinal degradation, formation of dissolved organic matter, and sorption processes to soil minerals. The interplay of all of these processes lead to different fractions of soil organic matter with different source materials, chemical and physical structural composition, and functions. Many of these fractions are well defined in terms of the process that leads to their formation and in terms of their functions, e.g., dissolved organic matter, free and occluded organic particulate organic matter, mineral-associated organic matter. In contrast, neither humification is an adequate description of specific processes, nor is humic substances a useful soil organic matter pool. It is time to be more specific in the separation, identification, and denomination of distinct processes that are taking place in mineralization, alteration, and stabilization of organic matter in soils. Such vague and poorly defined terms such as humification and humic substances are of little help to improve our understanding of the functioning of organic matter cycling in soils and, therefore, should be abandoned.

References

Achard FC (1786) Chemische Untersuchungen des Torfs. Crell's Chem Ann 2:391–403

Aiken GR, McKnight DM, Wershaw RL, MacCarthy P (1985) Humic substances in soil, sediment and water. Wiley, New York, NY

Almendros G, Dorado J (1999) Molecular characteristics related to the biodegradability of humic acid preparations. Eur J Soil Sci 50:227–236

Baldock JA, Nelson PN (2000) Soil organic matter. In: Sumner MR (ed) Handbook of soil science. CRC Press, Boca Raton, pp B-25–B-84

Baldock JA, Oades JM, Vassallo AM, Wilson MA (1990) Solid state CP/MAS ^{13}C N.M.R. analysis of particle size and density fractions of a soil incubated with uniformly labelled ^{13}C-glucose. Aust J Soil Sci 28:193–212

Benzing-Purdie L, Ripmeester J, Preston C (1983) Elucidation of the nitrogen forms in melanoidins and humic acids by nitrogen-15 cross polarization-magic angle spinning nuclear magnetic resonance spectroscopy. J Agric Food Chem 31:913–915

Bird MI, Moyo C, Veenendaal EM, Lloyd J, Howard DM (1999) Stability of elemental carbon in a savanna soil. Global Biogeochem Cycles 13:923–932

Butler MJ, Day AW (1998) Fungal melanins: a review. Can J Microbiol 44:1115–1136

Christensen BT (1996) Carbon in primary and secondary organo-mineral complexes. In: Carter MR, Stewart BA (eds) Advances in soil science – Structure and organic matter storage in agricultural soils. CRC Lewis, Boca Raton, pp 97–165

DeLong EF, King LL, Massana R, Cittone H, Murray A, Schleper C, Wakeham SG (1998) Cyclic and acyclic dibiphytanyl ether lipids in marine psychrophilic crenarchaeotes: evolutionary and ecological implications. Appl Environ Microbiol 64:1133–1138

Derenne S, Largeau C, Casadevall E, Berkaloff C, Rousseau B (1991) Chemical evidence of kerogen formation in source rocks and oil shales via selective preservation of thin resistant outer walls of microalgae: origin of ultralaminae. Geochim Cosmochim Acta 55:1041–1050

Derenne PS, Knicker H, Largeau C, Hatcher P (1998) Timing and mechanisms of changes in nitrogen functionality during biomass fossilization. In: Stankiewicz BA, van Bergen PF (eds) Nitrogen-containing macromolecules in the bio- and geosphere. ACS Symposium Series 707. American Chemical Society, Washington, DC, pp 245–253

Evershed RP, Bland HA, van Bergen PF, Carter JC (1997) Volatile compounds in archaeological plant remains and the Maillard reaction during decay of organic matter. Science 278:432–433

Glaser B, Haumaier L, Guggenberger G, Zech W (1998) Black carbon in soils: the use of benzenecarboxylic acids as specific markers. Org Geochem 29:811–819

Glaser B, Balashov E, Haumaier L, Guggenberger G, Zech W (2000) Black carbon in density fractions of anthropogenic soils of the Brazilian Amazon region. Org Geochem 31:669–678

Gleixner G, Bol R, Balesdent J (1999) Molecular insight into soil carbon turnover. Rapid Comm Mass Spectrom 13:1278–1283

Gleixner G, Czimczik C, Kramer C, Lühker BM, Schmidt MWI (2001) Plant compounds and their turnover and stability as soil organic matter. In: Schulze E-D, Heimann M, Harrison S, Holland E, Lloyd J, Prentice IC, Schimel D (eds) Global biogeochemical cycles in the climate system. Academic Press, San Diego, pp 201–216

Gleixner G, Poirier N, Bol R, Balesdent J (2002) Molecular dynamics of organic matter in a cultivated soil. Org Geochem 33:357–366

Goldberg ED (1985) Black carbon in the environment. Wiley, New York

Gregorich G, Janzen HH (2000) Decomposition. In: Sumner ME (ed) Handbook of soil science. CRC Press, Boca Raton, pp C-107–C-119

Guggenberger G (2002) Organic matter structure and characterization. In: Lal R (ed) Encyclopedia of soil science. Marcel Dekker, New York, pp 929–935
Guggenberger G, Frey SD, Six J, Paustian K, Elliott ET (1999) Bacterial and fungal cell-wall residues in conventional and no-tillage agroecosystems. Soil Sci Soc Am J 63:1188–1198
Haider K (1992) Problems related to the humification processes in soils of the temperate climate. In: Bollag J-M, Stotzky G (eds) Soil biochemistry, vol 7. Marcel Dekker, New York, pp 55–94
Haider K, Martin JP (1975) Decomposition of specifically carbon-14 labelled benzoic and cinnamic acid derivatives in soil. Soil Sci Soc Am Proc 39:657–662
Hamer U, Marschner B, Brodowski S, Amelung W (2004) Interactive priming of black carbon and glucose mineralisation. Org Geochem 35:823–830
Harborne JB (1997) Role of phenolic secondary metabolites in plants and their degradation in nature. In: Cadisch G, Giller KE (eds) Driven by nature: Plant litter quality and decomposition. CAB International, Wallingford, pp 67–74
Haumaier L, Zech W (1995) Black carbon – possible source of highly aromatic components of soil humic acids. Org Geochem 23:191–196
Hayes MHB, MacCarthy P, Malcolm R, Swift RS (1989) Humic substances II. In search of structure. Wiley-Interscience, Chichester
Hedges JI (1978) The formation and clay mineral reactions of melanoidins. Geochim Cosmochim Acta 42:69–76
Hedges JI (1988) Polymerization of humic substances in natural environments. In: Frimmel FH, Christman, RF (eds) Humic substances and their role in the environment. Wiley, Chichester, pp 45–58
Hedges JI, Eglinton G, Hatcher PG, Kirchman DL, Arnosti C, Derenne S, Evershed RP, Kögel-Knabner I, de Leeuw JW, Littke R, Michaelis W, Rullkötter J (2000) The molecularly-uncharacterized component of nonliving organic matter in natural environment. Org Geochem 31:945–958
Herring JR (1985) Charcoal fluxes into sediments of the North Pacific Ocean: The cenozoic record of burning. Geophys Monogr 32:419–442
Kirk TK, Farrel RL (1987) Enzymatic combustions – the microbial degradation of lignin. Annu Rev Microbiol 41:465–505
Knicker H (2000) Biogenic nitrogen in soils as revealed by solid-state carbon-13 and nitrogen-15 nuclear magnetic resonance spectroscopy. J Environ Qual 29:715–723
Knicker H, Hatcher PG (1997) Survival of protein in an organic-rich sediment – possible protection by encapsulation in organic matter. Naturwissenschaften 84:231–234
Knicker H, Hatcher PG (2001) Sequestration of organic nitrogen in the sapropel from Mangrove Lake, Bermuda. Org Geochem 32:733–744
Knicker H, Scaroni AW, Hatcher PG (1996) ^{13}C and ^{15}N NMR spectroscopic investigation on the formation of fossil algal residues. Org Geochem 24:661–669
Kögel-Knabner I (1993) Biodegradation and humification processes in forst soils. In: Bollag J-M, Stotzky G (eds) Soil biochemistry, vol 8. Marcel Dekker, New York, pp 101–137
Kögel-Knabner I (2000) Analytical approaches for characterizing soil organic matter. Org Geochem 31:609–625
Kögel-Knabner I (2002) The macromolecular organic composition of plant and microbial residues as inputs to soil organic matter. Soil Biol Biochem 34:139–162
Khanna M, Yoder M, Calamai L, Stotzky G (1998) X-ray diffractometry and electron microscopy of DNA from *Bacillus subtilis* bound on clay minerals. Sci Soils 3:1
Kuhlbusch TAJ, Crutzen PJ (1995) Toward a global estimate of black carbon in residues of vegetation fires representing a sink of atmospheric CO_2 and a source of O_2. Global Biogeochem Cycles 9:491–501

Largeau C, Derenne S, Casadevall E, Kadouri A, Sellier N (1986) Pyrolysis of immature torbanite and of the resistant biopolymer (PRB A) from extant algae *Botryoccus braunii*; mechanism of formation and structure of torbanite. Org Geochem 10:1023–1032

Luis P, Walther G, Buscot F (2004) Diversity of laccase gene from basidiomycetes in a forest soil. Soil Biol Biochem 36:1025–1036

Madigan MT, Martinko JM, Parker J (1997) Brock – Biology of microorganisms. 8th edn. Prentice Hall, Upper Saddle River

Martin JP, Haider K (1986) Influence of mineral colloids on turnover rates of soil organic carbon. In: Huang PM, Schnitzer M (eds) Interactions of soil minerals with natural organics and microbes, Vol 17. Soil Science Society of America, Madison, pp 283–304

Möller A, Kaiser K, Amelung W, Niamskul C, Udomsri S, Puthawong M, Haumaier L, Zech W (2000) Forms of organic C and P extracted from tropical soils as assessed by liquid-state ^{13}C- and ^{31}P NMR spectroscopy. Aust J Soil Res 38:1017–1035

Nagata T, Kirchman DL (1996) Bacterial degradation of protein adsorbed to model submicron particles in seawater. Mar Ecol Prog Ser 132:241–248

Oades JM (1989) An introduction to organic matter in mineral soils. In: Dixon JB, Weed SB (eds) Minerals in soil environments, 2nd ed. Soil Science Society of America, Madison, pp 89–159

Ogren JA, Charlson RJ (1983) Elemental carbon in the atmosphere: cycle and lifetime. Tellus 35:241–254

Olah G-M, Reisinger O, Kilbertus G (1978) Biodégradation et humification. Atlas ultrastructural. Presses de l'Úniversité Laval, Quebec

Paul EA, Clark FE (1996) Soil microbiology and biochemistry, 2nd edn. Academic Press, San Diego

Piccolo A (2002) The supramolecular structure of humic substances: a novel understanding of humus chemistry and implications in soil science. Adv Agron 75:57–134

Preston CM (1996) Applications of NMR to soil organic matter analysis: history and prospects. Soil Sci 161:144–166

Saiz-Jimenez C (1996) The chemical structure of humic substances: recent advances. In: Piccolo A (ed) Humic substances in terrestrial ecosystems. Elsevier, Amsterdam, pp 1–44

Skjemstad JO, Clarke P, Taylor JA, Oades JM, McClure SG (1996) The chemistry and nature of protected carbon in soil. Aust J Soil Res 34:251–271

Sommers LE, Gilmour CM, Wildung RE, Beck SM (1981) The effect of water potential on decomposition processes in soils. In: Parr JF, Garnder WR, Elliott LF (ed) Water potential relations in soil microbiology. Soil Science Society of America Special Publication 9, Madison, pp 97–117

Stevenson FJ (1994) Humus chemistry. Genesis, composition, reactions. 2nd edn. Wiley, New York

Stotzky G, Gallori E, Khanna M (1996) Transformation in soil. In: Akkermans ADL, van Elsas JD, de Bruijn FJ (eds.) Molecular microbial ecology manual. Kluwer, Dordrecht, pp 1–28

Swift RS (2001) Sequestration of carbon by soil. Soil Sci 166:858–971

Tegelaar EW, de Leeuw JW, Saiz-Jimenez C (1989) Possible origin of aliphatic moieties in humic substances. Sci Total Environ 81/82:1–17

Wolters V (2000) Invertebrate control of soil organic matter stability. Biol Fertil Soils 31:1–19

Zech W, Kögel-Knabner I (1994) Patterns and regulation of organic matter transformation in soils: litter decomposition and humification. In: Schulze ED (ed) Flux control in biological systems: from the enzyme to the population and ecosystem level. Academic Press, New York, pp 303–334

5 Importance of Microorganisms for Soil Aggregation

Jean-Luc Chotte[1]

1 Introduction

Martin et al. (1955, quoted by Allison 1968) defined an aggregate as a "naturally occurring cluster or group of soil particles in which the forces holding the particles together are much stronger than the forces between adjacent aggregates". Brewer (1964) considered an aggregate as having an identifiable morphological boundary which distinguishes it from its neighbors. There have been many major papers describing of the ways in which these aggregates are formed. These models describe the mechanisms that form and stabilize aggregates: the orientation of the mineral particles, the attractions between these particles, the interactions between these particles and the soil microorganisms and the production of aggregating agents by these microorganisms (Emerson 1959; Harris et al. 1963; Dexter 1988 cited by Oades and Waters 1991).

In general, these papers support a hierarchic model for the aggregation process. Oades and Waters (1991) tested this model on three types of soil (Mollisol, Alfisol, Oxisol). In soils (Mollisol, Alfisol) where organic matter plays a role in the stabilization of aggregates greater than 250 μm, the hierarchic model was confirmed. However, in Oxisols, the role of iron oxides in cementing mineral particles minimizes the importance of the hierarchic model in aggregation. Nevertheless, it is evident that most papers describing the formation and stabilization of aggregates and the determining factors controlling the process have been strongly influenced by the hierarchic model of aggregation.

The most important strategies proposed for maintaining and improving soil fertility are those which target the physical properties of the soil. The abundance and stability of the aggregates are critical for several soil functions.

- Plant growth (Hamblin 1985; Letey 1985),
- Resistance to erosion (Le Bissonnais and Arrouays 1997; Barthès and Roose 2002)

[1]Laboratoire d'Ecologie Microbienne des Sols, UR Ibis R083, Centre ISRA-IRD Bel Air, BP 1386 Dakar, Sénégal, e-mail: Jean-Luc.Chotte@ird.sn, Tel: +221-8493308, Fax: +221-8321675

Soil Biology, Volume 3
Microorganisms in Soils: Roles in Genesis and Functions
(ed. by F. Buscot and A. Varma)

- Soil organic matter turnover (Feller and Beare 1997; Chotte et al. 1998; Puget et al. 2000; Six et al. 2001),
- The abundance, activity and diversity of organisms, both mesofauna (Quénéhervé and Chotte 1996) and microflora (Elliott 1986; Gupta and Germida 1988; Kabir et al. 1994; Chotte et al. 2002).

Consequently, it is still very important to determine the factors controlling the processes of formation and stabilization of aggregates. The importance of soil microorganisms for soil aggregation has been regularly reviewed (Griffiths 1965; Lynch and Bragg 1985; Robert and Chenu 1992; Oades 1993). This paper summarizes the literature and reports new findings.

2 Evidence of the Role of Soil Microorganisms

Many methods have been used to characterize the particle structure of a soil (reviewed by Diaz-Zorita et al. 2002). There is no agreed standard method. The most commonly used methods, however, use a sieving process to separate the soil particles physically. The sieving may be carried out on a soil sample maintained at its natural humidity level, on a soil sample rewetted by capillary action, on a soil sample immersed in water or on a soil sample dried and then rewetted. In these methods, the air trapped in the pores by the water and rapid wetting exerts a force on the solid mass, causing it to disperse. If the soil is rewetted slowly, the air can escape and so there is less force than when the soil is rewetted rapidly. It should be noted that some authors recommend more vigorous methods, such as using agitator beads, which are particularly useful for studying microaggregates in 2:1 clay soils (e. g., Vertisols; Jocteur Monrozier et al. 1991). There are a number of papers in the literature which show the effect of increasingly vigorous dispersion on the destruction of macroaggregates (> 250 µm; Cambardella and Elliott 1986; Elliott 1986; Bossuyt et al. 2001). Therefore, the method used for separating the aggregates must be specified when describing the role of soil microorganisms in aggregation.

In a recent paper, Bossuyt et al. (2001) have observed that the formation of macroaggregates (> 2000 µm) in an incubated soil with the addition of organic residues in the presence of a fungicide is significantly less than in the same soil without the fungicide. On the other hand, introducing a bactericide did not cause a reduction in the quantity of macroaggregates with respect to the control soil. This observation confirms the results in other papers and shows the role of fungi in the stabilization of aggregates (Tisdall and Oades 1982; Lynch and Bragg 1985; Miller and Jastrow 1990; Beare et al. 1997; Tisdall et al. 1997; Jastrow et al. 1998). In a laboratory study carried out using an apparatus specially designed to measure the effect of a mycorrhizal fungus (*Glomus mossae*) and the effect of the host plant's roots (*Sorghum bicolor*) separately on soil aggregation, Andrade et al. (1998) showed that more aggregates were formed when both the roots and the fungus were present (the effect of the mycorrhizosphere). The hyphae on their own (hyphosphere) had less effect than roots on their own (rhizosphere).

Although these studies illustrated the effect of fungi, other studies have shown that there is a limit to the effect of fungi on aggregation. Beyond a certain limit, further fungal growth has no further effect on aggregation (Bossuyt et al. 2001). A possible explanation of this observation may be deduced from the work of Rillig and Steinberg (2002). In this study, the growth of mycorrhizal fungus (*Glomus intraradices*) hyphae and the production of a metabolite involved in the formation and stabilization of aggregates were monitored. The study was carried out in two different artificial media using glass beads of different diameters. The large beads (710–1180 μm) simulated an aggregated soil, while the small beads (< 106 μm) simulated a nonaggregated soil. In the nonaggregated soil, which was a suboptimal growth environment for the fungus, there was little hyphal growth, but large quantities of the aggregating metabolite were produced. On the other hand, in the aggregated soil, which was more favorable for fungal growth, the opposite results were obtained with the metabolism of the fungus oriented to the production of hyphae. In the suboptimal growth environment, the fungus devotes a part of its metabolism to the production of an "organic cement" to reorganize its local environment and improve its growth conditions. Once optimal conditions have been established, fungal growth becomes more important and there is no further effect on aggregation. Fungal growth can be determined by measuring the hyphae (Baath 1988; Dighton and Kooistra 1993) or the quantity of ergosterol (Ruzicka et al. 2000). A recently established method uses monoclonal antibodies to detect hyphae using a microscope (Caesar-TonThat et al. 2001).

In a large number of field studies, the impact of microorganisms on aggregation has been compared for different types of tillage. In a study carried out over six sites along a climatic gradient (from 380 to 1140 mm/year rainfall), with conventional tillage and no-tillage agroecosystems at each site, Frey et al. (1999) found an increase in the fungal biomass and an increase in the median diameter of the aggregates in the soil. The bacterial biomass did not seem to be affected by the tillage. However, statistical analysis of the soil by the authors using the humidity of the soil as a co-variable showed that in this case the soil structure did not have an effect on the fungal biomass. This result can be partly explained by the different metabolisms of fungi and bacteria which adapt differently to changes in the humidity of the soil. Fungi are known to have a metabolism more suitable than that of bacteria for colonizing the dry environments (surface of litter, pores between aggregates, etc.) found in no-tillage soils (Holland and Coleman 1987; Beare et al. 1992). The metabolism of soil microorganisms in the formation and stabilization of aggregates is, therefore, particularly important.

3 Microbial Metabolites Responsible for Soil Aggregation

The papers cited above show the role played by soil microorganisms in the aggregate formation and stabilization processes. However, the nature of the determining microbial factors is not always clearly identified. In a paper on the effect of mycorrhizal fungi on aggregation, Degens et al. (1996) were unable to distinguish the respective roles of fine roots and fungal hyphae on the mechanical stabilization (enmeshing) of soil particles. Baldock and Kay (1987) suggested that the structural stability of soils was strongly dependent on a fraction of the total carbon content of the soil, without giving details of this carbon pool. Other authors (Haynes and Swift 1990; Gijsman and Thomas 1995; Baldock and Kay 1997) have reported positive correlations between the carbohydrate content and the stability of the aggregates. This correlation, however, does not prove that carbohydrates have a functional role in the aggregation process: Kouakoua et al. (1997) reported that removing the carbohydrates did not alter the stability of the aggregates. In a study monitoring the formation of aggregates by microorganisms in response to the addition of starch, Guggenberger et al. (1999) found a very rapid increase in the total mass of the largest aggregates (between 250 and 800 µm) in the first 4 days. They also measured a high level of fungal growth in these aggregates, whereas the microbial biomass was mostly found in the smaller aggregates. After 4 days, the measurements showed a significant reduction in the fungal biomass without, however, a reduction in the mass of the aggregates. It would appear that while the presence of fungal hyphae provides physical binding for the particles, it cannot wholly explain the formation and stabilization of aggregates.

The following paragraphs review papers that provide some information about the nature of the microbial metabolites involved in aggregation.

3.1 Polysaccharides

The importance of polysaccharides in the aggregation process has been recognized for several years (review by Robert and Chenu 1992). These papers show that the most active polysaccharides in the formation of aggregates are bacterial metabolites and exudates from roots. The effectiveness of polysaccharides stems from their macromolecular structure which enables them to be adsorbed on clay particles and also to glue particles together. The effect of polysaccharides on mineral particles is usually studied in laboratories. Chenu (1993) showed that the water retention capacity of clay particles is strongly influenced by adding polysaccharides. The author showed that the strength of this effect depends on the type of polysaccharide and the properties of the clay particles. Adsorption by clays results in a modification of the macro-structure of the clay particles which is more significant for kaolinites than for smectites. Electron microscope images and polysaccharide labeling techniques (Thiery 1967) show the role played by polysaccharides in the stabilization of microscopic aggregates (Foster 1981; Feller et al. 1991; Chotte et al. 1994). For the soil as a whole, the importance of polysaccharides in the aggregation process is shown by the correlation between the quantity of aggregates that are stable in water and the quantity of polysaccharides. Many extraction methods have been used (Cheshire 1979), but the correlation between stable aggregates and polysaccharides is strongest when the polysaccharides are extracted using hot water or dilute acids (Haynes et al. 1991; Angers et al. 1993). In a study on a silty soil (Typic Hapludalf), Puget et al. (1999) showed the predominance of plant carbohydrates in the largest aggregates. Furthermore, this study showed that the clay and silt in the stable aggregates $>50\ \mu m$ had high concentrations of microbial carbohydrates. These results confirm the hypothesis that the stability of the aggregates is partly due to the polysaccharides produced by microorganisms decomposing the vegetal residues trapped in the aggregates.

3.2
Glomalin

The characterization of arbuscular mycorrhizal fungi has shown the presence of a large quantity of an organic compound secreted by the hyphae (Wright and Upadhyaya 1996). This compound, which is insoluble in water, has aroused the interest of scientists because of its apparent recalcitrance. It is, therefore, thought to play a key role in soil stabilization. The early work concentrated on developing an extraction and identification method based on the methods used for proteins (Wright and Upadhyaha 1998). The method relies on the extractability and solubility of the fractions of this fungal metabolite. Normally two fractions, whose solubility depends on the concentration of the solvent, are separated. One fraction is called easily extractable glomalin (EEG) and the fraction extracted with the highest concentration of solvent is called total glomalin (TG). Glomalin is a glycoprotein having N-linked oligosaccharides and tightly bound iron. It is insoluble and possibly hydrophobic in its native state (Wright et al. 1996). Glomalin can be detected in situ using immunofluorescence. Wright (2000) used a monoclonal antibody produced from *Glomus intraradices* and indirect immunofluorescence to detect the glomalin on the surface of the hyphae and soil aggregates. This work showed that glomalin is detected on the mycorrhizal hyphae and on the surface of the aggregates. A number of studies have confirmed the role of this fungal metabolite in the formation and stabilization of aggregates (Wright and Upadhyaya 1998; Bird et al. 2002; Rillig and Steinberg 2002). One recent study has paid particular attention to defining the roles of the fungal hyphae and the glomalin. When glomalin is extracted, its properties will certainly change. It is, therefore, difficult to assess its role in the formation and stabilization of aggregates. In their paper, Rillig et al. (2002) compared the effect of several mycotrophic plants on aggregation. They used a path-model previously used to test for causal interaction among biological factors on soil stabilization (Jastrow et al. 1998). They concluded that the production of glomalin by the hyphae had a stronger effect than the direct action of enmeshing. This conclusion does not conflict with the findings that the hyphae have an effect on the formation of aggregates (Tisdall and Oades 1982). However, another property can be used to distinguish these two factors: the degradation time. The turnover of glomalin is between 6 and 42 years which is much longer than for the hyphae (Rillig et al. 2001). This difference has been confirmed by a laboratory study showing that after a 150-day incubation the concentration of glomalin was reduced by barely 25%, whereas the hyphae lost nearly 60% of their length after the host was removed. This difference could partly explain the seasonal variations of these two fungal factors in the aggregation. However, the conclusions of these papers, which clearly show the positive role played by glomalin in the formation and stabilization of the aggregates, must be treated with caution; no positive correlation for glomalin was found in a high carbonate soil, for which the carbonates were the primary cement for the aggregation (Rillig et al. 2003).

3.3 Lipids

The role played by lipids in the aggregation processes is less well documented than for other compounds (Capriel et al. 1990; Dinel et al. 1991). Monreal et al. (1995) compared the abundance of aggregates that were stable in water and the concentration of organic compounds (carbohydrates, lignin, lipids) in two agroecosystems (wheat fallow rotation versus continuous wheat). The results showed that the abundance of aggregates $> 250\,\mu m$

strongly correlated with the lipid concentration. The presence of lipids in the aggregates improves their resistance to slaking as lipids are hydrophobic (Paré et al. 1999).

4 Manipulation of Microbially Mediated Processes to Improve Soil Aggregation

Manipulating soil microorganisms or using their metabolites to improve the structure of soils has been investigated for a long time (Taylor and Baldbridge 1954, quoted by Robert and Chenu 1992). The most common methods are the modification of the microbial community associated with a plant, organic amendments and the introduction of microorganisms.

4.1 The Rhizosphere Microbial Community

The potential for improving the physical properties of soils provided by the plant rhizosphere and its microbial community has been exploited for a long time. This improvement has been described in recent papers which underline the continuing usefulness of this approach (Haynes and Beare 1997; Caravaca et al. 2002a,b, 2003). These papers show clearly that the presence of plants with their natural or introduced symbiotic fungal community encourages the formation of stable macroaggregates. However, the physical properties of soils can be improved by using nonmycotrophic plants. Haynes and Beare (1997) observed the same increase in the quantity of stable aggregates in the presence of mycotrophic and nonmycotrophic plants. The stimulation of saprophytic fungal communities by the root exudates partly explains this result.

4.2 Organic Residues

A wide range of organic materials has been studied to investigate the impact on aggregation (Tisdall et al. 1978; Schlecht-Pietsch et al. 1994; Roldan et al. 1996; Paré et al. 1999; Sonneleitner et al. 2003). These papers show that organic amendments have a positive effect on aggregation. In a study carried out on tropical soils, Spaccini et al. (2003) showed that organic amendments proved to be an effective strategy for improving the structure of soils with low stability. However, it should be noted that this positive effect varies depending on the type of clay (Denef et al. 2002). This study showed that the increase in stable aggregates (> 250 µm), after the addition of exogenous organic matter, is greater in a soil where the clay content is a mixture of 2:1 and 1:1 clays in comparison with the increase in soil with 1:1 clays. In a mixture of 2:1 and 1:1 clays, it is probable that electrostatic forces are added to the effects of the microbial metabolites from the decomposition of the added organic matter in forming and stabilizing the aggregates. To study the effect of the biochemical quality of the organic materials and the effect of the presence of an exogenous source of mineral nitrogen, Bossuyt et al. (2001) monitored the formation of aggregates stable in water during incubation in a laboratory. In this test, the authors compared the effect of a high nitrogen organic material (C/N:19.7) with a low nitrogen material (C/N:108). During the first 15 days of incubation, no significant difference was recorded between the quantity of stable aggregates (> 250 µm) formed in these two experiments. It should be noted that adding a source of mineral nitrogen caused a reduction in the quantity of aggregates formed (low quality amendment). This result may be explained by the lower microbial activity, hence the lower level of secretion of microbial metabolites, recorded in this experiment.

4.3 Inoculation with Microorganisms

The properties of polysaccharides are such that many studies have been carried out on the feasibility of exploiting them. Most studies have concentrated on the selection and inoculation of the microorganisms that produce polysaccharides. The following papers are of note.

- Amellal et al. (1999) on the effects of inoculating the soil with a bacterium selected from the wheat rhizosphere for its ability to produce exopolysaccharides,
- Alami et al. (1997) on the effects of a rhizobium obtained from the sunflower rhizosphere,
- Rogers and Burns (1994) on the inoculation of soils with cyanobacteria.

These laboratory experiments show the advantages of such microbiological management of the physical properties of soils.

5 Conclusion

A large number of papers are devoted to characterizing and understanding the aggregate formation–stabilization processes. Most of the papers concern the nature of the microbiological factors. They support the role, over and over again, of soil microorganisms. They emphasize the interactions between these microbial factors in aggregation and reveal the importance of complex systems with more than one agent and one process:

- bacteria versus fungi biomass,
- the growth of fungi and their dependence on roots and plant debris for their energy source,
- the importance of fungal and bacterial metabolites and exudates in adsorptive reactions with mineral particles,
- the distribution of these metabolites through the soil pore systems

It is all an energy-driven process. Therefore, any strategy aims at improving the impact of soil microorganism on soil structure should take into account the allocation of carbon inputs that fuel these biological processes.

Acknowledgements. The author expresses his gratitude to Dr M. Oades for his helpful comments

References

Alami Y, Achouak W, Marol C, Heulin T (1997) Rhizosphere soil aggregation and plant growth promotion of sunflowers by an exopolysaccharide-producing *Rhizobium* sp. strain isolated from sunflower roots. Appl Environ Microbiol 66(8):3393–3398
Allison FE (1968) Soil aggregation-zone facts and fallacies as seen by a microbiologist. Soil Sci 106:136–143
Amellal N, Bartoli F, Villemin G, Talouizte A, Heulin T (1999) Effects of inoculation of EPS-producing *Pantoea agglomerans* on wheat rhizosphere aggregation. Plant Soil 211:93–101
Andrade G, Mihara KL, Linderman RG, Bethlenfalvay GJ (1998) Soil aggregation status and rhizobacteria in the mycorrhizosphere. Plant Soil 202:89–96
Angers DA, N'dayegamiye A, Côté D (1993) Tillage-induced differences inorganic matter of particle-size fractions and microbial biomass. Soil Sci Soc Am J 57:512–516
Baath E (1988) Autoradiographic determination of metabolically-active fungal hyphae in forest soil. Soil Biol Biochem 20:123–125
Baldock JA, Kay BD (1987) Influence of cropping history and chemical treatments on the water stable aggregation of a silt loam soil. Can J Soil Sci 67:501–511
Barthès B, Roose E (2002) Aggregate stability as an indicator of soil susceptibility to runoff and erosion; validation at several levels. Catena 47:133–149
Beare MH, Parmelee RW, Hendrix PF, Cheng W (1992) Microbial and faunal interactions and effects on litter nitrogen and decomposition in agroecosystems. Ecol Monogr 62:569–591

Beare MH, Reddy MV, Tian G, Srivastava SC (1997) Agricultural intensification, soil biodiversity and agroecosystem function in the tropics: the role of decomposer biota. Appl Soil Ecol 6:87–108

Bird SB, Herrick JE, Wander MM, Wright SF (2002) Spatial heterogeneity of aggregate stability and soil carbon in semi-arid rangeland. Environ Pollut 116:445–455

Bossuyt H, Denef K, Six J, Frey SD, Merckx R, Paustian K (2001) Influence of microbial populations and residues quality on aggregate stability. Appl Soil Ecol 16:195–208

Brewer R (1964) Fabric and mineral analysis of soils. Wiley, New York

Caesar-TonThat TC, Shelver WL, Thorn RG, Cochran VL (2001) Generation of antibodies for soil aggregating basidiomycete detection as an early indicator of trends in soil quality. Appl Soil Ecol 18:99–116

Cambardella CA, Elliott ET (1993) Methods for physical separation and characterization of soil organic matter fractions. Geoderma 56:4449–4457

Capriel P, Beck T, Borchet H, Harter P (1990) Relationship between soil aliphatic fraction extracted with supercritical hexane, soil microbial biomass, and soil aggregate stability. Soil Sci Soc Am Proc 54:415–420

Caravaca F, Barea JM, Figueroa D, Roldan A (2002a) Assessing the effectiveness of mycorrhizal inoculation and soil compost addition for enhancing reafforestation with *Olea europea* subsp. *sylvestris* through changes in soil biological and physical parameters. Appl Soil Ecol 20:107–118

Caravaca F, Garcia C, Hernandez MT, Roldan A (2002b) Aggregate stability changes after organic amendment and mycorrhizal inoculation in the afforestation of a semiarid site with *Pinus halepensis*. Appl Soil Ecol 19:199–208

Caravaca F, Alguacil MM, Figueroa D, Barea JM, Roldan A (2003) Re-establishment of *Retama sphaerocarpa* as target species for reclamation of soil physical and biological properties in a semiarid Mediterranean area. For Ecol Manage 182:49–58

Chenu C (1993) Clay-or-sand polysaccharide associations as models for the interface between micro-organisms and soil: water-related properties and microstructure. Geoderma 56:143–156

Cheshire MV (1979) Nature and origin of carbohydrates in soils. Academic Press, London

Chotte JL, Villemin G, Guilloré P, Jocteur Monrozier L (1994) Morphological aspects of microorganism habitats in a vertisol. In: Ringrose-Voase AJ, Humphreys GS (eds) Proceedings of the IXth International Working Meeting on Soil micromorphology, vol 22. Developments in Soil Science, Townsville, pp 395–403

Chotte JL, Ladd JN, Amato M (1998) Sites of microbial assimilation and turnover of 14C soluble and particulate substrates decomposing in a clay soil. Soil Biol Biochem 30:205–218

Chotte JL, Schwartzman A, Bailly R, Jocteur Monrozier L (2002) Changes in bacterial communities and Azospirillum diversity in a tropical soil under 3 years and 19 years natural fallow assessed by soil fractionation. Soil Biol Biochem 34:1083–1092

Degens BP, Sparling GP, Abbott LK (1996) Increasing the length of hyphae in a sandy soil increases the amount of water-stable aggregates. Appl Soil Ecology 3:149–159

Denef K, Six J, Bossuyt H, Frey SD, Elliott ET, Merckx R, Paustian K (2001) Influence of dry-wet cycles on the interrelationship between aggregate, particulate organic matter, and microbial community dynamics. Soil Biol Biochem 33:1599–1611

Denef K, Six J, Merckx R, Paustian K (2002) Short term effects of biological and physical forces on aggregate formation in soils with different clay mineralogy. Plant Soil 246:185–200

Dexter AR (1988) Advances in characterization of soil structure. Soil Tillage Res 11:199–238

Diaz-Zorita M, Perfect E, Grove JL (2002) Disruptive methods for assessing soil structure.

Soil Tillage Res 64:3–22
Dighton J, Kooistra M (1993) Measurement of proliferation and biomass of fungal hyphae and roots. Geoderma 56:317–330
Dinel H, Schnitzer M, Mehuys GR (1991) Soil lipids: origin, nature, content, decomposition and effect on soil physical properties. In: Stotzky G (ed) Soil biochemistry, vol 6. Dekker, New York, pp 397–430
Elliott ET (1986) Aggregate structure and carbon, nitrogen and phosphorus in native and cultivated soils. Soil Sci Soc Am J 50:627–633
Emerson WW (1959) The structure of soil crumbs. J Soil Sci 10:235–244
Feller C, Beare MH (1997) Physical control of soil organic matter dynamics in the tropics. Geoderma 79:69–116
Feller C, François C, Villemin G, Portal JM, Toutain F, Morel JL (1991) Nature des matières organiques associées aux fractions argileuses d'un sol ferrallitique. CR Acad Sci 312:1491–1497
Foster R (1981) Localization of organic materials in situ in ultrathin sections of natural soil fabrics using cytochemical techniques. In: Bisdom EBA (ed) Working group on submicroscopy, Wageningen, The Netherlands, pp 309–319
Frey SD, Elliott ET, Paustian K (1999) Bacterial and fungal abundance and biomass in conventional and no-tillage agroecosystems along two climatic gradients. Soil Biol Biochem 31:573–585
Gijsman AJ, Thomas RJ (1995) Aggregate size distribution and stability of an Oxisol under legume-based and pure grass pastures in the Eastern Colombian savannas. Aust J Soil Res 33:153–165
Griffiths E (1965) Microorganisms and soil structure. Biol Rev 40:129–142
Guggenberger G, Elliott ET, Frey SD, Six J, Paustian K (1999) Microbial contributions to the aggregation of a cultivated grassland soil amended with starch. Soil Biol Biochem 3:407–419
Gupta V, Germida JJ (1988) Distribution of microbial biomass and its activity in different soil aggregate size classes as affected by cultivation. Soil Biol Biochem 20:777–786
Hamblin A (1985) The influence of soil structure on water movement, crop growth, and water uptake. Adv Agron 38:95–158
Harris RF, Allen ON, Chesters G, Attoe OJ (1963) Evaluation of microbial activity in soil aggregate stabilization and degradation by the use of artificial aggregates. Soil Sci Soc Am Proc 27:542–545
Haynes RJ, Swift RS (1990) Stability of soil aggregates in relation to organic constituents and soil water content. J Soil Sci 41:73–83
Haynes RJ, Beare MH (1997) Influence of six crop species on aggregate stability and some labile organic matter fractions. Soil Biol Biochem 29:1647–1653
Haynes RJ, Swift RS, Stephen RC (1991) Influence of mixed cropping rotations (pasture-arable) on organic matter content, stable aggregation and clod porosity in a group of soils. Soil Tillage Res 19:77–87
Holland EA, Coleman DC (1987) Litter placement effects on microbial and organic matter dynamics in an agrosystem. Ecology 68:425–433
Jastrow JD, Miller RM, Lussenhop J (1998) Contributions of interacting biological mechanisms to soil aggregate stabilization in restored prairies. Soil Biol Biochem 30:905–916
Jocteur Monrozier L, Ladd JN, Fitzpatrick RW, Foster RC, Raupach M (1991) Components and microbial biomass content of size fractions in soils of contrasting aggregation. Geoderma 49:37–62
Kabir M, Chotte JL, Rahman M, Bally R, Jocteur Monrozier L (1994) Distribution of soil fractions and location of soil bacteria in a vertisol under cultivation and perennial grass. Plant Soil 163:243–255

Kouakoua E, Sala GH, Barthès B, Larre-Larrouy MC, Albrecht A, Feller C (1997) La matière organique soluble à l'eau chaude et la stabilité de l'agrégation. Aspects méthodologiques et application à des sols ferrallitiques du Congo. Eur J Soil Sci 48:239–247
Lavalle P, Spain AV (2001) Soil ecology. Kluwer, Dordrecht
Le Bissonnais Y, Arrouays D (1997) Aggregate stability and assessment of soil crustabilty and erodibility. 2 Application to humic loamy soils with various organic carbon contents. Eur J Soil Sci 48:39–48
Letey J (1985) Relationship between soil physical properties and crop production. In: Stewart BA (ed) Advance in soil science. Springer, Berlin Heidelberg New York, 277 pp
Lynch JM, Bragg E (1985) Microorganisms and soil aggregate stability. Adv Soil Sci 2:134–170
Miller RM, Jastrow JD (1990) Hierarchy of root and mycorrhizal fungal interactions with soil aggregation. Soil Biol Biochem 22:579–584
Monreal CM, Schnitzer M, Schulten HR, Campbell CA, Anderson DW (1995) Soil organic structures in macro and microaggregates of a cultivated brown chernozem. Soil Biol Biochem 27:845–853
Oades JM (1993) The role of biology in the formation, stabilization and degradation of soil structure. Geoderma 56:377–400
Oades JM, Waters AG (1991) Aggregate hierarchy in soils. Aust Soil Res 29:815–828
Paré T, Dinel H, Moulin AP, Townley-Smith L (1999) Organic matter quality and structural stability of a Black Chernozemic soil under different manure and tillage practices. Geoderma 91:311–326
Puget P, Angers DA, Chenu C (1999) Nature of carbohydrates associated with water-stable aggregates of two cultivated soils. Soil Biol Biochem 31:55–63
Puget P, Chenu C, Balesdent J (2000) Dynamics of soil organic matter associated with particle-size fractions of water-stable aggregates. Eur J Soil Sci 51:595–605
Quénéhervé P, Chotte JL (1996) Distribution of nematodes in vertisol aggregates under a permanent pasture in Martinique. Appl Soil Ecol 4:193–200
Rillig MC, Steinberg PD (2002) Glomalin production by an arbuscular mycorrhizal fungus: a mechanism of habitat modification? Soil Biol Biochem 34:1371–1374
Rillig MC, Wright SF, Nichols KA, Schmidt WF, Torn MS (2001) Large contribution of arbuscular mycorrhizal fungi to soil carbon pools in tropical forest soils. Plant Soil 233:167–177
Rillig MC, Wright SF, Eviner T (2002) The role of arbuscular mycorrhizal fungi and glomalin in soil aggregation: comparing effects of five plant species. Plant Soil 238:325–333
Rillig MC, Fernando TM, Lamit LJ (2003) Microsite differences in fungal hyphal length, glomalin and soil aggregate stability in semiarid Mediterranean steppes. Soil Biol Biochem 35:1257–1260
Robert M, Chenu C (1992) Interactions between soil minerals and microorganisms. In: Stotzky G, Bollag JM (eds) Soil biochemistry, vol 7. Dekker, New York, pp 307–404
Rogers SL, Burns RG (1994) Changes in aggregate stability nutrient status, indigenous microbial populations, and seedling emergence following inoculation with *Nostoc muscorum*. Biol Fertil Soils 18:219–215
Roldan A, Albaladejo J, Thornes JB (1996) Aggregate stability changes in a semiarid soil after treatment with different organic amendments. Arid Soil Res Rehabil 10:139–148
Ruzicka S, Edgeerton D, Norman M, Hill T (2000) The utility of ergosterol as a bioindicator of fungi in temperate soils. Soil Biol Biochem 32:989–1005
Schlecht-Pietsch S, Wagner U, Anderson TH (1994) Changes in composition of soil polysaccharides and aggregate stability after carbon amendments to different textured soils. Appl Soil Ecol 1:145–154
Six J, Carpentier A, Kessel CV, Merckx R, Harris D, Horwath WR, Lüscher A (2001) Impact of elevated CO_2 on soil organic matter dynamics as related to changes in aggregate turnover and residue quality. Plant Soil 234:27–36

Sonnleitner R, Lorbeer E, Schinner F (2003) Effects of straw, vegetable oil and whey on physical and microbiological properties of a chernozem. Appl Soil Ecol 22:195–204
Spaccini R, Mbagwu JSC, Igwe CA, Conte P, Piccolo A (2003) Carbohydrates and aggregation in lowland soils of Nigeria as influenced by organic inputs. Soil Tillage Res 75:161–172
Thiery JP (1967) Mise en évidence des polysaccharides sur coupes fines en microscopie électronique. J Microsc 6:987–1017
Tisdall JM, Oades JM (1982) Organic matter and water stable aggregates in soils. J Soil Sci 33:141–163
Tisdall JM, Cockroft B, Uren NC (1978) The stability of soil aggregates as affected by organic materials, microbial activity and physical disruption. Aust J Soil Res 10:9–18
Tisdall JM, Smith SE, Rengasamy P (1997) Aggregation of soil by fungal hyphae. Aust J Soil Res 35:55–60
Wright SF (2000) A fluorescent antibody assay for hyphae and glomalin from arbuscular mycorrhizal fungi. Plant Soil 226:171–177
Wright SF, Upadhyaya A (1996)Extraction of an abundant and unusual protein from soil and comparison with hyphae protein of arbuscular mycorrhizal fungi. Soil Sci 161:575–586
Wright SF, Upadhyaya A (1998) A survey of soils for aggregate stability and glomalin, a glycoprotein produced by hyphae of arbuscular mycorrhizal fungi. Plant Soil 198:97–1007
Wright SF, Franke-Snyder M, Morton JB, Upadhyaya A (1996) Time-course study and partial characterization of a protein on hyphae of arbuscular mycorrhizal fungi during active colonization of roots. Plant Soil 181:193–203
Wright SF, Upadhyaya, A, Buyer JS (1998) Comparison of N-linked oligosaccharides of glomalin from arbuscular mycorrhizal fungi and soils by capillary electrophoresis. Soil Biol Biochem 30:1853–1857

Part III
Microorganisms and Biogeochemical Processes in Soils

6 Microbial Energetics in Soils

Oliver Dilly[1]

1 Introduction

The ultimate source of energy for life on Earth is the sun. Plants and photoautotrophic microorganisms preserve energy from sunlight in carbonic compounds derived largely from atmospheric carbon dioxide. In ecosystems, less than 3% of the radiation of the sun is generally fixed as the net primary production which may be harvested or disposed as litter to soil. In natural ecosystems, the microbiota in soil obtain energy mainly from dead organic matter derived from plant litter or from inputs of the living soil biomass (roots, animals and microorganisms) throughout the season. Along with chemoautotrophic microbes, the heterotrophic soil microbiota ensure the liberation of nutrients and, thus, is of great relevance for ecosystem functioning.

The energy of a green photon is 238 kilojoule (kJ) per mole. An alternative unit of energy is calorie; 1 kcal mol^{-1} is equal to 4.184 kJ mol^{-1}. Adenosine 5'-triphosphate (ATP) as a universal currency of energy has a usual energy content of about 30.55 kJ mol^{-1} (Ziegler 1983). The complete oxidation of 1 mol of glucose produces 36–38 ATP. In contrast, the average energy of each vibrational degree of freedom in a molecule is much smaller, 2.5 kJ mol^{-1} at 25 °C. This amount of energy is much less than that needed to dissociate covalent bonds, e. g. 199 kJ mol^{-1} for a C–C bond. Hence, the covalent framework of biomolecules is stable in the absence of enzymes and inputs of energy. On the other hand, non-covalent bonds in biological systems typically have an energy of only a few kilojoules per mole, so that thermal energy is enough to make and break them (Madigan et al. 2003).

Microbial energetics and all other forms of metabolism are driven by the Gibbs free-energy yield derived from ATP. ATP is heterotrophically generated by fermentation or respiration. The latter requires terminal electron acceptors (e. g. molecular oxygen, nitrate, sulphate, ferric iron, carbon dioxide, or molecular hydrogen), and produces greater amounts of ATP per unit of substrate. Fermentation is less feasible in the presence of either

[1]Institut für Bodenkunde, Universität Hamburg, Allende-Platz-2, 20146 Hamburg, Germany, e-mail: o.dilly@ifb.uni-hamburg.de, Tel: +49-40-428382010, Fax: +49-40-428382024

Soil Biology, Volume 3
Microorganisms in Soils: Roles in Genesis and Functions
(ed. by F. Buscot and A. Varma)

highly oxidised or highly reduced substrates, and may rise toxic products (simple organic acids) that can eventually impede the process. The energy in soil is preserved in both organic and inorganic components enabling the microbial communities to sustain catabolic and anabolic processes.

The way in which energy and matter pass through the ecosystem is clearly important in relation to soil fertility. If it were possible to control parts of the pathway either by suppressing or encouraging the energy transformation, it might be possible to increase food production. The amount of energy diverted into crop plants, animals and decomposition varies from one soil to another and if we are to exploit our environment to the full, these processes must be established.

2
Soil, Energy and Microorganisms

The soil is the epidermis of the earth and represents the interface between atmosphere, lithosphere, hydrosphere and biosphere. This interface shows several horizons related to both local abiotic conditions (climatic and geological) and biotic factors. Soil types and soil-specific characteristics adjust during succession and organic matter is generally accumulated particularly during the early stages of soil development. Long-term rates of accumulation of organic carbon in Holocene-age soils ranged in most areas between 1 and 2 g $C\,m^{-2}\,year^{-1}$, but can be up to 12 g $C\,m^{-2}\,year^{-1}$ in temperate and boreal forests (Schlesinger 1997).

The organic matter represents the main source of energy in soil. Furthermore, it plays a major role in soil structure and, thus, has a great impact on water penetration and preservation, root development, and erosion resistance. It also stores major nutrients such as nitrogen, sulphur and phosphorus and many minor elements. The nutrients are structural constituents or adsorbed via exchange capacity.

Organic matter content and distribution vary widely among soils (Schlesinger 1997). Per surface area, the highest accumulation of up to 70 kg $C\,m^{-2}$ to a depth of 1 m, occurs in Histosols of highly productive swamps and marshes where decomposition is inhibited by a lack of O_2 (Table 1). Chernozems, Kastanozems, Greyzems and Phaeozems below grassland, developing under climatic conditions with a wet-dry season and containing humus-stabilising Ca^{2+}, have higher organic carbon levels than the equivalent boreal and temperate forests, although the carbon input, as shown by the plant primary production of 6 mg $C\,ha^{-1}\,year^{-1}$, is on average only half of the 12 mg $C\,ha^{-1}\,year^{-1}$ in the forest. Despite high total carbon stocks in tropical soils, tropical forests, with an annual average net production of 19 mg $C\,ha^{-1}\,year^{-1}$, have less soil organic carbon (C_{org}) per hectare than

Table 1. Global distribution of plant biomass and soil organic carbon

Ecosystem type	Soil organic carbon[a] ($kg\,C\,m^{-2}$)	World area[a] ($ha \times 10^8$)	Total world soil organic carbon[a] ($Mg\,C \times 10^9$)	Plant biomass[b] ($Mg\,C\,ha^{-1}$)	Net primary production[b] ($Mg\,C\,ha^{-1}\,year^{-1}$)
Tropical forest	10.4	24.5	255	19	19
Temperate forest	11.8	12	142	12	12
Boreal forest	14.9	12	179		
Woodland and shrubland	6.9	8.5	59		
Tropical savannah	3.7	15	56		
Temperate grassland	19.2	9	173	0.7	6
Tundra and alpine	21.6	8	173		
Desert scrub	5.6	18	101	0.01	1
Extreme desert, rock and ice	0.1	24	3		
Cultivated	12.7	14	178	7	6
Swamp and marsh	68.6	2	137		30
		147	1456		

[a] After Schlesinger (1997)
[b] After Paul and Clark (1989)

temperate forests, with a productivity of $12\,mg\,C\,ha^{-1}\,year^{-1}$. In tropical soils, the organic matter can distributed deep throughout the profile in humic Nitisols (ISSS/ISRIC/FAO 1998).

Soil contains a complex array of compounds derived from various origins and differing in energetic value (Table 2). Readily available components exuded by roots such as organic acids or amino acids can be mineralised rapidly and complex substrate spectrum in litter, fulvic and humic acids are transformed more slowly. Substrate complexes in an ecosystem such as leaf litter are enriched by lignin content and a lignin/N or lignin/cellulose ratio enables the separation of readily available and recalcitrant substrates. Their transformation is studied by using nylon bags filled with litter and buried in the field (Berg and McClaugherty 2003). The degradation of these substrate complexes in soil is generally slow with half-lives between 0.2 and 5 years, which is controlled by water availability, temperature, substrate quality and the structure and activity of the microbial communities. Xenobiotic organic contaminants such as naphthalene, phenanthrene, anthracene and benzpyrene have a much longer half-life of 2–17 years (Litz 1992).

The soil microbiota is generally substrate limited under natural conditions which is reflected by the strong increase in microbial respiratory activity after the addition of readily available C such as glucose (Fig. 1). Such an increase in activity of heterotrophic organisms by the substrate

Table 2. Free energy of formation for a few compounds of biological interest and turnover time of organic compounds in soil

Compound	Formula	Mol wt.	Turnover rate k* (years)	kJ mol^{-1}	Respiratory quotient (mol CO_2 mol^{-1} O_2)
Water	H_2O	18	–	–237.2[a]	–
Carbon dioxide	CO_2	44	–	–394.4[a]	–
Oxygen	O_2	16	–	0[a]	–
Hydrogen	H_2	2	–	0[a]	–
Glucose (respiration)	$C_6H_{12}O_6$	180	0.003	–917.3[a]	1.0
Glucose (combustion)	$C_6H_{12}O_6$	180	–	–2817[d,e]	1.0
Oxalic acid (combustion)	$C_2O_4H_2$	90	–	–251[d]	4.0
Acetate	C_2OH_4	52	–	–369.4[a]	0.8
Palmitin (combustion)	$C_{16}H_{32}O_2$	256	–	–10037[d]	0.7
Octane	$C_8H_{18}(l)$	118	–	–5470[e]	0.7
Methane	CH_4	16	–	–50.8[a]	0.5
Methane (combustion)	CH_4	16	–	–890.3[e]	0.5
Methanol	CH_4O	32	–	–175.4[a]	0.67
Nitrous oxide	N_2O	44	–	+104.2[a]	–
Ammonium	NH_4^+	18	–	–79.4[a]	0
Cellulose	$n(C_6H_{12}O_6)$	$> 400,000$	0.07	–	1.0
Hemicellullose	$n(C_5H_{10}O_6)$	–	0.07	–	–
Lignin	–		1.5	–	< 0.909
Fats	$C_{57}H_{110}O_6(l)$	890	–	–75520[e]	0.7
Litter (*Alnus* leaves)	–	–	0.6–1.0[f]	–	–
Litter (*Secale* straw)	–	–	0.5[c]	–	–
Fulvic acids	–	–	1[c]	–	–
Humic acids and humins	–	–	10[c]	–	< 0.909

* k Decomposition constant for the estimation of the first order decomposition rate with the formula $X_t = X_o\,e^{-kt}$ (Berg and McClaugherty 2003)
[a] After Madigan et al. (2003)
[b] After Killham (1994)
[c] After Schlesinger (1997)
[d] After Ziegler (1983)
[e] Energy content of fuels (www.chemistry.usna.edu/mavapps/Fuels/default.htm)
[f] After Dilly and Munch (1996)

addition demonstrates the significant potential of heterotrophic activity in soil.

The energy yield is generally lower for chemoautotrophs than for heterotrophs, e. g. nitrifiers gain approximately 272 kJ or 8.8 ATP molecules per mole NH_4^+ for *Nitrosomonas* and 76 kJ or 2.5 ATP molecules per mole NO_2^- for *Nitrobacter*. The recovery of this energy by the microbes ranges from

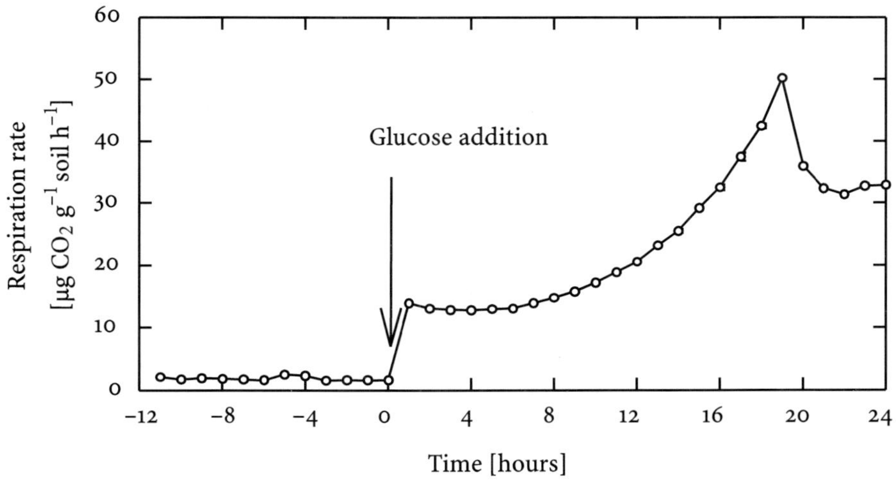

Fig. 1. Microbial respiration typical for a topsoil with and without addition of glucose (5 mg g^{-1} soil); Arenosol under crop rotation in the Bornhöved Lake district, Germany, with pH [H_2O] 6.4 and 16 mg C_{org} g^{-1} soil at 22 °C and 40–70% WHC

about 5–15% (Tate 2000). Thus, large quantities of NH_4^+ or NO_2^- must be oxidised for each C fixed by the microbes. Ammonium oxidisers typically oxidise between 14 and 70 NH_4^+-N per C incorporated into cellular biomass, whereas between 76 and 135 NO_2^--N must be oxidised for the similar task. In contrast, a mole of glucose may, under optimal conditions, yield an aerobic microbe 38 ATP or 1161 kJ or somewhat lower due to limited energetic efficiency (Table 2). This partly explains why autotrophic nitrifiers grow relatively slowly in the soil and even in laboratory culture where growth conditions can be made much closer to the optimal. However, the slow natural generation times for nitrifying bacteria in the order of 20–40 h, coupled with their small numbers in most soils, can give a very misleading impression of their vital contribution to nitrogen cycling and to soil ecology.

3 Microbial Communities

An enormous density and diversity of organisms can be detected in a very small soil compartment. The soil microbial biomass (C_{mic}) consists mainly of fungi and bacteria. It is assumed that several thousands of species are present in 1 g of soil (Torsvik et al. 1990).

Based on system-theoretic concepts, the spectrum of organisms can be distinguished in K- and r-strategists (Fig. 2). The classical thesis of Wino-

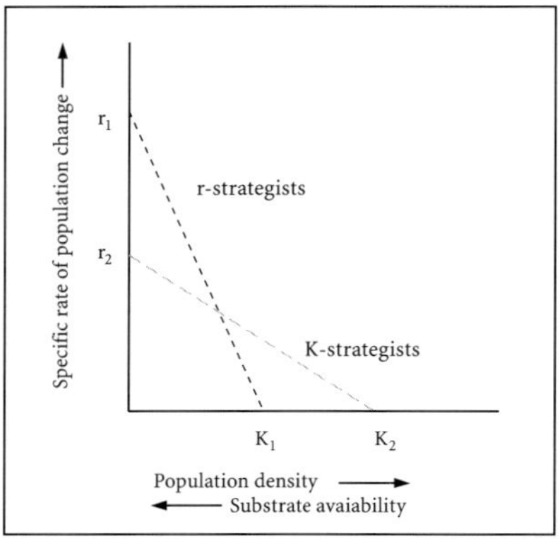

Equation for growth of populations:

$$dN\,dt^{-1}\,N^{-1} = r - (r\,N\,K^{-1})$$

where

$dN\,dt^{-1}\,N^{-1}$ = specific rate of population increase
N = population density as either numbers or biomass
r = rate of increase in biomass or population
K = carrying capacity of the environment

r-strategists

- Small organisms
- High qCO_2 values
- Rhizosphere, root tip
- Opportunistic to variable environmental conditions such as fresh plant material and dead animals
- 'Stress'

K-strategists

- Large organisms
- Low qCO_2 values
- Bulk soil
- Related to humic substances

Fig. 2. Characteristics of r- and K-strategists adopted for the soil microbiota. (Modified after Odum 1985; Atlas and Bartha 1998)

gradsky (1924) separates them into similarly autochthonous organisms associated with the organic matter in the soil and zymogenous organisms associated with freshly incorporated nutrients, respectively. The two groups have different growth behaviour and cell size with reference to available substrates and other environmental conditions (Cherepnev et al. 2003). In contrast to K-strategists, r-strategists are considered as being small and having a high metabolic quotient (qCO_2, defined as the respiration rate per unit of biomass), are predominant at root surfaces, respond rapidly to changing environmental conditions and are finally, more resistant to stress.

Microbial communities present in soil need to effectively adjust their performance to environmental factors by modifying biomass, community structure and/or activity rate (Kjøller and Struwe 1992). In addition to temperature and water availability, energy and nutrients supplied by heterotrophic plants are of great relevance for the soil microbial activity. Cheng et al. (1996) demonstrated that the presence of available substrates as reflected by the carbon availability index decreases along a transect from root surface to bulk soil. They defined the carbon availability index as basal respiration rate related to substrate-induced respiration. The carbon

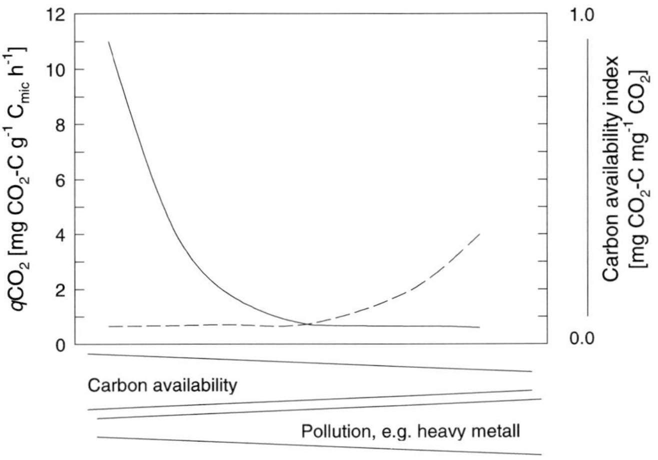

Fig. 3. Response of the metabolic activity as reflected by the metabolic quotient and carbon availability with reference to both carbon availability (*solid line*) and stressor such as pollution with heavy metal (*dashed line*); the metabolic quotient (qCO_2) is the ratio between microbial respiration and microbial biomass and the carbon availability index indicates the quotient of respiration without and with glucose addition

availability index corresponds to the qCO_2 if microbial biomass is determined by using substrate-induced respiration and soil is analysed without preconditioning for the stabilisation of the respiration rate. The qCO_2 is affected by both the presence of available substrate and the degree of soil pollution (Fig. 3).

4 Microbial Metabolism in Soil

4.1 Catabolism

Microbial metabolic activity in soil is frequently determined based on the catabolic respiratory activity since respiration is a key process in the global carbon cycle and of crucial importance in the partitioning of energy in soil. Aerobic respiratory processes consume O_2 and liberate CO_2 and lead to a loss of C from ecosystems to the atmosphere. In soil, organic matter, readily available litter and plant exudates are transformed and nutrients are mineralised mainly via respiratory processes of the microbiota. The CO_2 evolution from soil has recently been the focus of research because soil C storage, sequestration or liberation is considered to be related to climate change.

Under aerobic conditions, respiration encompasses the glycolysis and the subsequent oxidative decarboxylation of pyruvate feeding into the citric acid cycle coupled to a respiratory chain. When all these reactions are fully involved and alternative electron acceptors like NO_3^-, Fe^{3+}, ... are hardly used, the respiratory quotient (RQ) for substrates such as glucose is 1. Under these conditions, degradation of the substrate is considered as complete since a number of moles of CO_2 are consumed as O_2 are evolved. However, soil contains a wide spectrum of substrates that may be transformed and immobilised, and their complete oxidation may also be retarded by environmental and nutritional factors. Thus, soil management may modify the respiratory quotient.

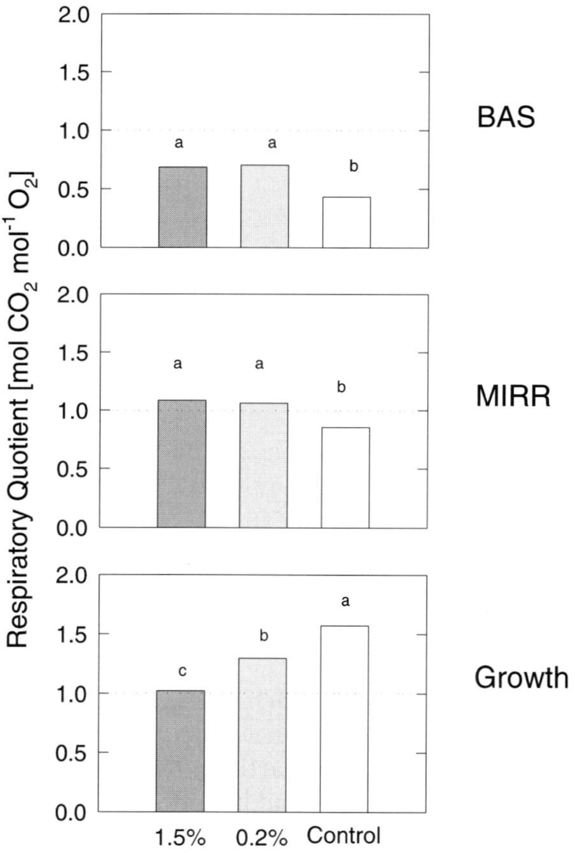

Fig. 4. Respiratory quotient of the topsoil from an eutric Arenosol under crop rotation in the Bornhöved Lake district, Germany, during basal respiration (*BAS*) and substrate-induced respiration from 0–4 h (*MIRR* maximal initial respiratory response) and 4–24 h (*Growth*); substrate rate was 5 mg glucose g^{-1} soil, soil was preconditioned with 1.5 and 0.2% oat straw for 46 days at 22 °C

The respiratory quotient is frequently lower than 1 during basal metabolism (Fig. 4), indicating that O_2 uptake is generally higher than CO_2 evolution. This may be related to the current degradation of lipids, proteins, lignin and humic acids and, furthermore, nitrification and methane oxidation in soil (Dilly 2003; Table 2).

Glucose application generally leads to the induction and de-repression of metabolic processes and to the stimulation of microbial growth and enzyme activity. RQ values approach 1 during the first 4 h after application, suggesting complete carbohydrate oxidation. Afterwards, RQ values were frequently approximately 1.3 between 4 and 24 h after substrate addition (Dilly 2001a). The presence of ammonia and nitrate may additionally modify the RQ value (Dilly 2003). The enhanced RQ value between 4 and 24 h after glucose addition indicates that the soil microbial communities may have, to a larger extent, performed glycolysis, the hexose monophosphate shunt, and Entner-Doudoroff pathway for biosynthetic purpose, the pyruvate decarboxylation and tricarboxylic acid cycle to obtain precursor metabolites for growth and low rates of oxidative phosphorylation via the respiration chain for ATP production. In addition to anabolic processes, anaerobic conditions and the mineralisation of organic acids leads to RQ values higher than 1.

Figure 4 shows that RQ values are modified when soil is preconditioned with oat straw. The available substrates enhanced the RQ value during basal metabolism and maximal initial respiratory response (MIRR) and reduced the RQ value during growth. RQ values approaching 1 during microbial growth indicated that glucose is then oxidised to a higher degree.

4.2
Anabolism

Although it is difficult to estimate the growth rate of microorganisms in soil, some idea of maximum growth rates can be obtained from data on energy input, community size and metabolic activity. It is generally accepted that the energy input in soil is sufficient to allow microorganisms to divide only a few times a year. Gray and Williams (1971) assumed for a forest soil that even if all the energy was utilised by bacteria, which is a minor component of the soil communities in terms of biomass, they could only divide once every 16 h. Other estimations for the microbial turnover under field conditions ranged much lower with values from 0.5–4 years (Joergensen 1995).

Anabolic processes occur when substrate availability is higher than catabolic requirements. Such conditions occur in the rhizosphere and in fresh organic residues that are therefore extensively colonised by both fungi and bacteria.

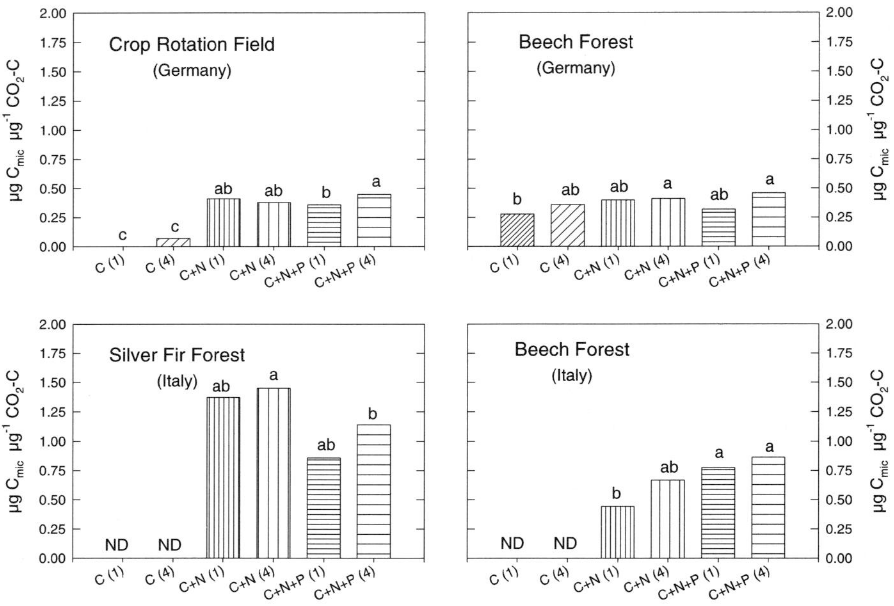

Fig. 5. Anabolic efficiency of the microbiota in four soils after the addition of glucose (*C*) and supplement N and N + P applied at one (1) or four (4) rates on four successive days; *different letters* indicate significant differences when applying the paired t-test, $p < 0.05$. (After Dilly 2001b)

The anabolic efficiency can be defined as (1) the yield coefficient, which is the C assimilated per C consumed or (2) the respiratory coefficient, which is the C assimilated per C respired. Since C assimilated and C respired are easy to estimate via growth (change in biomass content) and respiratory activity, the respiratory coefficient is uncomplicated for the quantification of the anabolic efficiency (Dilly 2001b). The anabolic efficiency can vary significantly and this is most likely attributed to the soil and nitrogen and phosphorus availability. The comparison of four soils showed that the lowest values occurred in the soils containing a low native microbial biomass and highest in soils having a higher biomass (Fig. 5). Sole C addition given in one dose resulted in an anabolic efficiency of almost zero in an arable soil, and supplemented N and N + P enhanced it. In a forest soil, the microbial C-use efficiency was considerably higher when C + N and C + N + P were applied.

At the same rate of application, the efficiency was lower in soils with a smaller native biomass than in soils with a larger biomass, which concurs with studies of Bremer and van Kessel (1990) and Witter and Kanal (1998). Witter and Kanal (1998) observed further that the ratio of biomass incorporated-to-respired glucose-C decreased with the rate of glucose application. Surprisingly, a trend of higher anabolic efficiency can even be observed when the substrate was added in one dose as in the silver fir forest soil with highest native biomass (Fig. 5).

Respiratory coefficient values of approximately 0.75 were frequently recorded in the literature for incubation times of 5 h and up to 5 days (summarised by Joergensen 1995). The values in Fig. 5, which were determined after 7 days, ranged from 0.0 to 1.45 $\mu g\ C_{mic}\ \mu g^{-1}\ CO_2 - C$ and thus represent a broad spectrum of microbial responses to the nutrient addition in agricultural and forest soils of different quality.

4.3 Soil Organic C, Microbial C and Biological Active C and Interactions with N

In comparison to agricultural or peat topsoils that contain in the range of 10 to 500 mg organic C g^{-1} soil, the basal mineralisation rate varies on average between 0.5 and 300 $\mu g\ CO_2$-C g^{-1} soil h^{-1} under optimised conditions with 40–70% water holding capacity (WHC) and 22 °C. Topsoils regularly contain a microbial biomass from 50 to 5000 $\mu g\ C_{mic}\ g^{-1}$ soil. Thus, the microbial quotient can be modified from 0.5 and 10 mg $C_{mic}\ g^{-1}\ C_{org}$ and the metabolic quotient from 0.5 and 10 mg CO_2-C $g^{-1}\ C_{mic}\ h^{-1}$ (Fig. 4). Interaction between stable and labile carbon and nitrogen pools can be evaluated by looking at the soil C/N and C_{mic}/N_{mic} ratio and the biologically active C and N and, furthermore, the qCO_2 and qN_{min} (Dilly et al. 2003).

5 Holistic Approaches to Evaluate Energetic Strategies of Soil Microbial Communities

Microorganisms dominate the biological component in most soils and respond rapidly to changing environmental conditions. These organisms are essential for many soil functions, but, in turn, their capabilities are controlled by environmental constraints.

The contents of soil organic matter (SOM) and microbial biomass (C_{mic}), both key factors of soil quality, are often closely correlated (Elliott 1997). However, Anderson and Domsch (1989) showed that the ratio between microbial biomass and organic matter content is adjusted to soil management in agricultural ecosystems, being higher in crop rotation systems and after application of organic fertiliser. Anderson and Gray (1991) observed higher values of up to 4% briefly after organic fertilisation, dropping afterwards to a site-specific and perhaps soil-quality characteristic level. In litter, values up to 8% can be found (Dilly and Munch 1996). In addition, extensive root densities frequently present in grassland soils may stimulate the development of the microbial biomass via exudation of readily available C compounds and root litter (Cheng et al. 1996; Grayston et al. 1996). In contrast, careless cultivation and stress factors reduce microbial biomass more rapidly than organic matter content, leading to low C_{mic}/C_{org} values (Sparling 1997).

The C_{mic}/C_{org} ratio, named 'microbial quotient' by Sparling (1992), is considered an indicator for biological activity and accumulation of organic matter in soil (Sparling 1992; Swift and Wommer 1993; both in Elliott 1994). Thus, the soil microbial biomass content is regulated by the site-, soil- and management-specific quantity and quality of the organic substrate (Sparling 1992). The C_{mic}/C_{org} ratio integrates the quality of soil properties looking at the extent of microbial colonisation. High values indicate that the biotope favours the establishment and energetic metabolism of many microorganisms.

The efficiency of soil microorganisms for transforming energy sources controls microbial growth for which the qCO_2 is used as an indirect easily determinable indicator. This indicator evaluates, with the C_{mic}/C_{org} ratio, the specific C mineralisation rate and the eco-physiological status of the soil microbiota (Insam et al. 1996), the succession stage (Insam and Haselwandter 1989), reflects the current energetic maintenance requirement and catabolic metabolism (Anderson 1994), and refers so far to the efficiency of the microbial metabolism (Wardle and Ghiani 1995). The term 'eco-physiology' is used here to evaluate the microbial biomass as a single organism with reference to its environment.

The maintenance requirement of actively metabolising microbial communities ranged in a similar order of magnitude to the qCO_2 value (Anderson and Domsch 1985a). In contrast, the maintenance requirement of dormant organisms is more than ten-fold lower (Anderson and Domsch 1985b). Therefore, the qCO_2 is used for the estimation of the maintenance requirement of the soil microbiota (Joergensen 1995) although there may be some variation as to the precise maintenance carbon requirement (Anderson and Domsch 1985a).

Under unfavourable conditions, the organisms require more energy to sustain the biomass, therefore, qCO_2 values are enhanced and the carbon is lost. High qCO_2 values indicate stress such as high heavy metal availability (Fließbach et al. 1995). The qCO_2 value is also enhanced when the SOM contains high amounts of readily available compounds (Fig. 3; Cheng et al. 1996; Dilly and Munch 1996). Consequently, the qCO_2/C_{org} ratio refers to the interrelationship between C-use efficiency and quality of the available organic matter in soil.

Figure 6 demonstrates the concept for the use of the qCO_2/C_{org} ratio: the value increases and simultaneously the C-use efficiency declines when half

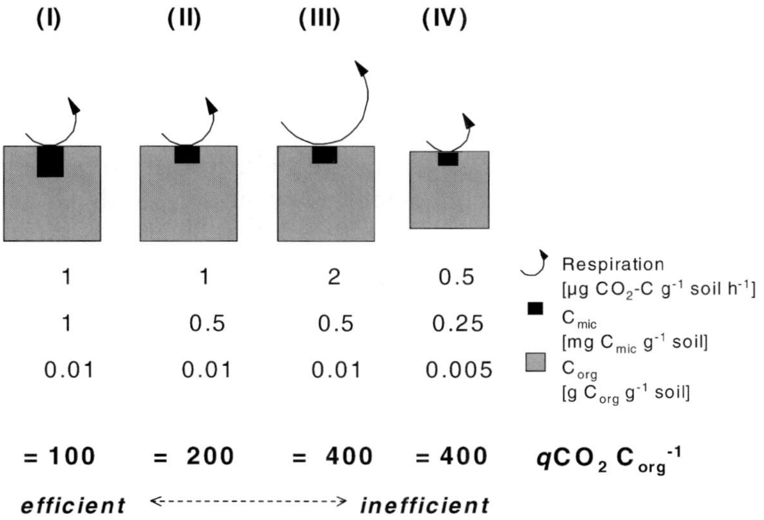

Fig. 6. Illustration of varying ratios between microbial respiration rate, C_{mic} and C_{org} content. (After Dilly et al. 2001)

of the microbial biomass and a non-modified respiration rate are supported by the same amount of organic matter (I, II). The increasing respiration rate leads to the increasing quotient indicating enhanced C-use inefficiency (III). This is unfavourable from an energetic perspective for soil microbial communities as C for growth is lost. When the respiration rate, C_{mic} and C_{org} content decrease proportionally, the qCO_2/C_{org} ratio increases (IV). Therefore, a scarce substrate level determines the efficiency value and, thus, limited substrate will be evaluated. To summarise, the qCO_2/C_{org} ratio considers the following interrelations: (1) higher respiration leads generally to increased inefficiency, (2) higher supported biomass enhances the efficiency, (3) more available substrate can support more organisms or biomass and enables higher activity (respiration), and (4) low quantity and quality of substrates are considered in particular. Since variations in microbial characteristics were frequently explained by changes in the C_{org} content, the C_{mic}/C_{org} and qCO_2/C_{org} ratios are structurally important indicators of the interdependence between microbial communities and organic matter and the energetic eco-physiological status as referring to different C levels. Although factors such as substrate quality, the water and oxygen supply and temperature in the respective biotope were not explicably taken into consideration, the three ratios should lead to structural and system-theoretical conclusions concerning the interaction between microbial communities and organic matter in soils under different management and land use (Dilly et al. 2001).

6 Conclusions

Microbial communities in soil represent a high diversity and density of biotic interactions. Fungi and bacteria dominate the microbial biomass and the potential activity is generally restricted by low nutrient availability in soil. Fresh plant residues and exudates provide the main internal source of nutrients in natural ecosystems. Substrate input rapidly stimulates catabolic and anabolic processes leading to high metabolic and microbial quotients. The respiratory quotient is additionally increased under growth conditions. The interaction between respiration, microbial C and organic C content in soil is discussed with reference to an integrative energetic indicator for soil organic matter quality.

Acknowledgements. The author is grateful for the excellent technical assistance of Sabine Splitzer, Sandra Wulff, Imke Meyer and Anne-Maria Schlöbohm and for financial support by the German Research Foundation (DFG; project no. MU 831/12–2).

References

Anderson T-H (1994) Physiological analysis of microbial communities in soil: application and limitation. In: Ritz K, Dighton J, Giller KE (eds) Beyond the biomass. Wiley, Chichester, pp 67–76

Anderson T-H, Domsch K-H (1985a) Maintenance requirement of actively metabolizing microbial populations under in situ conditions. Soil Biol Biochem 17:197–203

Anderson T-H, Domsch K-H (1985b) Determination of ecophysiological maintenance carbon requirement of soil microorganisms in a dormant state. Biol Fertil Soil 1:81–89

Anderson T-H, Domsch K-H (1989) Ratios of microbial biomass carbon to total organic carbon in arable soils. Soil Biol Biochem 21:471–479

Anderson T-H, Gray TRG (1991) The influence of soil organic carbon on the microbial growth and survival. In: Wilson ES (ed) The impact on agriculture and the environment. Adv Soil Organ Matter Res Spec Publ 90:253–266

Atlas RM, Bartha R (1998) Microbial ecology: Fundamentals and Applications. Addison-Wesley, Reading

Berg B, McClaugherty C (2003) Plant litter. Decomposition, humus formation, carbon sequestration. Springer, Berlin Heidelberg New York

Bremer E, van Kessel C (1990) Extractability of microbial ^{14}C and ^{15}N following addition of variable rates of glucose and $(NH_4)_2SO_4$ to soil. Soil Biol Biochem 22:707–713

Cheng W, Zhang Q, Coleman DC, Carroll CR, Hoffmann CA (1996) Is available carbon limiting microbial respiration in the rhizosphere? Soil Biol Biochem 28:1283–1288

Cherepnev GV, Abreimova YV, Yakovleva GY, Kurinenko BM (2003) The morphological and physiological differences between fast- and slow-growing *Escherichia coli* cells. Microbiology 72:238–239

Dilly O (2001a) Microbial respiratory quotient during basal metabolism and after glucose amendment in soils and litter. Soil Biol Biochem 33:117–127

Dilly O (2001b) Metabolic and anabolic responses of four arable and forest soils to nutrient addition. J Plant Nutr Soil Sci 164:29–34

Dilly O (2003) Regulation of the respiratory quotient of soil microbiota by availability of nutrients. FEMS Microb Ecol 43:375–381

Dilly O, Munch JC (1996) Microbial biomass content, basal respiration and enzyme activities during the course of decomposition of leaf litter in a black alder (*Alnus glutinosa* (L.) Gaertn.) forest. Soil Biol Biochem 28:1073–1081

Dilly O, Winter K, Lang A, Munch JC (2001) Energetic eco-physiology of the soil microbiota in two landscapes of southern and northern Germany. J Plant Nutr Soil Sci 164:407–413

Dilly O, Blume H-P, Sehy U, Jimenez M, Munch JC (2003) Variation of stabilised, microbial and biologically active carbon and nitrogen in soil under contrasting land use and agricultural management practices. Chemosphere 52:557–569

Elliott ET (1994) The potential use of soil biotic activity as an indicator of productivity, sustainability and pollution. In: Pankhurst CE, Doube BM, Gupta VVSR, Grace PR (eds) Soil biota. Management in sustainable farming systems. CSIRO, Melbourne, Australia, pp 250–256

Elliott ET (1997) Rationale for developing bioindicators of soil health. In: Pankhurst C, Doube BM, Gupta VVSR (eds) Biological indicators of soil health. CAB International, Wallingford, pp 49–78

Fließbach A, Martens R, Reber HH (1994) Soil microbial biomass and microbial activity in soils treated with heavy metal contaminated sewage sludge. Soil Biol Biochem 26:1201–1205

Gray TRG, Williams ST (1971) Soil microorganisms. Oliver and Boyd, Edinburgh

Grayston SJ, Vaugham D, Jones D (1996) Rhizosphere carbon flow in trees, in comparison with annual plants: the importance of root exudation and its impact on microbial activity and nutrient availability. Appl Soil Ecol 5:29–56

Insam H, Haselwandter K (1989) Metabolic quotient of the soil microflora in relation to plant succession. Oecologia 79:174–178

Insam H, Hutchinson TC, Reber HH (1996) Effects of heavy metal stress on the metabolic quotient of the soil microflora. Soil Biol Biochem 28:491–494

ISSS/ISRIC/FAO (1998) World reference base for soil resources. World soil resources, report 84. FAO, Rome

Joergensen RG (1995) Die quantitative Bestimmung der mikrobiellen Biomasse in Böden mit der Chloroform-Fumigations-Extraktions-Methode. Gött Bodenkd Ber 104:1–229

Killham K (1994) Soil ecology. Cambridge University Press

Kjøller A, Struwe S (1992) Functional groups of microfungi in decomposition. In: Carroll, GC, Wicklow DT (eds) The fungal community. Dekker, New York, pp 619–630

Litz N (1992) Organische Verbindungen. In: Blume, H-P (ed) Handbuch des Bodenschutzes. Bodenökologie und -belastung. Vorbeugende und abwehrende Schutzmaßnahmen. Ecomed, Landsberg, pp 353–399

Madigan MT, Martinko JM, Parker J (2003) Brock. Biology of microorganisms. Prentice Hall. Pearson Education, London

Odum EP (1985) Trends expected in stressed ecosystems. BioScience 35:419–422

Paul EA, Clark FE (1989) Soil microbiology and biochemistry. Academic Press, San Diego

Schlesinger WH (1997) Biogeochemistry. An analysis of global change. Academic Press, San Diego

Sparling GP (1992) Ratio between microbial biomass carbon to soil organic carbon as a sensitive indicator of changes in soil organic matter. Aust J Soil Res 30:195–207

Sparling GP (1997) Soil microbial biomass, activity and nutrient cycling as indicators of soil health. In: Pankhurst C, Doube BM, Gupta VVSR (eds) Biological indicators of soil health. CAB International, Wallingford, pp 97–119

Swift MJ, Woomer P (1993) Organic matter and the sustainability of agricultural systems: definition and measurement. In: Mulongoy K, Merckx R (eds) Soil organic matter dynamics and sustainability of tropical agriculture. Wiley-Sayce Co-Publication, IITA/KU Leuven, pp 3–18
Tate RL (2000) Soil microbiology. Wiley, New York
Torsvik V, Goksøyr J, Daae FL (1990) High diversity in DNA of soil bacteria. Appl Environ Microbiol 56:782–787
Wardle DA, Ghani A (1995) A critique of the microbial metabolic quotient (qCO_2) as a bioindicator of disturbance and ecosystem development. Soil Biol Biochem 27:1601–1610
Winogradsky MS (1924) Sur la microflore autochtone de la terre arable. Comptes Rendus 178:1236–1239
Witter A, Kanal E (1998) Characteristics of the soil microbial biomass in soils from a long-term field experiment with different levels of C input. Appl Soil Ecol 10:7–49
Ziegler H (1983) Physiologie des Stoff- und Energiestoffwechsels. In: Von Denffer D, Ziegler H, Ehrendorfer F, Bresinsky A (eds) Lehrbuch der Botanik für Hochschulen. Fischer, Stuttgart, pp 216–483

7 Role of Microorganisms in Carbon Cycling in Soils

Ellen Kandeler[1], Michael Stemmer[2], Martin H. Gerzabek[2]

1 Introduction

Soils have a large number of essential functions, some of them for the environment (protection function) and others for human or animal nutrition (production function). Most soil functions are significantly influenced by the quantity and quality of soil organic matter (SOM). This factor is essential for soil organisms and their diversity, plant nutrition, water holding capacity, aggregate stability and erosion control. More recently, the role of soil organic carbon within the global C cycle has received increasing interest (e.g., Percival et al. 2000; Bernoux et al. 2002). Quantification of possible measures to increase C stocks in soils – no or minimum tillage (e.g., Sainju et al. 2002), conversion of arable land into pasture or forest and changes in crop rotations (e.g., Blair and Crocker 2000; Gregorich et al. 2001) – is currently being intensively pursued. Changes in organic C dynamics in soils are intimately connected with or even driven by changes in microbial activities. In the past, biotic processes of decomposition were investigated at the molecular, organismal and community levels (see reviews by Sinsabaugh et al. 2002 and Tate 2002). Although the importance of soil microorganisms for global C cycling is well known, only a few researchers have attempted to combine the chemical and microbiological views of the C cycling (Kandeler et al. 2001). The present review, therefore, aims to link recent data on distribution and quality of carbon sources with the functions of the soil microbial community in C cycling. This paper also elucidates whether environmental changes (soil management and climate change) modify microbial resources and their decomposers.

[1]Institute of Soil Science, University of Hohenheim, 70599 Stuttgart, Germany, e-mail: kandeler@uni-hohenheim.de
[2]Institute of Soil Research, University of Agricultural Sciences, 1180 Vienna, Austria, e-mail: michael.stemmer@boku.ac.at, martin.gerzabek@boku.ac.at

Soil Biology, Volume 3
Microorganisms in Soils: Roles in Genesis and Functions
(ed. by F. Buscot and A. Varma)

2 Carbon Sources

Soil organic carbon (C_{org}) is the largest pool within the terrestrial carbon cycle. The annual carbon turnover through the terrestrial biosphere amounts to about 60 Gt (Schlesinger 1997; Fig. 1), which is around 9% of the atmospheric carbon pool (Esser 1990). Microbial C, measured by means of fumigation incubation, fumigation extraction or substrate-induced respiration, typically amounts to about 100–1000 $\mu g\,g^{-1}$ in arable soils and about 500–10,000 $\mu g\,g^{-1}$ in forest soils. The values rapidly decline with increasing soil depth and show distinct seasonal cycles (Table 1). The highest microbial biomass can be found in organic-C-rich litter layers of forests and boreal grasslands, whereby fungi biomass dominates there (Schimel et al. 1999). Independent of the considerably variable organic C content of soils, microbial biomass C generally corresponds to about 0.9–6% of total organic C, with a mean value around 2–3%. This indicates a relatively close relationship between microbial C and available C and N sources in soils. Turnover of microbial biomass C is high and varies between 0.2 and 3.9 a^{-1}, with mean values of about 0.5–1.5 (McGill et al. 1986; Kandeler and Böhm 1996). Raubuch and Joergensen (2002) reported turnover rates in a forest soil of about 2.7 a^{-1}, which corresponds to a C flux through the microbial biomass of 540 $kg\,ha^{-1}\,a^{-1}$.

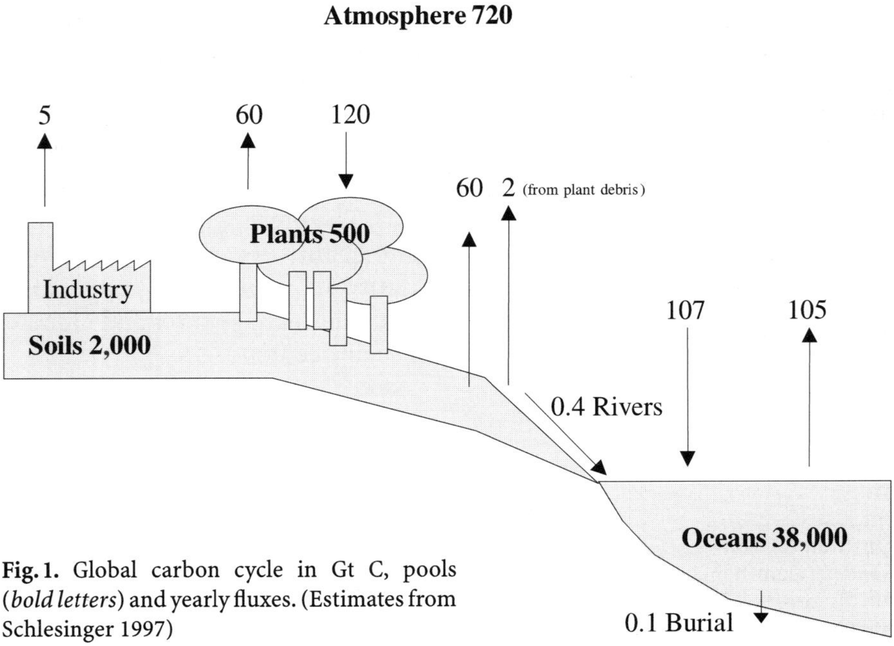

Fig. 1. Global carbon cycle in Gt C, pools (*bold letters*) and yearly fluxes. (Estimates from Schlesinger 1997)

Table 1. Microbial C content (C_{mic}) of soils from different ecosystems. Microbial biomass was measured by the fumigation extraction or substrate-induced respiration method

Ecosystem	Horizon (cm)	C_{mic} (µg/g)	C_{org} (%)	Reference
Arable soils	0–20	70– 720[a]	1.0–3.8	Beck et al. (1997)
	0–15	420– 980	2.5–5.5	Franzlübbers et al. (2000)
	0– 5	250– 1080	3.0–6.0	Hu et al. (1997)
Grassland	Litter	9650	2.1	Ross et al. (1996)
	0–10	2670	2.9	
	10–20	1120	1.9	
	20–30	470	2.0	
	F	25900[a]	ND	Johnson et al. (1998)
	0– 7	6600–15700[a]	ND	
	0–10	1120– 1510	ND	Bardgett et al. (1999)
Temperate forest	0–10	420– 1770[a]	0.9–2.6	Beck et al. (1997)
	0– 5	800– 1670	1.2–6.0	Hu et al. (1997)
	5–10	160– 430	ND	
	5–10	370– 600	ND	
	Litter	10830	2.3	Ross et al. (1996)
	FH	7670	2.0	
	0–10	1670	2.1	
	10–20	730	1.8	
	20–30	260	1.0	
Subtropical forest	0–10	330– 1090	0.9–1.8	Maithani et al. (1996)
	10–20	200– 790	0.7–1.3	
Tropical forest	0–10	210– 490	2.1–3.4	Salamanca et al. (2002)
Tropical forest	0–12	170– 860	0.5–1.1	Waldrop et al. (2000)
Tropical forest	0–10	950– 1970	1.4–2.9	Cleveland et al. (2004)
Tundra	0– 5	2990–13900	2.1–3.6	Weixin and Ross (1993)
Alpine meadows	0–10	1000– 2750	1.7–2.8	Zeller et al. (2001)
Boreal forest	Litter	2500– 6000	ND	Schimel et al. (1999)
Semiarid	0–15	120– 330	3.2–4.8	Kanchikerimath and Singh (2001)
Arid/Mediterranean	0–20	30– 700	0.8–8.0	García et al. (1994)

ND, Not determined
[a] Results based on the measurement of substrate-induced respiration

In principle, soil organic carbon is derived from autotrophic organisms – higher plants and autotrophic microorganisms. Dead plant and microbial material are the most important energy sources for soil microorganisms. The organic debris is either mineralized, i. e., completely decomposed to CO_2, or humified. The latter process leads to soil organic matter fractions of widely varying turnover half-lives, from several decades to thousands of years. Note that mineralization and humification are parallel processes. Decomposition of soil organic material is a dynamic, stepwise process. Fresh material (dead plant material) is partly decomposed, usually with a release of carbon dioxide and other inorganic compounds. This altered material is again available for further microbial attack. A small portion is converted to complex and more stable material. Per definition, the soil humus pool is divided into nonhumic and the more stable humic substances (Baldock and Nelson 2000). The first pool comprises identifiable organic compounds including polysaccharides and sugars, proteins and amino acids, lipids and lignin. The humic substances are divided into humic acids, organic materials soluble in alkaline solution, which precipitate by acidification, fulvic acids, which remain soluble when acidified, and humins, which are insoluble in alkaline solution.

3
Spatial Distribution and Protection of Carbon Sources

As organic matter input occurs both on the soil surface (litter and manures) and within the soil profile (root exudates and dead root material), the three-dimensional spatial distribution of organic carbon sources at the micro-, meso- and macroscales varies widely between soils of different origin, and according to land use and management. The depth distribution of organic carbon in soils is influenced by various factors, especially by (1) tillage, which leads to more or less unified C_{org} concentrations within the tillage depth and increases soil aeration; (2) the depth of the rooting zone, as plants transfer 20–30% of the assimilates into the soils (Kuzyakov and Domanski 2000); (3) the weathering status of the soil and thus the depth of penetrable soil horizons; (4) the activity of the soil meso- and macrofauna; (5) the chemical soil properties, as high clay contents favor the formation of organo-mineral complexes; (6) soil erosion and colluvial enrichment; and (7) soil management including historical land use (e.g., plaggen-manuring (North Germany), terra preta (Amazonia) caused unusually high organic carbon contents (Springob and Kirchmann 2002).

Figure 2 shows the high variability of C_{org} depth distributions within different soil types. Tropical soils (Lepto-/Niti-/Ferralsol and Fluvi-/Acrisol) show increasing organic C contents down to 2 m and more. Especially the deeply weathered soils store considerable amounts of C_{org} in deeper layers. Plant cover may change the C stocks drastically. Figure 2 gives an example from a chronosequence investigated in the Amazon region. A Latosol that had already been under grassland for 81 years had significantly more C_{org} than under forest. A Planosol from a foothill in the Alps contained the highest amount of C_{org} of all the presented soils within the first 100 cm of its profile due to the unfavorable mineralization conditions (wet, periodically

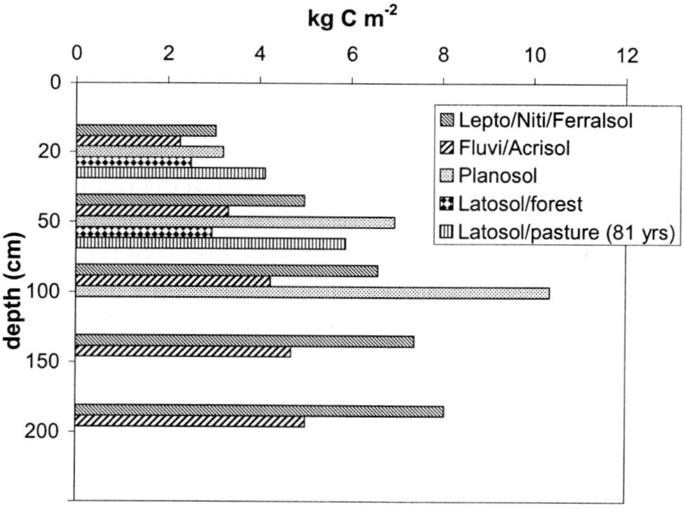

Fig. 2. Depth distribution of organic carbon stocks in a range of different soil profiles: Lepto/Niti/Ferralsol and Fluvi/Acrisol: mean values of three or two agriculturally used soils in Thailand, calculated from the excursion guide, 17th World Congress of Soil Science, southern Thailand tour; Planosol: mean values of four soil profiles under pasture at 2300 m a.s.l. in the Grossglockner massif, calculated from Gerzabek et al. (2004); Latosol: data from Neill et al. (1996, modified)

frozen subsoil). The high C_{org} stocks in deeper layers are due to colluvial materials on which this latter soil developed.

4 Spatial Distribution of Soil Microorganisms and Their Activities

During the last decade, several attempts were made to investigate the distribution of soil microorganisms within their natural environment and to relate their abundance to the turnover of organic material. These studies demonstrate that the areas of high microbial abundance and activity are heterogeneously distributed within the soil matrix. In most soils worldwide, hot spots representing > 90% of the total biological activity are concentrated in < 10% of the total soil volume (Beare et al. 1995). For example, the rhizo-, drilo-, und detritusphere are important microhabitats representing such hot spots of microbial activity (Beare et al. 1995). In many cases, complex experimental designs and/or sampling strategies were necessary to obtain microscale soil samples for chemical and microbiological analyses. Tarafdar and Jungk (1987), Gahoonia and Nielsen (1991) and Tarfdar and Marschner (1994) used 0.1–0.2 mm slices of soil cores that were separated from the root mat by a 53-μm nylon mesh to investigate the nutrient uptake of plant and microbial processes in the rhizosphere. The abundance of rhizosphere microorganisms and their activities decreased within a range of several millimeters to levels found in the bulk soil (Kandeler et al. 2002). The bacterial community composition in the rhizosphere of maize differed significantly from that of the unplanted control even up to 5 mm away from the mesh. Since about 75% of a total of 29 different bands yielded by the denaturing gradient gel electrophoresis (DGGE) were found in both planted and unplanted treatments, most soil bacteria appear to be ubiquitous. The extension of the rhizosphere depended on the excretion of easily degradable organic substances by roots, mass flow and the diffusion of dissolved organic substances used as substrates by soil microorganisms. In addition, microscale slices (0.2 mm) obtained by a freezing microtome were used to characterize gradients of dissolved organic matter and soil microbial processes at the soil–litter interface (Kandeler et al. 1999). The scale of the soil–litter interface ranged from 1.1–1.3 mm, in which the gradients of protease, xylanase and invertase activities followed an exponential function [$y = c + \exp(b_0 + b_1x_1 + b_2x_2)$]. The authors explained their results by the high local release of substrates driving C and N turnover within the 1–2 mm from the surface of the litter. The investigations on the drilosphere involved larger scales than those on the rhizo- and detritusphere. For example, Tiunov and Scheu (1999) showed that organic carbon and total nitrogen increased in burrow walls of *Lumbricus terrestris* L. by factors of 1.8–3.5 and 1.3–2.2 at distances of 0–4 and 8–12 mm, respectively, from the burrow. The high specific respiration (qO_2) and the fast growth response to nutrient additions indicated that the microbial community in the burrow walls contained a larger fraction of metabolically active microorganisms, which had adapted to the continuous resource additions by earthworm feces and mucus. The authors concluded that these burrows are stable microhabitats which sustain a large and active microbial community.

Compartmentation of substrates and soil microorganisms occurs not only at the scale of several millimeters as shown above for the rhizosphere, drilosphere and soil–litter interface, but also at an even smaller scale. Much effort was therefore made to understand the distribution of soil microorganisms and the mechanisms driving C turnover within different microcompartments of soils. On a small scale, microbial attack of soils depends on the chemical composition, C/N ratio, humification state, and physical position of the organic substrates within the soil matrix (Golchin et al. 1995). In natural soils, the activity of microorganisms and plant roots and the formation of organo-mineral complexes lead to aggregation of soil particles. This influences soil carbon storage and dynamics and a range of other soil properties (Oades and Waters 1991). Many authors (Jocteur Monrozier et

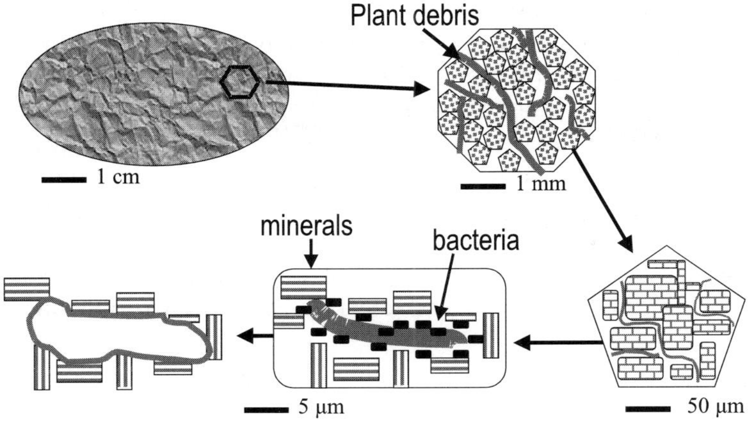

Fig. 3. Hierarchy of soil aggregates (Oades and Waters 1991, modified). The graphs shows: the bulk soil (*above, left*), a macroaggregate consisting of smaller aggregates including plant debris (*above, right*), an intermediately sized aggregate composed of microaggregates (*below, right*), a microaggegate showing minerals and bacteria associated with plant debris (*below, center*) and a microaggregate with stabilized soil organic matter (*below, left*)

al. 1991; Schulten et al. 1993; Desjardins et al. 1994; Guggenberger et al. 1994) point out that C_{org} content increases with diminishing particle size, whereas the C/N ratio decreases. Jastrow et al. (1996) showed that organic matter recently introduced into soil is predominantly located in larger aggregates. Tisdall and Oades (1982) demonstrated that macroaggregates (> 250 µm) can be destroyed by agricultural practices, whereas microaggregates cannot. Figure 3 shows the concept of the aggregate hierarchy. (1) Plant residues are introduced into the soil. This young organic matter comes into contact with the mineral substances, e.g., through macrofaunal action, and will be mainly contained in large aggregates several millimeters in size. (2) During the decomposition of organic matter, the size of the organic particles decreases, and the stability of the remaining C_{org} increases. This C_{org} fraction is then located in smaller aggregates, which partly compose larger ones. (3) Finally, we observe small mineral–organic matter–microbe associations. After the death of the microbes, the mineral–organic matter complexes, that physically protect the organic carbon against further decomposition, remain. These associations are called microaggregates and are highly stable.

Physical separation of particle size fractions revealed that much of the soil microbial biomass is associated with the smaller fractions (fine silt and clay); the highly variable amount of microbial biomass within the coarse fractions strongly depends on the quantity and quality of the macroorganic matter located there (Jocteur Monrozier et al. 1991). Investigations on structural diversity of the microbial community using the PLFA pattern and 16S rRNA gene fragments gave strong evidence that the microbial biomass within the clay fraction was mostly due to bacteria. On the other hand, a high abundance of linoleic acid (18:2ω6c) in the coarse sand is attributed to a fungal-specific membrane component (Poll et al. 2003). Moreover, higher amounts of fungal hyphae, relative to other community components, were found in larger aggregates using a dry-sieving procedure that yielded different aggregate-size classes (5.0–2.0, 2.0–1.0, 1.0–0.5, 0.5–0.25, 0.25–0.1 and < 0.1 mm; Schutter and Dick 2002). In general, coarse and fine sand fractions of soils were characterized by a rather simple bac-

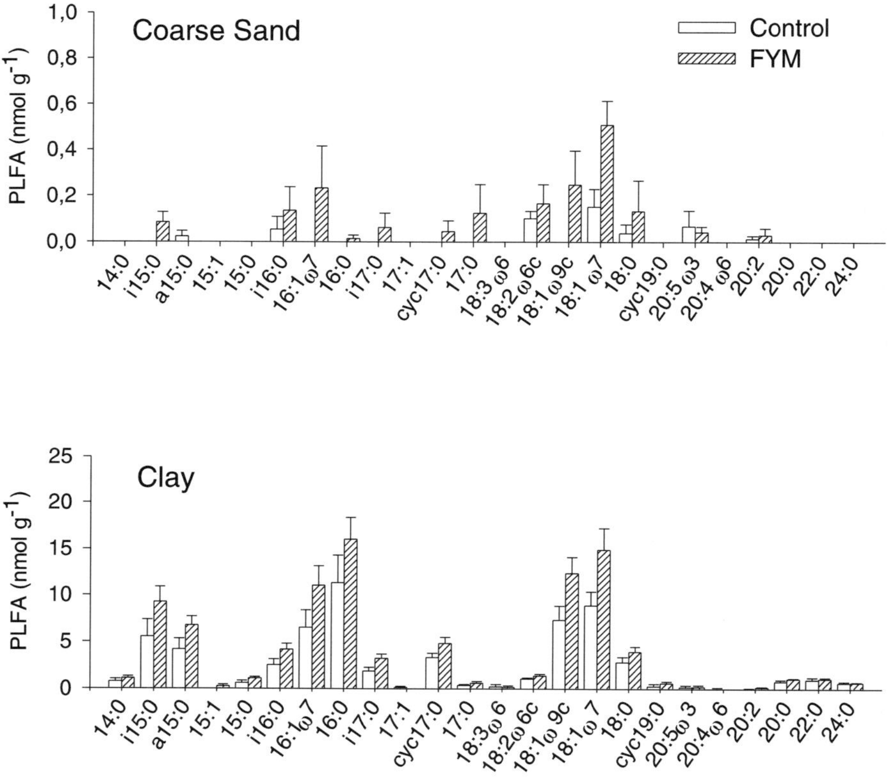

Fig. 4. Impact of farmyard manure (*FYM*) over 120 years to PLFA-pattern in the coarse sand and clay fraction of a Luvic Phaeozems (FAO). The *x-axis* represents single phospholipid fatty acids. Fatty acids are described as follows: the number of carbons in the chain is followed by a *colon*, then by the number of unsaturations. *ω* precedes the number of C atoms between the terminal double bond and the methyl end of the molecule. *i* iso methyl branching (second C from the methyl end); *a* anteiso methyl branching (third C from the methyl end); *cyc* cyclopropane ring. Data are given as means of three replicate samples, *bars* indicate standard error. (Modified after Poll et al. 2003)

terial microbial community [low number and abundance of phospholipid fatty acids (PLFA)], whereas smaller fractions tended to have a highly complex bacterial community using the heterogeneous substrates available in organo-mineral associations (Fig. 4). The amendment of farmyard manure with narrow C/N changed the microbial community structure mainly in the coarse sand fraction. The drop in fungal abundance here may be due to the lower C/N ratio of their substrates, which favored bacterial growth (Eiland et al. 2001). In contrast, typical Gram-negative bacteria PLFA biomarkers (monounsaturated plus cyclopropyl fatty acids) were detectable in the farmyard manure (FYM)-amended soil (Fig. 4). Several studies of soil PLFA have documented an increase in monounsaturated fatty acids with enhanced availability of organic substrates and manures (Bossio and Scow 1998; Peacock et al. 2001).

5
Microorganisms and Enzymes Involved in C Cycling

Decomposition of plant litter is a complex ecological process involving interactions of many taxa, spanning much of the range of biotic diversity (Sinsabaugh et al. 2002). Fungi as well as bacteria are key decomposers of the large amounts of substrates deposited on top of and below the ground each year. State-of-the-art reviews have recently been published on fungal communities involved in decomposition (Kjøller and Struwe 2002), and on microbial dynamics of litter decomposition (Sinsabaugh et al. 2002). Biotic processes of decomposition were investigated at three levels of resolution: at the molecular level, the topics of interest were plant fiber structure and enzymological characteristics of degradation. At the organismal level, the focus was on functional gene analyses, regulation of enzyme expression and growth kinetics, whereas at the community level, research concentrated on metabolism, microbial successions and competition between microbial and faunal communities. These three levels must be integrated to fully understand microbial litter decomposition. Sinsabaugh et al. (2002) explained the complex decomposition process as a successional loop in which the substrate selects the microbial community, which then produces extracellular enzymes that degrade and modify the substrate, which in turn drives community succession (Fig. 5; Sinsabaugh et al., 2002 modified).

The close link between litter decomposition, litter quality and enzyme activities is underlined in studies using different types of litter decomposed in the same environment (Sinsabaugh and Linkins 1988; Luxhøi et al. 2002). These and other studies showed the relationship between changes in plant litter quality during and the activity of enzymes involved in carbon cycling. The rapid mass loss within the first 10–18 days is due to leaching

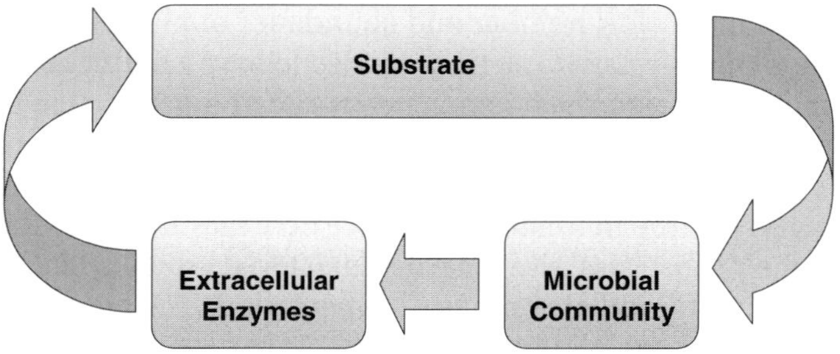

Fig. 5. Loop interaction chain between substrate, microbial community and their production of extracellular enzymes as rate-controlling agents. (Modified after Sinsabaugh et al. 2002)

and metabolism of soluble components by enzymes (invertase and alpha-glucosidase). In the later stage, xylanase, β-1,4-exoglucanase (exocellulase) and β-1,4-endoglucanase (endocellulase) are responsible for decomposing accessible polysaccharides. Phenoloxidase activity tends to increase with lignin-humus content (Sinsabaugh et al. 2002). The most important biotic processes are substrate selection by the microbial community and physiological regulation of enzyme release by soil microorganisms.

6 Dynamics of Organic Matter Decomposition in Agroecosystems

Soil management and agricultural practices significantly influence microbial decomposition of organic carbon in soil. In Austria, Gerzabek et al. (2003) evaluated C-inventories of different agricultural soils (Table 2). Soil use classes can be ranked according to their average C_{org} stocks (0–

Table 2. Weighted median values of organic carbon contents (t C/ha) and stored C_{org} amounts (Mt C) in Austria's agriculturally used soils. (Gerzabek et al. 2003)

Depth (cm)	0–20 $t\,C\,ha^{-1}$	20–50 $t\,C\,ha^{-1}$	0–20 Mt C	20–50 Mt C	Area 10^3 ha
Cropland	41.3± 5.0[a]	18.2	57.8	25.5	1397
Grassland	60.5± 9.1	20.5	56.7	19.3	938
Extensive grassland/alpine meadows	91.8±15.1	27.2	92.3	27.3	1005
Vineyards	39.3± 5.5	18.3	2.1	1.0	52.5
Orchards/domestic gardens	57.0±14.6	21.0	1.5	0.5	27.1
Sum			210.4	73.6	3420

[a] ±0.5 × AAD (%) Absolute average deviation from median

50 cm) as follows: vineyards ~ cropland < orchards/garden land < intensive grassland < extensive grassland. The obtained ranking clearly reflects the balance between organic matter input by plant residues and possibly organic manures, and mineralization. In general, most soil carbon is stored in topsoils (0–20 cm) in all land-use classes. The lowest C_{org} contents in topsoil layers are observed in intensively tilled systems (cropland and vineyards), the highest in extensive grasslands, which in most cases are characterized by slow mineralization of organic matter.

Comparing the subsoil C inventories under different land uses reveals smaller differences than for the topsoil layers. This indicates that subsoil C is less influenced by the present land use than the topsoil. Franzlübbers (2002) has recently suggested using the C stratification, which is also a common phenomenon in natural ecosystems, as an indicator for soil ecosystem functioning and especially for detecting management-induced changes. Calculating the stratification ratio (quotient of C_{org} inventories in the 0–20 and 20–50 cm layers) yielded the following ranking: vineyards (2.15) < cropland (2.27) < orchards/garden land (2.71) < intensive grassland (2.95) < extensive grassland (3.75). Land-use classes exhibiting the lowest ratios seem to have the largest potential for SOM buildup. Of course, this approach requires taking into account the site-specific potential of a soil to store organic carbon, governed, e.g., by the clay content, which influences the size of the physically protected organic matter.

A major factor influencing the C_{org} contents of agricultural soils is the application of organic residues that differ in their stability against decomposition. Figure 6 presents an example from the Ultuna long-term agricultural field experiment in Sweden (Uppsala). This experiment was established in 1956 on a clay loam (Eutric Cambisol) and is based on application of equal amounts (approx. 2000 kg C_{org} ha^{-1} $year^{-1}$), but different compositions of organic carbon. Therefore, changes in organic carbon contents in topsoil (0–20 cm) can be directly related to the stability of the applied organic matter. Figure 6 shows the development of C_{org} stocks in four selected treatments over the last 42 years. The bare fallow plot lost approximately one third of its initial C_{org} content due to mineralization. Root and stubble input (cereals, rape and fodder beet) in the non-nitrogen-treated plot resulted in slightly higher C_{org} contents as compared to bare fallow. The third and fourth treatments involved adding organic materials of quite high stability, such as animal manure and peat that are less available for microbial degradation and mineralization and increased the C_{org} level considerably. Note that changes in humus contents are still ongoing and no equilibrium between C_{org} input and mineralization has been observed in the Ultuna experiment until now. According to Fig. 6, most of the C_{org} introduced into the A_p horizon was stored in the silt-sized fraction, and only little reached the clay fraction during the decomposition process. Relative contributions

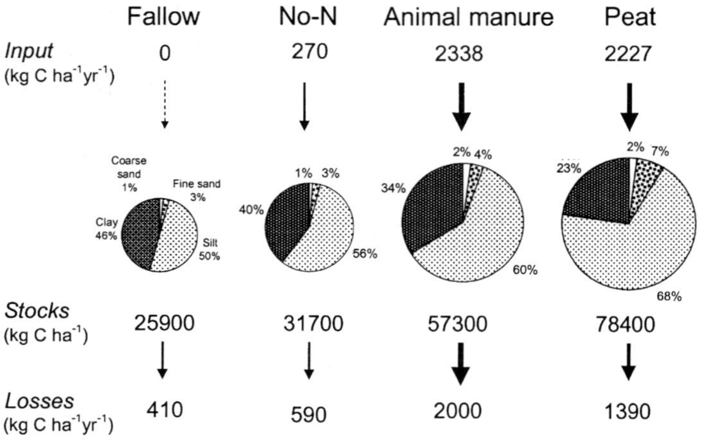

Fig. 6. Response of carbon pools to four selected treatments in the Ultuna experiment: carbon input, C partitioning in particle size fractions, C stocks and C losses as observed until 1998 (Kirchmann et al. 2004); the area of the circles reflects the size of the carbon stocks in the 0–20 cm layer

of C_{org} in silt increased from 52% in fallow plots to nearly 70% in peat plots, while C_{org} in clay-sized particles (2–0.1 µm) decreased from 36% (fallow) to 19% (peat). The proportions of C_{org} derived from organic amendments decreased from sand to silt and clay fractions. Organic carbon in the latter fraction consisted of approximately one third amendment-C (animal manure, peat), whereas in sand-sized particles, most C_{org} originated from organic amendments (Gerzabek et al. 2001). The silt-sized fraction can therefore be regarded as a medium-term C sink due to the buildup of stable microaggregates. The clay fraction contains the most stable and oldest C_{org} fraction (Buyanovsky et al. 1994).

Modeling of C_{org} decomposition and stabilization has been attempted in several models of differing complexity, from simple single-component models using one exponential decay term to quite sophisticated models considering various soil C pools. Most of today's multicomponent models are based on results from ^{14}C-labeling studies and long-term field experiments (Jenkinson and Rayner 1977). The two most widely used models are the CENTURY (Parton et al. 1987) and the Rothamsted model (Jenkinson and Rayner 1977). Both have recently yielded excellent results for different long-term experiments (Falloon and Smith 2002). Both also take into account two litter fractions of different decomposability, and three to four C pools of different stability. The size of the stable pool is frequently determined by the clay contents. The ROTHC-26.3 model considers detailed meteorological input, soil clay content, soil depth, soil cover, monthly input of plant residues and farmyard manure (if applicable). It also applies an estimate of the decomposability of the incoming plant material, the DPM/RPM ratio (decomposable plant material vs. resistant plant material; Coleman and Jenkinson 1999).

7 Soil Organic Matter, Below-Ground Processes and Climate Change

Since the beginning of industrialization, global concentration of atmospheric carbon dioxide (CO_2) has risen from about 280 to 365 ppm. During the last decade, the concentration has risen by an average 1.5 ppm/year (IPCC 2001). Direct effects of elevated CO_2 on soil organisms are unlikely, because CO_2 concentrations in soils are already 10–50 times higher than in the atmosphere (Lamborg et al. 1983). Nevertheless, much effort was put into resolving the issues whether soil microorganisms and their activities contribute to controlling the response of plant communities to increasing atmospheric carbon dioxide concentrations and play an important role in sequestering extra carbon in soils (Jones et al. 1998; Kampichler et al. 1998; Niklaus et al. 2003), as well as into determining the effects that soil microorganisms may have on elevating temperatures (Bardgett et al. 1999). These studies provide evidence that changes in the activity of microbial communities due to environmental changes can have lasting effects on ecosystem functioning (Kampichler et al. 1998; Zak et al. 2000; Ebersberger et al. 2003). Several mechanisms were discussed by which the elevation of atmospheric CO_2 might influence soil microbial communities and processes by plant-mediated effects. In general, elevated CO_2 stimulates plant photosynthesis, thus increasing assimilation, net primary productivity and below-ground allocation of assimilates. Root biomass increases, the root-to-shoot ratio widens, more fine roots are produced and their turnover is enhanced (Rogers et al. 1994). The chemistry of green leaves is altered: generally, nitrogen is depleted and C/N widens (Norby et al. 2001). To investigate the hypothesis that the turnover rates of fast-cycling carbon and/or the structure of the soil microbial community could change under elevated CO_2, an open-top chamber experiment on the shortgrass steppe in northern Colorado was established in 1997 (see Fig. 7). The experiment comprised nine experimental plots: three chambered plots maintained at the present CO_2 level of 360 µmol/mol (ambient treatment), three maintained at 720 µmol/mol CO_2 (elevated treatment), and three unchambered plots of equal ground area served as controls to monitor the chamber effect. In the first 3 years, doubling the CO_2 increased the functional diversity of the microbial community in the top 5 cm of the soil, measured on the basis of seven different enzyme activities involved in C, N, P and S cycling. Since

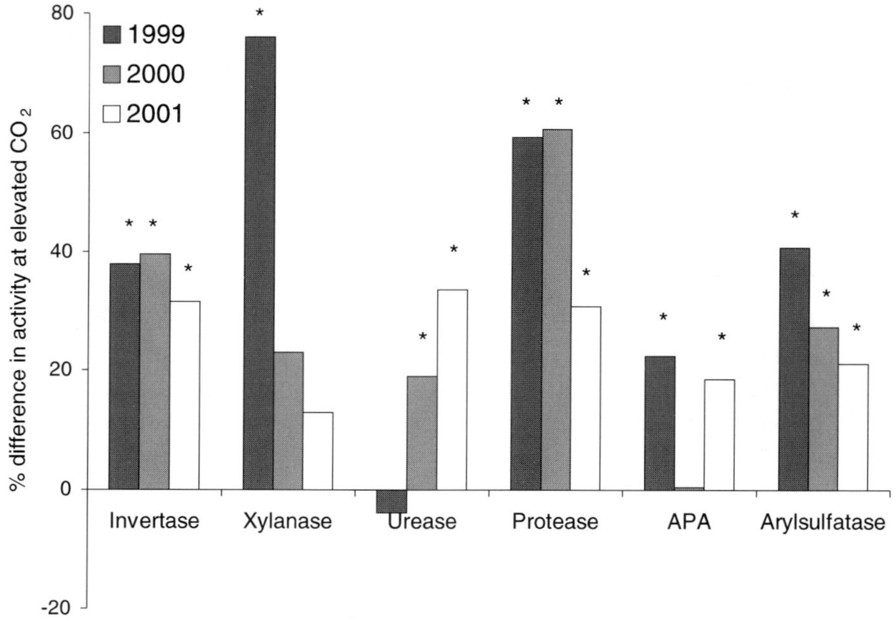

Fig. 7. Response of soil enzyme activities involved in C, N, P and S-cycling to the elevation of atmospheric carbon dioxide in a shortgrass steppe (Fort Collins, Colorado). Results are given as a percentage of the soil samples (0–5 cm) of the ambient treatment. (Kandeler et al., unpubl.)

changes in the functional diversity of soil microorganisms could only be detected in the top 5 cm of soils under elevated CO_2, the soil microorganisms probably make use of the improved supply of substrates at the soil surface of the shortgrass steppe. An alternative mechanism would be based on possible changes in soil moisture regimes under climate change. Elevated CO_2 reduces stomatal conductance of plants, boosting water use efficiency. At a plant community level, this reduces transpiration and increases soil water content. Higher soil water contents under elevated CO_2 are a common observation in CO_2 enrichment studies in grasslands (Körner 2000). Soil matric potential is an important control of soil microbial activity, directly through osmosis, as well as indirectly by controlling the nutrient supply (Killham 1994). Up to a certain threshold, soil moisture promotes soil microbes and their activity.

In conclusion, since changes in microbiological properties due to environmental changes may be manifested over a shorter time scale than changes in chemical soil properties (e.g., soil organic matter; Christensen 1996; Beyer et al. 1999), current concepts of soil monitoring include biological properties of soils (Stork and Dilly 1998; Wirth 1999; Tscherko and Kandeler 2000).

References

Baldock JA, Nelson PN (2000) Soil organic matter. In: Sumner ME (ed.) Handbook of soil science. CRC Press, Boca Raton, pp B-25–B-84

Bardgett RD, Lovel RD, Hobbs PJ, Jarvis SC (1999) Seasonal changes in soil microbial communities along a fertility gradient of temperate grasslands. Soil Biol Biochem 1999:1021–1030

Bardgett RD, Kandeler E, Tscherko D, Hobbs PJ, Jones TH, Thompson LJ, Bezemer TM (1999) Below-ground microbial community development in a high temperature world. Oikos 85:193–203

Beare MH, Coleman DC, Crossley DA Jr, Hendrix PF Odum EP (1995) A hierarchical approach to evaluating the significance of soil biodiversity to biogeochemical cycling. Plant Soil 170:5–22

Beck T, Joergensen RG, Kandeler E, Makeschin F, Nuss E, Oberholzer HR, Scheu S (1997) An inter-laboratory comparison of ten different ways of measuring soil microbial biomass. Soil Biol Biochem 29:1023–1032

Bernoux M, da Conceição Santana Carvalho M, Volkoff B, Cerri CC (2002) Brazil's carbon stocks. Soil Sci Soc Am J 66:888–896

Beyer L, Sieling K, Pingpank K (1999) The impact of a low humus level in arable soils on microbial properties, soil organic matter quality and crop yield. Biol Fertil Soils 28:156–161

Blair N, Crocker GL (2000) Crop rotation effects on soil carbon and physical fertility of two Australian soils. Aust J Soil Res 38:71–84

Bossio DA, Scow KM (1998) Impacts of carbon and flooding on soil microbial communities: phospholipid fatty acid profiles and substrate utilization patterns. Microb Ecol 35:265–278

Buyanovsky GA, Aslam M, Wagner GH (1994) Carbon turnover in soil physical fractions. Soil Sci Soc Am J 58:1167–1173

Christensen BT (1996) Matching measurable soil organic matter fractions with conceptual pools in simulation models of carbon turnover: revision of model structure. In: Powlsen DS, Smith P, Smith JU (eds) Evaluation of soil organic matter models using existing long-term datasets. NATO ASI series: global environmental change. Springer, Berlin Heidelberg New York, pp 143–160

Cleveland CC, Townsend AR, Ley RE, Constance BC, Schmidt SK (2004) Links between soil microbial dynamics and phosphorus availability in two tropical rain forests of varying P fertility in Costa Rica. Soil Biol Biochem (in press)

Coleman K, Jenkinson DS (1999) ROTHC-26.3 – a model for the turnover of carbon in soil; model description and windows used guide. IACR-Rothamsted, Harpenden, UK

Desjardins T, Andreux F, Volkoff B, Cerri CC (1994) Organic carbon and ^{13}C contents in soils and soil size-fractions, and their changes due to deforestation and pasture installation in eastern Amazonia. Geoderma 61:103–118

Ebersberger D, Wermbter N, Kandeler E (2004) Effects of long-term CO_2 enrichment on microbial community structure in calcareous grassland. Plant Soil 264:313–323

Eiland F, Klamer M, Lind AM, Leth M, Bååth E (2001) Influence of initial C/N ratio on chemical and microbial composition during long term composting of straw. Microb Ecol 41:272–280

Esser G (1990) Modelling global terrestrial sources and sinks of CO_2 with special reference to soil organic matter. In: Bouwman AF (ed) Soils and the greenhouse effect. Wiley, Chichester, pp 247–261

Falloon P, Smith P (2002) Simulating SOM changes in long-term experiments with RothC and CENTURY: model evaluation for a regional scale application. Soil Use Manage 18:101–111

Franzlübbers AJ (2002) Soil organic matter stratification ratio as an indicator of soil quality. Soil Tillage Res 66:95–106

Franzlübers AJ, Stuedemann JA, Schomberg HH, Wilkinson SR (2000) Soil organic C and N pools under long-term pasture management in the Southern Piedmont USA. Soil Biol Biochem 32:469–478

Gahoonia TS, Nielsen NE (1991) A method to study rhizosphere processes in thin soil layers of different proximity to roots. Plant Soil 135:143–146

García C, Hernández T, Costa F (1994) Microbial activity in soils under Mediterranean environmental conditions. Soil Biol Biochem 26:1185–1191

Gerzabek MH, Haberhauer G, Kirchmann H (2001) Soil organic matter pools and ^{13}C natural abundances in particle size fractions of a long-term agricultural field experiment receiving organic amendments. Soil Sci Soc Am J 65:352–358

Gerzabek MH, Haberhauer G, Stemmer M, Klepsch S, Hauhold E (2004) Fate of nitrogen in an alpine grassland ecosystem. Biogeochemistry (in press)

Gerzabek MH, Strebl F, Tulipan M, Schwarz S (2003) Quantification of carbon pools in agriculturally used soils of Austria by use of a soil information system as basis for the Austrian carbon balance model. In: Smith CAS (ed) Soil organic carbon and agriculture: developing indicators for policy analysis. Proceedings of the OECD-Expert Meeting on Organic carbon indicators for agriculturally used land, 15–18 Oct 2002, Ottawa/Canada p 73–78, Agriculture and Agri-Food Canada, Ottawa and Organisation of Economic Co-operation and Development, Paris, 329 pp

Golchin A, Oades JM, Skjemstad JO, Clarke P (1995) Structural and dynamic properties of soil organic matter as reflected by ^{13}C natural abundance, pyrolysis mass spectrometry and solid-state ^{13}C NMR spectrometry in density fractions of an oxisol under forest and pasture. Aust J Soil Res 33:59–76

Gregorich EG, Drury CF, Baldock JA (2001) Changes in soil carbon under long-term maize in monoculture and legume-based rotation. Can J Soil Sci 81:21–31

Guggenberger G, Christensen BT, Zech W (1994) Land-use effects on the composition of organic matter in particle-size separates of soil. I. Lignin and carbohydrate signature. Eur J Soil Sci 45:449–458

Hu S, Coleman DC, Caroll CR, Hendrix PF, Beare MH (1997) Labile soil carbon pools in subtropical forest and agricultural ecosystems as influenced by management practices and vegetation types. Agric Ecosyst Environ 65:69–78

IPCC (2001) IPCC third assessment report – climate change 2001. The Scientific Basis Technical Summary, Geneva

Jastrow JD, Boutton TW, Miller RM (1996) Carbon dynamics of aggregate-associated organic matter estimated by carbon-13 natural abundance. Soil Sci Soc Am J 60:801–807

Jenkinson DS, Rayner JH (1977) The turnover of soil organic matter in some of the Rothamsted classical experiments. Soil Sci 123:298–305

Jocteur Monrozier LJ, Ladd JN, Fitzpatrick RW, Foster RC, Raupach M (1991) Components and microbial biomass content of size fractionations in soils of contrasting aggregation. Geoderma 49:37–62

Johnson D, Leake JR, Lee JA, Campell CD (1998) Change in soil microbial biomass and microbial activities in response to 7 years simulated pollutant nitrogen deposition on a heathland and two grasslands. Environ Pollut 103:239–250

Jones TH, Thompson LJ, Lawton JH, Bezemer TM, Bardgett RD, Blackburn TM, Bruce KD, Cannon PF, Hall GS, Jones CG, Kampichler C, Kandeler E, Ritchie DA (1998) Impacts of rising atmospheric CO_2 on soil biota and processes in model terrestrial ecosystems. Science 280:411–413

Kampichler C, Kandeler E, Bardgett RD, Jones TH, Thomson LJ (1998) Impact of elevated atmospheric CO_2 concentration on soil microbial biomass and activity in a complex, weedy field model ecosystem. Global Change Biol 4:335–346

Kanchikerimath M, Singh D (2001) Soil organic matter and biochemical properties after 26 years of maize-wheat-cowpea cropping as affected by manure and fertilization in a Cambisol in semiarid region of India. Agric Ecosyst Environ 86:155–162
Kandeler E, Böhm KE (1996) Temporal dynamics of microbial biomass, xylanase activity, N-mineralization and potential nitrification in different tillage systems. App Soil Ecol 4:181–191
Kandeler E, Luxhøi J, Tscherko D, Magid J (1999) Xylanase, invertase and protease activities at the soil-litter interface of a sandy loam. Soil Biol Biochem 31:1171–1179
Kandeler E, Tscherko D, Stemmer M, Schwarz S, Gerzabek MH (2001) Organic matter and soil microorganisms – investigations from the micro- to the macro-scale. Bodenkultur 52:117–131
Kandeler E, Marschner P, Tscherko D, Gahoonia TS, Nielsen NE (2002) Structural and functional diversity of soil microbial community in the rhizosphere of maize. Plant Soil 238:301–312
Killham K (1994) Soil ecology. Cambridge Univ Press, Cambridge, p 242
Kirchmann H, Haberhauer G, Kandeler E, Sessitsch A, Gerzabek MH (2004) Effects of level and quality of organic matter input on carbon storage and biological activity in soil – synthesis of a long-term experiment. Global Biogeochem. Cycles (in press)
Kjøller AH, Struwe S (2002) Fungal communities, succession, enzymes, and decomposition. In: Burns R, Dick RP (eds) Enzymes in the environment – activity, ecology, and applications. Dekker, New York, pp 267–284
Körner C (2000) Biosphere responses to CO_2 enrichment. Ecol Appl 10:1590–1619
Kuzyakov Y, Domanski G (2000) Carbon input into the soil. Review. J Plant Nutr Soil Sci 163:421–431
Lamborg MR, Hardy RWF, Paul EA (1983) Microbial effects. In: Lemon ER (Ed) CO_2 and plants, the response of plants to rising levels of atmospheric carbon dioxide. Westview Press, Boulder, pp 131–176
Luxhøi J, Magid J, Tscherko D, Kandeler E (2002) Dynamics of invertase, xylanase and coupled quality indices of decomposing green and brown plant residues. Soil Biol Biochem 34:501–508
Maithani K, Tripathi RS, Arunachalam A, Pandey HN (1996) Seasonal dynamics of microbial biomass C, N and P during regrowth of a disturbed subtropical humid forest in north-east India. App Soil Ecol 4:31–37
McGill W, Cannon K, Robertson J, Cook F (1986) Dynamics of soil microbial biomass and water soluble organic C in Breton L after 50 years of cropping to two rotations. Can J Soil Sci 66:1–19
Neill C, Frey B, Melillo JM, Steudler PA, Moraes JFL, Cerri CC (1996) Forest- and pasture-derived carbon contributions to carbon stocks and microbial respiration of tropical pasture soils. Oecologia 107:113–119
Niklaus PA, Spinnler D, Körner C (1998) Soil moisture dynamics of calcareous grassland under elevated CO_2. Oecologia 117:201–208
Niklaus PA, Alphei J, Ebersberger D, Kampichler C, *Kandeler E*, Tscherko D (2003) Six years of in situ CO_2 enrichment evoke changes in soil structure and soil biota of nutrient-poor grassland. Global Change Biol 9:585–600
Norby RJ, Cotrufo MF, Ineson P, O'Neill EG, Canadell JG (2001) Elevated CO_2, litter chemistry, and decomposition: a synthesis. Oecologia 127:153–165
Oades JM, Waters AG (1991) Aggregate hierarchy in soils. Aust J Soil Res 29:815–825
Parton WJ, Schimel DS, CV Cole, Ojima DS (1987) Analysis of factors controlling soil organic matter levels in Great Plains grasslands. Soil Sci Soc Am J 51:1173–1179
Peacock AD, Mullen MD, Ringelberg DB, Tyler DD, Hedrick DB, Gale PM, White DC (2001) Soil microbial community responses to dairy manure or ammonium nitrate applications. Soil Biol Biochem 33:1011–1119

Percival HJ, Pargitt RL, Scott NA (2000) Factors controlling soil carbon levels in New Zealand grasslands: is clay content important? Soil Sci Soc Am J 64:1623–1630
Poll C, Thiede A, Wermbter N, Sessitsch A, Kandeler E (2003) Micro-scale distribution of microorganisms and microbial enzyme activities in a soil with long-term organic amendment. Eur J Soil Sci 54:715–724
Raubuch M, Joergensen RG (2002) C and net N mineralisation in a coniferous forest soil: the contribution of the temporal variability of microbial C and N. Soil Biol Biochem 34:841–849
Rogers HH, Runion GB, Krupa SV (1994) Plant responses to atmospheric CO_2 enrichment with emphasis on roots and the rhizosphere. Environ Pollut 83:155–189
Ross DJ, Tate KR, Feltham CW (1996) Microbial biomass, and C and N mineralization, in litter and mineral soil of adjacent montane ecosystems in a southern beech (Nothofagus) forest and a tussock grassland. Soil Biol Biochem 28:1613–1620
Sainju UM, Singh BP, Whitehead WF (2002) Long-term effects of tillage, cover crops, and nitrogen fertilization on organic carbon and nitrogen concentrations in sandy loam soils in Georgia, USA. Soil Tillage Res 63:167–179
Salamanca EF, Raubuch M, Joergensen RG (2002) Relationships between soil microbial indices in secondary tropical forest soils. App Soil Ecol 21:211–219
Schimel JP, Gulledge JM, Clein-Curley JS, Lindstrom JE, Braddock JF (1999) Moisture effects on microbial activity and community structure in decomposing birch litter in the Alaskan taiga. Soil Biol Biochem 31:831–838
Schlesinger WH (1997) Biogeochemistry: an analysis of global change. Academic Press, San Diego
Schulten H-R, Leinweber P, Sorge C (1993) Composition of organic matter in particle-size fractions of an agricultural soil. J Soil Sci 44:677–691
Schutter ME, Dick RP (2002) Microbial community profiles and activities among aggregates of winter fallow and cover-cropped soil. Soil Sci Soc Am J 66:142–153
Sinsabaugh RL, Linkins AE (1988) Adsorption of cellulase components by leaf litter. Soil Biol Biochem 20:927–932
Sinsabaugh RL, Carreiro MM, Alvarez S (2002) Enzyme and microbial dynamics of litter decomposition. In: Burns RG, Dick RP (eds) Enzymes in the environment – activity, ecology, and applications. Dekker, New York, pp 249–265
Springob G, Kirchmann (2002) C-rich sandy Ap horizons of specific historical land-use contain large fractions of refractory organic matter. Soil Biol Biochem 34:1571–1581
Stork R, Dilly O (1998) Maßstabsabhängige räumliche Variabilität mikrobieller Bodenkenngrößen in einem Buchenwald. Z Pflanzenernähr Bodenkd 161:235–242
Tarafdar JC, Jungk A (1987) Phosphatase activity in the rhizosphere and its relation to the depletion of soil organic phosphorus. Biol Fertil Soils 3:199–204
Tarafdar JC, Marschner H (1994) Phosphatase activity in the rhizosphere of VA-mycorrhizal wheat supplied with inorganic and organic phosphorus. Soil Biol Biochem 26:387–395
Tate RL III (2002) Microbiology and enzymology of carbon and nitrogen cycling. In: Burns RG, Dick RP (eds) Enzymes in the environment – activity, ecology, and applications. Dekker, New York, pp 227–135
Tisdall JM, Oades JM (1982) Organic matter and water stable aggregates in soils. J Soil Sci 33:141–163
Tiunov AV, Scheu S (1999) Microbial respiration, biomass, biovolume and nutrient status in burrow walls of *Lumbricus terrestris* L. (Lumbricidae). Soil Biol Biochem 31:2039–2048
Tscherko D, Kandeler E (2000) Classification and monitoring of soil microbial biomass, N-mineralization and enzyme activities to indicate environmental changes. Bodenkultur 50:215–226
Waldrop MP, Balser TC, Firestone MK (2000) Linking microbial community composition to function in a tropical soil. Soil Biol Biochem 32:1837–1846

Weixin C, Ross AV (1993) Measurement of microbial biomass in arctic tundra soils using fumigation-extraction and substrate-induced respiration procedures. Soil Biol Biochem 25:135–141

Wirth SJ (1999) Soil microbial properties across an encatchment in the moraine, agricultural landscape of Northeast Germany. Geomicrobiol J 16:207–219

Zak DR, Pregitzer KS, King JS, Holmes WE (2000) Elevated atmospheric CO_2, fine roots and the response of soil microorganisms: a review and hypothesis. New Phytologist 147:201–222

Zeller V, Bardgett RD, Tappeiner U (2001) Site and management effects on soil microbial properties of subalpine meadows: a study of land abandonment along a north-south gradient in the European Alps. Soil Biol Biochem 33:639–649

8 Contribution of Bacteria to Initial Input and Cycling of Nitrogen in Soils

Laurent Philippot[1], J.C. Germon[1]

1 Introduction

Nitrogen was discovered by the Scottish chemist and physician Daniel Rutherford in 1772 by removing oxygen and carbon dioxide from air. At the same time, the French chemist, Antoine Laurent Lavoisier, isolated what we would call nitrogen and named it *azote*, meaning *without life* since it did not support life or combustion. However, nitrogen is the fourth most common element in many biomolecules, which are essential for life, being outranked only by carbon, hydrogen and oxygen. Thus, nitrogen is found in amino acids that form proteins and in the nucleoside phosphates of nucleic acids. The cycle of nitrogen in soil has been studied more extensively than any other nutrient cycle. Nevertheless, despite decades of investigations by microbiologists of the different steps of the nitrogen cycle, many of these are still poorly understood or quantified.

One striking aspect of the N cycle is the coexistence in nature of different oxidation states of the nitrogen atom ranging from reduced compounds, e.g., −3 as in ammonia to fully oxidized state, e.g., +5 as in nitrate. The conversion between the different forms of nitrogen is mediated by processes performed by soil microorganisms. Together these processes form the global nitrogen cycle and living organisms are essential for maintaining the balance between nitrogen's reduced and oxidized forms.

The aim of this chapter is to provide an introductory-level survey of the main microbial processes that participate in the N cycle, to present recent data on the diversity and the distribution of the bacteria involved in these different processes, and to give an estimation of N fluxes within the N cycle.

[1]Microbiologie et Geochimie des Sols, INRA-University of Burgundy, 17 rue Sully BP 86510, 21065 Dijon Cedex, France, e-mail: Laurent.Philippot@dijon.inra.fr

Soil Biology, Volume 3
Microorganisms in Soils: Roles in Genesis and Functions
(ed. by F. Buscot and A. Varma)

2 Nitrogen Transformations in the Soil

With 79% of gaseous N_2 and a total mass of 3.8×10^{21} g of nitrogen atoms, the atmosphere is the largest reservoir of nitrogen. However, this N_2 is metabolically unavailable to the higher plants which do not have the ability to break the triple covalent bond of N_2. Nitrogen from the air enters the nitrogen cycle through the action of several unique types of microorganisms that can convert N_2 gas to inorganic forms available to plants. This conversion of molecular nitrogen into the available form is known as nitrogen fixation. Nitrogen is the nutrient element most frequently found to be limited to plants because of its continual loss by denitrification, soil erosion, leaching, chemical volatilization, etc. Therefore, biological nitrogen fixation (BNF) is important in agriculture because it provides a source of fixed nitrogen for the growth of plants that does not require fossil fuels for production. Biological nitrogen fixation is mainly performed by microorganisms called diazotrophs and can be represented by the following equation, in which 2 moles of ammonia are produced from 1 mole of nitrogen gas:

$$N_2 + 8H^+ + 8e^- + 16ATP \rightarrow 2NH_3 + H_2 + 16ADP + 16Pi \,.$$

The product of nitrogen fixation is generally used directly by free nitrogen-fixing bacteria or exported to the plants by symbiotic nitrogen-fixing bacteria. Exported nitrogen is used by plants to synthesize organic nitrogen compounds. These organic nitrogen compounds are used by animals as a nitrogen source. After the death of an organism (plant, animal, fungi, bacteria etc.) organic nitrogen, i. e., proteins, amino acids and nucleic acids, is degraded by microorganisms into ammonia. This decomposition of dead organic matter by saprophytic microorganisms is termed ammonification or mineralization. Urea and uric acid excreted by living animals are also mineralized by microorganisms in the soil. Once ammonia is produced, it can be (1) fixed by clays or by soil organic matter, (2) volatilized as NH_3, (3) assimilated by plants and microorganisms or finally (4) converted to nitrate by highly specialized bacteria during a two-step process: nitrification.

In the first step of nitrification, ammonia is oxidized into nitrite via NH_2OH:

$$NH_3 + O_2 + 2H^+ + 2e^- \rightarrow NH_2OH + H_2O \rightarrow NO_2^- + 5H^+ + 4e^- \,.$$

In the second step, nitrite is oxidized into nitrate without detectable intermediates:

$$NO_2^- + H_2O \rightarrow NO_3^- + 2H^+ + 2e^- \,.$$

The conversion by nitrification of the relatively immobile nitrogen form NH_4^+ to the highly mobile form NO_3^- provides opportunities for nitrogen losses from soil. Thus, nitrification in soil can lead to losses of nitrogen by runoff and leaching which can result in an accumulation of nitrate in lakes, groundwater and rivers. Alternatively, nitrification is of significance in aerobic sewage treatment in which ammonium from ammonification is converted to nitrate. The nitrate formed by nitrification can be assimilated by plant roots and bacteria or used as a terminal electron acceptor by microorganisms when oxygen is limited. Therefore, nitrification can be considered to be central for the flow, transfer or loss of nitrogen in soil. Due to the fact that the end product of nitrification is NO_3^- – the precursor for pathways of major losses of nitrogen – it has been suggested that the only realistic approach for controlling the nitrogen cycle in an agricultural ecosystem is to inhibit this process (Hauck 1983).

The reduction of nitrate into gaseous nitrogen in an anaerobic environment is performed by a four-step reaction respiratory process: denitrification.

Denitrification is a microbial respiratory process in which soluble nitrogen forms, nitrate and nitrite, are reduced into gas, nitric oxide, nitrous oxide and dinitrogen when oxygen is limited:

$$NO_3^- \rightarrow NO_2^- \rightarrow NO \rightarrow N_2O \rightarrow N_2 \,.$$

The importance of denitrification in the nitrogen cycle is depicted in Figure 1: it is the main biological process responsible for the return of fixed nitrogen to the atmosphere, thus completing the N cycle. Denitrification is also of interest for several reasons. First, reduction of nitrate to gaseous nitrogen depletes the soil of nitrate and is thus responsible for the loss of an essential plant nutrient. It is also used to remove nitrate from water, accumulated mainly as a result of agricultural nitrogen-fertilizer. Third, denitrification contributes to the modification of the global atmospheric chemistry, essentially through the greenhouse effect (Lashof and Ahuja 1990) and destruction of the Earth's ozone layer (Waibel et al. 1999) by emitting N_2O.

Alternatively, nitrate produced by nitrification can be also reduced into ammonium by either respiratory or dissimilatory processes:

$$NO_3^- \rightarrow NO_2^- \rightarrow NH_4^+ \,.$$

In respiratory nitrate reduction to ammonium, reduction of nitrite is coupled to the generation of an electrochemical proton gradient across the membrane. This reduction of nitrite to ammonia is performed without the release of intermediate products (Simon 2002). Similarly to denitrification, this respiratory nitrate ammonification is carried out under anaerobic conditions.

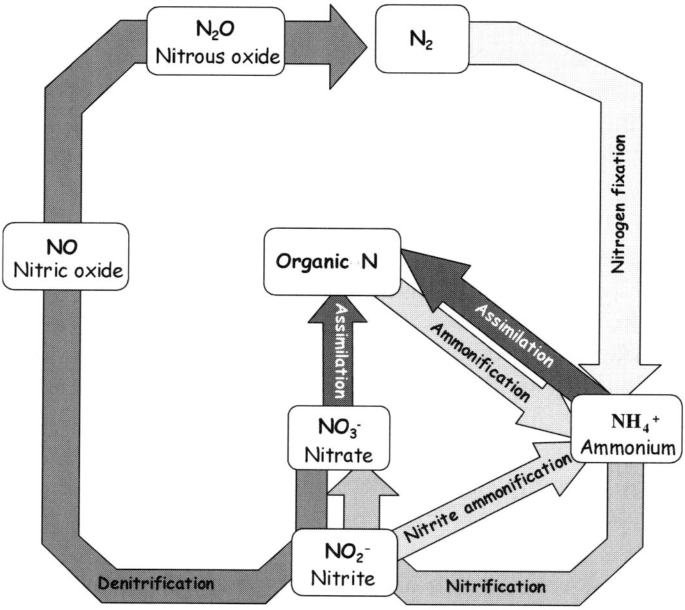

Fig. 1. Overall transformation in the N cycle

In contrast, in dissimilatory nitrate reduction to ammonium (DNRA), nitrite is used as an electron sink to regenerate NAD^+ and to detoxify the nitrite that accumulated in nitrate-reducing cells (Moreno-Vivian and Ferguson 1998). Only the first step of this ammonification, reduction of nitrate to nitrite, is coupled to energy production in most microorganisms.

The reduction of nitrite to ammonia can be considered as a short circuit that bypasses denitrification and nitrogen fixation (Cole and Brown 1980). By transforming nitrate into ammonia, a less mobile nitrogen form, dissimilative reduction of nitrate to ammonia, plays a central role in preventing leaching and pollution of the groundwater.

3 Bacteria Involved in the Nitrogen Cycle

3.1 Nitrogen-Fixing Bacteria

Nitrogen fixation is limited to prokaryotes. Bacteria which are capable of fixing nitrogen are classified as nonsymbiotic or symbiotic, depending on the required involvement of one or more than one organism, respectively, in the process. There is a great diversity of metabolic types of free-living, nitrogen-fixing bacteria which are obligate anaerobes, facultative aerobes or aerobes. This includes species of an increasing number of genera such as *Azotobacter*, *Azospirillum*, *Beijerinckia*, *Chromotium*, *Clostridium*, *Desulfovibrio*, *Klebsiella*, *Paenibacillus*, *Pseudomonas*, *Rhodopseudomonas*, *Rhodosospirillum*, *Thiobacillus*. Species of cyanobacteria and actinobacteria have also been found to fix nitrogen. Although the rates of nitrogen fixation by free-living bacteria are relatively low, these bacteria are widespread in soil and some may be rhizosphere-associated.

Most of the legumes have the ability to establish a dinitrogen-fixing association with bacteria defined as rhizobia. Rhizobia are characterized by their ability to form a defined nodule on the root or on the stem of a leguminous plant. Rhizobia are common saprophytic bacteria that are, with very few exceptions, unable to fix nitrogen in their free-living state. Rhizobia are composed of species belonging to the genera *Rhizobium*, *Bradyrhizobium*, *Sinorhizobium*, *Mesorhizobium*, *Azorhizobium* and *Allorhizobium* and represent different phyletic groups within the *α-Proteobacteria*. They represent only a small fraction of the soil microflora with densities varying widely in a range of 10^2–10^4 bacteria g^{-1} soil (Amarger 2001). However, the density of rhizobia can reach up to 10^7 bacteria g^{-1} soil under the host plant.

Symbiotic nitrogen fixation is not restricted to legumes. Some shrub and tree species form a symbiosis with a nitrogen-fixing Actinomycete called *Frankia*. Characteristic families of dicotyledonous plants forming symbioses with *Frankia* are Betulaceae, Casuarinaceae, Coriariaceae, Datiscaceae, Eleagnaceae, Myriacaceae, Rhamnaceae and Rosaceae. In addition, some lichenous fungi, liveworts, pteridophytes, gymnosperms and angiosperms are able to establish symbioses with nitrogen-fixing cyanobacteria *Nostoc* and *Anabaena*. Cyanobacteria and rhizobial symbiosis are more common in tropical and subtropical soils, whereas actinomycete symbioses are more frequent in temperate and circumpolar regions.

Mechanisms of N_2 fixation appear to be quite similar in most of the nitrogen-fixing bacteria. Thus, the fixation of dinitrogen is catalyzed by the nitrogenase enzyme complex consisting of two conserved proteins: the MoFe dinitrogenase encoded by the *nifD* and *nifK* genes and the Fe dinitrogenase reductase encoded by the *nifH* gene. Because the nitrogenase complex is extremely sensitive to O_2 inactivation, the fixation of N_2 must occur in environments or cells with very low O_2 partial pressure. The study of diversity of uncultivated diazotrophs has been largely done by focusing on 16S rRNA from phylotypes that are known to fix N_2 (Simonet et al. 1991), or on nitrogenase genes (Ueda et al. 1995; Widmer et al. 1999; Lovel et al. 2001; Poly et al. 2001). Thus, the database for the *nifH* gene has become one of the largest functional gene datasets after that of the 16S rRNA. Results of these numerous studies on the distribution of *nifH* phylotypes among habitats indicated that there are some distinct patterns of diazotrophs across environments suggesting that their distribution can be predicated on the basis of the habitat characteristics (Zehr et al. 2003).

3.2 Nitrifiers

The pioneering work of Winogradsky established that nitrification is performed by chemolithotrophic bacteria using the oxidation reactions to assimilate CO_2. Nitrifying bacteria are classified into two groups, based on their ability to oxidize ammonia to nitrite or nitrite to nitrate. Oxidation of ammonia to nitrite via hydroxylamine is performed by the five genera of specialized bacteria *Nitrosomonas*, *Nitrosococcus*, *Nitrosospira*, *Nitrosolobus* and *Nitrosovibria*. Recently, based on 16S rRNA homology, Head et al. (1998) have proposed to group *Nitrosolobus* and *Nitrosovibrio* in the genus *Nitrospira*. The oxidation of ammonia to hydroxylamine is catalyzed by the ammonia monooxygenase (Amo), which can be inhibited by either competitively or by covalently binding the substrate to its active site. For instance, acetylene, which is used to estimate nitrification in soils, inhibits ammonia oxidation at a concentration between 0.1 and 10 Pa (Berg et al. 1982). Oxidation of hydroxylamine to nitrite is catalyzed by the hydroxylamine oxidoreductase. All known terrestrial autotrophic ammonia oxidizers belong to the genus *Nitrosomonas* and *Nitrosospira* which form a monophyletic cluster within the β subclass of *Proteobacteria*. Due to the difficulty to isolate ammonia oxidizers in pure culture, nearly all the recent studies on the ecology of this functional community in the environment are based on direct molecular approaches using primers targeting the 16S rRNA or *amoA* encoding the ammonia oxygenase (Rotthauwe et al. 1997; Stephen et al. 1998; Kowalchuk et al. 2000). The monophyletic nature of these bacteria in terrestrial environments has facilitated these molecular biological approaches. However, knowledge of the ammonia-oxidizing bacteria in terrestrial environment is scarce because they have been best characterized in a bioreactor and other systems used for the treatment of nitrogen-rich waste (Kowalchuk and Stephen 2001). The number of autotrophic ammonia oxidizing bacteria has been estimated using the most probable number (MPN) method in several acid soils revealing densities between 10^2 and 10^4 bacteria g^{-1} soil (DeBoer et al. 1992; Pennington and Ellis 1993; Hastings et al. 2000). Higher densities of 10^5–10^7 are commonly observed in agricultural soils (Bruns et al. 1999). Cultivation-independent quantification of ammonia oxidizing bacteria in soils by competitive PCR revealed similar densities (Bruns et al. 1999; Mendum et al. 1999). In addition to chemolithotrophic nitrifiers, a number of bacteria and fungi are capable of heterotrophic nitrification. In contrast to autotrophic nitrification, this process is not linked to cellular growth. Of the heterotrophs capable of nitrification, the fungi are considered to be the most numerous and efficient. However, many of the approaches to study heterotrophic nitrification have been performed in pure culture systems and the significance of heterotrophic nitrification in soils still needs to be determined (Stams et al. 1990; DeBoer and Kowalchuk 2001).

Four genera of bacteria oxidizing nitrite to nitrate have been described to date: *Nitrobacter* (*α-Proteobacteria*), *Nitrococcus* (*γ-Proteobacteria*), *Nitrospina* (*δ-Proteobacteria*), *Nitrospira* (*Nitrospira* group). *Nitrobacter* was considered the dominant or even the sole nitrite oxidizer in soils, but evidence has been recently given that *Nitrospira* is also a common soil bacterium (Bartosch et al. 2002). Studies of the ecology of nitrite-oxidizing bacteria are scarce and are mainly specific of the genus *Nitrobacter*. Quantification of *Nitrobacter* in forest soil indicated densities of 10^4–10^5 bacteria g^{-1} soil (Degrange et al. 1998).

3.3 Nitrate Reducers, Denitrifiers and Nitrite Ammonifiers

The utilization of nitrate as an electron acceptor when oxygen is limiting is the first step of denitrification, respiratory nitrate reduction to ammonia and dissimilatory nitrate reduction to ammonia. Nitrate-reducing bacteria are widespread in the environment and, in contrast to nitrifiers, belong to most of the prokaryotic families. Actually, in the N cycle, there is no transformation carried out by a wider diversity of microorganisms than nitrate reduction. The proportion of total bacteria capable of reducing nitrate is estimated between 10–50%. In a culture-based study, Shirey and Sextone (1989) found that the majority of nitrate-reducing bacteria in abandoned and reclaimed mine soils were *Streptomyces*, *Bacillus* and Enterobacteriacea strains. Nitrate-reducing bacteria can possess either the membrane-bound or the periplasmic nitrate reductase or two types of nitrate reductase. Both the *napA* and *narG* genes encoding the catalytic subunit of the periplasmic and membrane-bound nitrate reductase, respectively, have been exploited as molecular markers to study the nitrate reductase community in the environment (Flanagan et al. 1999; Philippot et al. 2002). The comparison of *narG* sequences from different soils has revealed a much greater diversity of the nitrate-reducing community than previously appreciated (Flanagan et al. 1999; Philippot et al. 2002).

Virtually all bacteria that are able to reduce nitrate are also able to reduce nitrite. Thus, nitrite is used as an electron acceptor and reduced into gaseous nitrogen by denitrifying bacteria. Denitrifiers are primarily heterotrophs characterized by two criteria (1) a growth yield increase proportional to the amount of N oxide present and (2) a reduction of nitrite mainly into N_2O or N_2(Tiedje 1988; Mahne and Tiedje 1995). Denitrification is present in many prokaryotic families such as Thermoproteaceae, Cytophagaceae, Corynebacteriaceae, Streptomycineae, Bacillaceae, Rhodospirillaceae, Rhodobacteraceae, Rhizobiaceae, Burkholderiaceae, Nitrosomonadaceae, Neisseriaceae, and Pseudomonaceae. In contrast to nitrate reducers, only 0.1–5% of the soil bacteria are denitrifiers, representing densities of 10^5–10^6 CFU g^{-1} soil (Cheneby et al. 2000). Dominant denitrifiers in soils were first identified as the *Pseudomonas* and *Alcaligenes* species (Gamble et al. 1977). However, in more recent studies, Weier and MacRae (1992) and Cheneby et al. (2000) found a broader diversity of dominant denitrifiers in soils with isolates falling into five groups related to the genera *Pseudomonas, Ralstonia-Burkholderia, Xanthomonas-Frateuria, Bacillus,* and *Streptomyces-Arthrobacter*. Most of the denitrification genes have been exploited to design primers to target the denitrifying bacteria in the environment (Braker et al. 1998; Scala and Kerkhof 1998; Hallin and Lindgren 1999; Rosch et al. 2002; Braker and Tiedje 2003), revealing in most cases a higher diversity of this functional community than previously observed. A recent study has indicated that differences in the composition of the denitrifying community in soil can be of great importance for regulating in situ N emissions (Holtan-Hartwig et al. 2000).

It is interesting to note that the denitrification capacity can also be present in nitrogen-fixing bacteria and in nitrifiers. Thus, denitrification is very common in rhizobia species such as *Bradyrhizobium japonicum* and *Rhizobium fredii* (O'Hara and Daniel 1985; Tiedje 1988). In nitrifier denitrification, the oxidation of NH_3 to NO_2^- is followed by the reduction of NO_2^- to N_2O and/or N_2. This sequence of reactions is probably carried out by one group of microorganisms, namely autotrophic NH_3-oxidizers such as *Nitrosomonas europaea*.

Respiratory nitrite reducers can be distinguished from denitrifiers by the production of ammonium instead of gaseous nitrogen. Moreover, the reduction of nitrite to ammonia is catalyzed by a periplasmic cytochrome *c* nitrite reductase, NrfA which is completely different from the NO-producing copper and cd_1 denitrifying nitrite reductases. Respiratory nitrite ammonification is present in the γ, δ, and ε subclasses of the Proteobacteriaceae with species of *Wolinella, Sulfurospirillum, Campylobacter, Desulfovibrio* and *Escherichia*. Nonetheless, many more bacteria are probably able to carry out respiratory nitrite ammonification. Thus, analysis of complete bacterial genome sequences revealed the presence of predicted proteins in *Porphiromonas* (Bacteroides), *Carboxydothermus* (Gram-positive), *Pasteurella multicoda* and *Salmonella tiphy* and *thy-*

phimurium (*γ-Proteobacteria*) exhibiting identity with Nrf proteins (Simon 2002).

Dissimilative nitrate reduction to ammonia (DNRA) is typical for the Enterobacteriaceae, but is also present in the *Bacillus*, *Pseudomonas*, and *Neisseria* species (Tiedje 1988). This nitrite reduction to ammonia, which is used to detoxify the nitrite accumulating in nitrate-reducing bacteria and does not generate a proton motive force, is not a respiratory process (Moreno-Vivian and Ferguson 1998). Enzymes which catalyze nitrite dissimilation are of two types: hexa-haem *c* type cytochromes located in the periplasm and cytoplasmic flavoproteins with sirohaem as the prosthetic group (Cole 1990). Data on the ecology of DNRA microorganisms in soils are scarce and it is often assumed that conditions in soil are more favorable for denitrification than DNRA (Tiedje 1988). However, in a recent study, Yin et al. (2002) have shown that DNRA can account for up to 15% of the reduced ^{15}N-labeled nitrate added to soil and up to 21% of isolates, being able to growth anaerobically, were identified as DNRA bacteria.

4 Nitrogen Fluxes

Microbial transformations in the N cycle affect N bioavailability in soil by conditioning plant growth and N storage in biosphere. Organic compounds represent only a small part (0.0007%) of the total nitrogen present in terrestrial ecosystems and the atmosphere. These mechanisms regulate

Table 1. Global fluxes in nitrogen cycle. (Adapted from Tiedje 1988)

	Tg N year^{-1}
N_2 biological fixation	
Soil	139–180
Oceans	1–130
Total biological fixation	140–310
Industrial fixation	80– 85
Combustion and atmospheric reactions	24– 30
Total N Inputs	290–430
Denitrification	
Soil	105–185
Oceans	25–250
Total denitrification	140–445
Storage in sediments	10
Total N Outputs	140–445
Net	−150 to +15

losses of nitrogen compounds generating pollution in the environment and more specifically NH_4^+, NO_3^- and NO_2^- in surface water, and NH_3, NO_x and N_2O in the atmosphere.

The flux data adapted from Tiedje (1988) highlight the uncertainties in global estimations of N inputs and outputs and confirm that the questions on a general equilibrium in global N cycle are unresolved (Table 1).

4.1 Biological Nitrogen Fixation

According to Newton (2000), 65% of total nitrogen fixed is provided by biological nitrogen fixation (BNF), 25% by industrial fertilizer production and 10% by abiotic, natural processes (lightning, combustion and volcanic activity). BNF in the terrestrial ecosystem is in the order of 90–130 Tg N $year^{-1}$ (Galloway 1998). However, the quantities of biologically fixed nitrogen vary with the microflora involved and the kind of association with plants, and also with environmental conditions. BNF by free bacteria in cultivated soil from temperate areas does not exceed several kg N ha^{-1} $year^{-1}$. BNF by endophytic diazotroph bacteria colonizing sugarcane rhizosphere can reach 170 kg N ha^{-1} $year^{-1}$ in Brazil (Baldani et al. 2000). In water-submerged rice fields, BNF is due to free cyanobacteria and diazotrophic bacteria colonizing rice roots. Fixed nitrogen can vary from several tens to 150 kg N ha^{-1} $year^{-1}$ (NRC, 1979, mentioned by Atlas and Bartha 1993) and could provide the equivalent to 20% of the total nitrogen incorporated into the plant (Baldani et al. 2000). Similar orders are observed with the symbiotic association between a cyanobacterium (*Anabeana*) and a plant (*Azolla*) used as green manure in tropical areas (Vance 1998).

However, in the terrestrial habitat, the symbiotic fixation of nitrogen by rhizobia accounts for the largest contribution of combined nitrogen. The rates of nitrogen fixation by symbiotic rhizobia is often two or three orders of magnitude higher than rates exhibited by free-living nitrogen-fixing bacteria in soil. Thus, BNF obtained with the symbiotic associations between rhizobia and leguminous plants varies from several tens to 350 kg N ha^{-1} $year^{-1}$ depending on the associations. BNF obtained with nonleguminous angiosperms producing symbiosis with *Frankia* is between 15 and 77 kg N ha^{-1} $year^{-1}$ for *Casurarina equisetifolia* (Dommergues 1997), 29 and 117 kg N ha^{-1} $year^{-1}$ for *Alnus nepalensis* (Shrama 1993) and 18 kg N ha^{-1} $year^{-1}$ for *Myrica faya* (Vitousek and Walker 1989).

An estimation of BNF by human-induced cultivation of rice and legumes ranges between about 30 and 50 Tg N $year^{-1}$ (Galloway 1998). Furthermore, BNF efficiency depends on mineral nitrogen availability and is tremendously diminished by N fertilizer application (Waterer et al. 1994; Vance 1998).

4.2 Nitrogen Mineralization

N mineralization is an organic matter hydrolysis that induces ammonium formation, a toxic compound which is evacuated out of the cells as ammonium salt, urea or uric acid. Mineralization concerns fresh as well as humified organic matter. Studies using ^{15}N have allowed to define a net mineralization as a result of a bulk mineralization and a simultaneous microbial immobilization. The respective importance of both transformations principally depends on the C and N bioavailability.

In soils where mineralization is the dominant process, mineral N is produced in accordance with a first-order kinetics,

$$dN/dT = k(N_0 - N) \quad \text{or} \quad N = N_0(1 - e^{-kt}) .$$

N is the mineralized nitrogen at t, N_0 is the initial pool of mineralizable N and k is the constant of rate varying with the parameters of the environment. In regions with temperate weather the annual mineralization rate is around 1–2% of the humified organic compounds generating several tens to 100 kg N ha^{-1} year^{-1} of mineral N.

4.3 Nitrification

The soil nitrification rate refers to the production of ammonium rapidly transformed into nitrate. Available nitrogen is the most important factor limiting nitrification rates in terrestrial ecosystems (Roberston 1982; Donaldson and Henderson 1990), and soil nitrification appears to be a substrate-limited reaction. Soil ammonium storage is considered as an index of defective nitrification due to limiting conditions in the environment (insufficient aeration, acidic soil, low temperature). It is well known that nitrifiers can grow in pure culture in relatively limiting conditions of aeration, pH and temperature. Nevertheless, it is accepted that nitrification in soil proceeds under a much broader range of conditions than the range that might be predicted from the knowledge of the physiology of the bacteria involved (MacDonald 1986). In ammonium-enriched soils due to nitrogenous fertilizers or a strong organic matter mineralization, nitrification can be inhibited by a low concentration in free ammonia or nitrous acid in a close relationship with the soil pH (Anthonisen et al. 1976).

In order to adjust nitrate availability to plant requirements and to reduce nitrogen losses by leaching or by denitrification, many authors have tried to manage the nitrification rate with the help of nitrification inhibitors: the most frequently used products are nitrapyrine (2-chloro-6-(trichloromethyl)pyridine) and dicyandiamide (DCD). However, numerous studies have shown that nitrification inhibitor efficiency largely varies and it is difficult to draw a conclusion on a beneficial and reliable effect on nitrogen balance and fertilizer efficiency (Keeney 1986).

Nitrification is also a source of the gaseous nitrogen oxides N_2O and NO directly or indirectly involved in the greenhouse effect; the 100-year global warming potential of N_2O is about 300 times as strong as that of carbon dioxide with a lifetime of approximately 120 years (IPCC 2001) and N_2O emissions are taken into account in the global balance of greenhouse gas. N_2O is produced by nitrifying bacteria during ammonium oxidation into nitrite, on the one hand, and by nitrite reduction into nitrogen gas in oxygen limiting conditions, i. e., nitrifier denitrification, on the other. In the absence of denitrification in different soils, Garrido et al. (2002) assert that N_2O emissions are in direct relation to the amounts of nitrified nitrogen with a variable proportion between 0.03 and 1% depending on the soils. Under similar conditions, NO production varied from 0 to 2.5% of nitrified nitrogen depending on the soils and the water potential. The significance of N_2O loss via nitrifier denitrification in soils is still a matter of speculation and varies between insignificant amounts (Robertson and Tiedje 1987) to 30% of the total N_2O production (Webster and Hopkins 1996).

4.4 Dissimilatory Nitrate Reduction to Ammonium

Dissimilatory nitrate reduction to ammonium is developed under anaerobic conditions and can be considered as an alternative mechanism to denitrification, allowing nitrogen to be maintained in soil as an available form for the vegetation. Several studies have tried to evaluate the respective contribution of denitrification and DNRA during nitrate reduction in soils in relation to different conditions, particularly oxygen and carbon availability. They demonstrated that the indigenous labile soil carbon is the key factor influencing the partitioning of nitrate reduction. In cultivated and normally aerated soils kept under anaerobic conditions, DNRA corresponds to only a few percent of denitrified nitrogen, while it can reach 15% of reduced nitrate in soils naturally rich in biodegradable organic C (Yin et al. 2002). DNRA can be increased strongly in soil receiving fermentable C source with a DNRA/denitrification ratio increasing with the biodegradable-C/N-NO_3^- ratio (Fazzolari et al. 1998). Moreover, DNRA can be a N_2O source in the environment (Tiedje 1988): in a pure culture of a dissimilatory nitrate reducer into ammonium (*Enterobacter amnigenus*), Fazzolari et al. (1990) obtained a quantitative nitrite reduction into 79.7% as NH_4^+ and 15.4% as N_2O, highlighting the possible N_2O production in this way.

4.5 Denitrification

Denitrification allows the return of nitrate to the atmosphere: a durable nitrate accumulation in soil and water is an indicator of an unbalanced nitrogen cycle or a necessary lag time for re-establishing a new balance after disturbance. Denitrification occurs in soils containing nitrate and submitted to anaerobiosis phases. It occurs only in the presence of a reducing compound source, generally organic matter whose determining effect has been widely demonstrated (Beauchamp et al. 1998).

Denitrification measurement in soils presents a very large variability in time and space and is a consequence of the heterogeneous distribution of soil structure and organic matter. In cultivated soils, denitrifying activity measured in normally aerated field conditions is generally several orders lower than the potential activity estimated in laboratory optimal conditions, illustrating the driving rule of environmental parameters on its expression. In situ denitrification in cultivated soils under temperate weather does not exceed several $kg\,N\,ha^{-1}\,year^{-1}$ and is not in accordance with the deficiency in the nitrogen balance observed under N fertilization. However, denitrification can be tremendously increased in soils enriched in organic matter and temporarily submitted to anoxic conditions. It can reach several tens of $kg\,N\,ha^{-1}\,year^{-1}$ in fertilized grasslands and range from 100 to $200\,kg\,N\,ha^{-1}\,year^{-1}$ in field conditions under vegetable crop production in fine textured soils with high organic matter content in regions with hot or tropical weather (Ryden and Lund 1980). These nitrogen losses are important in riparian areas; they can be over several tens to $100\,kg\,N\,ha^{-1}\,year^{-1}$ (Pinay et al. 1993).

Many statistical or mechanistic models have been proposed to forecast denitrification in soils. Potential activities established on undisturbed soil cores and corrected by different functions taking into account the driving effect of environmental parameters (water-filled pore space, nitrate content and temperature) allow an interesting approach for evaluating denitrification losses for several months in field conditions (Hénault and Germon 2000).

Denitrification was initially considered as the main mechanism generating N_2O and more attention was paid to nitrification. Different authors tried to estimate the respective contributions of nitrification and denitrification to these emissions, indicating that both mechanisms can be alternatively the dominant processes depending on the environmental conditions (Stevens et al. 1997). The denitrification part in N_2O emissions was determined in several studies: in their review Pratt et al. (1997) consider that 5–10% of denitrified nitrogen is on average emitted as N_2O while Aulack et al. (1992) mention that the proportion of N_2O in gaseous denitrification products can vary from 0 to 100%. However, the specific contribution for a given soil can be less variable (Germon and Jacques 1990). Hénault et al. (2001) determined this contribution by comparing denitrification and N_2O emission kinetics in different situations. They showed that the transient accumulation of N_2O during denitrification in laboratory conditions appeared to be a relevant indicator of soil in situ N_2O emissions and suggested an empirical index. Recent studies underlined that the quality of the denitrifying community (size, composition, enzyme induction) must be taken into account when trying to understand and to model N_2O field fluxes from soils (Holtan-Hartwig et al. 2000; Cavigelli and Robertson 2001).

In conclusion, soil microorganisms carry out processes that are important to ecosystem function, such as maintaining soil fertility by cycling the nitrogen. With the development of molecular genetic tools in microbial ecology, rapid progress has been made in the field especially for bacteria contributing to the N cycle (Bothe et al. 2000; Kowalchuk and Stephen 2001; Zehr et al. 2003). However, formulating meaningful conclusions about the importance of diversity within these functional groups is still difficult and the relative contribution of the different bacterial populations to the N transformation processes remains unclear. This brings a new challenge for soil microbiologists, i.e., relating microbial diversity to ecosystem functioning.

References

Amarger N (2001) Rhizobia in the field. Adv Agron 73:109–167

Anthonisen A, Loehr R, Prakasam T, Srinath E (1976) Inhibition of nitrification by ammonia and nitrous acid. J Water Pollut Control Fed 48:835–852

Atlas R, Bartha R (1993) Microbial ecology. Benjamin/Cummings , Redwood City, CA, 563 pp

Aulack M, Doran J, Mosier A (1992) Soil denitrification: significance, measurements and effect of management. Adv Soil Sci 18:1–57

Baldani J, Oliveira A, Guimaraes S, Baldani V, Reis F, Silva L, Reis V, Teixera K, Döberiner J (2000) Biological nitrogen fixation (BNF) in nonleguminous plants, the role of endophytic diatozotrophs. In: Pedrosa F, Hungria H, Yates G, Newton W (eds) Nitrogen fixation: from molecules to crop productivity. Kluwer, Dordrecht, pp 397–400

Bartosch S, Hartwig C, Spieck E, Bock E (2002) Immunological detection of *Nitrospira*-like bacteria in various soils. Microb Ecol 43:26–33

Beauchamp E, Trevors J, Paul J (1998) Carbon sources for bacterial denitrification. Adv Soil Sci 10:113–142
Berg P, Klemedtsson L, Roswall T (1982) Inhibitory effect of low partial pressures of acetylene on nitrification. Soil Biol Biochem 14:301–303
Bothe H, Jost G, Schloter M, Ward B, Witzel K (2000) Molecular analysis of ammonia oxidation and denitrification in natural environments. FEMS Microbiol Rev 24:673–690
Braker G, Tiedje JM (2003) Nitric oxide reductase (norB) genes from pure cultures and environmental samples. Appl Environ Microbiol 69:3476–3483
Braker G, Fesefeldt A, Witzel KP (1998) Development of PCR primer systems for amplification of nitrite reductase genes (*nirK* and *nirS*) to detect denitrifying bacteria in environmental samples. Appl Environ Microbiol 64:3769–3775
Bruns MA, Stephen JR, Kowalchuk GA, Prosser JI, Paul EA (1999) Comparative diversity of ammonia oxidizer 16S rRNA gene sequences in native, tilled, and successional soils. Appl Environ Microbiol 65:2994–3000
Cavigelli MA, Robertson GP (2001) Role of denitrifiers diversity in rates of nitrous oxide consumption in a terrestrial ecosystem. Soil Biol Biochem 33:297–310
Cheneby D, Philippot L, Hartmann A, Hénault C, Germon JC (2000) 16S rDNA analysis for characterization of denitrifying bacteria isolated from three agricultural soils. FEMS Microbiol Ecol 34:121–128
Cole JA (1990) Physiology, biochemistry and genetics of nitrate dissimilation to ammonia. In: Revsbech NP, Sorensen J (eds) Denitrification in soil and sediment. Plenum Press, New York, pp 57–76
Cole JA, Brown CM (1980) Nitrite reduction to ammonia by fermentative bacteria: a short circuit in the biological nitrogen cycle. FEMS Microbiol Lett 7:65–72
DeBoer W, Kowalchuk G (2001) Nitrification in acid soils: micro-organisms and mechanisms. Soil Biol Biochem 33:853–866
DeBoer W, Tietema A, Gunnewiek PJAK, Lannbroek HJ (1992) The chemolithotrophic ammonium-oxidizing community in a nitrogen-saturated acid forest soil in relation to pH-dependent nitrifying activity. Soil Biol Biochem 24:229–234
Degrange V, Coûteaux M, Anderson J, Berg M, Lensi R (1998) Nitrification and occurrence of *Nitrobacter* by MPN-PCR in low and high nitrifying coniferous forest soils. Plant Soil 198:201–208
Dommergues Y (1997) Contribution of actinorhizal plants to tropical soil productivity and rehabilitation. Soil Biol Biochem 6:931–941
Donaldson JM, Henderson GS (1990) Nitrification potential of secondary-succession upland oak forest. II. Regulation of ammonium-oxidizing bacteria populations. Soil Sci Soc Am J 54:898–902
Fazzolari E, Mariotti A, Germon J (1990) Nitrate reduction to ammonia: a dissimilatory process in *Enterobacter amnigenus*. Can J Microbiol 36:779–785
Fazzolari E, Nicolardot B, Germon J (1998) Simultaneous effects of increasing levels of glucose and oxygen partial pressures on denitrification and dissimilatory nitrate reduction to ammonium in repacked soil cores. Eur J Soil Biol 34:47–52
Flanagan D, Gregory L, Carter J, Karakas-Sen A, Richardson D, Spiro S (1999) Detection of genes for periplasmic nitrate reductase in nitrate respiring bacteria and in community DNA. FEMS Microbiol Lett 177:263–270
Galloway J (1998) The global nitrogen cycle: changes and consequences. Environ Pollut 102:15–24
Gamble T, Betlach H, Tiedje J (1977) Numerically dominant denitrifying bacteria from world soils. Appl Environ Microbiol 33:926–939
Garrido F, Hénault C, Gaillard H, Pérez S, Germon J (2002) N_2O and NO emissions by agricultural soils with low hydraulic potentials. Soil Biol Biochem 34:559–575

Germon J, Jacques D (1990) Denitrifying activity measurement by soil core method. Effect of depth and characterisation of N_2O/N_2 ratio in different soils. Mitt Dtsch Bodenkd Ges 60:51–58

Hallin S, Lindgren P (1999) PCR detection of genes encoding nitrite reductase in denitrifying bacteria. Appl Environ Microbiol 65:1652–1657

Hastings R, Butler C, Singleton I, Saunders J (2000) Analysis of ammonia-oxidizing bacteria populations in acid forest soil during conditions of moisture limitation. Lett Appl Microbiol 30:14–18

Hauck R (1983) Agronomic and technological approaches to minimize gaseous nitrogen losses from croplands. In: Freney JR, Simpson JR (eds) Gaseous loss of nitrogen from plant-soil systems. Nijhoff/Junk, The Hague, pp 285–312

Head IM, Saunders JR, Pickup RW (1998) Microbial evolution, diversity and ecology: a decade of ribosomal RNA analysis of uncultivated microorganisms. Microb Ecol 35:1–21

Hénault C, Germon J (2000) Nemis, a predictive model of denitrification on the field. Eur J Soil Sci 51:257–270

Hénault C, Chèneby D, Heurlier K, Garrido F, Perez S, Germon JC (2001) Laboratory kinetics of soil denitrification are useful to discriminate soils with potentially high levels of N_2O emission on the field scale. Agronomie 21:713–723

Holtan-Hartwig L, Dörsch P, Bakken L (2000) Comparison of denitrifying communities in organic soils: kinetics of NO_3^- and N_2O reduction. Soil Biol Biochem 32:833–843

IPCC (2001) Climate change, the scientific basis. Cambridge Univ Press, Cambridge, 881 pp

Keeney D (1986) Inhibition of nitrification in soils. In: Prosser J (ed) Nitrification. IRL Press, Oxford, pp 99–115

Kowalchuk G, Stephen J (2001) Ammonia-oxidizing bacteria: a model for molecular biology. Annu Rev Microbiol 55:485–529

Kowalchuk GA, Stienstra AW, Heilig GH, Stephen JR, Woldendorp JW (2000) Molecular analysis of ammonia-oxidising bacteria in soil of successional grasslands of the Dreentsche A (The Netherlands). FEMS Microbial Ecol 31:207–215

Lashof D, Ahuja D (1990) Relative contributions of greenhouse gas emissions to global warming. Nature 344:529–531

Lovel C, Friez M, Longshore J, Bagwell C (2001) Recovery and phylogenetic analysis of *nifH* sequences from diazotrophic bacteria associated with dead aboveground biomass of *Spartina alterniflora*. Appl Environ Microbiol 67:5308–5314

MacDonald R (1986) Nitrification in soil: an introductory history. In: Prosser J (ed) Nitrification. IRL Press, Oxford, pp 1–16

Mahne I, Tiedje JM (1995) Criteria and methodology for identifying respiratory denitrifiers. Appl Environ Microbiol 61:1110–1115

Mendum TA, Sockett RE, Hirsch PR (1999) Use of molecular and isotopic techniques to monitor the response of autotrophic ammonia-oxidizing populations of the beta subdivisons of the class Proteobateria in arable soils to nitrogen fertilizer. Appl Environ Microbiol 65:4155–4162

Moreno-Vivian C, Ferguson S (1998) Definition and distinction between assimilatory, dissimilatory and respiratory pathways. Mol Microbiol 29:661–669

Newton WE (2000) Nitrogen fixation in perspective. In: Pedrosa F, Hungria M, Yates G, Newton W (eds) Nitrogen fixation: from molecules to crop productivity. Kluwer, Dordrecht, pp 3–8

O'Hara G, Daniel R (1985) Rhizobial denitrification. Annu Rev Soil Biol Biochem 17:1–9

Pennington PI, Ellis RC (1993) Autotrophic and heterotrophic nitrification in acidic forest and native grassland soils. Soil Biol Biochem 25:1399–1408

Philippot L, Piutti S, Martin-Laurent F, Hallet S, Germon J (2002) Molecular analysis of the nitrate-reducing community from unplanted and maize planted soil. Appl Environ Microbiol 68:6121–6128
Pinay G, Rogues L, Fabre A (1993) Spatial and temporal pattern of denitrification in a riparian forest. J Appl Biol 30:581–591
Poly F, Ranjard L, Nazaret S, Gourbiere F, Monrozier L (2001) Comparison of *nifH* gene pools in soils and soil microenvironments with contrasting properties. Appl Environ Microbiol 67:2255–2262
Pratt P, Barber J, Corrin M, Goering J, Hauck R, Johnston H, Klute A, Knowles R, Nelson D, Pickett R, Stephens E (1997) Effect of increased nitrogen fixation on stratospheric ozone. Climate Change 1:109–135
Roberston GP (1982) Factors regulating nitrification in primary and secondary succession. Ecology 63:1561–1573
Robertson GP, Tiedje JM (1987) Nitrous oxide sources in aerobic soils: Nitrification, denitrification and other biological processes. Soil Biol Biochem 19:187–193
Rosch C, Mergel A, Bothe H (2002) Biodiversity of denitrifying and dinitrogen-fixing bacteria in an acid forest soil. Appl Environ Microbiol 68:3818–3829
Rotthauwe J-H, Witzel K-P, Liesack W (1997) The ammonia monoxygenase structural gene *amoA* as a functional marker: molecular fine-scale analysis of natural ammonia-oxidizing populations. Appl Environ Microbiol 63(12):4704–4712
Ryden J, Lund L (1980) Nature and extent of directly measured denitrification losses from some irrigated vegetable crop production units. Soil Sci Soc Am J 49:505–511
Scala D, Kerkhof L (1998) Nitrous oxide reductase (*nosZ*) gene-specific PCR primers for detection of denitrifiers and three *nosZ* genes from marine sediments. FEMS Microbiol Lett 162:61–68
Shirey JJ, Sextone AJ (1989) Denitrification and nitrate-reducing bacterial populations in abandoned and reclaimed minesoils. FEMS Microbiol Ecol 62:59–70
Shrama E (1993) Nutrient dynamics in Himalayan alder plantations. Ann Bot 72:329–336
Simon J (2002) Enzymology and bioenergetics of respiratory nitrite ammonification. FEMS Microbiol Rev 26:285–309
Simonet P, Grosjean M, Misra A, Nazaret S, Cournoyer B, Normand P (1991) *Frankia* genus-specific characterization by polymerase chain reaction. Appl Environ Microbiol 57:3278–3286
Stams A, Flameling E, Marnette E (1990) The importance of autotrophic versus heterotrophic oxidation of atmospheric ammonium in forest ecosystems with acid soil. FEMS Microbiol Ecol 74:337–344
Stephen JR, Kowalchuk GA, Bruns M-AV, McCaig AE, Philipps C, Embley TM, Prosser JI (1998) Analysis of β-subgroup proteobacterial ammonia oxidizer populations in soil by denaturing gradient gel electrophoresis analysis and hierarchical phylogenetic probing. Appl Environ Microbiol 64:2958–2965
Stevens R, Laughlin R, Burns L, Arah J, Hood R (1997) Measuring the contributions of nitrification and denitrification to the flux of nitrous oxide from soil. Soil Biol Biochem 29:139–151
Tiedje JM (1988) Ecology of denitrification and dissimilatory nitrate reduction to ammonium. In: Zehnder AJB (ed) Biology of anaerobic microorganisms. Wiley, New York, pp 179–244
Ueda T, Suga Y, Yahiro N, Matsugushi T (1995) Remarkable N_2 fixing bacterial diversity detected in rice roots by molecular evolutionary analysis of *nifH* gene sequences. J Bacteriol 177:1414–1417
Vance C (1998) Legume symbiotic nitrogen fixation: agronomic aspects. In: Spaink H, Kondorosi A, Hooykaas P (eds) The Rhizobiaceae; molecular biology of plant-associated bacteria. Kluwer, Dordrecht, pp 509–503

Vitousek P, Walker L (1989) Biological invasion by *Myrica faya* in Hawaii: plant demography, nitrogen fixation, ecosystems effects. Ecol Monogr 59:247–265

Waibel A, Peter T, Carslaw K, Oelhaf H, Wetzel G, Crutzen P, Poschl U, Tsias A, Reimer E, Fischer H (1999) Arctic ozone loss due to denitrification. Science 283:2064–2069

Waterer J, Vessey J, Stobbe E, Soper R (1994) Yield and symbiotic nitrogen fixation in a pea-mustard intercrop as influenced by N fertilizer addition. Soil Biol Biochem 26:447–453

Webster EA, Hopkins DW (1996) Contributions from different microbial processes to N_2O emission from soil under different moisture regimes. Biol Fertil Soils 22:3331–3335

Weier KL, MacRae IC (1992) Denitrifying bacteria in the profile of a brigalow clay soil beneath a permanent pasture and a cultivated crop. Soil Biol Biochem 29:919–923

Widmer F, Shaffer B, Porteous L, Seidler R (1999) Analysis of *nifH* gene pool complexity in soil and litter at a Douglas fir forest site in the Oregon cascade mountain range. Appl Environ Microbiol 65:374–380

Yin S, Chen D, Chen L, Eddis R (2002) Dissimilatory nitrate reduction and responsible microorganisms in two Chinese and Australian paddy soils. Soil Biol Biochem 34:1131–1137

Zehr J, Jenkins B, Short S, Stewart G (2003) Nitrogenase gene diversity and microbial community structure: a cross-system comparison. Environ Microbiol 5:539–554

9 Influence of Microorganisms on Phosphorus Bioavailability in Soils

Annette Deubel[1], Wolfgang Merbach[1]

1 Introduction

Phosphorus is the most important plant growth-limiting nutrient in soils besides nitrogen. The total phosphorus content of arable soils varies from 0.02–0.5% with an average of 0.05% in both inorganic and organic forms (Barber 1995). While most mineral nutrients in a soil solution are present in millimolar amounts, phosphorus is only available in micromolar quantities or less (Ozanne 1980). This low availability is due to the high reactivity of phosphorus with calcium, iron and aluminum. Alkaline soils contain different calcium phosphates like hydroxyapatite and fluoroapatite, while acidic soils include amorphous iron- and aluminum phosphates, variscite ($AlPO_4 \cdot 2H_2O$), strengite ($FePO_4 \cdot 2H_2O$) and similar minerals (Barber 1995). An important portion of inorganic soil phosphate is adsorbed onto iron and aluminum oxides and hydroxides, clay minerals and organic substances which contain iron or aluminum complexes. Organic phosphorus compounds have to be mineralized or enzymatically cleaved to become available to plants. All these forms are not directly available to plants.

As a result, higher plants depend on diffusion processes and a continuous release from insoluble sources to meet their phosphorus demand. As decomposers of organic matter as well as mobilizers of inorganic phosphates – in association and competition with higher plants – soil microorganisms influence the availability of phosphorus for plants to a great extent.

2 Microbial Effects on Rhizodeposition

As a result of the low phosphorus mobility in soils, plants only have access to phosphorus a few millimeters around their roots. This volume is highly affected by root deposits and intensively settled by microorganisms, which

[1]Martin-Luther University Halle–Wittenberg, Institute of Soil Science and Plant Nutrition, Adam-Kuckhoff Str. 17b, 06108 Halle, Germany, e-mail: merbach@landw.uni-halle.de, Tel: +49-345-5522421, Fax: +49-345-5527113

Soil Biology, Volume 3
Microorganisms in Soils: Roles in Genesis and Functions
(ed. by F. Buscot and A. Varma)

Table 1. Influence of water-soluble root exudates on double lactate (DL) soluble P 2 mm distant from the frontage of a loess soil block

Treatment	μg DL-P (g soil)$^{-1}$
Without exudates	125
3 days after trickling with 200 mg exudate, sterile	370
3 days after trickling with 200 mg exudate, nonsterile[a]	270
3 days after trickling with 320 mg exudate, nonsterile[b]	432

[a] After 3 days, 75% of ^{14}C was respired by microbes

[b] Plant roots release under nonsterile conditions approx. 60% more exudates in comparison with those grown under sterile conditions

use root-borne compounds as energy sources. The rhizosphere effect leads to an approximately tenfold increase in the microbial population density in comparison to bulk soil.

A wealth of articles describe the exudation of phosphorus-mobilizing substances by plant roots (Johnson et al. 1996; Zhang et al. 1997; Neumann and Römheld 1999; Gaume et al. 2001; Ishikawa et al. 2002). However, neutral sugars that sparingly affect P availability occupy the largest part of water-soluble root exudates of annual plants (Kraffczyk et al. 1984; Merbach et al. 1999; Gransee and Wittenmayer 2000). In addition, such easily decomposable substances are used by microorganisms within a short time.

We used water-soluble ^{14}C-labeled maize root exudates containing 80% sugars, 7% amino acids and amides, and 13% carboxylic acids to examine the phosphorus-mobilizing effect with respect to the concentration gradient in the rhizosphere. Freeze-dried excretions were dissolved in water and trickled onto the front of small soil blocks (20 × 20 × 10 mm Perspex boxes filled with loess soil, 1.43 g/cm^3, 36vol% water content). The blocks were sealed with paraffin to prevent water loss and incubated at 20 °C. After 3 days, the blocks were frozen and cut into slices with a cryomicrotome. Root deposits increased the double-lactate soluble P content of the soil particularly near the trickled area (Schilling et al. 1998). Table 1 gives a comparison between sterile (using ^{60}Co for soil sterilization) and nonsterile treatments.

Microbial decomposition of root exudates leads to a smaller increase in phosphorus solubility in comparison to the sterile treatment. If one takes into account that only one quarter of the radioactivity remains in the soil after 3 days, these regained substances, which include microbial metabolites, have a considerably higher specific phosphorus-mobilizing ability than the original plant-derived substances. In addition, microbial colonization increases the exudation rate of plants (Merbach and Ruppel

1992; Meharg and Killham 1995). In general, nonsterile growing plants can usually mobilize more P than sterile growing plants do. The following paragraph discusses P-mobilizing mechanisms in detail.

3 Mechanisms of Microbial Influence on Phosphorus Availability

3.1 Solubilization of Calcium Phosphates

Under alkaline conditions, soil phosphates are fixed in the form of different calcium phosphates, mainly apatites and metabolites of fertilizer phosphates. Their solubility increases with a decrease of soil pH.

Phosphorus-mobilizing microorganisms are ubiquitous in soils. Their detectable portion among the total microflora depends on soil characteristics as well as on the selection method used. A common simple test is the use of calcium phosphate-containing agar plates (Whitelaw 2000), on which phosphorus-mobilizing colonies produce clear zones. Another possibility is the visualization of a pH decrease using an indicator (Mehta and Nautyal 2001). Both methods assume that P release is mainly based on acidification of the nutrient medium or the soil. However, the decrease in pH is not always in the same correlation to the calcium phosphate solubilization by microorganisms.

To compare the results of different methods, we tested selected rhizosphere bacterial strains qualitatively on calcium phosphate agar plates as well as quantitatively in a liquid medium containing 200 $\mu g\ P\ ml^{-1}$ in the form of $Ca_3(PO_4)_2$, 1% glucose and 0.1% asparagine as C and N sources, respectively.

Only two of the eight strains showed clear zones on calcium phosphate agar and could be identified as P solubilizers (Table 2). However, seven of the eight strains mobilized significant amounts of tricalcium phosphate. Although some strains acidified the nutrient solution remarkably, we found no correlation between pH and P in solution. Hence, proton release cannot be the single mechanism of calcium phosphate mobilization. For this reason, we identified carboxylic anions which are produced under these conditions (Deubel et al. 2000). We found the following substances in the order of their amounts:

- D 5/23: succinate, hydroxyglutarate, adipate, lactate, ketogluconate
- PsIA12: succinate, lactate, malate, ketogluconate, galacturonate, citrate

Table 2. Phosphorus mobilizing ability of selected rhizosphere bacteria[a]

Strain	Clear zones on agar plates	Quantitative estimation in a standing liquid culture (μg P ml^{-1})	pH
D 5/23 *Pantoea agglomerans*	–	62.76	5.93
PsIA12 *Pseudomonas fluorescens*	+	44.09	4.77
CC 322 *Azospirillum* sp.	–	83.39	6.19
Mac 27 *Azotobacter chroococcum*	–	98.11	4.84
Ala 27 *Azotobacter chroococcum*	–	1.10	7.50
Msx 9 *Azotobacter chroococcum*	–	65.90	5.82
ER 3	+	75.48	5.32
ER 10	–	36.16	5.72

[a]Incubation time, 7 days at 28 °C; the P content of sterile controls was in the range of 2–5 µg/ml at a pH of 6.6–6.9 at the end of incubation time

- CC 322: gluconate, succinate, 2-ketoglutarate, ketogluconate
- Mac 27: citrate, malate, fumarate, succinate, lactate
- Msx 9: citrate, fumarate, malate, lactate, succinate
- ER 10: lactate, gluconate, malonate, citrate
- ER3: fumarate, isocitrate, lactate, malonate

Rhizosphere bacteria are able to produce a broad spectrum of potential phosphorus-solubilizing substances (Bajpai and Sundara Rao 1971; Banic and Dey 1981; Subba-Rao 1982). Whitelaw (2000) reviewed the production of oxalate, lactate, glycollate, citrate, succinate and tartrate by different P-mobilizing fungi. Carboxylic anions, which have a high affinity to calcium, solubilize more phosphorus than acidification alone (Staunton and Leprince 1996). Under the same conditions of the experiment above, the following P-concentrations were released from $Ca_3(PO_4)_2$ by:

- lactic acid: 126 µg P mg^{-1} = 11.35 µg P μmol^{-1}
- succinic acid: 178 µg P mg^{-1} = 21.02 µg P μmol^{-1}
- citric acid: 236 µg P mg^{-1} = 45.34 µg P μmol^{-1}

In some in vitro experiments, proton release seems to be the main mechanism of calcium phosphate mobilization (Illmer and Schinner 1995; Villegas and Fortin 2002). However, it should be taken into consideration that nutrient media often are unbuffered, while soils have effective buffer systems (Whitelaw 2000). Only the proton release in combination with microbial N_2 fixation really decreases the pH of alkaline soils. For that reason, carboxylic anions which mobilize calcium phosphates also in buffered systems probably have the highest efficiency under natural soil conditions. They often produce insoluble calcium compounds like calcium citrate or oxalate, which may prevent lysing zones on agar plates. Hence, these common tests can fail to detect the really effective microbial strains.

3.2 Mobilization of Iron- and Aluminum-Bound Phosphorus

An important potential P source of arable soils is the P fraction adsorbed on iron and aluminum compounds. Carboxylic anions are able to replace phosphate from sorption complexes by ligand exchange (Otani et al. 1996; Whitelaw 2000) and to chelate both Fe and Al ions associated with phosphate. Citrate for instance, is able to release phosphate from goethite (Geelhoed et al. 1999) or amorphous ferric hydroxides (Dye 1995). Chelation involves the formation of two or more coordinate bonds between a ligand molecule and a metal ion, thereby creating a ring structure complex (Whitelaw 2000). The ability of different carboxylic anions to desorb P generally decreases with a decrease in the stability constants of Fe (III)- or Al-organic acid complexes ($\log K_{Al}$ or $\log K_{Fe}$) in the following order: citrate > oxalate > malonate/malate > tartrate > lactate > gluconate > acetate > formiate (Ryan et al. 2001; Whitelaw 2000). The extent to which an organic acid is able to chelate metal cations is greatly influenced by its molecular structure, particularly by the number of carboxyl and hydroxyl groups. Tricarboxylic acids like citrate have a higher efficiency than dicarboxylic or monocarboxylic acids. Moreover, the location of groups (α- or β-hydroxy acid structures) can influence the stability of formed complexes.

To demonstrate the P-desorbing efficiency of organic compounds, we extracted P from different slightly acidic soils (pH 5–6) with water, glucose and ribose (important sugars in root exudates), gluconate, succinate, citrate and oxalate (also possibly released by plant roots directly or produced by microorganisms). Figure 1 shows the average of four soils.

Under nonsterile conditions, glucose and ribose released nearly twice the amount of P as water alone. A neutral gluconate solution was not more efficient than sugars, and succinate not more than water. In contrast, citrate and oxalate showed an enormous P-mobilizing ability. The P release with citrate strongly increased within 24 h, but we cannot say whether this is a chemical or microbial effect.

The pH value has controversial effects on P sorption. A proton release increases the availability of iron and aluminum and decreases the negative charge of adsorbing surfaces, which facilitates the sorption of the also negatively charged P ions. However, an acidification leads to an increase of $H_2PO_4^-$ ions in relation to HPO_4^{2-} ions, which have a higher affinity to

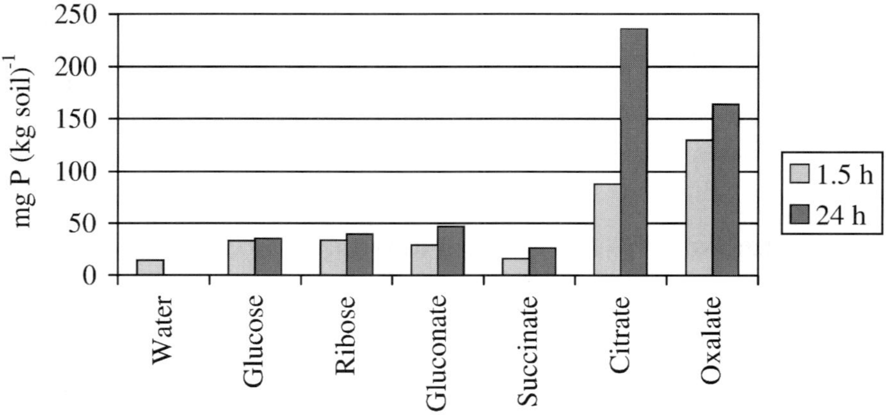

Fig. 1. Average of extractable P amounts with different sugars and carboxylic anions at pH 7 (2 g soil was extracted for 1.5 and 24 h with 100 ml solutions containing 5 g/l sugar or carboxylic acid)

reactive soil surfaces. Therefore, proton release can also decrease P sorption (Whitelaw 2000).

3.3 Influence on Phosphorus Diffusion

Carboxylic anions and higher molecular organic substances compete with phosphate for binding sites at adsorbing surfaces. This effect can improve the diffusion of phosphate ions through the leached zone around plant roots. While the effects of low-molecular compounds are well documented, lack of knowledge still exists concerning the role and composition of mucilages in the rhizosphere of higher plants, which may be of both root and bacterial origin. Mucilages are mainly extracellular polysaccharides containing, for instance, galactose, fucose and uronic acids (El-Shatnawi and Makhadmeh 2001). They include acidic groups and 1,2 diol polysaccharide residues. Bacterial mucilages, which are more complex than those of plant origin, also comprise proteins (Martens and Frankenberger 1991; Watt et al. 1993). Gaume et al. (2000) and Grimal et al. (2001) reported a decrease in P adsorption on goethite or ferrihydrite, respectively, by polygalacturonic acid, galacturonic acid and maize root mucilage. In addition to competition for the same adsorption sites, a coating of the minerals by mucilages is another possible explanation.

Furthermore, mucilages might play an important role in maintaining the root/soil contact in drying soils (Read et al. 1999). They affect the soil structure by gluing soil particles together, sometimes producing a sheath surrounding the roots (Watt et al. 1993; Gregory and Hinsinger 1999). Amellal et al. (1999) found a significant aggregation and stabilization of root-adhering soil by bacterial extracellular polysaccharides combined with an increase in aggregate mean weight, diameter, aggregate macro-porosity, adhering soil: root mass ratio, water-stable > 200 μm aggregates and 0.1–2 μm elementary clayey microaggregates. Watt et al. (1993) reported a 10% higher soil binding capacity of bacterial mucilages in comparison to plant released materials.

Phospholipides in root and bacterial mucilages are powerful surfactants that alter the interaction of soil solids with water and ions (Read et al. 2003). The occupation of adsorption sites as well as influences on aggregate stability and water balance can facilitate phosphorus diffusion processes.

3.4 Release of Phosphorus from Organic Sources

Between 20 and 85% of the total P in agricultural soils are present in the organic form, including inositol phosphate esters, phospholipids, nucleic acids, phosphate linked to sugars and derivatives of phosphoric acid (Tarafdar et al. 2001). Microorganisms mineralize organic materials like plant residues and organic manures and enable the nutrient cycling. Using an isotopic dilution technique, Oehl et al. (2001) found a daily mineralization of 1.7 mg P kg^{-1} in an organically fertilized loamy silt soil. This amount was approximately equivalent to soil solution P, indicating that mineralization is a significant process in delivering available P. Although soil solutions can include higher concentrations of organic than of inorganic phosphates, plants can acquire phosphorus only in inorganic form (Tarafdar et al. 2002). For this reason phosphatases which hydrolyse C–O–P ester bonds are very important for P nutrition. Depending on their pH optimum, acid and alkaline phosphatases can be determined. Phosphodiesterase, which is able to degrade nucleic acids, has not been extensively studied in soils (Dodor and Ali Tabatabai 2003). Acid phosphatases are released by plant roots as well as by microorganisms (Seeling and Jungk 1996; Yadaf and Tarafdar 2001), while alkaline phosphatases are probably mostly of microbial origin (Tarafdar and Claasen 1988). The largest portion of extracellular soil phosphatases is derived from the microbial population and strongly correlates with microbial biomass (Dodor and Ali Tabatabai 2003). Tarafdar et al. (2001) identified a better hydrolysis of lecithin and phytin by fungal acid phosphatases in comparison to those of plant origin. Glycerophosphate was equally hydrolyzed by both enzymes. Tarafdar et al. (2002) reported that different fungi released only 25% of their acid phosphatases extracellularly, but a 39 times higher extracellular phytase activity was noted in comparison with the one inside the fungal cells. This indicates that fungi especially can make this important organic P compound available. This corresponds with the results of Hayes et al. (2000) who reported a limited use of phytate-P by sterile growing plants.

4
Interactions Between Microorganisms and Higher Plants from Competition to Symbiosis

Interactions between microorganisms and higher plants extend from directly detrimental effects of plant pathogens to directly beneficial effects in the case of symbiosis. Rhizosphere microorganisms can be competitors for limited nutrients like P. As a result of higher nucleic acid contents, they have higher P concentrations than higher plants. Some microbes are able to incorporate high P amounts in the form of polyphosphates, an energy reservoir for limited oxygen conditions. Microorganisms often have a higher P uptake efficiency than plant roots. On the other hand, this incorporated P can become available to plants as the microbes die. Oberson et al. (2001) reported a rapid microbial P turnover under different land-use conditions.

Plant hormone production by rhizosphere microorganisms can influence root architecture, the development of root hairs and the affinity of roots for phosphate, indirectly affecting the P uptake.

In particular, different forms of mycorrhiza have great importance for the P nutrition of higher plants. Most agricultural crops are potential host plants for arbuscular mycorrhizal (AM) fungi. In addition to an exudation of carboxylates, phosphatases and plant hormones, mycorrhiza increase the exploitation of the soil volume by the hyphal network, which increases the active adsorption surface and spreads beyond the phosphate depletion zone (Lange Ness and Vlek 2000; Martin et al. 2001). Mycorrhizal hyphae have a higher affinity for phosphate as expressed in the Michaelis-Menten equation by a lower Km value and absorb P at lower solution concentrations than roots do (Lange Ness and Vlek 2000). AM fungi store phosphate in the form of orthophosphate, polyphosphate and organic P in their vacuoles and transfer it to the roots of the host plant (Ezawa et al. 2002).

5
Phosphorus-Mobilizing Microorganisms as Biofertilizers

A great interest in the use of microorganisms as biofertilizers exists especially in areas with a low P availability as a result of an unfavorable soil pH. In addition, inoculates are used to improve the fertilizer efficiency of rock phosphate (Goenadi et al. 2000; Reddy et al. 2002). Although mycorrhizal fungi are able to improve the phosphorus supply of higher plants more than other microbes, their use as biofertilizers is complicated because AM fungi, which are able to infect a lot of arable crops, are obligate symbionts. Up-to-date methods for an in vitro production of inocula do not exist. In addition, P fertilization depresses the formation of arbuscular mycorrhizas. Therefore, the use of AM inocula is mostly confined to horticulture and recultivation of mine areas. Much research focuses on bacteria and fungi which live in association with higher plants. Some bacterial inoculates can enhance root colonization by mycorrhizal fungi (Ratti et al. 2001).

Although many authors report a growth-promoting effect of phosphorus-solubilizing microorganisms (PSM; (Narula et al. 2000; Sundara et al. 2002)), results in the field are highly variable (Gyaneshwar et al. 2002). A yield increase is not always combined with higher P uptake (de Freitas et al. 1997; Deubel et al. 2002; Reyes et al. 2002). The varying success of PSM inoculations can be due to different reasons (Kucey et al. 1989; Gyaneshwar et al. 2002): (1) insufficient survival and colonization of inoculated strains, (2) competition with native microorganisms, (3) nature and properties of soils and plant varieties, (4) starvation of nutrients in the rhizosphere to produce enough organic acids to solubilize soil phosphates and (5) inability of PSMs to solubilize soil phosphates.

Because of the ubiquitous occurrence of phosphorus-mobilizing microbes, a yield increase by inoculation with additional strains may be beneficial if these organisms possess different growth-promoting abilities, for instance N_2 fixation, phytohormone production and phosphorus mobilization (Peix et al. 2001). One of the greatest problems is insufficient selection and test methods for phosphorus-mobilizing microorganisms. The selection on clear zones or a pH decrease in simple plate tests cannot reflect the real P binding capabilities under soil and rhizosphere conditions. A better possibility may be the selection of microbes, which can effectively use P adsorbed on goethite or other minerals (He et al. 2002). Because many more microbes have the ability to solubilize phosphates under special conditions than to colonize and promote growth of higher plants, it is useful to select first on the basis of growth-promoting abilities. A limited number of strains can then be tested for special properties. It must be taken into account that test conditions influence the results to a large extent. A given strain can respire a sugar to CO_2 under high oxygen supply, or produce carboxylic acids by fermentation or incomplete respiration if oxygen is limiting. Moreover, the type of C sources affects the production of microbial metabolites (Kim et al. 1998; Deubel et al. 2000).

We tested the influence of different sugars on the $Ca_3(PO_4)_2$-mobilizing efficiency of different bacterial strains. Figure 2 shows a comparison of glucose, which represents a large portion of water-soluble root exudates of

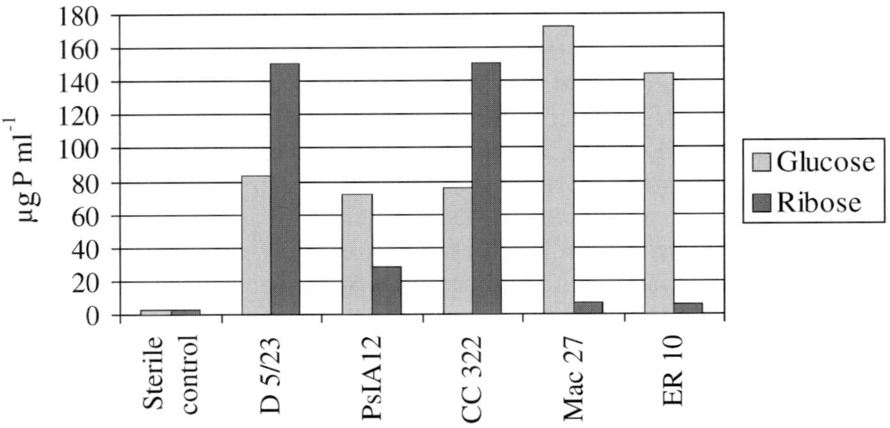

Fig. 2. P release from $Ca_3(PO_4)_2$ by different bacterial strains with glucose or ribose as C source (standing liquid culture 7 days at 28 °C, 4 mg C ml^{-1} as sugar, 200 µg P ml^{-1} as $Ca_3(PO_4)_2$)

well-nourished plants, with ribose, which is present in increasing amounts under P deficiency (Deubel et al. 2000).

Although all strains were able to grow with both sugars as the C source, *Pantoea* and *Azospirillum* increased P mobilization with ribose, while other strains released less or no phosphate with this sugar. These changes in P-mobilizing ability were combined with changes in the pattern of produced carboxylic acids. The same results were found in the response of different bacteria on synthetically mixed root exudates of well-nourished and P-deficient plants (Fig. 3).

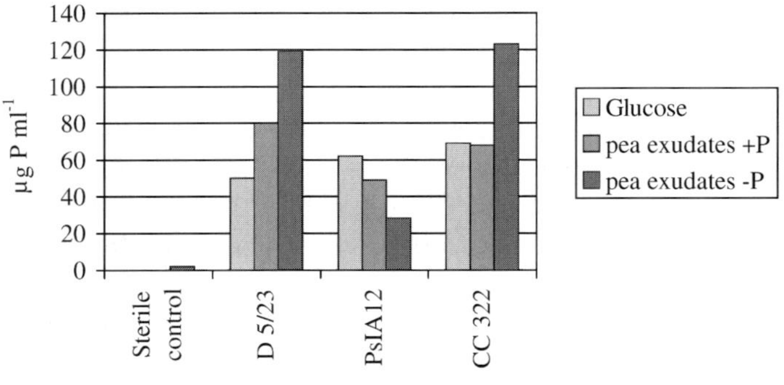

Fig. 3. Influence of synthetic sugar mixtures [in analogy to the saccharide portion of root deposits of *Pisum sativum* plants with P (pea exudates + P) and without P supply (pea exudates – P)] in comparison to standard medium (glucose) on the $Ca_3(PO_4)_2$ solubilizing ability of the bacterial strains D 5/23, PsIA12 and CC 322

Therefore, it is very important to consider the real conditions in the rhizosphere of potential host plants during the test procedure.

Another issue is the demand of C sources to produce carboxylic acids in sufficient amounts. A symbiotic relation with higher plants, for instance mycorrhiza or rhizobia symbiosis (Vance 2001), can best provide enough organic C compounds for P-mobilizing processes. On the other hand, only a few substances, particularly citrate and oxalate are effective in micromolar concentrations, which are realistic for rhizosphere conditions. Hence, it is useful to look for producers of such effective substances, for instance by genetic characterization (Igual et al. 2001).

One possibility to increase the P-mobilizing efficiency of microbial strains is the induction of mutations by UV light (Reyes et al. 2001), or chemical substances (Narula et al. 2000), as well as genetic manipulation (Gyaneshwar et al. 1998; Rodriguez et al. 2000). However, the chances and risks in the spreading of genetically modified microorganisms have to be weighed carefully.

Soil microorganisms have an enormous potential to improve phosphorus bioavailability. A better understanding of the interactions between different microorganisms as well as between microorganisms and higher plants, improved selection and test procedures and the development of culture methods for mycorrhizal fungi will help to realize this potential as biofertilizers.

6 Conclusions

Soil microorganisms, particularly the rhizosphere flora of higher plants, remarkably affect the phosphorus bioavailability in soils. Microbially derived carboxylic acids mobilize calcium phosphates as well as iron- and aluminum-bound phosphorus. Microbial mineralization of organic matter is essential for nutrient cycling in soils and phosphatases enhance the use of organic P compounds by higher plants. Plants, especially in nutrient-poor habitats like forest ecosystems, often depend on symbiotic relations with microorganisms like mycorrhizal fungi. However, rhizosphere flora also decomposes P-mobilizing substances derived from plant roots. Microorganisms can be powerful competitors for growth-limiting nutrients like P, but microbial turnover can also make P available for higher plants. The difficulty in quantifying all these complex and partially contrary processes is a substantial weak point in mathematical P-utilization models as well as in the use of P-mobilizing microbes as biofertilizers. The investigation of these complex effects with modern methods, which cover also the large majority of noncultivable microorganisms, is an important aim for further research.

References

Amellal N, Bartoli F, Villemin G, Talouizte A, Heulin T (1999) Effects of inoculation of EPS-producing *Pantoea agglomerans* on wheat rhizosphere aggregation. Plant Soil 211:93–101

Bajpai PD, Sundara Rao WVB (1971) Phosphate solubilizing bacteria. Soil Sci Plant Nutr 17:41–53

Banic S, Dey BK (1981) Phosphate-solubilizing microorganisms of a lateritic soil. I. Solubilization of inorganic phosphates and production of organic acids by microorganisms, isolated in sucrose calcium phosphate agar plates. Zentralbl Bakt Abt II 136:478–486

Barber SA (1995) Soil nutrient bioavailability. Wiley, New York

De Freitas JR, Banerjee MR, Germida JJ (1997) Phosphate solubilizing rhizobacteria enhance the growth and yield but not the phosphorus uptake of canola. Biol Fertil Soil 24:358–364

Deubel A, Gransee A, Merbach W (2000) Transformation of organic rhizodepositions by rhizosphere bacteria and its influence on the availability of tertiary calcium phosphate. J Plant Nutr Soil Sci 163:387–392

Deubel A, Narula N, Gransee A, Merbach W (2002) Einfluss einer Bakterieninokulation auf die N- und P-Ernährung junger Weizenpflanzen bei unterschiedlichen Düngungsstufen. In: Merbach W, Hütsch BW, Wittenmayer L, Augustin J (eds) Durchwurzelung, Rhizodeposition und Pflanzenverfügbarkeit von Nährstoffen und Schwermetallen. Teubner, Stuttgart, pp 31–35

Dodor DE, Ali Tabatabai M (2003) Effect of cropping systems on phosphatases in soils. J Plant Nutr Soil Sci 166:7–13

Dye C (1995) Effect of citrate and tartrate on phosphate adsorption by amorphous ferric hydroxide. Fertil Res 40:129–134

El-Shatnawi MKJ, Makhadmeh IM (2001) Ecophysiology of the plant-rhizosphere system. J Agron Crop Sci 187:1–9

Ezawa T, Smith SE, Smith FA (2002) P metabolism and transport in AM fungi. Plant Soil 244:221–230

Gaume A, Weidler PG, Frossard E (2000) Effects of maize root mucilage on phosphate adsorption and exchangeability on a synthetic ferrihydrite. Biol Fertil Soils 31:525–532

Gaume A, Mächler F, Frossard E (2001) Aluminium resistance in two cultivars of *Zea mays* L.: root exudation of organic acids and influence of phosphorus nutrition. Plant Soil 234:73–81

Geelhoed JS, van Riemsdijk WH, Findenegg GR (1999) Simulation of the effect of citrate exudation from roots on the plant availability of phosphate adsorbed on goethite. Eur J Soil Sci 50:379–390

Goenadi DH, Siswanto, Sugiarto Y (2000) Bioactivation of poorly soluble phosphate rocks with a phosphorus-solubilizing fungus. Soil Sci Soc Am J 64:927–932

Gransee A, Wittenmayer L (2000) Qualitative and quantitative analysis of water-soluble root exudates in relation to plant species and development. J Plant Nutr Soil Sci 163:381–385

Gregory PJ, Hinsinger P (1999) New approaches to studying chemical and physical changes in the rhizosphere: an overview. Plant Soil 211:1–9

Grimal JY, Frossard E, Morel JL (2001) Maize root mucilage decrease adsorption of phosphate on goethite. Biol Fertil Soils 33:226–230

Gyaneshwar P, Naresh Kumar G, Parekh LJ (1998) Cloning of mineral phosphate solubilizing genes from *Synechocystis* PCC 6803. Curr Sci 74:1097–1099

Gyaneshwar P, Naresh Kumar G, Parekh LJ, Poole PS (2002) Role of soil microorganisms in improving P nutrition of plants. Plant Soil 245:83–93

Hayes JE, Simpson RJ, Richardson AE (2000) The growth and phosphorus utilization of plants in sterile media when supplied with inositol hexaphosphate, glucose 1-phosphate or inorganic phosphate. Plant Soil 220:165–174

He ZL, Bian W, Zhu J (2002) Screening and identification of microorganisms capable of

utilizing phosphate adsorbed by goethite. Commun Soil Sci Plant Anal 33:647–663
Igual JM, Valverde A, Cervantes E, Velazquez E (2001) Phosphate-solubilizing bacteria as inoculants for agriculture: use of updated molecular techniques in their study. Agronomie 21:561–568
Illmer P, Schinner F (1995) Solubilization of inorganic calcium phosphates – solubilization mechanisms. Soil Biol Biochem 27:257–263
Ishikawa S, Adu-Gyamfi JJ, Nakamura T, Yoshihara T, Watanabe T, Wagatsuma T (2002) Genotypic variability in phosphorus solubilizing activity of root exudates by pigeonpea grown in low-nutrient environments. Plant Soil 245:71–81
Johnson JF, Vance CP, Allan DL (1996) Phosphorus deficiency in *Lupinus albus*. Plant Physiol 112:31–41
Kim KY, Jordan D, McDonald GA (1998) *Enterobacter agglomerans*, phosphate solubilizing bacteria, and microbial activity in soil: effect of carbon sources. Soil Biol Biochem 30:995–1003
Kraffczyk I, Trolldenier G, Beringer H (1984) Soluble root exudates of maize : Influence of potassium supply and microorganisms. Soil Biol Biochem 16:315–322
Kucey RMN, Jenzen HH, Leggett ME (1989) Microbially mediated increases in plant available phosphorus. Adv Agron 42:199–228
Lange Ness RL, Vlek PLG (2000) Mechanism of calcium and phosphate release from hydroxy-apatite by mycorrhizal hyphae. Soil Sci Am J 64:949–955
Martens DA, Frankenberger WT Jr (1991) Saccharide composition of exocellular polymers produced by soil microorganisms. Soil Biol Biochem 23:731–736
Martin FM, Perotto S, Bonfante P (2001) Mycorrhizal fungi: A fungal community at the interface between soil and roots. In: Pinton R, Varanini Z, Nannipieri P (eds) The rhizosphere. Dekker, New York, pp 263–296
Meharg AA, Killham K (1995) Loss of exudates from the roots of perennial ryegrass inoculated with a range of micro-organisms. Plant Soil 170:345–349
Mehta S, Nautyal CS (2001) An efficient method for qualitative screening of phosphate-solubilizing bacteria. Curr Microbiol 43:51–56
Merbach W, Ruppel S (1992) Influence of microbial colonization on $^{14}CO_2$ assimilation and amounts of root-borne ^{14}C compounds in soil. Photosynthetica 26:551–554
Merbach W, Mirus E, Knof G, Remus R, Ruppel S, Russow R, Gransee A, Schulze J (1999) Release of carbon and nitrogen compounds by plant roots and their possible ecological importance. J Plant Nutr Soil Sci 162:373–383
Narula N, Kumar V, Behl RK, Deubel A, Gransee A, Merbach W (2000) Effect of P-solubilizing *Azotobacter chroococcum* on N, P, K uptake in P-responsive wheat genotypes grown under greenhouse conditions. J Plant Nutr Soil Sci 163:393–398
Neumann G, Römheld V (1999) Root excretion of carboxylic acids and protons in phosphorus-deficient plants. Plant Soil 211:121–130
Oberson A, Friesen DK, Rao IM, Bühler S, Frossard E (2001) Phosphorus transformations in an oxisol under contrasting land-use systems: the role of the soil microbial biomass. Plant Soil 237:197–210
Oehl F, Oberson A, Sinaj S, Frossard E (2001) Organic phosphorus mineralization studies using isotopic dilution techniques. Soil Sci Soc Am J 65:780–787
Otani T, Ae N, Tanaka H (1996) Phosphorus (P) uptake mechanisms of crops grown in soils with low P status. II. Significance of organic acids in root exudates of pigeonpea. Soil Sci Plant Nutr 42:553–560
Ozanne PG (1980) Phosphate nutrition of plants – a general treatise. In: Khasawneh FH, Sample EC, Kamprath EJ (eds) The role of phosphorus in agriculture. American Society of Agronomy, Madison, pp 559–590
Peix A, Rivas-Boyero AA, Mateos PF, Rodriguez-Barrueco C, Martinez-Molina E, Velazquez

E (2001) Growth promotion of chickpea and barley by a phosphate solubilizing strain of *Mesorhizobium mediterraneum* under growth chamber conditions. Soil Biol Biochem 33:103–110

Ratti N, Kumar S, Verma HN, Gautam SP (2001) Improvement in bioavailability of tricalcium phosphate to *Cymbopogon martinii* var. *motia* by rhizobacteria, AMF and *Azospirillum* inoculation. Microbiol Res 156:145–149

Read DB, Gregory PJ, Bell AE (1999) Physical properties of axenic maize root mucilage. Plant Soil 211:87–91

Read DB, Bengough AG, Gregory PJ, Crawford JW, Robinson D, Scrimgeour CM, Young IM, Zhang K, Zhang X (2003) Plant roots release phospholipid surfactants that modify the physical and chemical properties of soil. New Phytol 157:315–326

Reddy MS, Kumar S, Babita K, Reddy MS (2002) Biosolubilization of poorly soluble rock phosphates by *Aspergillus tubingensis* and *Aspergillus niger*. Bioresource Technol 84:187–189

Reyes I, Baziramakenga R, Bernier L, Antoun H (2001) Solubilization of phosphate rocks and minerals by a wild-type strain and two UV-induced mutants of *Penicillium rugulosum*. Soil Biol Biochem 33:1741–1747

Reyes I, Bernier L, Antoun H (2002) Rock phosphate solubilization and colonization of maize rhizosphere by wild and genetically modified strains of *Penicillium rugulosum*. Microb Ecol 44:39–48

Rodriguez H, Gonzalez T, Selman G (2000) Expression of a mineral phosphate solubilizing gene from *Erwinia herbicola* in two rhizobacterial strains. J Biotechnol 84:155–161

Ryan PR, Delhaize E, Jones DL (2001) Function and mechanism of organic anion exudation from plant roots. Annu Rev Plant Physiol Plant Mol Biol 52:527–560

Schilling G, Gransee A, Deubel A, Lezovic G, Ruppel S (1998) Phosphorus availability, root exudates, and microbial activity in the rhizosphere. J Plant Nutr Soil Sci 161:465–478

Seeling B, Jungk A (1996) Utilization of organic phosphorus in calcium chloride extracts of soil by barley plants and hydrolysis by acid and alkaline phosphatases. Plant Soil 178:179–184

Staunton S, Leprince F (1996) Effect of pH and some organic anions on the solubility of soil phosphate: implications for P bioavailability. Eur J Soil Sci 47:231–239

Subba-Rao NS (1982) Advances in agricultural microbiology. Butterworth Scientific, London

Sundara B, Natarajan V, Hari K (2002) Influence of phosphorus solubilizing bacteria on the changes in soil available phosphorus and sugarcane and sugar yields. Field Crops Res 77:43–49

Tarafdar JC, Claasen N (1988) Organic phosphorus compounds as a phosphorus source for higher plants through the activity of phosphatases produced by plant roots and microorganisms. Biol Fertil Soils 5:308–312

Tarafdar JC, Yadav RS, Meena SC (2001) Comparative efficiency of acid phosphatase originated from plant and fungal sources. J Plant Nutr Soil Sci 164:279–282

Tarafdar JC, Yadav RS, Niwas R (2002) Relative efficiency of fungal intra- and extracellular phosphatase and phytase. J Plant Nutr Soil Sci 165:17–19

Vance CP (2001) Symbiotic nitrogen fixation and phosphorus acquisition. Plant nutrition in a world of declining renewable resources. Plant Physiol 127:390–397

Villegas J, Fortin JA (2002) Phosphorus solubilization and pH changes as a result of the interactions between soil bacteria and arbuscular mycorrhizal fungi on a medium containing NO_3 as nitrogen source. Can J Bot 80:571–576

Watt M, McCully ME, Jefree CE (1993) Plant and bacterial mucilages of the maize rhizosphere: comparison of their soil binding properties and histochemistry in a model system. Plant Soil 151:151–165

Whitelaw MA (2000) Growth promotion of plants inoculated with phosphate-solubilizing

fungi. Adv Agron 69:100–153
Yadaf RS, Tarafdar JC (2001) Influence of organic and inorganic phosphorus supply on the maximum secretion of acid phosphatase by plants. Biol Fertil Soils 34:140–143
Zhang FS, Ma J, Cao YP (1997) Phosphorus deficiency enhances root exudation of low-molecular weight organic acids and utilization of sparingly soluble inorganic phosphates by radish (*Raphanus sativus* L.) and rape (*Brassica napus* L.) plants. Plant Soil 196:261–264

Part IV
Biotic Interactions Involving Soil Microorganisms

10 Interactions Between Mycorrhizal Fungi and Bacteria to Improve Plant Nutrient Cycling and Soil Structure

Jose Miguel Barea[1], R. Azcón[1], C. Azcón-Aguilar[1]

1 Introduction

The stability and productivity of either agro-ecosystems or natural ecosystems largely depend on soil quality, therefore, the management of soil–plant systems must consider the maintenance of the quality and sustainability of soil resources (Altieri 1994). An important issue is that soil quality is fundamental not only to produce healthy crops or to ensure self-sustainability of the ecosystems, but also to prevent erosion and to minimize environmental impacts (Parr et al. 1992). Soil quality is determined by diverse chemical, physical and biological factors and their interactions. Thus, for the appropriate management of soil–plant systems, the understanding of how the physicochemical and biological (microbial) components function and interact, and how perturbations affect these interactions is critical (Kennedy and Smith 1995). Actually, perturbations of either agro- or natural ecosystems are known to disturb interactive processes which affect essential determinants of soil quality, such as soil structure, plant nutrient availability, organic matter content and/or microbial diversity and activity (Kennedy and Smith 1995). Therefore, the success of any restoration approach largely depends on the integration of management strategies addressed to optimize the interactions among the soil components improving its quality (Requena et al. 2001).

While many studies have been devoted in the past to investigate the physicochemical components of soil quality (Parr et al. 1992), the biological properties have received less attention. Nevertheless, evidence is accumulating to demonstrate that the maintenance of diverse and active soil microbial communities is fundamental to soil quality (Kennedy and Smith 1995; Barea 1997). Microbial activities are particularly relevant at the root–soil interfaces/interphases known as the rhizosphere where microorganisms, plant roots and soil constituents interact (Lynch 1990; Kennedy 1998; Werner 1998; Bowen and Rovira 1999). A key issue in rhizosphere

[1]Departamento de Microbiología del Suelo y Sistemas Simbióticos, Estación Experimental del Zaidín, CSIC Prof. Albareda 1, 18008 Granada, Spain, e-mail: josemiguel.barea@eez.csic.es, Tel: +34-958-181600, Fax: +34-958-129600

Soil Biology, Volume 3
Microorganisms in Soils: Roles in Genesis and Functions
(ed. by F. Buscot and A. Varma)

ecology is that not only the microbial and the physicochemical components interact, but also the microbial components themselves. In fact, soil microbial populations are involved in a network of interactions known to affect soil–plant developments, and some of the beneficial microbial activities can be exploited as a low-input biotechnology with regard to sustainability issues (Kennedy and Smith 1995). The root-associated microorganisms can be integrated into two main groups: saprophytes and plant symbionts. Both of them comprise detrimental, neutral and beneficial bacteria and fungi. Beneficial microorganisms are involved in many key ecosystem processes. Among them, the biogeochemical cycling of both inorganic and organic nutrients in soil and in the maintenance of soil structure are to be pointed out. Particularly, some interactions between bacteria and fungi are relevant to the sustainability of the agro- and natural ecosystems (Jeffries and Barea 2001; Barea et al. 2002a,b), therefore, discussion of these interactions is a topic of current interest.

2 Beneficial Bacteria and Fungi in Agro- and Natural Ecosystems

Three main types of beneficial microbes can be distinguished: (1) the saprophytic root colonist rhizosphere bacteria, the so-called plant growth-promoting rhizobacteria (PGPR), (2) the microsymbionts arbuscular mycorrhizal fungi and (3) the microsymbionts N_2-fixing bacteria. The PGPR are involved in nutrient cycling and plant protection against diseases (Kloepper 1994; Bashan 1999; Dobbelaere et al. 2001; Probanza et al. 2002), the N_2-fixing bacteria are responsible for the N inputs to the biosphere (Postgate 1998), and the arbuscular mycorrhizal fungi improve plant growth, nutrition and health (Smith and Read 1997).

The arbuscular mycorrhiza is a universal symbiosis which is established with more than 80% of plant species, including almost all major agricultural crops and herbaceous and shrub species in natural ecosystems (Barea et al. 1997). Other types of mycorrhizal associations (Smith and Read 1997) will not be considered here. The responsible fungi in the arbuscular mycorrhiza (hereafter mycorrhiza) belong to the order Glomales in the Zygomycetes, according to Morton and Redecker (2001), or to the new fungal phylum Glomeromycota, as proposed by Schüßler et al. (2001); a taxonomic issue still a matter of discussion.

The mycorrhizal fungi, after the biotrophic colonization of the root cortex, develop an external mycelium which is a bridge connecting the root with the surrounding soil microhabitats. The beneficial growth responses to mycorrhizal fungi is primarily attributed to an increased uptake of phosphate (and other nutrients) from the soil solution, in such a way that the external mycorrhizal mycelium is responsible for a major part of the P acquired by the plant (Joner et al. 2000). Because of the low mobility of phosphate ions, a phosphate-depletion zone develops around plant roots (Jeffries and Barea 2001), thus, the mycorrhizal activity, due to the ability of external hyphae to exploit soil volumes that are not accessed by non-mycorrhizal roots, is a critical issue for nutrient capture and cycling in soil–plant systems (Smith and Read 1997). In cooperation with other soil organisms, the external mycorrhizal mycelium forms water-stable aggregates necessary for good soil tilth (Miller and Jastrow 2000). Mycorrhizal fungi also improve plant health through increased protection against biotic and abiotic stresses (Jeffries et al. 2003).

Accordingly, some interactions between these beneficial microorganisms are particularly relevant to benefit plant fitness and soil quality (Barea et al. 2002a, b) and merit further considerations. The aim of this chapter is to analyze some key examples of these interactions involving fungi and bacteria which benefit two important properties of soil quality, i. e., nutrient mobilization and cycling and soil structure stabilization. Because almost all plant species growing in terrestrial ecosystems are mycorrhizal (Smith and Read 1997), mycorrhizal fungi will have a central position in the interactions to be studied. The mycorrhizosphere (Barea et al. 2002a, b) interactions related to nutrient cycling concern three types of soil bacteria: (1) plant symbiotic N_2-fixing rhizobial bacteria; (2) phosphate-solubilizing bacteria; and (3) phytostimulators *Azospirillum*. Representative experiments using isotope (^{15}N and ^{32}P) dilution techniques to ascertain the extent of soil–plant benefit from these interactions will be described. The activity of mycorrhizal fungi with regard to soil aggregation, known to occur in interaction with saprophytic microorganisms, including bacteria, will also be reviewed as these are fundamental for soil structure stabilization.

3 Interactions Between Mycorrhizal Fungi and Symbiotic N_2-Fixing Rhizobial Bacteria

Nitrogen fixation is a key factor in biological productivity, it being accepted that more than 60% of the N-input to the plant community has a biological origin, and that half of this input is due to the symbiotic plant–bacteria systems, particularly those involving legumes (Postgate 1998). The bacterial partners in the symbiotic relationships with legume species belong to the genera *Rhizobium*, *SinoRhizobium*, *BradyRhizobium*, *MesoRhizobium*, and *AzoRhizobium*, collectively termed as *Rhizobium* or rhizobia. Their association with legume roots leads to formation of N_2-fixing nodules

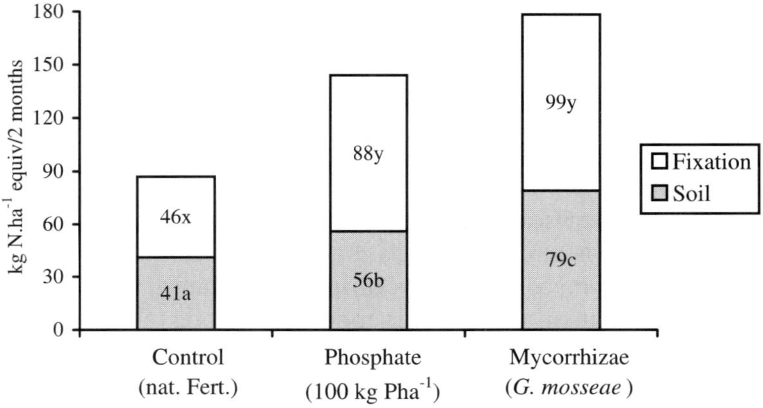

Fig. 1. Sources of N for a legume plant either inoculated with mycorrhizal fungi and *Rhizobium*, given extra phosphate, or the corresponding controls. For each response variable, means ($n = 5$), not sharing a letter in common differ significantly ($p = 0.05$) from each other according to Duncan's multirange test. (Barea et al. 1987)

(Spaink et al. 1998). The significance of this symbiosis is strengthened by the importance of legumes for food production, forage, green manure, and horticulture (Postgate 1998). Legumes also play a fundamental role in natural ecosystems (Jeffries and Barea 2001). The agronomic significance of the symbiotic N_2 fixation in legumes has been demonstrated in many experiments as reviewed by Vance (2001).

Arbuscular mycorrhiza is one of the most efficient ecological factors improving growth and N content in legumes (Barea and Azcón-Aguilar 1983). However, it was questioned whether the extra N found in nodulated legumes, as a mycorrhizal response, is due to increased N_2fixation, as expected because of the phosphate supply by mycorrhizal system, taking into account the importance of P supply for N_2-fixation (Barea et al. 1992), or to an increased N uptake from soil by the mycorrhizal hyphae network. The latter alternative was supported by the finding that mycorrhizal hyphae were demonstrated to absorb, transport and utilize NH_4^+ (Ames et al. 1983). As will be expanded below, Barea et al. (1987) demonstrated by using ^{15}N-based techniques under field conditions that both mechanisms, improvement of N_2 fixation and increased N uptake from soil, operate. These conclusions are graphically recorded in Fig. 1. Further studies using ^{15}N and compartmented plant-growing systems, corroborated that the mycorrhizal hyphae colonizing the ^{15}N-labeled soil in a root-free compartment are able to increase ^{15}N uptake (Tobar et al. 1994a, b).

The use of ^{15}N-labeled soils has been the basis of several experiments aimed at ascertaining the effect of microbial interactions on N cycling (Barea et al. 1987, 2002c). Because some key experiments using ^{15}N-labeled soils will be described here, some brief comments are necessary to help understand the fundamentals of the ^{15}N-aided methodologies (Danso 1986, 1988). A basic, widely accepted, assumption is that when a plant is confronted with two or more sources of a nutrient element, the nutrient uptake from each source is proportional to their available amount (Zapata 1990).This implies a situation in which a soil [having the natural abundance of available N (0.366% ^{15}N/99.634% ^{14}N)], if it is supplemented with a ^{15}N-labeled fertilizer (with a % ^{15}N atomic excess), reaches a "constant" $^{15}N/^{14}N$ ratio in the N available to any plant. Consequently, different plant species, either fixing or nonfixing N_2 and growing on such a ^{15}N-labeled soil, will incorporate N from the soil with a similar $^{15}N/^{14}N$ ratio. For N_2-fixing plants, however, another N source is available; i.e., the N from the atmosphere, in which the $^{15}N/^{14}N$ ratio is that of the natural abundance. Thus, the N_2-fixing plant lowers the ratio of $^{15}N/^{14}N$ by incorporating atmospheric N. This is the basis to assess the N_2-fixing capacity of a fixing system, as discussed below. The $^{15}N/^{14}N$ ratio is a yield-independent parameter with regard to both biomass and N accumulation (Danso 1986, 1988).

Based on these concepts, methodologies using ^{15}N-labeled fertilizers were proposed to measure N_2fixation by *Rhizobium*-legume symbiosis in the field (Danso 1986, 1988). As indicated before, for quantitative estimates, a nonfixing (reference) crop is needed to assess the $^{15}N/^{14}N$ ratio in the target soil that was labeled with a small amount of ^{15}N-containing fertilizer. A lower $^{15}N/^{14}N$ ratio in *Rhizobium*-inoculated legumes compared to those achieved by the reference nonfixing crop is usually found under field conditions. The most effective *Rhizobium* strain will induce the greatest lowering of the $^{15}N/^{14}N$ ratio. Therefore, methodologies based on ^{15}N-labeled soils offer the only direct approach to distinguish the relative contribution of the three N sources to legume plants, i.e., soil, fertilizer and atmosphere. The ^{15}N methodology is the only direct procedure which gives a truly integrated measurement of N_2 fixation over a growing period (Danso 1986, 1988).

A further advantage of ^{15}N methodologies is the possibility of assessing the influence on N_2fixation of a given treatment distinct from soil or fertilizer N supply (Danso 1986, 1988). The effect of mycorrhizal inoculation at improving N_2 fixation in legume-*Rhizobium* associations was investigated (Barea et al. 1987) with this technique under field conditions. It was found that mycorrhizal inoculation of the target legume enhanced dry matter yield, N concentration and total N yield. The use of ^{15}N allowed one to distinguish the source of N in plant tissues and it was realized that both the amount of N derived from soil and from fixation were higher in mycorrhiza-inoculated plants than in either phosphate-added or nontreated controls. These findings demonstrated that mycorrhizal fungi acted both by a P-mediated mechanism to improve N_2 fixation by the rhizobial microsymbiont, and by enhancing N uptake from soil. Thus, mycorrhizas were found to improve two processes of great physiological and ecological importance.

Recent studies based on the application of ^{15}N dilution techniques have further corroborated that mycorrhizal fungi and *Rhizobium* interact to increase N_2 fixation in legumes under greenhouse (Toro et al. 1998) and field (Barea et al. 2002c) conditions, and also support an important ecological role of these microbial interactions in contributing to N cycling in ecosystems.

Because legumes are known to enrich their rhizospheric soil with N from fixation (at natural abundance in the $^{15}N/^{14}N$ ratio), these plants "dilute" the $^{15}N/^{14}N$ ratio in ^{15}N-labeled soils. Nonfixing plants growing nearby will tap this N from the legume rhizosphere, resulting in a lowered $^{15}N/^{14}N$ ratio compared to nonfixing plants not associated to legumes. This can be measured as "N transfer from fixation" (Danso 1986, 1988). Mycorrhizal inoculation has been demonstrated to increase such N transfers (Barea et al. 1989; Requena et al. 2001).

While the mycorrhizal activity in improving N_2-fixation represents a considerable contribution to N inputs in legume species, this increase was interpreted as an indication that the mycorrhizal mycelium uses N sources less available to nonmycorrhizal plants (Barea et al. 1987). In other words, it seems that the available N pool in soils is greater for mycorrhizal than for nonmycorrhizal plants. To investigate mycorrhiza contribution to N acquisition by nonfixing plants, the ^{15}N isotope was used to measure the apparent plant-available N pool size, i. e., the A_N value of the soil (Zapata 1990) for mycorrhizal and nonmycorrhizal plants. The A_N value is calculated on the basis of the ratio of the plant N derived from the soil (^{14}N) to the plant N derived from a ^{15}N-labeled fertilizer. The A_N value, which determines the available amount of N for a given plant, is an inherent, yield-independent, property of the soil, constant for any one set of experimental conditions (Zapata 1990). However, if a given treatment is able to induce changes in the N uptake pattern allowing roots to use N forms (^{14}N) different from those without treatment, the A_N value would be expected to increase. This is due to the apparent 'constancy' of the $^{14}N/^{15}N$ ratio in plants actually depending on the same $^{14}N/^{15}N$ pools in soil under different treatments. The experiments carried out by Barea et al. (1991) demonstrated that the A_N value for plant inoculated with mycorrhizal fungi is higher than for the noninoculated controls. This demonstrated that the mycorrhizal mycelium is accessing N forms less available for nonmycorrhizal plants.

From the point of view of the soil quality improvement, the experiments of Requena et al. (2001) illustrate that target mycorrhizosphere interactions influence accumulation in N content in soils surrounding dually inoculated legume plants. These field trials, which will be described below in Section 6

(Interactions Improving Soil Structure Stabilization), investigated several aspects related to changes in the physicochemical soil properties in soils around transplanted target legumes, as a result of the mycorrhizal and *Rhizobium* inoculation. The experiments, actually revegetation trials, were carried out in a degraded shrub ecosystem, and evidenced a long-term improvement of N content in soils by dually inoculated transplanted target legume species (Requena et al. 2001). Using ^{15}N, it was found that mycorrhizal inoculation increased N_2 fixation by the shrub legume and contributed to N accumulation in the soil around the target legumes. In addition, the use of ^{15}N allowed the authors to demonstrate that mycorrhizal inoculation benefited "N transfer" from the rhizosphere of the nodulated legumes to nonfixing plant species associated in the natural succession. These facts actually indicated microbiologically driven N-cycling activities vital for restoration of a degraded ecosystem.

4 Interactions Between Mycorrhizal Fungi and Phosphate-Solubilizing Bacteria

Among the microbiological processes involved in nutrient cycling, those responsible for increasing the phosphate availability in soils are particularly relevant (Kucey et al. 1989; Richardson 2001). Two general types of processes have been described: those known to promote solubilization of nonavailable P-sources in soils and those known to improve plant uptake of the already solubilized phosphate. P-solubilization is carried out by a great number of saprophytic bacteria and fungi acting on sparingly soluble soil phosphates, mainly by chelation-mediated mechanisms (Kucey et al. 1989; Whitelaw 2000). Improvement of phosphate uptake by plants is typically carried out by mycorrhizal fungi (Smith and Read 1997). As indicated before, the external mycelium of mycorrhizal fungi acts as a bridge between roots and the surrounding soil microhabitats, which give access to the phosphate ions from soil solution beyond the phosphate-depletion zone surrounding the roots. Therefore, it is accepted that, by linking the biotic and geo-chemical portions of the soil ecosystem, the mycorrhizal fungi cannot only contribute to P capture and supply, but also affect P cycling rates and patterns in both agro- and natural ecosystems (Jeffries and Barea 2001).

The possibility that extracellular phosphatases from mycorrhizal hyphae could contribute to mineralization of organic P, an important nutrient pool in soil, was investigated. Using a compartmented growth system, Joner and Jakobsen (1995) applied ^{32}P-labeled organic matter to assess whether this P source might be used by mycorrhizal hyphae developing in the root-free compartment. They concluded that the mycorrhizal fungi did not mineralize organic P, however, they contribute to a closer P cycling and avoid P fixation upon mineralization by associate saprophytic microorganisms. A recent review (Joner et al. 2000) confirmed that mycorrhizal fungi do not play a significant role in P mineralization.

Because of the importance of microbial solubilization processes in soil ecosystems, inoculation of phosphate-solubilizing bacteria was assayed (Kucey et al. 1989; Barea et al. 2002c). For bacterial solubilization of phosphates in soil to be effective, several factors must operate. First of all, phosphate-solubilizing bacteria inoculants must be established in the root-associated soil habitats. This is why it has been recommended to select the inoculated phosphate-solubilizing bacteria from the subset of rhizobacteria populations (Glick 1995). In addition, the role of such inoculated bacteria for the P supply seems limited, because of the transient nature of the released compounds and their possible re-fixation on their way to the root surface (Kucey et al. 1989). However, it was proposed that if the solubilized phosphate ions were taken up by a mycorrhizal mycelium, this synergistic microbial (mycorrhizosphere) interaction should improve the P acquisition by the plant (Barea et al. 1983, 2002c). A great amount of information has been accumulated, and some key studies and concepts will be summarized here. In particular, the information will concentrate on studies which included the application of less expensive, but poorly reactive in nonacidic soils, rock phosphate as P source, and on the use of ^{32}P-based methodologies to ascertain the source of the P taken up by plants, as affected by mycorrhiza and bacteria treatments (Barea et al. 2002c). The fundamentals of ^{32}P-based methodologies to investigate P cycling in soil–plant systems will be briefly summarized before describing selected experiments.

Since radioactive P (^{32}P) has been used for evaluating the exchange rates governing phosphate equilibrium between the solution and the solid phases of soil (Fardeau 1993), ^{32}P-based techniques were recommended to measure P availability in rock phosphates (Zapata and Axmann 1995). Labeling of the so-called isotopically exchangeable soil P is carried out with phosphate ions labeled with ^{32}P, it being assumed that all 'labile' P, and only this fraction, may attain isotopic exchange within a short-term experimental period (Fardeau 1993). The isotopic composition, or "specific activity" (SA), i.e., the ^{32}P/^{31}P ratio, is then determined in the plant tissues. The SA in plants growing in ^{32}P-labeled soils is the basis for calculations to determine which of the P sources a plant is actually using (Zapata and Axmann 1995), therefore, it was assessed whether mycorrhizal and nonmycorrhizal plants use the same P sources (Raj et al. 1981). The general conclusion was that the SA in plants was not changed by mycorrhiza inoculation. However, experiments carried out in our laboratory (Toro et al. 1997, 1998; Barea et al. 2002c) further investigated whether coupled inoculation of phosphate-solubilizing bacteria and mycorrhizal fungi affect the SA of plants in rock phosphate-enriched and ^{32}P-labeled soils. These studies found that dual inoculation induced a lowering in the SA of the host plants indicating that these used extra ^{31}P solubilized from other not directly available P sources, either endogenous or added as rock phosphate.

A model experiment (Barea et al. 2002c) is summarized here to describe the phosphate-solubilizing bacteria x mycorrhiza interactions on P capture, cycling and supply. This experiment involved a factorial combination of four microbial and two chemical treatments. The microbial treatments were: (1) mycorrhiza inoculation; (2) phosphate-solubilizing rhizobacteria inoculation; (3) mycorrhiza plus bacteria dual inoculation; and (4) non-inoculated controls, exposed to the naturally existing mycorrhizal fungi and phosphate-solubilizing bacteria. The two chemical treatments were: (1) nonamended control without P application, and (2) rock phosphate application. Labeling was done by mixing the soil thoroughly with a solution containing ^{32}P phosphate ions. Plant seedlings were transplanted after soil labeling. The ^{32}P activity in the plant material was measured, and the specific activity of P was calculated by considering the radioactivity per amount of total P content in the plant (Zapata and Axmann 1995). Both rock phosphate addition and microbial inoculation improved biomass production and P accumulation in the test plants, with dual microbial inoculation as the most effective treatment. Independently of rock phosphate addition, mycorrhiza-inoculated plants showed a lower specific activity ($^{32}P/^{31}P$) than their comparable nonmycorrhizal controls. If the $^{32}P/^{31}P$ ratio in soil solution is uniform both spatially and temporally, similar SA in the plants should be assumed whether or not they are mycorrhiza-inoculated. However, in the reported experiments, the SA values were lower in mycorrhiza-inoculated plants than in the corresponding controls, particularly when they were inoculated with phosphate-solubilizing bacteria. This means that mycorrhiza-inoculated plants were taking soil P which is labeled differentially from that taken up by nonmycorrhiza-inoculated controls. The explanation could be that phosphate-solubilizing bacteria, either inoculated or naturally present, were effective in releasing ^{31}P from sparingly soluble sources, either from the soil components or from the added rock phosphate. This release of P ions would constitute a part of the total ^{31}P pool from which the mycorrhizal mycelium taps phosphate and transfers it to the plants. Such microbial activities could result in the lower SA in dually inoculated plants. It can, therefore, be concluded that the interactions between mycorrhizal fungi and phosphate-solubilizing bacteria display a fundamental role for P-cycling, a fact of considerable interest in ecosystems. These conclusions are graphically illustrated in Fig. 2.

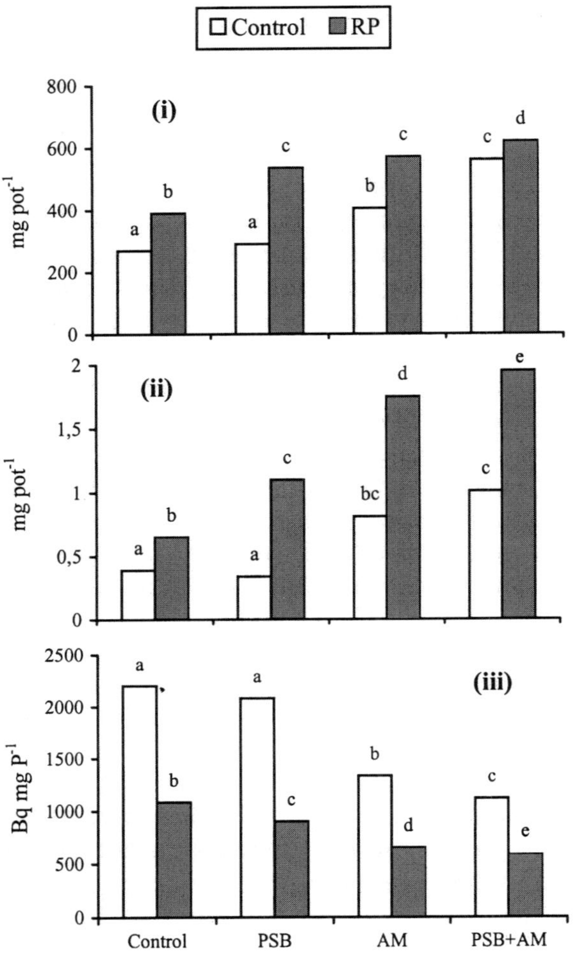

Fig. 2. Shoot dry weight (**i**); shoot P content (**ii**); and specific activity (**iii**) of alfalfa plants receiving several microbial inoculation treatments with and without rock phosphate (*RP*) application. For each response variable, means ($n = 5$), not sharing a letter in common differ significantly ($p = 0.05$) from each other according to Duncan's multirange test. (Barea et al. 2002c)

5 Interactions Between Mycorrhizal Fungi and Phytostimulators *Azospirillum* Bacteria

Bacteria belonging to the genus *Azospirillum* are known to promote plant development and yield under appropriate conditions (Okon 1994; Bashan 1999; Dobbelaere et al. 2001). *Azospirillum* species are free-living N_2-fixing rhizobacteria developing in close association with plant roots (Okon and Vanderleyden 1997). Early studies with *Azospirillum* spp. aimed to assess whether their effect on plant growth derives from the N_2-fixation ability. It was concluded that *Azospirillum* mainly acts by influencing the morphology, geometry and physiology of the root system rather than as N_2-fixing bacteria (see Dobbelaere et al. 2001). The changes in root architecture were based on the production of plant hormones by the bacteria (Dobbelaere et al. 1999). Because of these changes in rooting patterns, it was concluded that the effect of *Azospirillum* enhancing plant N acquisition was due to the improved root system through hormone production (Dobbelaere et al. 2001).

Positive interactions between mycorrhizal fungi and *Azospirillum* were to be expected (Azcón-Aguilar and Barea 1992). Diverse experiments were carried out, and the information reviewed (Volpin and Kapulnik 1994; Barea 1997) shows that *Azospirillum* enhances mycorrhizal formation and response and that, conversely, mycorrhizal fungi improve *Azospirillum* establishment. These results were corroborated by Vázquez et al. (2000) who further described an improvement in biomass and N accumulation in dually inoculated plants.

Because the increased N acquisition by *Azospirillum*-inoculated plants was attributed to the improvement of N uptake abilities by the root system, this effect is being investigated in interactions with mycorrhizal fungi by using ^{15}N methodologies (Collados and Barea, unpubl.). Both ammonium and nitrate ^{15}N-labeled salts were used as tracer, together with nitrification inhibitor. Preliminary results show that mycorrhizal inoculation improves N acquisition of *Azospirillum*-inoculated plants, and that the nitrate was the main N form used by such dually inoculated plants.

6
Interactions Improving Soil Structure Stabilization

Soil structure is defined as the arrangement of soil particles bound together in aggregates of different sizes and building pores (Tisdall 1996). Structure stability is the ability of the soil to maintain such an arrangement after being exposed to environmental stresses (Oades 1993). A well-aggregated soil structure is fundamental to ensure appropriate soil tilth, soil–plant water relations, water infiltration rates, soil aeration, root penetrability, organic matter accumulation, and to control erosion (Miller and Jastrow 2000). A critical aspect in soil structure stability is the formation of relatively water-stable aggregates. The contribution of mycorrhizal fungi to the formation and stabilization of soil aggregates has been demonstrated in several studies, as recently reviewed (Miller and Jastrow 2000). The main conclusions from this review are summarized below, before the report of a model experiment carried out in this laboratory (Requena et al. 2001).

According to Miller and Jastrow (2000), understanding the contributions of mycorrhizal fungi to the formation and stabilization of soil aggregates is necessary to realize the hierarchical nature of the mechanisms involved in aggregation. In the frame of this chapter, only the microbially mediated processes will be mentioned (see Miller and Jastrow 2000 for information on other factors). In a first stage, soil particles are bound together by bacterial products and by structures of saprophytic and mycorrhizal fungi into stable microaggregates (2–20 μm in diameter). These are bound by microbial products into larger microaggregates (20–250 μm in diameter), in which bacterial polysaccharides act as binding agents. Microaggregates are then bound into macroaggregates (> 250 μm in diameter), a process in which bacterial polysaccharides act as binding agents and the mycorrhizal mycelia contribute by increasing the size of macroaggregates. Such a mycorrhizal role is accounted for by the size, branching habits and three-dimensional structure of the external mycelium colonizing the soil surrounding the root (Miller and Jastrow 2000), an activity that can persists up to 22 weeks after the plant had died (Tisdall and Oades 1980).

The effect of the mycorrhizal mycelium in the formation of water-stable soil aggregates has been evidenced in different ecological situations (Andrade et al. 1995, 1998; Bethlenfalvay et al. 1999; Requena et al. 2001), and the involvement of glomalin, a glycoprotein produced by the external hyphae of mycorrhizal fungi, has been demonstrated (Wright and Upadhyaya 1998). Glomalin has been suggested to contribute to hydrophobicity of soil particles and, because of its glue-like hydrophobic nature, it also participates in the initiation and stabilization of soil aggregates (Miller and Jastrow 2000).

The importance of mycorrhizosphere interactions in improving soil structure was investigated in a revegetation experiment aimed at restoring a degraded Mediterranean ecosystem in southern Spain (Requena et al. 2001). This experiment used a shrub legume belonging to the natural succession and dual mycorrhizal and rhizobial inoculation, a biotechnology that has received considerable attention in the last decade (Herrera et al. 1993). The experiments carried out by Requena et al. (2001) aimed at assessing the long-term benefits of inoculation with these two types of plant microsymbionts not only on the establishment of the target legume species, but also on changes in key physicochemical soil properties known to affect soil quality (Kennedy and Smith 1995). In particular, the effect on soil structure, plant nutrient availability, organic matter content, microbial activity, etc., was analyzed as their degradation is often concomitant to disturbance of natural plant communities, as a result of the degradation/desertification processes (Jeffries and Barea 2001).

The experiments were carried out under field conditions in a representative area within a desertified semi-arid ecosystem in southeast Spain. The existing natural vegetation was a degraded shrubland, in which *Anthyllis cytisoides*, a drought-tolerant legume able to form symbioses with both rhizobial and mycorrhizal microsymbionts, was the dominant species (Requena et al. 1997). The experimental variables tested, as inoculants for *Anthyllis cytisoides* seedlings to be transplanted, involved three microsymbiont inoculation treatments, including: (1) an exotic mycorrhizal fungi from culture collection, (2) a mixture of five taxa of indigenous mycorrhizal fungi representing the natural abundance and diversity at the site, and (3) an indigenous rhizobial inoculum (Requena et al. 1997). A time-course (every 6 months) sampling was established over 5 years after transplanting. A long-term improvement in the physicochemical properties in the soil around *Anthyllis* plants inoculated with a mycorrhizal inoculum based on indigenous taxa was observed. The benefits included an increased content of both N and organic matter, and in the number of hydro-stable macroaggregates.

As described before, it can be assumed that the increase in N content in the rhizosphere of the legume can be accounted to an improvement in nodulation and N-fixation rates resulting from inoculation with mycorrhizal fungi (Barea et al. 1992), while increases in organic matter content and improvement in soil aggregation are due to the role of the mycorrhizal mycelium on soil structure stabilization (Miller and Jastrow 2000). Because rhizobial species are usually recognized for their ability to produce exo-polysaccharides (Spaink et al. 1998), it can be assumed that, given the

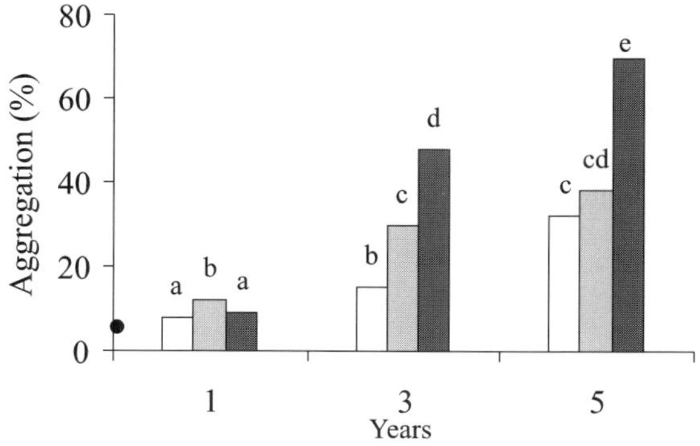

Fig. 3. Time-course changes in soil aggregation in the rhizosphere of field-established nodulated plants of *Anthyllis cytisoides* growing under natural conditions, either noninoculated with AMF (*white columns*), or inoculated with *G. intraradices* (*light gray columns*), or with native AMF (*dark gray columns*). For each response variable, means ($n = 5$), not sharing a letter in common differ significantly ($p = 0.05$) from each other according to Duncan's multirange test. Data points on the *y-axis* represent background values in the bare soil before transplanting. (Requena et al. 2001)

importance of these bacterial metabolites in soil aggregation (Miller and Jastrow 2000), rhizobia would be contributing to aggregate formation in interaction with the co-inoculated mycorrhizal fungi. The effects of mycorrhizal inoculation on aggregate formation are graphically recorded in Fig. 3.

Inoculation with native mycorrhizal fungi also benefited plant growth, N fixation and N transfer and improved the N status of nonleguminous plants grown in association with the legumes. These effects were previously described for agricultural crops (Azcón-Aguilar et al. 1979), but the experiments by Requena et al. (2001), are the first demonstration of this phenomenon for natural plant communities in a semi-arid ecosystem. An important role of the mycotrophic shrub legumes as a source of mycorrhizal inoculum for the surrounding area and in improving N nutrition for non-N-fixing vegetation was also observed. The results of this study support the idea that the introduction of target indigenous species of plants, associated with a managed community of microbial symbionts, is a successful biotechnological tool to aid the integral recovery of desertified ecosystems, i. e., improvements in both plant development and soil quality are an initial step in the restoration of a self-sustaining ecosystem.

7 Conclusions

In summary, it can be concluded that arbuscular mycorrhizal fungi and specific rhizosphere bacteria interact to improve plant nutrient (mainly N and P) cycling, as evidenced by using isotope dilution approaches. Such microbial interactions also improve physicochemical soil properties, particularly aggregate formation. These microbial activities have been demonstrated to contribute to plant fitness and soil quality, critical issues for sustainable agricultural development and ecosystem functioning.

References

Altieri MA (1994) Sustainable agriculture. Encycl Agric Sci 4:239–247

Ames RN, Reid CPP, Porter LK, Cambardella C (1983) Hyphal uptake and transport of nitrogen from two ^{15}N-labelled sources by *Glomus mosseae*, a vesicular-arbuscular mycorrhizal fungus. New Phytol 95:381–396

Andrade G, Azcón R, Bethlenfalvay GJ (1995) A rhizobacterium modifies plant and soil responses to the mycorrhizal fungus *Glomus mosseae*. Appl Soil Ecol 2:195–202

Andrade G, Mihara KL, Linderman RG, Bethlenfalvay GJ (1998) Soil aggregation status and rhizobacteria in the mycorrhizosphere. Plant Soil 202:89–96

Azcón-Aguilar C, Azcón R, Barea JM (1979) Endomycorrhizal fungi and *Rhizobium* as biological fertilizers for *Medicago sativa* in normal cultivation. Nature 279:325–327

Azcón-Aguilar C, Barea JM (1992) Interactions between mycorrhizal fungi and other rhizosphere microorganisms. In: Allen MJ (ed) Mycorrhizal functioning. An integrative plant-fungal process. Chapman and Hall, New York, pp 163–198

Barea JM (1997) Mycorrhiza/bacteria interactions on plant growth promotion. In: Ogoshi A, Kobayashi L, Homma Y, Kodama F, Kondon N, Akino S (eds) Plant growth-promoting rhizobacteria, present status and future prospects. OCDE, Paris, pp 150–158

Barea JM, Azcón-Aguilar C (1983) Mycorrhizas and their significance in nodulating nitrogen-fixing plants. In: Brady NC (ed) Advances in agronomy, vol 36. Academic Press, New York, pp 1–54

Barea JM, Azcón R, Azcón-Aguilar C (1983) Interactions between phosphate solubilizing bacteria and VA mycorrhiza to improve plant utilization of rock phosphate in non-acidic soils. 3rd international congress on phosphorus compounds, Brussels, 4–6 Oct, pp 127–144

Barea JM, Azcón-Aguilar C, Azcón R (1987) Vesicular-arbuscular mycorrhiza improve both symbiotic N_2-fixation and N uptake from soil as assessed with a ^{15}N technique under field conditions. New Phytol 106:717–721

Barea JM, Azcón R, Azcón-Aguilar C (1989) Time-course of N_2-fixation (^{15}N) in the field by clover growing alone or in mixture with ryegrass to improve pasture productivity, and inoculated with arbuscular mycorrhizal fungi. New Phytol 112:399–404

Barea JM, Azcón-Aguilar C, Azcón R (1991) The role of VA mycorrhizae in improving plant N acquisition from soil as assessed with ^{15}N. In: IAEA, Naciones Unidas (eds) Stable isotopes in plant nutrition. IAEA, Vienna, pp 209–216

Barea JM, Azcón R, Azcón-Aguilar C (1992) Vesicular-arbuscular mycorrhizal fungi in nitrogen-fixing systems. In: Norris JR, Read D, Varma A (eds) Methods in microbiology, vol 24. Techniques for the study of mycorrhizae. Academic Press, London, pp 391–416

Barea JM, Azcón-Aguilar C, Azcón R (1997) Interactions between mycorrhizal fungi and rhizosphere micro-organisms within the context of sustainable soil–plant systems. In: Gange AC, Brown VK (eds) Multitrophic interactions in terrestrial systems. Backwell, Cambridge, pp 65–77

Barea JM, Azcón R, Azcón-Aguilar C (2002a) Mycorrhizosphere interactions to improve plant fitness and soil quality. Anton Leeuw 81:343–351

Barea JM, Gryndler M, Lemanceau Ph, Schüepp H, Azcón R (2002b) The rhizosphere of mycorrhizal plants. In: Gianinazzi S, Schüepp H, Barea JM, Haselwandter K (eds) Mycorrhiza technology in agriculture from genes to bioproducts. Birkhäuser, Basel, pp 1–18

Barea JM, Toro M, Orozco MO, Campos E, Azcón R (2002c) The application of isotopic ^{32}P and ^{15}N-dilution techniques to evaluate the interactive effect of phosphate-solubilizing rhizobacteria, mycorrhizal fungi and *Rhizobium* to improve the agronomic efficiency of rock phosphate for legume crops. Nutr Cycl Agroecosyst 63:35–42

Bashan Y (1999) Interactions of *Azospirillum* spp. in soils: a review. Biol Fertil Soils 29:246–256

Bethlenfalvay GJ, Cantrell IC, Mihara KL, Schreiner RP (1999) Relationships between soil aggregation and mycorrhizae as influenced by soil biota and nitrogen nutrition. Biol Fertil Soils 28:356–363

Bowen GD, Rovira AD (1999) The rhizosphere and its management to improve plant growth. Adv Agron 66:1–102

Danso SKA (1986) Review, estimation of N_2 fixation by isotope dilution: an appraisal of techniques involving ^{15}N enrichment and their application. Soil Biol Biochem 18:243–244

Danso SKA (1988) The use of ^{15}N enriched fertilizers for estimating nitrogen fixation in grain and pasture legumes. In: Beck DP, Materon L (eds) Nitrogen fixa-

tion by legumes in Mediterranean agriculture. ICARDA, Martinus Nijhoff, Dordrecht, pp 345–358
Dobbelaere S, Croonenborghs A, Thys A, van de Broek A, Vanderleyden J (1999) Phytostimulatory effect of *Azospirillum brasilense* wild type and mutant strains altered in IAA production on wheat. Plant Soil 212:155–164
Dobbelaere S, Croonenborghs A, Thys A, Ptacek D, Vanderleyden J, Dutto P, Labandera-Gonzalez C, Caballero-Mellado J, Aguirre JF, Kapulnik Y, Brener S, Burdman S, Kadouri D, Sarig S, Okon Y (2001) Response of agronomically important crops to inoculation with *Azospirillum*. Aust J Plant Physiol 28:1–9
Fardeau JC (1993) Le phosphore assimilable des sols: sa représentation par un modèle fonctionnel á plusieurs compartiments. Agronomie 13:317–331
Glick BR (1995) The enhancement of plant growth by free-living bacteria. Can J Microbiol 41:109–117
Herrera MA, Salamanca CP, Barea JM (1993) Inoculation of woody legumes with selected arbuscular mycorrhizal fungi and rhizobia to recover desertified Mediterranean ecosystems Appl Environ Microbiol 59:129–133
Jeffries P, Barea JM (2001) Arbuscular mycorrhiza – a key component of sustainable plant–soil ecosystems. In: Hock B (ed) The Mycota, vol IX. Fungal associations. Springer, Berlin Heidelberg New York, pp 95–113
Jeffries P, Gianinazzi S, Perotto S, Turnau K, Barea JM (2003) The contribution of arbuscular mycorrhizal fungi in sustainable maintenance of plant health and soil fertility. Biol Fertil Soil 37:1–16
Joner EJ, Jakobsen I (1995) Growth and extracellular phosphatase activity of arbuscular mycorrhizal hyphae as influenced by soil organic matter. Soil Biol Biochem 27:1153–1159
Joner EJ, van Aarle IM, Votsatka M (2000) Phosphatase activity of extra-radical arbuscular mycorrhizal hyphae: a review. Plant Soil 226:199–210
Kennedy AC (1998) The rhizosphere and spermosphere In: Sylvia DM, Fuhrmann JJ, Hartel PG, Zuberer DA (eds) Principles and applications of soil microbiology. Prentice Hall, Upper Saddle River, New Jersey, pp 389–407
Kennedy AC, Smith KL (1995) Soil microbial diversity and the sustainability of agricultural soils. Plant Soil 170:75–86
Kloepper JW (1994) Plant growth-promoting rhizobacteria (other systems) In Okon Y (ed) *Azospirillum*/Plant Associations. CRC Press, Boca Raton, pp 111–118
Kucey RMN, Janzen HH, Leggett ME (1989) Microbiologically mediated increases in plant-available phosphorus. In: Brady NC (ed) Advances in agronomy, vol 42. Academic Press, New York, pp 199–228
Lynch JM (1990) The rhizosphere. Wiley, New York
Miller RM, Jastrow JD (2000) Mycorrhizal fungi influence soil structure. In: Kapulnik Y, Douds Jr DD (eds) Arbuscular mycorrhizas: physiology and function. Kluwer, Dordrecht, pp 3–18
Morton JB, Redecker D (2001) Two new families of *Glomales*, *Archaeosporaceae* and *Paraglomaceae*, with two new genera *Archaeospora* and *Paraglomus*, based on concordant molecular and morphological characters. Mycologia 93:181–195
Oades JM (1993) The role of biology in the formation, stabilization and degradation of soil structure. Geoderma 56:277–400
Okon Y (1994) *Azospirillum*/plant associations. CRC Press, Boca Raton
Okon Y, Vanderleyden J (1997) Root-associated *Azospirillum* species can stimulate plants. ASM News 63:364–370
Parr JF, Papendick RI, Hornick SB, Meyer RE (1992) Soil quality: attributes and relationship to alternative and sustainable agriculture. Am J Alternative Agric 7:2–3
Postgate JR (1998) Nitrogen fixation. Cambridge Univ Press, Cambridge

Probanza A, Lucas García JA, Ruiz Palomino M, Ramos B, Gutiérrez Mañero FJ (2002) *Pinus pinea* L. seedling growth and bacterial rhizosphere structure after inoculation with PGPR *Bacillus* (*B. licheniformis* CECT 5106 and *B. pumillus* CECT 5105). Appl Soil Ecol 20:75–84

Raj J, Bagyaraj DJ, Manjunath A (1981) Influence of soil inoculation with vesicular-arbuscular mycorrhiza and a phosphate-dissolving bacterium on plant growth and ^{32}P-uptake. Soil Biol Biochem 13:105–108

Requena N, Jimenez I, Toro M, Barea JM (1997) Interactions between plant-growth-promoting rhizobacteria (PGPR), arbuscular mycorrhizal fungi and *Rhizobium* spp. in the rhizosphere of *Anthyllis cytisoides*, a model legume for revegetation in Mediterranean semi-arid ecosystems. New Phytol 136:667–677

Requena N, Pérez-Solís E, Azcón-Aguilar C, Jeffries P, Barea JM (2001) Management of indigenous plant–microbe symbioses aids restoration of desertified ecosystems. Appl Environ Microbiol 67:495–498

Richardson AE (2001) Prospects for using soil microorganisms to improve the acquisition of phosphorus by plants. Aust J Plant Physiol 28:897–906

Schüßler A, Schwarzott D, Walker C (2001) A new fungal phylum, the Glomeromycota, phylogeny and evolution. Mycol Res 105:1413–1421

Smith SE, Read DJ (1997) Mycorrhizal symbiosis. Academic Press, San Diego

Spaink HP, Kondorosi A, Hooykaas PJJ (1998) The rhizobiaceae. Kluwer, Dordrecht

Tisdall JM (1996) Formation of soil aggregates and accumulation of soil organic matter. In: Carter MR, Stewart BA (eds) Structure and organic matter storage in agricultural soils. CRC Press, Boca Raton, pp 57–96

Tisdall JM, Oades JM (1980) The management of ryegrass to stabilize aggregates of red-brown earth. Aust J Soil Res 18:415–422

Tobar RM, Azcón R, Barea JM (1994a) Improved nitrogen uptake and transport from ^{15}N-labelled nitrate by external hyphae of arbuscular mycorrhiza under water-stressed conditions. New Phytol 126:119–122

Tobar RM, Azcón R, Barea JM (1994b) The improvement of plant N acquisition from an ammonium-treated, drought-stressed soil by the fungal symbiont in arbuscular mycorrhizae. Mycorrhiza 4:105–108

Toro M, Azcón R, Barea JM (1997) Improvement of arbuscular mycorrhizal development by inoculation with phosphate-solubilizing rhizobacteria to improve rock phosphate bioavailability (^{32}P) and nutrient cycling. Appl Environ Microbiol 63:4408–4412

Toro M, Azcón R, Barea JM (1998) The use of isotopic dilution techniques to evaluate the interactive effects of *Rhizobium* genotype, mycorrhizal fungi, phosphate-solubilizing rhizobacteria and rock phosphate on nitrogen and phosphorus acquisition by *Medicago sativa*. New Phytol 138:265–273

Vance CP (2001) Symbiotic nitrogen fixation and phosphorus acquisition. Plant nutrition in a world of declining renewable resources. Plant Physiol 127:390–397

Vázquez MM, César S, Azcón R, Barea JM (2000) Interaction between arbuscular mycorrhizal fungi and other microbial inoculants (*Azospirillum*, *Pseudomonas*, *Trichoderma*) and their effects on microbial population. Appl Soil Ecol 15:261–272

Volpin H, Kapulnik Y (1994) Interaction of *Azospirillum* with beneficial soil microorganisms. In: Okon Y (ed) *Azospirillum*/plant associations. CRC Press, Boca Raton, pp 111–118

Werner D (1998) Organic signals between plants and microorganisms. In: Pinton R, Varanini Z, Nannipieri P (eds) The rhizosphere: biochemistry and organic substances at the soil–plant interfaces. Dekker, New York

Whitelaw MA (2000) Growth promotion of plants inoculated with phosphate-solubilizing fungi. Adv Agron 69:99–151

Wright SF, Upadhyaya A (1998) A survey of soils for aggregate stability and glomalin, a glycoprotein produced by hyphae of arbuscular mycorrhizal fungi. Plant Soil 198:97–107

Zapata F (1990) Isotope techniques in soil fertility and plant nutrition studies. In: Hardarson G (ed) Use of nuclear techniques in studies of soil–plant relationships. IAEA, Vienna, pp 61–128

Zapata F, Axmann H (1995) ^{32}P isotopic techniques for evaluating the agronomic effectiveness of rock phosphate materials. Fertil Res 41:189–195

11 Mycorrhizosphere: Strategies and Functions

Bhoopander Giri[1], Pham Huong Giang[2], Rina Kumari[3], Ram Prasad[3], Minu Sachdev[3], Amar P. Garg[3], Ralf Oelmüller[4], Ajit Varma[1]

1 Introduction

Hiltner recognized the rhizosphere as the volume of soil in the immediate vicinity of the roots, which is predominantly affected by the activity of plants. The rhizosphere differs from the surrounding soil in most of the physico-chemical factors and a wide range of microorganisms colonizes this rhizosphere soil along with the rhizoplane (i. e., the root surface; Phillips et al. 2003). The number of these microorganisms per gram of soil is much larger in the rhizosphere compared to bulk soil. This increased microbial activity in the vicinity of roots can be ascribed to root exudates, sloughed senescent root cells and mucigel, which have been described as rhizodeposition (Mukerji et al. 1997; Bansal et al. 2000).

In nature, most of the actively absorbing rootlets form a symbiotic association with mycorrhizal fungi, which are ubiquitous soil inhabitants. The formation of symbiotic associations with mycorrhizae significantly changes the physiology and/or morphology of roots and plants in general, leading to altered root exudation (Bansal and Mukerji 1994). The changes in root exudates affect the microbial communities around the roots, leading to the formation of the "mycorrhizosphere" (Mukerji et al. 1997; Varma et al. 1999). The mycorrhizosphere is the zone of soil where the physical, chemical and microbiological processes are influenced by plant roots and their associated mycorrhizal fungi. A major difference in the rhizosphere around the nonmycorrhizal roots and mycorrhizosphere effect is the presence of extramatrical hyphae of mycorrhizal fungi. These extramatrical hyphae extend well beyond the roots into the bulk soil and are an impor-

[1]School of Life Science, Jawaharlal Nehru University, New Delhi 110067, India, e-mail: ajitvarma73@hotmail.com, Tel: +91-26704511, Fax: +91-26187338/26198234, and Amity Institute of Herbal and Microbial Studies, Sector 125, New Super Express Highway, Noida, India, Tel: 95120-2432400, Fax: 95120-2432200, e-mail: ajitvarma@aihmr.amity.edu
[2]International Centre for Genetic Engineering and Biotechnology (UNO, Triesta, Italy) New Delhi, India
[3]Ch. Charan Singh University, Meerut, Uttar Pradesh, India
[4] Institutes for General Botany and Plant Physiology, University of Jena, Dornburger Str. 159, 07743 Jena, Germany

Soil Biology, Volume 3
Microorganisms in Soils: Roles in Genesis and Functions
(ed. by F. Buscot and A. Varma)

tant source of carbon to the soil organisms (Schreiner and Bethlenfalvay 1995). The mycorrhizal hyphae increase the soil aggregation and in root association increase exudation, which favors the microbial growth (Schreiner and Bethlenfalvay 1995; Bansal and Mukerji 1996).

The mycorrhizosphere microbiota differs qualitatively as well as quantitatively from the rhizosphere of nonmycorrhizal plants. The soil microfauna influences the mycorrhiza formation as well as the host growth response (Fitter and Garbaye 1994). Many kinds of interactions occur between these microbial communities in the mycorrhizosphere and mycorrhizae. The interactions between the mycorrhizae and soil microorganisms may be mutualistic or competitive and they affect the establishment and functions of mycorrhizal symbionts as well as modify the interactions of the plant with other symbionts or pathogens in soil.

2 The Rhizosphere

The rhizosphere is the region in which materials released from the root, and root metabolic activities such as respiration, affect microbes (Table 1). Roots in the process of rhizodeposition release volatile, soluble, and particulate materials. The rhizosphere microbes, after their growth on these materials

Table 1. Various spheres and materials released in the soil

Terms	Definition
Rhizosphere	Region around the plant root where materials released from the root modify microbial populations and their activities
Endorhizosphere	Regions of the various cell layers of the root itself where microorganisms also colonize
Ectorhizosphere	An area surrounding the root and containing root hairs, plant and bacterial mucilage
Rhizoplane	Root surface that can be colonized by microorganisms
Mycorrhizosphere	The ectorhizosphere extends a substantial distance from the root with the development of mycorrhizal fungal associations. Materials released from the fungus increase the microbial populations and their activities around the fungal hyphae
Spermosphere	The region around the germinating seed
Rhizodeposition	Release of materials from roots
Exudates	Compounds of low molecular weight produced by plant cells and released into the root environment
Mucilages	Gelatinous organic materials released by the plant in the root cap region derived from the Golgi apparatus, polysaccharides hydrolysis, and epidermal materials

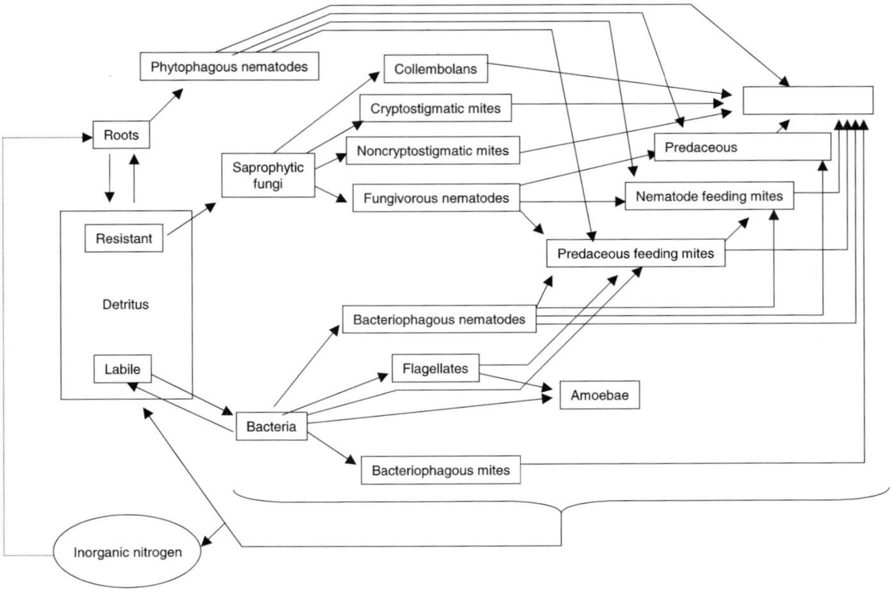

Fig. 1. Components of the rhizosphere food web. (Modified after Moore et al. 2003)

and their cellular turnover, release nutrients in forms which can be utilized by plants. Plants and their rhizospheres are found in soils in which the environment is primarily aerobic, and in many marine and freshwater environments in which oxygen is often limited. The rhizosphere encompasses not only the region of nutrient uptake by the roots, but also extends into the soil by the action of root products and the trophic interactions that are affected by these products (van der Putten et al. 2001). A growing root can reach the regions from the root tip to the crown, where different populations of soil biota have access to a continuous flow of organic substrates derived from the root. This infusion of organic substrates into the rhizosphere by plants explains why the biomass and activity of microbes and soil fauna are greater in the rhizosphere than the bulk soil (Parmelle et al. 1993; Bardgett et al. 1998). The root tip is the site of root growth and is characterized by rapidly dividing cells and secretions or exudates that lubricate the tip as it passes through the soil. The exudates and sloughed root cells provide carbon for bacteria and fungi, which in turn immobilize nitrogen and phosphorus. Further up the root is the region of nutrient exchange, characterized by root hairs and lower rates of exudation which stimulate additional microbial growth (Bringhurst et al. 2001).

The food web that develops within the rhizosphere is complex (Fig. 1), consisting of multiple assemblages of species that are supported by roots and their by-products. These assemblages are dubbed as the root, bacterial, and fungal energy channels. Live roots form the basis of the root energy

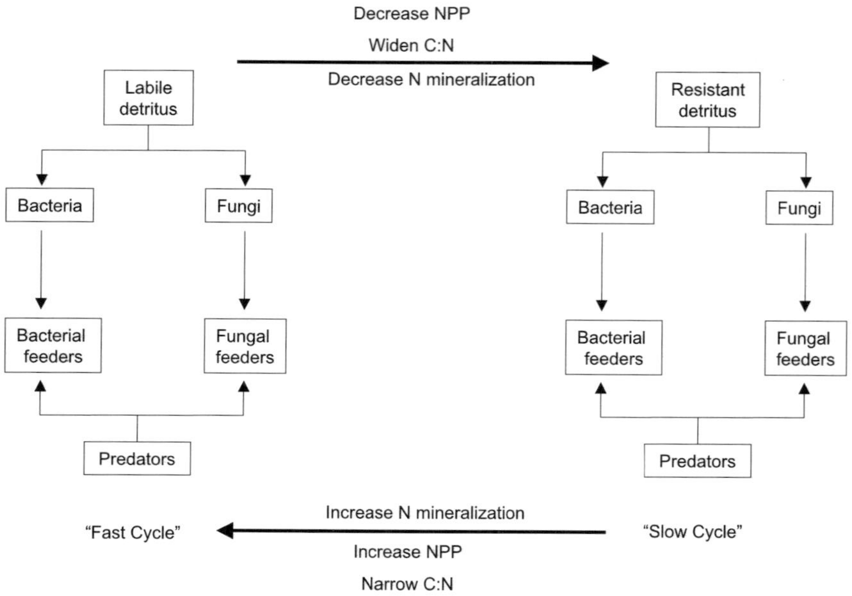

Fig. 2. Events in the evolution of the rhizosphere. (Modified after Phillips et al. 2003)

channels. The root energy channel consists of root-feeding insects and nematodes, and microbes that engage in symbiotic relationships with plant roots (mycorrhizal fungi, rhizobia, *Frankia*). Detritus forms the basis of the bacterial and fungal channels. The bacterial energy channel is composed of bacteria, protozoa, rotifers, nematodes, and a few arthropods. The fungal energy channel largely consists of saprophytic fungi, arthropods, and nematodes.

Soil saprophytic bacteria that compose most of the microbial biomass in the rhizosphere are aquatic organisms and are more efficient in using the more labile root exudates than saprophytic fungi. In contrast, fungi are better adapted to utilize more resistant root cells and substrates than are bacteria (Lynch 1990). The bacterial energy channel represents a "fast cycle", while the fungal energy channel represents a "slow cycle" (Fig. 2).

A common suite of nematode and arthropod predators links the root, bacterial, and fungal energy channels. The linkages between the energy channels tend to be weak at the trophic levels occupied by roots, bacteria and fungi, and strongest at the trophic levels occupied by predatory mites (Moore et al. 2003). The strength of the linkages between energy channels and the dominance of a given energy channel vary with the type of ecosystem, changes with disturbance, and affects nutrient turnover rates (Fig. 2). The fungal energy channel tends to be more dominant in systems where the ratio of carbon to nitrogen (C:N) is high while the bacterial channel is more dominant in systems with narrow C:N ratios (Moore et al. 2003).

3
Evolution of the Rhizosphere

Plants surely encountered microorganisms in primordial soil as they moved from aquatic to terrestrial environments (Fig. 3). Geochemical evidence for microorganisms exists from 2600 million years ago (m.y.a.) and bacterial fossils dating back 1200 m.y.a. are known (Horodyski and Knauth 1994; Watanabe et al. 2000). Although true roots with vascular tissue appeared

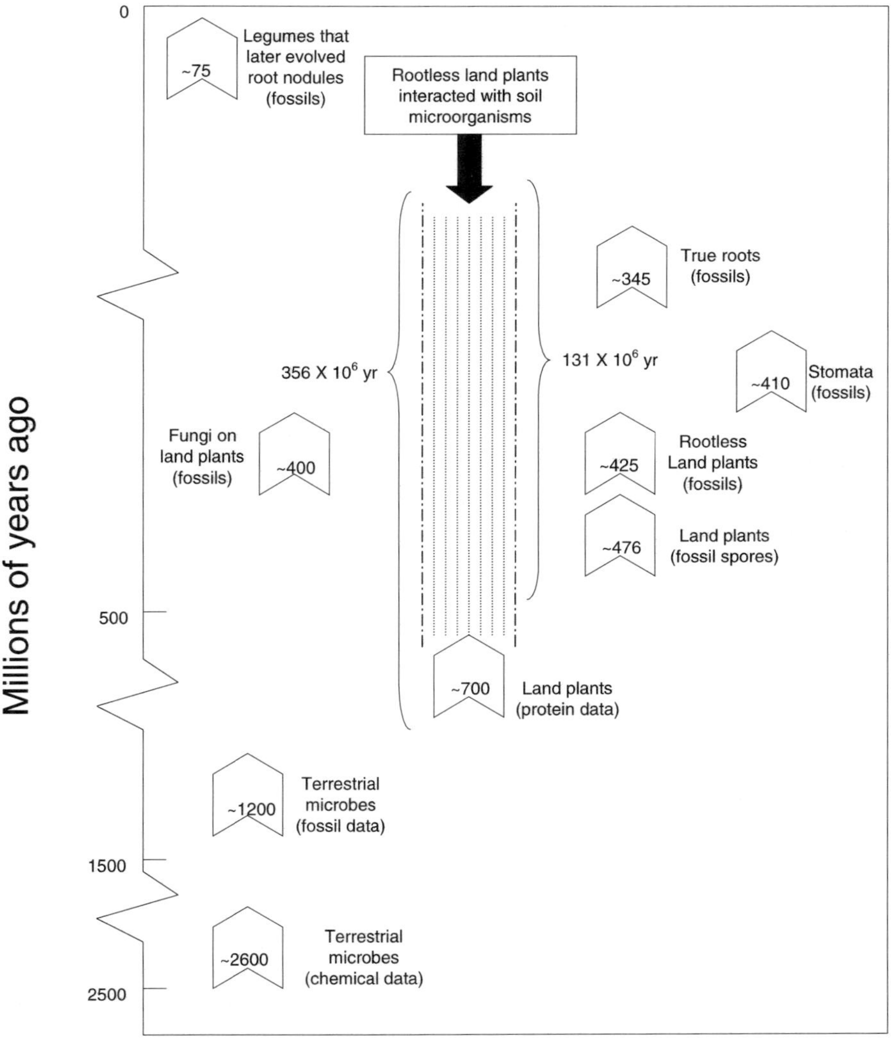

Fig. 3. Bacterial energy channels representing fast slow cycles. (Modified after Moore et al. 2003)

perhaps 345 m.y.a. (Stewart and Rothwell 1993), early terrestrial plants had a variety of underground structures, including stems and rhizoidal appendages (Raven and Edwards 2001), which were beset early on by bacteria and from at least 400 m.y.a. by fungi (Taylor et al. 1995). Stomata were present approximately 410 m.y.a. (Edwards et al. 1998), and thus water movement through the evolving soil food web towards early terrestrial plant tissues probably predated roots.

Phillips et al. (2003) pointed out that all plants in the early terrestrial environment interacted with microorganisms. Those relationships in primordial soil predated vascular roots by 131–355 million years, depending on whether one documents the beginning of interactions by plant microfossils (Kenrick and Crane 1997), or by estimates based on protein data (Heckman et al. 2001). It is often thought that the complex rhizobial symbiosis with legumes evolved a mere 75 million years after the Caesalpiniodeae group of legumes appeared; either estimate offers sufficient time for simpler mutualisms to develop (Phillips et al. 2003).

One cannot assess the extent to which primitive plants resisted microbial attacks, but the presence of their reasonably intact, fossilized remains shows that some protective mechanisms existed. Thus, it is reasonable to suggest that populations of epiphytic and endophytic microorganisms were an accepted fact of life for early land plants. The chemical residues of those microbial populations, as well as any signals released among the microorganisms must have been in close contact with early land plants. Under such conditions, a sifting of water-soluble microbial products for potentially important data on the water and mineral content of nearby environments probably occurred (Phillips et al. 2003).

4
Anatomy of the Root Through the Eyes of a Microbiologist

Vascular plants are widely distributed over the world. They are one of the most important links which humans have to nature. The vast majority of our food and fiber are directly derived from plants. Although it often is not evident, plant roots and their surrounding microbes (the rhizosphere) are important wherever plants are found: forests, grasslands, tundra, deserts, and wet areas such as marshes and mangrove swamps. The root of these plants is divided into three zones: (1) zone of cell division or meristematic activity; (2) zone of cell elongation and (3) zone of cell maturation. Roots grow by the activity of apical meristems, which also form a root cap distally. The root cap is a dynamic, specialized organ that facilitates root penetration of soil, senses threats as well as bounty, and responds by transmitting signals that alter growth patterns. The root supports a unique modified

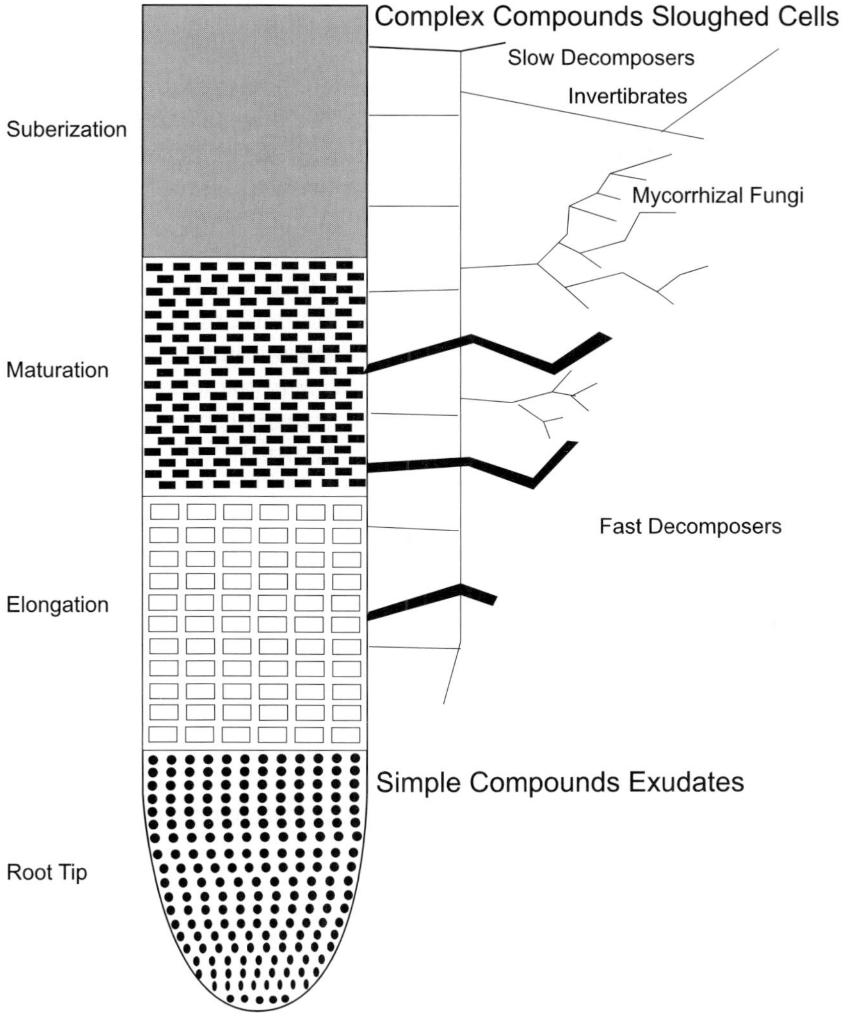

Fig. 4. Structure of mycorrhizal fungi in relationship to the structure of the root and associated root organisms

microbial community in an environment termed the rhizosphere, the region influenced by the root and its activities. Microbes directly colonize root surfaces and are also found under them, creating additional unique environments for microbes (Fig. 4).

The term rhizosphere, which has been used for 100 years, is critical to understanding how plants interact with their environment. In essence, the microbes in the rhizosphere provide the critical link between plants, which require inorganic nutrients, and the environment, which contains the nutrients, but often in organic and largely inaccessible forms.

Microbes also colonize the plant root surface (the rhizoplane). The colonization of the rhizoplane by microbes can involve specific attachment mechanisms. For *Agrobacterium thaliana*, which forms tumors in susceptible plants, this involves a two-step process of (1) loose binding to the cell surface and (2) the synthesis of cellulose fibrils by the bacterium. This results in binding of the bacteria to the plant root surface. If mutant bacteria are used which do not have these attachment characteristics, they will not bind to the root surface.

Plants also have other microbes with which they develop unique relationships in the root environment, including the symbiotic nitrogen-fixing bacteria such as *Rhizobium*. These bacteria form nodules on susceptible legumes, and the fixation of nitrogen by filamentous bacteria of the genus *Frankia*, an association that occurs with a wide range of shrubs and woody plants. Another important group of microbes, which form direct associations with plants, includes the mycorrhizae or "fungus roots", which occur in a wide variety of plants, considered to be one of the oldest plant-microbe associations. The nitrogen-fixing bacteria and the mycorrhizae form structures within the plant root, indicating these physiologically active relationships. The mycorrhizal hyphal network, supported by carbon derived from the plant, also releases organic carbon. Microbes grow around the mycorrhizal hyphae.

5
Production of Chemical Compounds in the Rhizosphere by Plant Roots

A general thought is that aerial parts (stem and leaves) contain greater biomass than root. This impression is misleading. For many plants the root:shoot ratio is such that more of the plant mass is in the roots than in stems and leaves. The materials released by the plants include a wide variety of organic compounds (Table 2). The types of these substances are constantly changing due to a wide range of plant and environment-related factors. These factors can include temperature and moisture stress, fertilizer additions, herbage removal (both above- and below-ground) changes in sunlight, herbicide additions, plant age, and other changes in the plant's environment. The materials lost from plant roots can be 30–40% of the carbon fixed through photosynthesis.

The fine hairs are a critical part of the root system (Fig. 5). These can be rapidly shed when environmental conditions become less suitable for plant growth. Cortical and epidermal cells, called mucilages, and soluble metabolic products (amino acids, sugars, organic acids, etc.), described as exudates, are also released. In addition, a variety of gaseous metabolites flow

Table 2. Compounds released by plant roots in the process of rhizodeposition

Compound	Exudate components
Sugars	Glucose, fructose, sucrose, maltose, galactose, rhamnose, ribose, xylose, arabinose, raffinose, oligosaccharide
Amino compounds	Asparagine, α-alanine, glutamine, aspartic acid, leucine/isoleucine, serine, γ-aminobutyric acid, glycine, cystine/cysteine, methionine, phenylalanine, tyrosine, threonine, lysine, proline, tryptophane, β-alanine, arginine, homoserine, cystathionine
Organic acids	Tartaric, oxalic, citric, malic, proponic, butyric, succinic, fumaric, glycolic, valeric, malonic
Fatty acids and sterols	Palmitic, stearic, oleic, linoleic, linolenic acids, cholestrol, campesterol, stigmasterol, sitosterol
Growth factors	Biotin, thiamine, niacin, pantothenate, choline, inositol, pyridoxine, ρ-aminobenzoic acid, N-methyl nicotinic acid
Nucleotides, flavonines and enzymes	Flavonine, adenine, guanine, uridine/cytidine, phosphatase, invertase, amylase, protease, polygalacturonase
Miscellaneous compounds	Auxins, scopoletin, fluorescent substances, hydrocyanic acid, glycosides, saponin (glucosides), organic phosphorus compounds, nematode-cyst or egg-hatching factors, nematode attractants, fungal mycelium growth stimulants and inhibitors, zoospore attractants

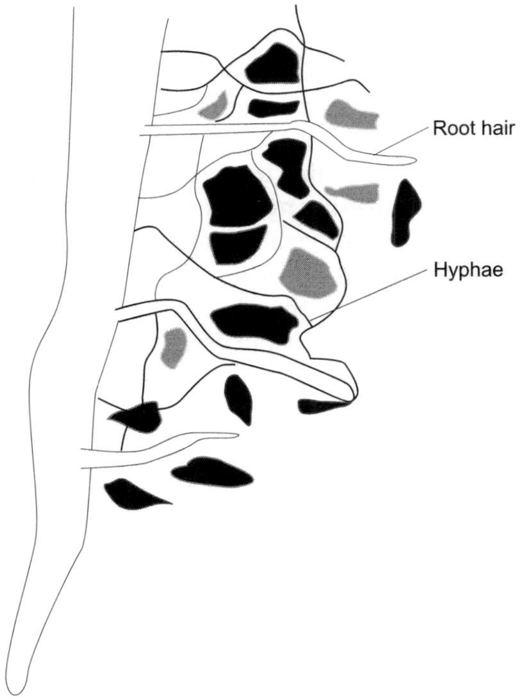

Fig. 5. Root hairs that assist the plant in exploring resources present in soils

from the roots. The release of these different materials is described as the process of rhizodeposition. When the mucilages combine with microbes, soil colloids, and soil organic matter, mucigels are formed which cover and protect the root tip.

6 Microbial Diversity in the Rhizosphere

The rhizosphere is a "cloud" of microbes which literally surrounds plant roots and is vital for the plant's survival and growth. Plant roots create new environments for microbes due to the increased levels of nutrients; microbial populations increase, often by 1000–10,000-fold, and marked changes in the composition of the microbial community will also occur (Table 3), as indicated by the rhizosphere:soil (R:S) ratio for a soil. The number and types of microbes often increase along the root away from the tip of the plant root. The plant roots also respire (use oxygen), which changes the environment of the rhizosphere microbes.

The microbial community, which develops in this changed rhizospheric environment will face additional challenges; many of the materials released from roots do not contain sufficient nitrogen, and sometimes phosphorus, to allow rapid microbial growth. This situation limits both the plant and the associated rhizosphere microbes.

The plant has an increasing demand for inorganic nutrients, which are often not available at a sufficient rate. The rhizosphere contains a wide variety of free-living and symbiotic nitrogen-fixing bacteria (Table 3), which make a major contribution to meet this demand, but at a high energetic cost for the plant. The filamentous fungi, including the free-living and mycorrhizal types, also play a unique role in making nutrients available to the plant which cannot be provided by most bacteria. The filamentous fungi in the rhizosphere have an extensive hyphal network. With this hyphal network, they can utilize carbon derived from the plant while obtaining their nitrogen and other limiting resources from outside the immediate root zone.

Table 3. Microbial diversity of major groups in the rhizospheric and nonrhizospheric soils

Organism	Rhizosphere soil (microbes/g dry soil)	Nonrhizosphere soil (microbes/g dry soil)	R:S ratio
Bacteria	1200×10^6	53×10^6	23
Actinomycetes	46×10^6	7×10^6	7
Fungi	12×10^5	1×10^5	12
Algae	5×10^3	27×10^3	0.2

Free-living nitrogen-fixing bacteria, including the genera *Azotobacter*, *Azospirillum*, and *Azoarcus* are abundant in the rhizosphere. In the presence of nitrogen-free or lower nitrogen-content substrates released from the root, these bacteria play an important role. They carry out associative nitrogen fixation and thus provide nutrient for the plant. The rhizosphere community not only has bacteria and fungi but also contains protozoans and nematodes. These consumers feed on the nutrient-rich bacteria and fungi, leading to more rapid turnover of the microbes, which leads to an accelerated release of nutrients for plant use.

7
What Are Mycorrhizal Fungi?

Mycorrhizae provide an intimate link between the soil environment and the functional nutrient-absorbing system of the plant. The modification of plant roots by symbiotic fungi into the distinct structures characteristic of mycorrhizae results in a unique and intriguing component of the rhizosphere (Fig. 6). Since the first published description of a mycorrhizal

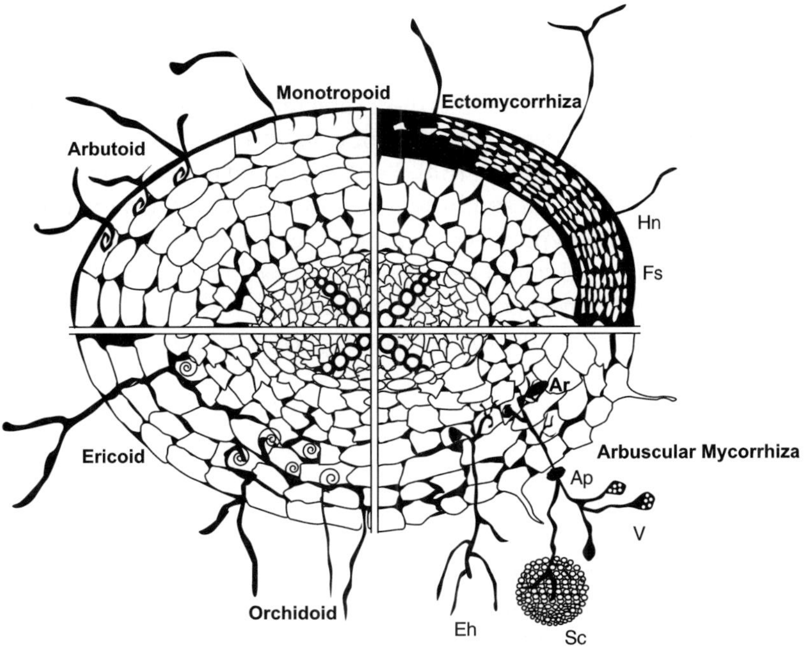

Fig. 6. Various kinds of mycorrhizal colonization in the root of the vascular plant (*Fs* fungal sheath, *Eh* extramatrical hyphae, *Hn* Hartig's , *V* vesicle, *Ar* arbuscule, *Sc* sporocarp, *Ap* appresorium)

association by Frank, scientists continue to be challenged by the role of mycorrhizae in the ecological and physiological context of plants. Although most research and observations on mycorrhizae have been concerned with nutrient uptake by mycorrhizae, especially immobile elements such as phosphorous, there is an increasing awareness of their potential importance in many diverse aspects of a plant's ability to grow and survive in natural and man-altered environments.

8 Types of Mycorrhizal Fungi

Over the years, seven types of mycorrhizae have come into general use on the basis of morphology and anatomy, but also of either host plant taxonomy or fungal taxonomy (Srivastava et al. 1996; Smith and Read 1997). These are: ectomycorrhiza, endomycorrhiza or arbuscular mycorrhiza, ericoid mycorrhiza, arbutoid mycorrhiza, monotropoid mycorrhiza, ect-endomycorrhiza and orchidaceous mycorrhiza.

8.1 Ectomycorrhiza

The ectomycorrhizae (ECM) are sometimes referred to as "sheathing" mycorrhizae because of the distinct presence of a sheath or mantle of fungal mycelium that covers the absorbing root. ECM are found almost exclusively on woody perennials. The plant symbionts include both Gymnosperms and Angiosperms. There is no hyphal penetration of cells. Fungal hypha is generally separate. A distinct Hartig's net is present between the cells. Hartig's net is a plexus of fungal hyphae between epidermal and cortical cells. It provides a large surface area for the interchange of nutrients between the host and the fungi.

8.2 Arbuscular Mycorrhiza

The term refers to the presence of intracellular structures – vesicles and arbuscules – that form in the root during various phases of development (Fig. 7). These mycorrhizae are the most commonly occurring group since they occur on a vast taxonomic range of plants, both herbaceous and woody. The plant symbionts range from Bryophytes to Angiosperms. There is no fungal sheath. Aseptate hyphae enter the root cortical cells and form

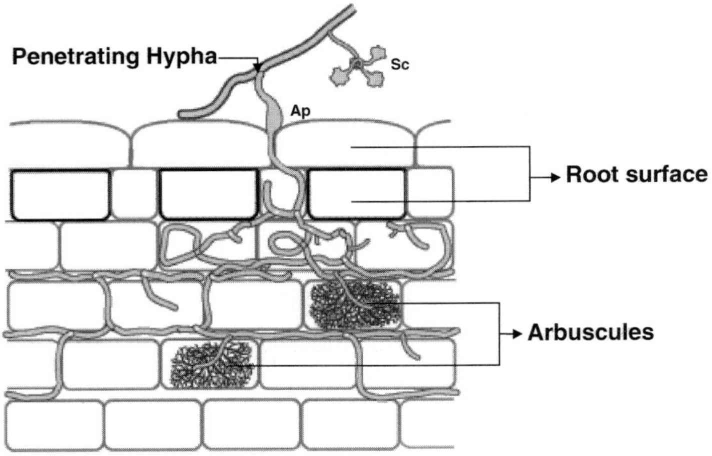

Fig. 7. Colonization of AM fungi in the root of the vascular plant. Arbuscules are the sites for bidirectional flux. Transfer of photosynthates from root to soil and nutrients from soil to root (*Ap* appresorium, *Sc* sporocarp)

characteristic vesicles and arbuscules. The plasmalemma of the host cell invaginates and encloses the arbuscules.

8.3 Ericoid Mycorrhiza

The ericoid mycorrhizae are endomycorrhizae in the general sense, since the fungal symbiont penetrates the cortical cell wall and invaginates the plasmalemma. Infection of each cortical cell takes place from the outer cortical wall; lateral spread from cell to cell does not occur. Infected cells appear to be fully packed with fungal hyphae. The mycorrhizae do not form a sheath although a loose weft of hyphae around the root can sometimes be observed. The functional life of the association in epidermal cells may be short-lived, being only a matter of weeks in *Rhododendron*. In the ericoid mycorrhizae, the host cell dies as the association disintegrates, thereby restricting the functional life (i. e., nutrient absorption) of these epidermal cells to the period prior to breakdown of the infected cell.

8.4 Arbutoid Mycorrhiza

The arbutoid mycorrhizae have characteristics found in both ECM and other endomycorrhizae. Intracellular penetration of cortical cells and formation of a sheath can occur, and a Hartig's net is present. A feature distinguishing them from ericoid mycorrhizae is the presence of the dolipore septate in internal hyphae. It appears from most reports that the fungal associate in arbutoid mycorrhizae is a basidiomycete.

8.5 Monotropoid Mycorrhiza

This term is applied specifically to mycorrhizae that are observed on the achlorophyllous plants in the family Monotropaceae. These mycorrhizae are very similar to the ECM and form a distinct sheath and Hartig's net. However, they exhibit a distinctive type of intracellular penetration in cortical cells that is unlike other endomycorrhizal types. The fungus forms a fungal peg, which invaginates the cell wall.

8.6 Ect-endomycorrhiza

Ect-endomycorrhiza are only formed with genera in the Pinaceae. These mycorrhizae form a Hartig's net in the cortex of the root, but develop little or no sheath. Intracellular penetration of cortical cells takes place, and thus they are similar to the arbutoid type. Ect-endomycorrhizae in Pinaceae seem to be limited to forest nurseries and are formed by a group of fungi called E-strain. These fungi are most likely to be the imperfect stage of ascomycetes; they may cause ect-endomycorrhizae in some tree species and ECM in other tree species.

8.7 Orchidaceous Mycorrhiza

The fungal association is of the endomycorrhizal type, where the fungus penetrates the cell wall and invaginates the plasmalemma and forms hyphal coil within the cell. Once the plant is invaded, spread of the fungus may occur from cell to cell internally. The internal hyphae eventually collapse or are digested by the host cell. Since the symbiosis forms an external network of hyphae, it would seem probable that the fungal hyphae function in nutrient uptake as with other mycorrhizae and that the coarse root system of orchids would be supplemented by the increased absorbing surface area of the hyphae (Smith and Read 1997). A number of basidiomycetes genera have been shown to be involved in the symbiosis, although many reports on the isolation of the symbiotic fungus from the roots of orchids have placed the symbionts in the form genus *Rhizoctonia* when the perfect stage was not known or the isolate was not induced to fruit in culture.

9 Functions of Mycorrhizal Fungi

In terrestrial ecosystems, arbuscular mycorrhizal (AM) fungi make various promises (Table 4) where the organic detritus–decomposer pathway accounts for the majority of energy flow and nutrient turnover. Microflora coupled with microfauna in the soil are the major components of both biomass and activity affecting nutrient availability. Although soil bacteria and fungi generally immobilize mineral nutrients as carbon is consumed and thereby compete with plants for macronutrients, mycorrhizal fungi,

Table 4. Advantages of AM fungi

Promotes plant growth	Maintain plant and soil health
Bio-protection against root diseases (bacteria, fungi and nematodes)	Plant production with reduced fertilizers and pesticides
Nutrient acquisition	Plant size or biomass
Improved soil-root contact	Influence population dynamics of soil flora
Symbiosis alters host water relations	Revegetation of landscape or contaminated soils
Symbiosis alters root length, root architecture and root/shoot ratio	Biological hardening of tissue culture-raised plants
Alters rate of water movement into, through and out of host plants	Effects on tissue hydration and leaf physiology
Postpones leaf dehydration	Alters leaf osmotic potential
Alters the number of photosynthetic units	Photosynthetic storage and export rates
Dissimilar symplastic solute pools	More effective scavenging of soil water
Effects on osmotic adjustment	Drought responses
Altered transpiration rates	Stomatal conductance to water vapor
Intrinsic leaf hydraulic or biochemical properties	Osmoprotection of enzymes
Altered nodule number and their activity	Enhanced P acquisition
Altered total protein	Altered morphological and phenological effects
Altered leaf abscission	Altered leaf drop, necrosis and senescence
Altered leaf movements	Altered wilting of leaves
Altered recovery from wilting	Provide salt tolerance to plant

because of their unique carbon strategies, can efficiently couple soil mineralization and nutrient uptake by the plant roots. In many cases, the mycorrhizal system actually "bridges" across the rhizosphere, and provides an organic link between the root and the bulk soil. In addition, AM fungi help the plant to cope with various kinds of stress such as soil pH, heavy metals, soil salinity, and water and drought stresses. A brief account of the functioning of AM fungi under these stresses is described in this chapter.

9.1 Arbuscular Mycorrhizal Fungi in Relation to Soil pH

Soil pH crucially affects development of mycorrhizal fungi by affecting the solubility of several compounds. Most of the phosphate exists as insoluble complexes of Al and Fe at acidic pH and at alkaline pH, phosphate exists as insoluble complexes of Ca and Mg while maximum solubility of phosphate occurs at neutral pH. However, inorganic forms are still largely insoluble. Many metals are insoluble under alkaline edaphic conditions, but highly soluble at acidic conditions.

It has been well demonstrated that mycorrhizal fungi vary in their tolerance of soil pH. Some grow only in low pH soils, whereas others grow after modifying the soil pH with a certain amount of lime (Giri et al. 2003a). A few mycorrhizae have a tendency to grow at the pH from which they have been isolated (Giri and Mukerji 2003).

Soil pH is an important factor in studying the ecology of endomycorrhizal fungi. Low soil pH has a profound effect on the movement and uptake of P. Rhizosphere acidification affects fungal soil plant nutrient supply mechanisms (Giri et al. 2003a). Gillespie and Pope (1991) reported that the P diffusion rate increased with an increase in the acidity of the soil. *Glomus* sp. extracted from a neutral soil grew best at pH 7, while its growth was less at acidic and alkaline pH. Relative tolerance of *Glomus mosseae*, *G. fasciculatum*, and *G. macrocarpum* to graded pH levels (7.8–10.5) and their influence on P uptake in *Prosopis juliflora* were evaluated by Sidhu and Behl (1997). They found that an increase in pH adversely affected growth, biomass and P concentration in seedlings. Chlamydospore formation by all three AM fungi decreased with an increase in the rhizosphere soil pH. However, application of AM resulted in a significant increase in seedling root and shoot length, collar diameter and biomass production at high pH levels. Biomass production of mycorrhizal seedlings grown at pH 10.5 was equivalent to that of uninoculated plants at pH 7.8. In AM-colonized roots P concentration increased by 68% at pH 7.8 and 40% at pH 10.5. *Glomus fasciculatun* originally isolated from a site having a high pH (9.2) showed relatively high tolerance to a pH ranging between 8.5–10.5 compared to other species as exhibited by a higher degree for chlamydospore formation and root colonization (Sidhu and Behl 1997).

Soil pH and available soil nutrients have a cumulative effect on the efficiency of AM fungi on plant growth. Habte and Soedarjo (1996) reported that increased P concentration leads to high soil pH and reduced available Mn concentration in soil. This is probably due to the precipitation of cations directly by an excess of phosphate, which induced an elevation in pH that increased plant growth.

9.2 Arbuscular Mycorrhizal Fungi in Relation to Heavy Metal Stress

Metal toxicity in soil may be induced by the discharge of sewage or industrial pollutants into the soil, or due to the presence of excessive quantities of metals in certain categories of soils. In soil they are present as free ions, soluble metal complexes, exchangeable metal ions, organically bound metals, precipitated or insoluble complexes such as oxides, carbonate and hydroxides or they may form part of the structure of silicate minerals (indigenous soil content). The toxicity of metals in the soil depends on their bioavailability, which may be defined as the availability of metals to be transferred from a soil compartment to living organisms. High metal concentration in soil is not only toxic to plants, but also affects germination and growth of soil microorganisms. Amongst the myriad of soil microorganisms, mycorrhizal fungi are considered as integral functioning parts of plant roots and the fungi involved provide a direct link between soil and plant roots. The influence of mycorrhizal fungi on plant nutrition is greater for elements with narrow diffusion zones around plant roots such as P (Smith and Read 1997; Giri et al. 1999).

There are only a few reports concerning the interaction of arbuscular mycorrhiza with metals. Nonetheless, it is apparent that soils high in available metals do provide a habitat for specific AM fungi, which do provide some degree of protection to the host plant from toxic metals by restricting the uptake of metals to the plant or tolerating these themselves. *Medicago sativa* grown in mine spoils and waste sediments containing high amount of Zn and Cd showed significant mycorrhizal colonization although the number of AM fungal spores was lower than in an adjacent soil not altered by mining activity (Diaz and Honrubia 1993). *Agropyron trachycaulum* showed considerable mycorrhizal colonization in a subalpine coal mine spoil and on oil sand tailings amended with peat containing AM fungal propagules, but colonization was absent in the same species cultivated in the unamended oil sand tailings, revealing them to be devoid of arbuscular mycorrhizal fungi. These results indicated that poor or absent mycorrhizal propagules in some of the mine spoils may have resulted in nonmycorrhizal colonization. However, mycorrhizal rather than nonmycorrhizal grasses could colonize polluted mining sites, suggesting that heavy metal tolerance was due to mycorrhizal association (Shetty et al. 1994).

In *Festuca* and *Calamagrostis epigejos*, mycorrhizal colonization was observed when grown in coastal dunes contaminated by atmospheric deposition from a blast furnace (Duke et al. 1986). Similarly, extensive colonization of mycorrhizal fungi was observed in *Agrostis capillaris* in a Zn- and Cd-contaminated site (Griffioen et al. 1994). Experiments conducted on three populations of *A. capillaris* using a sandy soil contaminated with smelter and limestone-derived clay with or without metals of natural origin, however, did not show a significant difference in mycorrhizal root colonization between these populations.

In *Albizzia amara* a high level of colonization by AM fungi was observed in agricultural soil and fly ash severely contaminated with heavy metals (Giri 2001). *Oxalis acetosella* grown in low pH soil polluted with Cd, Zn, and Pb showed even higher mycorrhizal colonization (Turnau et al. 1996). These results further substantiate the fact that mycorrhizal fungal colonization has the potential to tolerate heavy metals particularly in the case of those AM fungi which originated from metal-contaminated sites. In contrast, there are some reports demonstrating inhibition of mycorrhizal colonization in the presence of metals of different origin (Chao and Wang 1991; Weissenhorn et al. 1995; Vidal et al. 1996; Joner et al. 2000).

Experiments conducted on maize in the metal-contaminated soil showed that mycorrhizal colonization either increased plant biomass and decreased Cd, Cu, Zn and Mn concentrations in shoot- and root-tissues, or had no effect on growth and metal uptake depending on root density, plant growth conditions, and mycorrhizal inoculum (Weissenhorn et al. 1995). Loth and Hofner (1995) observed that mycorrhizal colonization increased uptake of Cu, Zn and Cd in *Avena sativa* roots from a highly contaminated soil, but reduced translocation to the aerial part. Weissenhorn and Leyval (1995) reported a higher uptake of metal by mycorrhizal plants under high metal concentration. It was found that *G. fasciculatum* reduced the negative effect of Zn on plant growth. However, they did not report on the effect of mycorrhiza on the Zn concentration in shoot and root tissues.

The effect of AM fungi on plant metal uptake also depends on soil pH. With increasing soil pH, (diethylenetriaminepentaacetate (DTPA)-extractable metals decrease, but at the same time AM fungi increased Cd, Zn and Mn uptake in the shoots of *Medicago sativa*. At a lower soil pH, mycorrhizal colonization decreased metal uptake. In both cases, mycorrhizal colonization enhanced alfalfa growth. It is unfortunate that most of the studies on heavy metal tolerance/uptake by mycorrhizal fungi have been performed in pots where it is not possible to separate the effect of the fungus and of the host plant on the mobilization and uptake of metals (Leyval et al. 1997). To differentiate between fungus and host plant effects, plant containers with different compartments separating root and extraradical hyphae of *G. mosseae* from a sandy loam to clover was undertaken. In mycorrhizal clover, uptake of Cd increased by 55% in comparison to nonmycorrhizal plants. In the same plant, Burkert and Robson (1994) also reported transport of Zn in extraradical hyphae from a sandy soil. In the bean plants, a higher amount of Cd, Zn and Cu was transferred by mycorrhizal extraradical hyphae.

The experiments carried out on the capacities of extramatrical hyphae to bind Cd and Zn have shown that AM fungal mycelium has a high metal sorption capacity and a cation exchange capacity (CEC) comparable to other microorganisms. Metal sorption by AM fungi was rapid and appeared mainly to be due to passive adsorption. It was also noticed that the highest adsorption took place in a metal-tolerant *G. mosseae* isolate and intermediate for fungus isolated from a soil treated with metal contaminated waste. *G. mosseae* absorbed ten times more metals than the commonly used biosorption organism *Rhizopus arrhizus* (Joner et al. 2000). These results confirm the AM involvement in plant protection against excess heavy metal uptake and more binding capacity of mycorrhizal fungi than others.

9.3
Arbuscular Mycorrhizal Fungi in Relation to Soil Salinity

Soil salinity is a problem of great concern. About one third of the world's irrigated land and half of the land in arid, semi-arid and tropical regions is not in use due to salinity (Juniper and Abbott 1993; Briccoli-Bati et al. 1994; Giri and Chamola 1999; Giri et al. 2002, 2003a, b; Giri and Mukerji 2003). Ten million hectares of irrigated agricultural land is abandoned annually. In India alone, 75 Mh of land has lost its fertility because of the deposition of salts. Most of the areas of Uttar Pradesh (Indo-gangatic plain), Rajasthan and Haryana are adversely affected due to a high salt concentration and have lost their fertility (Giri et al. 2000). Thus, there is an urgent need for improving such degraded wastelands to combat the increasing pressure of the alarming rise in population on agriculture by low input technologies. Several microorganisms are known to have the ability to tolerate high salt concentrations. They can survive under a wide salinity range. Among these microorganisms, mycorrhizal fungi are of growing concern (Table 5). Several field experiments have demonstrated the impacts of high salt concentration on AM fungi. Arbuscular mycorrhizal fungi-colonized plants established and survived better in soils with an electrical conductivity of 10 dS/m or higher (Al-Karaki et al. 2001; Giri et al. 2003a).

Table 5. Plants and AM fungi that tolerate high salinity levels

Plants	AM fungi	EC	References
Bell pepper	*Glomus fasciculatum*	1–12	Hirrel and Gerdemann (1980)
Bell pepper	*Gigaspora margarita*	1–12	Hirrel and Gerdemann (1980)
Onion	*G. fasciculatum*	1–12	Hirrel and Gerdemann (1980)
Onion	*Gigaspora margarita*	1–12	Hirrel and Gerdemann (1980)
Indian ricegrass	*Entrophospora infrequences*	1.6–2.0	Stahl and Williams (1986)
Indian ricegrass	*G. fasciculatum*	1.6–2.0	Stahl and Williams (1986)
Indian ricegrass	*Glomus microcarpum*	1.6–2.0	Stahl and Williams (1986)
Indian ricegrass	*Glomus mosseae*	1.6–2.0	Stahl and Williams (1986)
Yellow sweetclover	*E. infrequences*	1.6–2.0	Stahl and Williams (1986)
Yellow sweetclover	*G. fasciculatum*	1.6–2.0	Stahl and Williams (1986)
Yellow sweetclover	*G. microcarpum*	1.6–2.0	Stahl and Williams (1986)
Yellow sweetclover	*G. mosseae*	1.6–2.0	Stahl and Williams (1986)
Big sagebrush	*G. fasciculatum*	0.6	Stahl et al. (1988)
Big sagebrush	*G. microcarpum*	0.6	Stahl et al. (1988)
Big sagebrush	*E. infrequences*	2.6–3.8	Stahl et al. (1988)
Big sagebrush	*G. fasciculatum*	2.6–3.8	Stahl et al. (1988)
Big sagebrush	*G. macrocarpum*	2.6–3.8	Stahl et al. (1988)
Big sagebrush	*G. microcarpum*	2.6–3.8	Stahl et al. (1988)
Big sagebrush	*G. mosseae*	2.6–3.8	Stahl et al. (1988)
Acacia auriculiformis	*G. fasciculatum*	1–10	Giri et al. (2003b)
Acacia auriculiformis	*G. macrocarpum*	1–10	Giri et al. (2003b)
Sesbania aegyptiaca	*G. macrocarpum*	1–5	Giri and Mukerji (2003)
Sesbania grandiflora	*G. macrocarpum*	1–5	Giri and Mukerji (2003)

A high salt concentration inhibits the germination of AM fungal spores as well as the growth of hyphae, resulting in decreased growth and development of AM fungal density in soil (Juniper and Abbott 1993; Al-Karaki and Clark 1998; Al-Karaki et al. 2001). McMillen et al. (1998) found that an increasing concentration of NaCl inhibits either the hyphal growth or the infectivity of hyphae and AM colonization of plant roots. This may be due to the adverse effect of NaCl on the hyphal growth as well as the altered supply of carbohydrates from the plant to the fungus.

In our laboratory, the effect of AM fungi *Glomus fasciculatum* and *G. macrocarpum* was investigated on growth, development and nutritional responses of *Acacia auriculiformis*, under nursery and field conditions (Giri et al. 2003a). Plants were grown under different salinity levels imposed by 3, 5 and 10 dS/m solutions of 1 N NaCl. Both AM fungi protected the host plant against the detrimental effects of salinity. The extent of AM response on growth as well as root colonization varied with their species and salinity levels. Mycorrhiza-inoculated plants accumulated greater amounts of P and K, while Na uptake was lowered as salinity increased. Greater nutrient acquisition, change in root morphology and electrical conductivity of soil in response to AM colonization were observed during the course of the study and may be the possible mechanisms protecting the plant from salt stress (Giri et al. 2003a). Inoculation of *Sesbania grandiflora* and *Sesbania aegyptiaca* with AM fungus *Glomus macrocarpum* had a significant increase in growth and biomass production (Giri and Mukerji 2003). Under saline conditions, *Sesbania* spp. had a higher amount of Mg and reduced Na content in shoot tissues; the increased Mg uptake and reduced sodium uptake helped in chlorophyll synthesis. AM fungus also increased the establishment and survival of tree plants. Both the tree species were highly dependent on *G. macrocarpum* (Giri and Mukerji 1999; Giri et al. 1999).

The response of mycorrhizal and nonmycorrhizal *Olea europaea* under saline conditions with or without supplemental Ca resulted in less depolarization, at the cellular level (cell transmembrane electropotential), in roots of mycorrhizal than nonmycorrhizal plants. Supplemental Ca in the saline treatments had a protective effect on membrane integrity canceling or reducing the differences in depolarization between mycorrhizal and nonmycorrhizal plants. Mycorrhizal roots accumulated greater quantities of Na, K, and Ca and exhibited a lower K:Na ratio, but in leaves, mycorrhizal plants had a greater K:Na ratio than nonmycorrhizal plants (Rinaldelli and Mancuso 1996).

Mycorrhizal colonization brought about a noticeable improvement in salt-tolerance in olive plants, which was clearly demonstrated by trends in impedance parameters (Mancuso and Rinaldelli 1996). Uncer saline conditions, the electrical impedance parameters in shoots and leaves of olive plants were studied to understand variations in extracellular resistance, intracellular resistance and the state of the membrane in mycorrhizal and nonmycorrhizal plants. There was a reduction in extra- and intracellular resistance for nonmycorrhizal plants with increased NaCl concentration (Mancuso and Rinaldelli 1996). Ezz and Nawar (1994) found that sour orange seedlings irrigated with water containing 450 ppm salt reduced total leaf chlorophyll, peroxidase activity, starch and total carbohydrate concentration in leaves and roots. Inoculation with *Glomus intraradices* increased total chlorophyll, polyphenol activity, leaf and root sugars, and carbohydrate concentrations, but peroxidase activity was not altered.

9.4 Arbuscular Mycorrhizal Fungi in Relation to Water and Drought Stress

Arbuscular mycorrhizal fungi often results in altered rates of water movement into, through and out of host plants, with consequent effects on tissue hydration and leaf physiology (Smith and Read 1997; Auge 2000, 2001). The notation that AM effects on water relations were mainly nutritional in nature was prevalent for several years, i. e., the behavior of mycorrhizal and nonmycorrhizal plants altered because plants differed in size or tissue P concentrations. Various reports indicated that water relations and gas exchange of soybean could be affected by AM symbiosis independently of P nutrition (Bethlenfalvay et al. 1988; Auge 2001).

A few studies have shown important AM effects on stomata conductance and water potential of host plants (Gupta 1991; Koide 1993; Auge 2000). These studies suggested that such alterations in mycorrhizal plants are due to hormonal involvement, more effective scavenging of soil water, possibly through improved soil/root contact, stimulation of gas exchange through increased sink strength with possible effects on osmotic adjustment, and contributions of soil hyphae to water absorption (for more details, see Auge 2001). Many workers have studied water transport in terms of hydraulic conductivity of the root. Koide (1993) suggested that hydraulic conductivity is generally not improved by AM symbiosis in the absence of AM-induced growth or P effects. In fact, hydraulic conductivity was lower in mycorrhizal than in nonmycorrhizal roots when plants of similar size were examined (Graham et al. 1987; Auge 2001). In studies comparing AM and non-AM plants of either dissimilar size or tissue P concentrations, hydraulic conductivity was usually higher in AM than in non-AM roots (Cui and Nobel 1992), but not always (Syvertsen and Graham 1990). *Glomus fasciculatum* colonization increased whole plant, soil-to-root and root-to-leaf hydraulic conductance in *Bouteloua* (Allen et al. 1981; Allen 1982) and decreased soil-to-plant hydraulic conductance in *Bromus* relative to similarly sized nonmycorrhizal plants. AM hyphae were reported to enhance water uptake in sunflower and cowpea and lettuce, but not in clover or couchgrass or wheat (Faber et al. 1990; Tarafdar 1995; Auge 2001).

Arbuscular mycorrhiza symbiosis usually increased host growth rates during drought by affecting nutrient acquisition and possibly hydration. AM fungi have also typically increased water use efficiency and colonization by different fungi has affected water use efficiency differently (Simpson and Daft 1990). AM effects on host growth during drought are often related to improved P acquisition, as the availability of P in soils is reduced by soil drying. Copper and zinc concentrations were each higher in leaves of drought-affected mycorrhizal than nonmycorrhizal plants (Subramanian and Charest 1995; Giri and Chamola 1999).

Under drought conditions, inoculation of soybean plants with the AM fungus *Glomus mosseae* enhanced nodule dry weight and increased its leghemoglobin and protein contents as well as the nodule activity, thus helping to alleviate drought-induced nodule senescence in legume plants (Porcel et al. 2003). Drought considerably enhanced oxidative damage to lipids and proteins in nodules of nonmycorrhizal plants, whereas mycorrhizal treatments were protected against oxidative damage. Therefore, the alleviation of oxidative damage in nodules of AM plants has been suggested as an important mechanism involved in the protective effects of the AM symbiosis against premature nodule senescence (Ruiz-Lozano et al. 2001).

10 The Mycorrhizosphere

The rhizosphere is defined as the narrow zone (1–2 mm) of soil around the plant roots which is influenced by the presence and activity of the root. This area has the largest microbial activity of soil since it represents an important source of nutrients and physical support for many microorganisms (Weller and Thomashow 1994; Varma et al. 1999). The constant release of exudate, cell debris, mucilage or lysates provides the nutrient requirements of most saprophytic microbes. The root itself is a perfect niche for some symbiotic microorganisms such as *Rhizobium* or mycorrhizal fungi, and for other microorganisms intimately associated with the roots such as the so-called plant growth promoting rhizobacteria (PGPRs). The rhizosphere is a dynamic environment where microbial interactions take place constantly and may significantly affect plant growth and health. The microbial impact on plant growth is called the rhizosphere effect and is due to the microbial production of plant hormones, enzymes, or changes in the nutrient availability for the plant (Azcon-Aguilar and Barea 1992).

The rhizosphere of mycorrhized plants is very different from the rhizosphere of the same plant when nonmycorrhized. First, the prolongation of the root absorption ability in the form of fungal external mycelium increases the nutrient depletion area surrounding the roots. Secondly, the pattern of root exudation of mycorrhizal plants is altered and consequently, the physico-chemical soil properties, such as pH, moisture, nutrient content, organic matter or soil aggregation are normally modified around the mycorrhizal roots. The mycorrhizae act in modifying the nutrient availability for the rest of the rhizospheric microorganisms, and also in providing an additional ecological niche for these microorganisms. All these crucial changes due to mycorrhizal formation have caused some authors rename

Table 6. Synergistic interactions between mycorrhizal fungi and other rhizosphere microorganisms. (Modified from Bansal et al. 2000)

Mycorrhizal fungi	Rhizosphere microorganisms
Glomus fasciculatum	*Frankia*
G. fasciculatum	*Streptomyces cinnamomeous*
Glomus mosseae	*Rhizobium leguminosarum*
G. versiforme	*R. loli*
Glomus fasciculatum	*Beijerinckia*
G. fasciculatum	*Azotobacter chroococcum*
G. fasciculatum	*Azospirillium brasilense*
G. versiforme	*Corynebacterium*
Glomus versiforme	*Pseudomonas* sp.
Endogone sp.	*Agrobacterium* sp.
	Pseudomonas sp.
G. macrocarpum	*Bacillus megaterium*
	Pseudomonas fluorescence
Glomus macrocarpum	*Cladosporium* sp.
	Gliocladium virens
G. intraradices	*Fusarium oxysporium* f.sp. *chrysanthemi*

Table 7. Antagonistic interactions between mycorrhizal fungi and rhizosphere microorganisms. (Modified after Bansal et al. 2000)

Mycorrhizal fungi	Soil microorganisms
Glomus mosseae	*Phytophthora nicotianae* var. *parasitica*
G. macrocarpum	*Fusarium* sp.
G. intraradices	*Fusarium oxysporium* f.sp. *radicic-lycopersici*
G. intraradices	*Ordium lini*
G. fasciculatum	*Pythium ultimum*
G. fasciculatum	*Aphanomyces euteiches*
Glomus sp.	*Verticillium albo-atrium*
Glomus sp.	*Rhizoctonia solani*
G. fasciculatum	*Phytophthora perasitica*
G. fasciculatum	*P. fragariae*
G. etunicatum	*P. fragariae*
G. fistulosum	*Meloidogyne hapla*
Gigaspora margarita	*Pratylenchus vulnus*
G. intraradices	*Pratylenchus vulnus*
G. etunicatum	*Rhodopholus similis*
G. manihotis	*M. incognita*
G. mosseae	*Azotobacter chrocococcum*

this area as the mycorrhizosphere effect (Linderman 1988). The microbial components of this mycorrhizosphere have normally an increased activity which, in turn, may affect plant growth and health, and also modify the behavior of its fungal counterpart. These complex interactions are crucial to understanding the hidden world under the soil surface. Interactions among mycorrhizal fungi and other soil microorganisms are reciprocal, i. e., mycorrhizal fungi affect the other microbes and other microbes in turn affect the mycorrhizal association. The interaction can be synergistic and antagonistic, and is summarized in Tables 6 and 7, respectively.

11 Interactions in the Mycorrhizosphere

11.1 Interactions at the Pre-Symbiotic Stage

During the pre-symbiotic phase, the AM fungi do not interact actively with the rest of the soil microbiota since their saprophytic growth is very limited and mainly supported by the endogenous lipid reserve of the spore. Nevertheless, the positive influence of certain microorganisms on AM fungal germination has been reported. The presence of certain contaminant bacteria in the germination media accelerated the germination of AM fungal spores (Tylka et al. 1991; Carpenter-Boggs et al. 1995), although there are very few reports showing a net increase in the final number of germinated spores due to any microbial interaction. The addition of antibiotics to prevent bacterial contamination of the spores also inhibits germination of *Glomus versiforme*. Some of the bacteria involved in this effect are *Pseudomonas* and *Cornybacterium*. Soil fungi have also been shown to exert a beneficial effect on AM fungal spore germination. *Trichoderma* sp. hastened spore germination of *Glomus mosseae* in water agar (Calvet et al. 1992). In addition to the stimulatory effect on germination, most microorganisms tested have been shown to exert a stimulatory effect on hyphal growth, branching pattern, vegetative spore or auxiliary cell formation of AM. These effects are species-specific since not all bacterial or fungal treatments can induce them, and the magnitude of the effects is also dependent on the specific interaction.

Soil microorganisms also have detrimental effects on AM fungi. Thus, it has been observed that two AM fungi, *Glomus etunicatum* and *Glomus mosseae*, were only able to germinate in a certain soil after disinfection, showing a fungistatic ability exerted by some soil microbes. Moreover, the fungistasis was restored when the original nonsterile sieved soil was added to the pasteurized soil. Wilson et al. (1989) showed that this fungistasis could be due to the competition for P because it was reverted by addition of P to the soil. In addition, the spores and the emerging hyphae can contribute to microbial nutrition since they are sometimes observed to be parasitized by other fungi, actinomycetes or amoeboid organisms (Paulitz and Linderman 1991; Requena et al. 1996).

A very novel and attractive topic in the context of mycorrhiza-bacterial interactions was raised with the recent disclosure of the existence of an endosymbiotic bacteria living inside AM fungal spores. The discovery of the presence of these bacteria-like organisms (BLOs) in the cytoplasm of the spore was already reported in 1973, but the impossibility of cultivating them hampered their study (Perotto and Bonfante 1997). Recently, by using the PCR technique, it has been possible to amplify bacterial DNA from spores of *Gigaspora margarita*, and to determine their taxonomic position inside the genus *Burkholderia* using 16S rDNA primers (Bianciotto et al. 1996). So far, the role of BLOs is unknown, but current research on the topic seems to indicate a possible implication in the nitrogen metabolism (Perotto and Bonfante 1997).

11.2 Interactions at the Post-Symbiotic Stage

Given the obligate biotrophism of AM fungi, it is logical that most of the interactions between this group of fungi and the rest of the soil microbiota take place during their symbiotic phase. This fact hinders the analysis of such interactions in detail, because many of the effects and also the causes involved are mediated through the plant. It has been described that the root colonization rates of AM fungi can be improved by the presence of certain microorganisms, such as *Azospirillum*, *Rhizobium*, *Acetobacter, Pseudomonas* (Staley et al. 1992; Barea et al. 2002a). Some microbes have been shown to induce specifically the arbuscular formation, while having a slight, or no effect on the total percentage of root colonization (arbuscules + internal mycelium; Gryndler et al. 1995). It is not well established, however, whether these effects take place through a direct stimulation of AM fungi or via the stimulation of the root exudate production, which is strongly correlated to mycorrhiza colonization (Barea et al. 2002a). The extracellular polysaccharides (EPS) from *Rhizobium meliloti* were able to increase the formation of mycorrhizae in alfalfa plants, probably due to the ability of the EPS to increase exudation rates of their specific legume. Nevertheless, there are also many other beneficial microbial effects on the symbiosis with AM fungi, which do not necessarily correlate with an increase in the root colonization rate (Requena et al. 1997, 2001).

Moreover, in the rhizosphere AM fungi interact with various types of other soil bacteria. These interactions can be considered as: (1) interaction with plant growth-promoting rhizobacteria (PGPRs); (2) interaction with plant symbiotic N_2-fixing rhizobacteria; (3) interaction with phosphate-solubilizing bacteria and (4) interaction with soil-borne pathogens.

11.3 Interactions Between Arbuscular Mycorrhizal Fungi and Plant Growth-Promoting Rhizobacteria

The plant growth-promoting rhizobacteria (PGPRs) are involved in the nutrient cycling and the protection of the plant against plant diseases (Dobbelaere et al. 2001; Barea et al. 2002a; Probanza et al. 2002). After the biotrophic colonization of the root cortex, the arbuscular mycorrhizae (AM) develop an external mycelium which is a bridge connecting the root with the surrounding soil microhabitats. Mycorrhizal fungi in the mycorrhizosphere interact with PGPRs. Their activity in soil affects the populations of PGPRs in the rhizosphere both quantitatively and qualitatively. AM fungus, *Glomus fasciculatum*, inoculation showed a positive influence on the population of actinomycetes in the tomato rhizosphere. The same effect was observed after inoculation with *Azotobacter chroococcum*. The population of fluorescent pseudomonads was reduced significantly after inoculation of cucumber seedlings with *Glomus intraradices*, but not after inoculation with *Glomus etunicatum*. *Glomus fasciculatum* in the rhizosphere of sweet corn and clover reduced the population of *Streptomyces* sp. and chitinase-producing actinomycetes. Interaction studies with bacteria have indicated the longer survival of bacteria in the rhizosphere of the mycorrhizal root. Mycorrhizal fungi positively influence the survival of *Azotobacter paspali* in the rhizosphere of *Paspalum notatum* (Barea et al. 1998, 2002b). Secilia and Bagyaraj (1987) reported a higher bacterial population and number of nitrogen fixers, *Streptomyces* and *Pseudomonas solanacearum* in the rhizosphere of AM fungal-colonized plants. Moreover, it has been reported that plants in the presence of bacteria and AM fungi produced more phytohormones. These hormones and growth-promoting substances have a direct effect on the root biomass of the plant and germination, penetration, and establishment of AM fungi (Barea et al. 2002b).

11.4 Interactions Between Arbuscular Mycorrhizal Fungi and N_2-Fixing Bacteria

Nitrogen fixation is a key factor in biological productivity, it being accepted that more than 60% of the N-input to the plant community has a biological origin, with half of this input due to the symbiotic plant–bacteria systems, particularly those involving legumes (Postgate 1998). The bacterial partner in the symbiotic relationship with legume species belongs to the genera *Rhizobium*, *SinoRhizobium*, *Bradyrhizobium*, *MesoRhizobium*, and *AzoRhizobium*, collectively known as *Rhizobium* or rhizobia, which interact with legume roots, leading to the formation of N_2-fixing nodules (Spaink et al. 1998).

Arbuscular mycorrhiza is one of the most efficient ecological factors in improving growth and N content in legumes (Barea et al. 2002b). *Rhizobium*-associated plants are usually mycorrhizal. The mycorrhizal and *Rhizobium* symbiosis usually acts synergistically on infection rate, mineral nutrition and growth of the plant. AM fungi improve P uptake in conditions where N and P are limited. The higher P concentration in the plant benefits the bacterial symbiont and nitrogenase functioning, leading to increased nitrogen fixation, which in turn promotes root and mycorrhizal development. The AM fungus-mediated enhancement of N_2 fixation can be reduced, but either or both their mean weight and specific nitrogenase activity may increase in legumes. Symbiotic N_2 fixation depends on an adequate steady supply of P to the root nodules. AM fungi play important roles in improving growth, nodulation and N_2 fixation by legume crops symbiotic with *Rhizobium* spp. (Barea et al. 1993). Certain rhizobial strains improve processes involved in AM formation, i. e., spore germination, mycelial growth from the mycorrhizal propagules and "entry point" formation on the developing root system of the common host legume plant (Barea et al. 1996).

The use of ^{15}N-labeled soils has ascertained the effect of microbial interactions on N_2 cycling (Barea et al. 2002b). This methodology confirmed that mycorrhizae improve nitrogen nutrition in crop plants by facilitating the use of certain nitrogen forms that are difficult for nonmycorrhizal plants to exploit (Barea et al. 1993). Measurements of the $^{15}N/^{14}N$ ratio in plant shoots indicated enhancement of the N_2-fixation rates in *Rhizobium*-inoculated mycorrhizal plants, relative to that achieved by the same *Rhizobium* strain in nonmycorrhizal plants (Toro et al. 1997). In addition, mycorrhizae can indirectly enhance biological N_2 fixation, hence facilitating nitrogen input into the plant–soil system, thereby participating in N_2 cycling.

11.5 Interactions Between Arbuscular Mycorrhizal Fungi and Phosphate-Solubilizing Bacteria

The synergistic interaction of phosphate-solubilizing bacteria (PSB) and mycorrhizal fungi is well reported (Toro et al. 1997; Jeffries and Barea 2001; Barea et al. 2002a; Jeffries et al. 2003). Phosphate-solubilizing bacteria survived longer around mycorrhizal than nonmycorrhizal roots and acted synergistically with the mycorrhiza to increase plant growth, especially where rock phosphate was added to the soil. Two general types of microbiologically mediated processes for increasing the phosphate availability in soil have been described: those known to promote solubilization of non-available P sources in soil, yielding available phosphate ions, and those known to improve plant uptake of the already solubilized phosphate. The first type of activity is carried out by a great number of both saprophyte bacteria and fungi acting on sparingly soluble phosphates in soil, mainly by chelation-mediated mechanisms (Whitelaw 2000), while activity to improve the phosphate uptake properties of the host plants is typically carried out by mycorrhizal fungi (Smith and Read 1997). The external mycelium of mycorrhizal fungi acts as a bridge connecting the root with the surrounding soil microhabitats to extract phosphate ions from soil solution beyond the phosphate-depletion zone surrounding the roots, and transfer these ions to the plant. Thus, by linking the biotic and geochemical proportions of the ecosystem, the mycorrhizal fungi can contribute to P capture and supply, and P cycling (Jeffries and Barea 2001).

The interaction of PSB and AM fungi on plant use of soil P sources having low bioavailability has been evaluated in the soil microcosm using ^{32}P isotope (Toro et al. 1997). The rhizobacteria acted as 'Mycorrhiza-Helper Bacteria', promoting AM establishment by either the indigenous or inoculated AM fungi, while AM formation increased the size of the PSB population. The dual inoculation of PSB and AM fungi significantly increased microbial biomass and N and P accumulation in plant tissues. It was observed from the isotope dilution approach that the mycorrhizal and bacterized plants were using P sources otherwise unavailable to the plant (Barea et al. 2002b). Sharif (1999) studied the interactions among *Bacillus megaterium* var *phosphaticum* and *Glomus manihot* and their effects on growth and N and P uptake of pearl millet. P uptake by plants was significantly increased by the combined inoculation of *G. manihot* and *B. megaterium* var *phosphaticum*. Seed inoculation with *Pseudomonas striata* and *Glomus fasciculatum*, *G. mosseae* and *Gigaspora margarita* resulted in increased root biomass and P uptake in soybean. In a dual inoculation of PSB and AM fungi, the PSB rendered more P soluble, while mycorrhizae enhanced P uptake. The combined inoculation of *Glomus macrocarpum* and *Bacillus polymyxa* resulted in more fruit production due to their synergistic effect of P supply (Chandraghatgi and Sreenivasa 1995).

11.6 Interactions Between Arbuscular Mycorrhizal Fungi and Soil-Borne Pathogens

Many studies have demonstrated that AM fungi inhibit growth of the soil-borne pathogens (Al-Raddad and Adhmad 1995; Azcon-Aguilar and Barea 1996; Mukerji et al. 1997; Sharma et al. 1998; Sharma and Mukerji 1999; Mukerji 1999; Joseph and Sivaprasad 2000; Singh et al. 2000). Since AM fungi are established in the roots of host plants, research on the mycorrhizae and disease incidence has been concentrated on disease caused by soil-borne pathogens only. In the rhizosphere AM fungi occupy a unique ecological position as they are partly inside and partly outside the host thus, root-borne pathogens could directly interact with AM fungi in the mycorrhizosphere. A summary of the AM fungi and soil-borne pathogens is given in the Tables 8 and 9.

12 Conclusion

Microbial survival and reproductive success in many systems require colonization of a surface and/or integration into a biofilm community. Success in a community context requires morphological, physiological, and genetic attributes that have only recently been explored. The development of multicellular biofilm communities represents the interplay of many factors including specific cell–cell interactions and, in many cases, metabolic communications. Microbial interactions enable a variety of microorganisms to coexist in environments in which individual organisms cannot survive. Typically, these communities consist of various microbial aspects with different metabolic activities and nutritional requirements. Particularly within a biofilm, temporal and spatial formation of chemical microzones, positioning of syntrophic partners and establishment of complimentary metabolic pathways may all occur. Interaction between different species and populations is often characterized by close but, in general, poorly understood interdependencies.

Predation may also affect microbial activity. The role of protozoa in regulating population numbers in the microbial community is well recognized. This leads to increased mineralization of carbon, phosphorus, and nitrogen as a result of predation. Collembola are established mycophagous and after mites and nematodes they are among the soil's most abundant microfauna. They distribute the mycorrhizosphere flora.

Currently, there is considerable resistance to the use of chemical insecticides, pesticides, herbicides, weedicides and fungicides and fertilizers, be-

Table 8. Influence of VAM fungi on the control of nematodes. (Modified after Singh et al. 2000)

Host	Nematodes	AM fungi	Effect on host
Allium cepa	*Meloidogyne hapla, Meloidogyne incognita*	*Glomus etunicatum*	Tolerance of plants against nematodes increased
Avena sativa	*Meloidogyne incognita*	*Glomus mosseae*	Inhibitory effect on the disease incidence and development
Citrus limon	*Radopholus citrophilus*	*Glomus intraradices*	Larger shoot and root weights, lower nematode population densities
Citus sp.	*Radopholus citrophilus*	*Glomus etunicatum*	Increased host tolerance to nematodes
Citus sp.	*Radopholus citrophilus*	*Glomus intraradices*	Increased host tolerance to nematodes
Citus sp.	*Tylenchulus semipenetrans*	*Glomus fasciculatum*	Growth enhanced
Citus sp.	*Tylenchulus semipenetrans*	*Glomus mosseae*	Growth enhanced
Cydonia oblonga	*Pratylenchus vulnus*	*Glomus intraradices*	Growth favored and protection against nematodes
Daucus carota	*Meloidogyne hapla*	*Glomus mosseae*	Inhibitory effect on disease development
Elettaria cardomomum	*Meloidogyne incognita*	*Glomus fasciculatum, Gigaspora margarita*	Improved plant growth and reduced nematode population
Glycine max	*Heterodera glycines*	*Gigaspora margarita*	Growth enhanced
Glycine max	*Meloidogyne incognita*	*Gigaspora heteogama*	Growth enhanced
Glycine max	*Meloidogyne incognita*	*Gigaspora margarita*	Growth enhanced
Glycine max	*Meloidogyne incognita*	*Gigaspora margarita*	Growth enhanced
Gossypium hirsutum	*Aphelenchus avenae*	*Gigaspora margarita*	Growth stimulated
Gossypium hirsutum	*Meloidogyne incognita*	*Glomus fasciculatum*	Reduction in egg and nematode number/g of root
Lycopersicon esculentum	*Meloidogyne incognita*	*Glomus fasciculatum*	Reduction in number and size of root galls produced by nematodes
Lycopersicon esculentum	*Meloidogyne incognita*	*Gigaspora margarita*	Growth inhibited
Lycopersicon esculentum	*Meloidogyne javanica*	*Glomus mosseae*	Suppression of gall index and the average number of galls per root system
Lycopersicon esculentum	*Rotylenchus reniformis*	*Glomus fasciculatum*	Reduced juvenile penetration and development of nematode
Musa acuminata	*Radopholus similis*	*Glomus fasciculatum*	Increased length, dry and fresh weight
Nicotiana tabacum	*Meloidogyne incognita*	*Glomus mosseae*	Improved plant growth with lower incidence of disease
Nicotiana tabacum	*Meloidogyne incognita*	*Glomus fasciculatum*	

Table 9. Influence of AM fungi on the control of fungal pathogens. (Modified after Singh et al. 2000)

Host	Disease	Pathogen	AM fungi	Effect on host
Allium cepa	White rot	*Sclerotium cepivorum*	*Glomus* sp.	Delayed disease epidemic and increased the yield by 22%
Asparagus officinalis	Root rot	*Fusarium oxysporum*	*Glomus fasciculatum*	Lower incidence of disease
Cajanus cajan	Pigeon pea blight	*Phytophthora drechsleri* f.sp. *cajani*	*Gigaspora calospora*	Inhibitory effect on the development of disease and improved plant growth
Capsicum frutescens	Wilt	*Fusarium oxysporum*	*Glomus mosseae*	No effect
Chamaecyparis lawsoniana	Root rot	*Phytophthora cinnamomi*	*G. mosseae*	Delayed onset of disease incidence and the population of pathogens plant growth
Cicer arietinum	Wilt	*Fusarium oxysporum*	*G. mosseae*	No effect on wilt incidence
Citus sp.	Wilt	*Phytopthora parasitica*	*Gi. margarita, Glomus constrictum*	Significant increase in root dry weight and better growth of plants
Cuminum cyminum	Wilt	*Fusarium oxysporum* f.sp. *cumini*	*Acaulospora laevis, Gigaspora calospora*	Increased nutrient uptake and reduced disease severity
Glycine max	Root rot	*Fusarium solani, Macrophomina phaseolina, Rhizoctonia solani*	*Glomus mosseae*	Reduced disease symptoms
Glycine max	Root rot	*Phytopthora megasperma* var. *sojae*	*Glomus macrocarpum* var. *geosporum*	Increased severity of disease, internal stem discoloration
Hevea brasiliensis	Leaf blight	*Microcyclus ulei*	*Glomus etunicatum*	Increased resistance to leaf blight, lesion size and production of spores of the pathogen significantly lower
Hordeum vulgare	Root rot	*Bipolaris sorokiniana*	*Glomus* sp.	Severity of disease reduced

Linum usitatissimum	Wilt	*Fusarium oxysporum*, f.sp. *oidium lini*	*Glomus intraradices*	AM plants showed increased tolerance to *F. oxysporum*
Lucerne sp.	Wilt	*Fusarium oxysporum* f.sp. *medicagnisis*, *Verticilium aldoatrrum*	*Glomus fasciculatum*, *Glomus mosseae*	Lower incidence of wilt and increase in plant growth
Lycopersicon esculentum	Root rot	*Fusarium oxysporum* f.sp. *radices-lycopersici*	*Glomus intraradices*	Reduction in disease incidence and the population of pathogen and improvement in plant growth
Lycopersicon esculentum	Wilt	*Fusarium* sp.	*Glomus mosseae*	Reduced wilt to 11% as against 45% in nonmycorrhizal plants
Lycopersicon esculentum	Corky root disease	*Pyrenochaeta lycipersici*	*Glomus caledonium*	Disease index lower and root growth better
Lycopersicon esculentum	Root rot	*Phytopthora nicotiana* var. *parasitica*	*Glomus mosseae*	Decreased root necrosis
Nicotiana tabacum	Root rot	*Thielaviopsis basicola*	*Glomus microcarpum*	Reduction in the number of pathogen propagules
Nicotiana tabacum	Root rot	*Thielaviopsis basicola*	*Glomus monosporum*	Better tolerance to pathogen, higher root and leaf dry weight
Phaseolus aureus	Root rot	*Macrophomina phaseolina*	*G. mosseae*	Disease incidence reduced
Pisum sativum	Root rot	*Aphanomyces euteiches*	*G. fasciculatum*	Infection suppressed
Vicia faba	Wilt	*Fusarium oxysporum*	*G. mosseae*, *Gigaspora calospora*	More incidence of disease when plants has more AM infection
Vigna unguiculata	Root rot	*Macrophomina phaseolina*	*Glomus etunicatum*	Disease incidence reduced

cause of their hazardous influence on the environment, and on soil, plant, animal, and human health. Hence, the use of biofertilizers and biocontrol agents is recommended in practical agriculture, forestry, horticulture and flori-culture. There are a large number of bacteria and fungi that solubilize unavailable forms of phosphate and make it available for plant growth. Among them, mycorrhizae form symbiotic associations with the roots of plants and help in the uptake of phosphorus from the labile pool.

Colonization by mycorrhizal fungi alters the physiology, morphology, and nutritional status of the host–soil biota and structure. There is no host–plant or host–soil specificity, but some plant–fungus and some soil–fungus combinations are more effective than others. In all ecosystems, mycorrhizae link plants and soil, and that coupling influences most of the dynamics that occur in the mycorrhizosphere. Research efforts on mycorrhizosphere are fragmented, data synthesis and modeling on the ecosystem level are lacking. There is no information on the genetic potential of mycorrhizal fungi and very little on the associated plant growth promoting rhizobacteria (PGPRs) to tolerate environmental or cultural conditions to modulate host–plant and soil responses. In damaged or disturbed coupling, there is a need to develop an understanding and technology to recouple plants and soil with mycorrhizae, emulating the balance that occurs in undisturbed ecosystems and returning our crop production systems to a level of sustainability that allows for reduced inputs of chemicals.

Our knowledge of the regulation of a specific process may be detailed, but our understanding of its role in microbial survival and proliferation in natural systems is limited. Interpreting how heterotrophic microorganisms respond to and benefit from community growth in the natural environment, as well as the underlying molecular biological mechanisms, awaits application of the range of techniques now available. The science of the mycorrhizosphere is expanding fast and will soon connect the mycorrhizae with other sciences of plant systematics, ecology, and physiology, molecular biology, horticulture, agronomy, soil science, climatology, and certainly, plant pathology.

Acknowledgements. The authors thank the Council of Scientific and Industrial Research, Department of Science and Technology and University Grant Commission for financial support.

References

Al-Karaki GN, Clark RB (1998) Growth, mineral acquisition and water use by mycorrhizal wheat grown under water stress. J Plant Nutr 21:263–276

Al-Karaki GN, Hammad R, Rusan M (2001) Response of two tomato cultivars differing in salt tolerance to inoculation with mycorrhizal fungi under salt stress. Mycorrhiza 11:43–47

Allen MF (1982) Influence of vesicular-arbuscular mycorrhizae on water movement through *Bouteloua gracilis* (H.B.K.) Lag ex Steud. New Phytol 91:191–196

Allen MF, Smith WK, Moore TS Jr, Christensen M (1981) Comparative water relations and photosynthesis of mycorrhizal and non-mycorrhizal *Bouteloua gracilis* H.B.K. New Phytol 88:683–693

Al-Raddad A, Adhmad M (1995) Interaction of *Glomus mosseae* and *Paecilomyces lilacinus* on *Meloidogyne javanica* of tomato. Mycorrhiza 5:233–236

Auge RM (2000) Stomatal behavior of arbuscular mycorrhizal plants. In: Kapulnik Y, Douds D (eds) Mycorrhizal symbiosis: molecular biology and physiology. Kluwer, Dordrecht, pp 201–237

Auge RM (2001) Water relations, drought and vesicular-arbuscular mycorrhizal symbiosis. Mycorrhiza 11:3–42

Azcon-Aguilar C, Barea JM (1992) Interaction between mycorrhizal fungi and other rhizosphere microorganisms. In: Allen MJ (ed) Mycorrhizal functioning: an integrative plant-fungal process. Chapman and Hall, New York, pp 163–198

Azcon-Aguilar C, Barea JM (1996) Arbuscular mycorrhizas and biological control of soil born-plant pathogens: an overview of the mechanisms involved. Mycorrhiza 6:457–464

Bansal M, Mukerji KG (1994) Positive correlation between root exudation and VAM induced changes in rhizosphere mycoflora. Mycorrhiza 5:39–44

Bansal M, Mukerji KG (1996) Root exudates in rhizosphere biology. In: Mukerji KG, Singh VP (eds) Concepts in applied microbiology and biotechnology. Aditya Book, New Delhi, India, pp 98–120

Bansal M, Chamola BP, Sarwer N, Mukerji KG (2000) Mycorrhizosphere: interaction between rhizosphere microflora and VAM fungi. In: Mukerji KG, Chamola BP, Singh J (eds) Mycorrhizal biology. KA/PP, New York, pp 143–152

Bardgett RD, Wardle DA, Yeates GW (1998) Linking above-ground and below-ground food webs: how plant responses to foliar herbivory influences soil organisms. Soil Biol Biochem 30:1867–1878

Barea JM, Azcon R, Azcon-Aguilar C (1993) Mycorrhiza and crops. In: Ingram DS, Williams PH (eds) Advances in plant pathology: mycorrhiza synthesis. Academic Press, London, pp 391–416

Barea JM, Tober RM, Azcon-Aguilar C (1996) Effect of genetically modified *Rhizobium meliloti* inoculant on the development of arbuscular mycorrhiza, root morphology, nutrient uptake and biomass accumulation in *Medicago sativa* L. New Phytol 134:361–369

Barea JM, Andrade G, Bianciotto V, Dowling D, Lohrke S, Bonfante P, O'Gara F, Azcon-Aguilar C (1998) Impact on arbuscular mycorrhiza formation of *Pseudomonas strains* used as inoculant for the biocontrol of soil-borne plant fungal pathogens. Appl Environ Microbiol 64:2304–2307

Barea JM, Azcon R, Azcon-Aguilar C (2002a) Mycorrhizosphere interactions to improve plant fitness and soil quality. Anton Leeuw 81:343–351

Barea JM, Toro M, Orozco MO, Campos E, Azcon R (2002b) The application of isotopic (^{32}P and ^{15}N) dilution techniques to evaluate the interactive effect of phosphate solubilizing rhizobacteria, mycorrhizal fungi, and *Rhizobium* to improve agronomic efficiency of rock phosphate for legume crops. Nutr Cycl Agroecosyst 63:35–42

Bethlenfalvay GJ, Brown MS, Ames RN, Thomus RS (1988) Effect of drought on host and

endophyte development in mycorrhizal soybeans in relation to water use and phosphate uptake. Plant Physiol 72:565–571
Bianciotto V, Bandi C, Minerdi D, Sironi M, Tichy HV, Bonfante P (1996) An obligately endosymbiontic mycorrhizal fungus itself harbors obligately intracellular bacteria. Appl Environ Microbiol 62:3005–3010
Briccoli-Bati C, Rinaldi R, Tocci C, Sirianni T, Tannotta N (1994) Influence of salty water irrigation on mycorrhizae of young olive trees in containers. Acta Hortic 356:218–220
Bringhurst RM, Cardon ZG, Gage DJ (2001) Galactosides in the rhizosphere: utilization by *Sinorhizobium meliloti* and development of a biosensor. Proc Natl Acad Sci USA 98:4540–4545
Burkert B, Robson A (1994) ^{65}Zn uptake in subterranean clover (*Trifolium subterraneum* L.) by three vesicular arbuscular mycorrhizal fungi in a root-free sandy soil. Soil Biol Biochem 26:1117–1124
Calvet C, Barea JM, Pera J (1992) In vitro interaction between the vesicular-arbuscular mycorrhizal fungus *Glomus mosseae* and some saprophytic fungi isolated from organic substrates. Soil Biol Biochem 24:775–780
Carpenter-Boggs L, Loynachan TE, Stahl PD (1995) Spore germination *of Gigaspora margarita* stimulated by volatiles of soil-isolated actinomycetes. Soil Biol Biochem 27:1445–1451
Chandraghatgi SP, Sreenivasa MN (1995) Possible synergistic interaction between *Glomus macrocarpum* and *Bacillus polymyxa* in chilli. In: Adholeya A, Singh S (eds) Proceeding of 3rd National Conference on Mycorrhiza. TERI, New Delhi, pp 180–183
Chao CC, WangYP (1991) Effects of heavy metals on vesicular arbuscular mycorrhizae and nitrogen fixation of soybean in major soil groups of Taiwan. J Chin Agric Chem Soc 29:290–300
Cui M, Nobel PS (1992) Nutrient status, water uptake and gas exchange for three desert succulents infected with mycorrhizal fungi. New Phytol 122:645–649
Diaz G, Honrubia M (1993) Infectivity of mine soils from Southeast Spain. II. Mycorrhizal population levels in spoilt sites. Mycorrhiza 4:85–88
Dobbelaere S, Croonenborghs A, Thys A, Ptacek D, Vanderleyden J, Dutto P, Labander-Gonzalez C, Caballero-Mellado J, Aguirre JF, Kapulnik Y, Brener S, Burdman S, Kadouri D, Sarig S, Okon Y (2001) Response of agronomically important crops to inoculation with *Azosprillum*. Aust J Plant Physiol 28:1–9
Duke ER, Johnson CR, Koch RE (1986) Accumulation of phosphorus, dry matter and betaine during NaCl stress of split-root citrus seedlings colonized with vesicular-arbuscular mycorrhizal fungi on zero, one or two halves. New Phytol 104:583–590
Edwards D, Kerp H, Hass H (1998) Stomata in early land plants: an anatomical and ecophysiological approach. J Exp Bot 49:255–278
Ezz T, Nawar A (1994) Salinity and mycorrhizal association in relation to carbohydrate status, leaf chlorophyll and activity of peroxidase and polyphenol oxidase enzymes in sour orange seedlings. Alexandria J Agric Res 39:263–280
Faber B, Zasoske R, Munns D, Shackel K (1990) A method for measuring hyphal nutrient and water uptake in mycorrhizal plants. Can J Bot 69:87–94
Fitter AH, Garbaye J (1994) Interaction between mycorrhizal fungi and other soil organisms. Plant Soil 159:123–132
Gillespie AR, Pope PE (1991) Consequences of rhizosphere acidification on deliver and uptake kinetics of soil phosphorus. Tree Physiol 8:195–204
Giri B (2001) Mycorrhization in afforestation of stressed habitats. PhD Thesis, Department of Botany, University of Delhi, Delhi, India
Giri B, Chamola BP (1999) Vesicular arbuscular mycorrhizal fungi under salinity and

drought conditions. In: Tewari JP, Lakhanpal TN, Singh J, Gupta R, Chamola BP (eds) Advances in microbial biotechnology. APH Publ, New Delhi, pp 422–427

Giri B, Mukerji KG (1999) Improved growth and productivity of *Sesbania grandiflora* Pers under salinity stress through mycorrhizal technology. J Phytol Res 12:35–38

Giri B, Mukerji KG (2003) Mycorrhizal inoculant alleviates salinity stress in *Sesbania aegyptiaca* Pres and *S. grandiflora* Pres under field conditions: evidence for improved magnesium and decreased sodium uptake. Mycorrhiza (published online 23 Oct 2003)

Giri B, Kaur M, Mukerji KG (1999) Growth responses and dependency of *Sesbania aegyptiaca* on vesicular arbuscular mycorrhiza in salt stressed soil. Ann Agric Res 20:109–112

Giri B, Kapoor R, Mukerji KG (2000) *Sesbania aegyptiaca* Pers seedlings response to VA mycorrhization in two types of soil. Phytomorphology 50:327–332

Giri B, Kapoor R, Mukerji KG (2002) VAmycorrhizal/VAM technology in establishment of plants under salinity stress conditions. In: Mukerji KG, Manoharachari C, Chamola BP (eds) Techniques in mycorrhizal studies. Kluwer, Dordrecht, pp 313–327

Giri B, Kapoor R, Lata A, Mukerji KG (2003a) Preinoculation with arbuscular mycorrhizae helps *Acacia auriculiformis* in a degraded Indian wasteland soil. Comm Soil Sci Plant Anal [coming in vol 35 issue (1&2)]. Dekker, New York

Giri B, Kapoor R, Mukerji KG (2003b) Influence of arbuscular mycorrhizal fungi and salinity on growth, biomass, and mineral nutrition of *Acacia auriculiformis*. Biol Fertil Soils 38:170–175

Graham JH, Syvertsen JP, Smith ML (1987) Water relation of mycorrhizal and phosphorus-fertilized non-mycorrhizal *Citrus* under drought stress. New Phytol 105:411–419

Griffioen WAJ, Ietswaart JH, Ernst HO (1994) Mycorrhizal infection of an *Agrostis capillaries* population on a copper contaminated soil. Plant Soil 158:83–89

Gryndler M, Vejsadova H, Vosatka M, Catska V (1995) Influence of bacteria on vesicular-arbuscular mycorrhizal infection of maize. Folia Microbiol 40:95–99

Gupta RK (1991) Drought response in fungi and mycorrhizal plants. In: Arora DK (ed) Hand book of applied mycology, soil and plants. Dekker, New York, pp 55–75

Habte H, Soedarjo M (1996) Response of *Accacia mangium* to vesicular arbuscular mycorrhizal inoculation, soil pH and soil P concentration in an oxisol. Can J Bot 74:155–161

Heckman DS, Geiser DM, Eidell BR, Stauffer RL, Kardos NL, Hedges SB (2001) Molecular evidence for the early colonization of land by fungi and plants. Science 293:1129–1133

Hirrel MC, Gerdemann JW (1980) Improved growth of onion and bell pepper in saline soils by two vesicular arbuscular mycorrhizal fungi. Soil Sci Soc Am J 44:654–655

Horodyski RJ, Knauth LP (1994) Life on land in the Precambrian. Science 263:494–498

Jeffries P, Barea JM (2001) Arbuscular mycorrhiza – a key component of sustainable plant-soil ecosystem. In: Hock B (ed) The Mycota, vol IX. Fungal association. Springer, Berlin Heidelberg New York, pp 95–113

Jeffries P, Gianinazzi S, Perotto S, Turnau K, Barea JM (2003) The contribution of arbuscular mycorrhizal fungi in sustainable maintenance of plant and soil fertility. Biol Fertil Soil 37:1–16

Joner E, Roberto B, Leyval C (2000) Metal-binding capacity of arbuscular mycorrhizal mycelium. Plant Soil 226:227–234

Joseph PJ, Sivaprasad P (2000) The potential of arbuscular mycorrhizal associations for biocontrol of soil-borne diseases. In: Upadhyay RK, Mukerji KG, Chamola BP (eds) Biocontrol potential and its exploitation in sustainable agriculture. KA/PP, New York, pp 139–153

Juniper S, Abbott L (1993) Vesicular arbuscular mycorrhizas and soil salinity. Mycorrhiza 4:45–57

Kenrick P, Crane PR (1997) The origin and early evolution of plants on land. Nature 389:33–39

Koide R (1993) Physiology of the mycorrhizal plants. Adv Plant Pathol 9:33–54
Leyval C, Turnau K, Haselwandter K (1997) Effect of heavy metal pollution on mycorrhizal colonization and function: physiological, ecological and applied aspects. Mycorrhiza 7:139–153
Lindermann RG (1988) Mycorrhizal interactions with the rhizosphere microflora. The mycorrhizosphere effect. Phytopathology 78:366–371
Loth FG, Hofner W (1995) Einfluss der VA-Mykorrhiza auf die Schwermetallaufnahmen von Hafer (*Avena sativa* L.) in Abhangigheit vom Kontaminationsgrad der Boden. Z Pflanzenernaehr Bodenkd 158:339–345
Lynch JM (1990) The rhizosphere. Wiley, New York
Mancuso S, Rinaldelli E (1996) Response of young mycorrhizal and non-mycorrhizal plants of olive tree (*Olea europaea* L.) to saline conditions. II. Dynamics of electrical impedance parameters of shoots and leaves. Adv Hortic Sci 10:135–145
McMillen BG, Juniper S, Abbott LK (1998) Inhibition of hyphal growth of vesicular arbuscular mycorrhizal infection from fungus spores in soil containing sodium chloride. Soil Biol Biochem 30:1639–1646
Moore JC, McCann K, Setala H, de Ruiter PC (2003) Top-down is bottom-up: dose predation in the rhizosphere regulate aboveground dynamics? Ecology 84:846–857
Mukerji KG (1999) Mycorrhiza in control of plant pathogens: molecular approaches. In: Mukerji KG, Chamola BP, Upadhyay RK (eds) Biotechnological approaches in biocontrol of plant pathogens. Kluwer/Plenum, New York, pp 135–156
Mukerji KG, Chamola BP, Sharma M (1997) Mycorrhiza in control of plant pathogens. In: Agnihotri VP, Sarbhoy A, Singh DV (eds) Management of threatening plant diseases of national importance. MPH, New Delhi pp 297–314
Parmelle RW, Ehrenfeld JG, Tate RA (1993) Effects of pine roots on microorganisms, fauna and nitrogen availability in two soil horizons of a coniferous spodosol. Biol Fertil Soils 15:113–119
Paulitz TC, Lindermann RG (1991) Mycorrhizal interactions with soil organisms. In: Arora DK, Mukerji KG, Knudsen GR (eds) Hand book of applied mycology, soil and plants. Marcel Dekker, New York, pp 77–129
Perotto S, Bonfante P (1997) Bacterial associations with mycorrhizal fungi: close and distant friends in the rhizosphere. Trends Microbiol 5:496–501
Phillips DA, Ferris H, Cook DR, Strong DR (2003) Molecular control points in rhizosphere food webs. Ecology 84:816–826
Porcel R, Barea JM, Ruiz-Lozana MJ (2003) Antioxidant activities in mycorrhizal soybean plants under drought stress and their possible relationship to the process of nodule senescence. New Phytol 157:135–143
Postgate JR (1998) Nitrogen fixation. Cambridge Univ Press, Cambridge
Probanza A, Lucas Garcia JA, Ruiz Palomino M, Ramos B, Gutierrez Manero FJ (2002) *Pinus pinea* L. seedling growth and bacterial rhizosphere structure after inoculation with PGPR *Bacilli* (*B. licheniformis* CECT 5106 and *B. pumillus* CECT 5105). Appl Soil Ecol 20:75–84
Raven JA, Edwards D (2001) Roots: evolutionary origins and biogeochemical significance. J Exp Bot 52:381–401
Requena N, Jimenez I, Barea JM (1996) Bacteria-mycorrhiza interactions in land restoration. In: Azcon-Aguilar C, Barea JM (eds) Mycorrhizas in integrated systems: from genes to plant development. Office for publications of the European committees, Luxemburg, pp 657–660

Requena N, Jimenez I, Toro M, Barea JM (1997) Interaction between plant growth promoting rhizobacteria (PGPR), arbuscular mycorrhizal fungi and *Rhizobium* spp. in the rhizosphere of *Anthyllis cytisoides*, a model legume for revegetation in Mediterranean semi-arid ecosystems. New Phytol 136:667–677

Requena N, Perez-Solis E, Azcon-Aguilar C, Jeffries P, Barea JM (2001) Management of indigenous plant-microbe symbiosis aids restoration of desertified ecosystem. Appl Environ Microbiol 67:495–498

Rinaldelli E, Mancuso S (1996) Response of young mycorrhizal and non-mycorrhizal plants of olive tree (*Olea europaea* L.) to saline conditions. I. Short-term electrophysiological and long-term vegetative salt effects. Adv Hort Sci 10:126–134

Ruiz-Lozano JM, Collados C, Barea JM, Azcon R (2001) Arbuscular mycorrhizal symbiosis can alleviate drought-induced nodule senescence in soybean plants. New Phytol 151:493–502

Schreiner RP, Bethlenfalvay (1995) Mycorrhizal interactions in sustainable agriculture. Crit Rev Biotechnol 15:271–285

Secilia J, Bagyaraj DJ (1987) Bacteria and actinomycetes associated with pot cultures of vesicular-arbuscular mycorrhizas. Can J Bot 33:1069–1073

Sharif M (1999) The interaction among phosphate solubilizing bacteria, AM fungus and associative N_2 fixing bacteria and their effects on growth and N and P uptake of pearl millet. Pakis J Soil Sci 16:53–62

Sharma M, Mukerji KG (1999) VA mycorrhizae by control of fungal pathogens. In: Singh J, Aneja RK (eds) From ethnomycology to fungal biotechnology-exploiting fungi from natural resources for novel products. KA/PP, New York, pp 185–196

Sharma M, Mittal N, Kumar RN, Mukerji KG (1998) Fungi: tool for plant diseases management. In: Varma A (ed) Microbes for health, wealth and sustainable environment. MPH, New Delhi, pp 101–154

Shetty KG, Banks MK, Hetrick BA, Schwab (1994) Biological characterization of a southeast Kansas mining site. Water Air Soil Pollut 78:169–177

Sidhu OP, Behl HM (1997) Response of three *Glomus* species on growth of *Prosopis juliflora* Swartz at high levels. Symbiosis 23:23–24

Simpson D, Daft MJ (1990) Interaction between water stress and different mycorrhizal inocula on plant growth and mycorrhizal development in maize and sorghum. Plant Soil 121:179–186

Singh R, Adholeya A, Mukerji KG (2000) Mycorrhiza in control of soil-borne pathogens. In: Mukerji KG, Chamola BP, Singh J (eds) Mycorrhizal biology. KA/PP, New York, pp 173–197

Smith SE, Read DJ (1997) Mycorrhizal symbiosis. Academic Press, San Diego, 605 pp

Spaink HP, Kondorosi A, Hooykass PJJ (1998) The rhizobiaceae. Kluwer, Dordrecht

Srivastava D, Kapoor R, Srivastava SK, Mukerji KG (1996) Vesicular arbuscular mycorrhiza: an overview. In: Mukerji KG (ed) Concepts in mycorrhizal research. Kluwer, Dordrecht, pp 1–39

Stahl PD, Williams SE (1986) Oil shale process water affects activity of vesicular-arbuscular fungi and *Rhizobium* 4 years after application to soil. Soil Biol Biochem 18:451–455

Stahl PD, Williams SE, Christensen (1988) Effect of native vesicular arbuscular mycorrhizal fungi after severe soil disturbance. New Phytol 110:347–353

Staley TE, Lawrence, EG, Nance EL (1992) Influence of a plant growth-promoting pseudomonad and vesicular-arbuscular mycorrhizal fungus on alfalfa and bridesfoot trefoil and nodulation. Biol Fertil Soils 14:175–180

Stewart WN, Rothwell GW (1993) Paleobotany and the evolution of plants, 2nd edn. Cambridge Univ Press, Cambridge

Subramanian S, Charest C (1995) Influence of arbuscular mycorrhizae on the metabolism of maize under drought stress. Mycorrhiza 5:273–278

Syvertsen JP, Graham JH (1990) Influence of vesicular arbuscular mycorrhizae and leaf age on net gas exchange of *Citrus* leaves. Plant Physiol 94:1424–1428

Tarafdar JC (1995) Role of a VAmycorrhizal fungus on growth and water relations in wheat in presence of organic and inorganic phosphate. J Indian Soc Soil Sci 43:200–204

Taylor TN, Remy W, Hass H, Kerp H (1995) Fossil arbuscular mycorrhizae from the early Devonian. Mycologia 87:560–573

Toro M, Azcon R, Barea JM (1997) Improvement of arbuscular mycorrhizal development by inoculation with phosphate-solubilizing rhizobacteria to improve rock phosphate bioavailability (^{32}P) and nutrient cycling. Appl Environ Microbiol 63:4408–4412

Turnau K, Kottke I, Dexheimer J (1996) Toxic element filtering in *Rhizopogon roseolus*/*Pinus sylvestris* mycorrhizas collected from calamine dumps. Mycol Res 100:16–22

Tylka GL, Hussey RS, Roncadori RW (1991) Axenic germination of vesicular-arbuscular mycorrhizal fungi: effects of selected *Streptomyces* species. Phytopathology 81:754–659

Van der Putten WH, Vet LEM, Harvey JA, Wackers FL (2001) Linking above-ground and below-ground multitrophic interactions of plants, herbivores, pathogens, and their antagonists. Trends Ecol Evol 16:547–554

Varma A, Verma A, Sudha S, Sahay N, Britta B, Franken P (1999) *Piriformospora indica* -- a cultivable plant growth promoting root endophyte with similarities to arbuscular mycorrhizal fungi. Appl Environ Microbiol 65:2741–2744

Vidal MT, Azcon-Aguilar C, Barea JM (1996) Effects of heavy metals (Zn, Cd and Cu) on arbuscular mycorrhiza formation. In: Azcon-Aguilar C, Barea JM (eds) Mycorrhiza in integrated systems: from genes to plant development. European Commission, EUR 16728, Luxembourg, pp 487–490

Watanabe Y, Martini JEJ, Ohmoto H (2000) Geochemical evidence for terrestrial ecosystem 2.6 billion years ago. Nature 408:574–578

Weissenhorn I, Leyval C (1995) Root colonization of maize by a Cd-sensitive and a Cd-tolerant *Glomus mosseae* and cadmium uptake in sand culture. Plant Soil 175:233–238

Weissenhorn I, Leyval C, Berthelin J (1995) Bioavailability of heavy metal and abundance of arbuscular mycorrhiza in a soil polluted by atmospheric deposition from a smelter. Biol Fertil Soils 19:22–28

Weller DM, Thomashow LS (1994) Current challenges in introducing beneficial microorganisms into the rhizosphere. In: Gara FO, Dowling DN, Boesten B (eds) Molecular ecology of rhizosphere microorganisms biotechnology and the release of GMOs. VCH Verlagsgesellschaft, Weinheim, pp 1–13

Whitelaw MA (2000) Growth promotion of plants inoculated with phosphate-solubilizing fungi. Adv Agron 69:99–151

Wilson GWT, Hetrick BAD, Kitt DG (1989) Suppression of vesicular-arbuscular mycorrhizal fungus spore germination by non-sterile soil. Can J Bot 67:18–23

12 Interactions Between Microorganisms and Soil Micro- and Mesofauna

Stefan Scheu, L. Ruess, Michael Bonkowski[1]

1 Introduction

The soil food web is extraordinarily complex; its trophic structure and the relationships between its components are still poorly understood. By recycling plant material and mineralizing nutrients therein, the belowground decomposer system provides the basis for soil fertility and plant life. Decomposition processes are dominated by microorganisms with their vast array of enzymes for the breakdown of organic matter. However, the microbial environment and, therefore, the activity and composition of the microbial soil community are strongly structured and influenced by an exceptionally diverse community of soil-dwelling invertebrates. Interactions between microorganisms and soil invertebrates include direct predator–prey relationships, but also indirect effects, such as competition for resources and habitat formation. Of fundamental importance for the type of interaction between microorganisms and soil invertebrates is the size of the contributors (Fig. 1). Bacterial consumers necessarily need to be small to be able to exploit the water-filled pores colonized by bacteria. The microfauna which is dominated by protozoa and nematodes may effectively prey upon bacteria as primary consumers and form part of one of the most important food web compartments in soils, the bacterial energy channel (Moore et al. 1988). Saprophytic fungi, the second major microbial primary decomposer group, are preyed upon by gnawing and piercing soil invertebrates, such as collembolans and oribatid mites, which form a part of the soil mesofauna. Fungi, fungal grazers and the associated predators constitute another major compartment of soil food webs, the fungal energy channel (Moore et al. 1988; de Ruiter et al. 2002). Macrofauna species such as earthworms, millipedes and isopods also ingest microorganisms, but in contrast to predator–prey interactions they predominantly affect microbial life indirectly by forming their habitat ('engineering' sensu Jones et al. 1994). However, habitat formation is not restricted to macrofauna species,

[1]Technische Universität Darmstadt, Fachbereich Biologie, Schnittspahnstr. 3, 64287 Darmstadt, Germany, e-mail: scheu@bio.tu-darmstadt.de, Tel: +49-6151-163006, Fax: +49-6151-166111

Soil Biology, Volume 3
Microorganisms in Soils: Roles in Genesis and Functions
(ed. by F. Buscot and A. Varma)

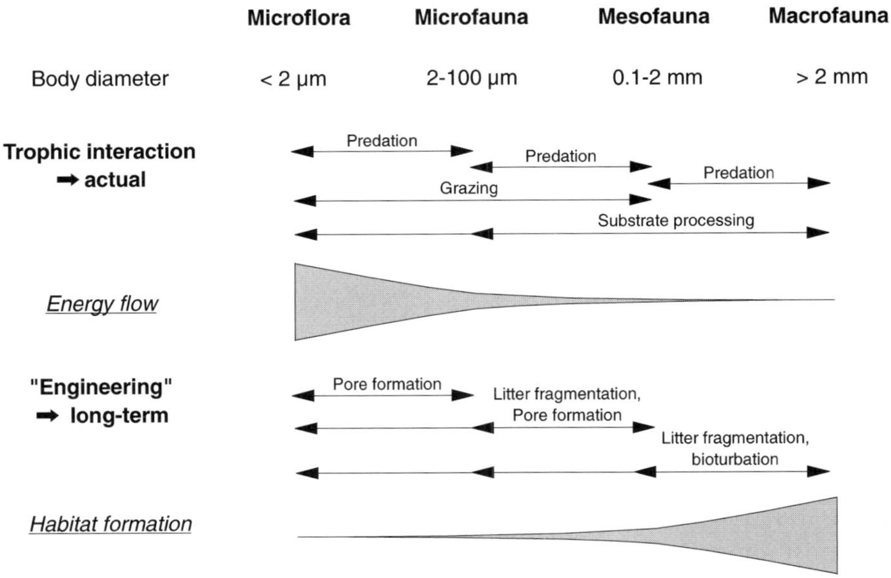

Fig. 1. Size-dependent interactions among soil organisms. Trophic interactions and interactions caused by "engineering" are separated; both are indicated by *arrows*. Note that trophic interactions and interactions caused by engineering are strongly size-dependent, but complement each other (*tapering* and *widening triangles*). Both function at different scales: trophic interactions drive the current energy flow, engineering sets the conditions for the existence of the soil biota community in the long term (from cheu and Setälä 2002)

mesofauna groups such as enchytraeids and collembolans also function as ecosystem engineers, particularly in more acidic soils.

Predator–prey interactions between small soil invertebrates and microorganisms directly modify the functioning of microbial communities and affect current microbial processes, whereas habitat formation by macro- and mesofauna is responsible for microbial community composition in the long term. Both the short-term modification of microbial activity and the structuring of microbial communities in the long term complement each other. Short-term effects rely on heavy grazing of microbes which implies that the consumers must be numerous in order to be significant. In contrast, even small numbers of larger soil invertebrates, which may only marginally affect current microbial metabolism, may be of essential importance for microbial community composition in the long term and, therefore, for basic ecosystem characteristics and processes.

In this chapter, we explore how soil micro- and mesofauna affect microbial activity and community composition. According to the two major plant resources entering the decomposer system, we distinguish between the plant detritus (litter) and root exudate-based food web compartments.

Finally, we explore how these interactions may be fed back to plants via modifying rhizosphere processes and the root environment.

In addition to the components mentioned above, the soil micro- and mesofauna include species which directly feed on living plant tissue, i. e., root feeders with associated antagonists and higher order predators (Moore et al. 1988; de Ruiter et al. 2002). They form a third, root-based energy channel which is an essential component of the belowground food web. However, it will not be considered in this review since we focus on animal–microbe interactions and their feedback to plant growth. A strict separation of the detritivore and root herbivore food web, however, is difficult since the root herbivore food chain is closely associated with, e. g., the fungal energy channel, where fungivore soil nematodes probably feed on both mycorrhizas and roots. In fact, from the plant perspective, both function in a very similar way.

2 Interactions in the Detritus Food Web

2.1 Structure of the Decomposer Animal Community

Soil animals are extremely densely packed: underneath one footprint of forest soil there may be billions of protozoa, thousands of nematodes, Collembola and mites and a large number of isopods, spiders, beetles, etc. Soil animals are not only extraordinarily numerous, they also are exceptional divers. Considering that consumers and resources did not coevolve because the basal food resource of the decomposer community is dead organic matter, this diversity is surprising (Anderson 1975; Giller 1996; Maraun et al. 2003). Until today, consumer-resource relationships in decomposer systems have been poorly understood. Most decomposer soil animals appear to be food generalists rather than specialists. Furthermore, by consuming dead organic matter, decomposer animals ingest food resources consisting not only of plant detritus, but also of bacteria, fungi, algae, protozoa and even multicellular organisms such as nematodes. The actual resources used by detritivores often remain obscure (Scheu and Setälä 2002).

Due to the difficulties of ascribing decomposer organisms to specific trophic levels, information on the structure of soil food webs is difficult to obtain. A methodology which improved this situation is the analysis of variation in the abundance of stable isotopes. Using δ^{15}N data, it has been recently shown that soil organisms form two clusters consisting of predators and microbi-detritivores which span a range equivalent to approximately four trophic levels (excl. detritus; Ponsard and Arditi 2000;

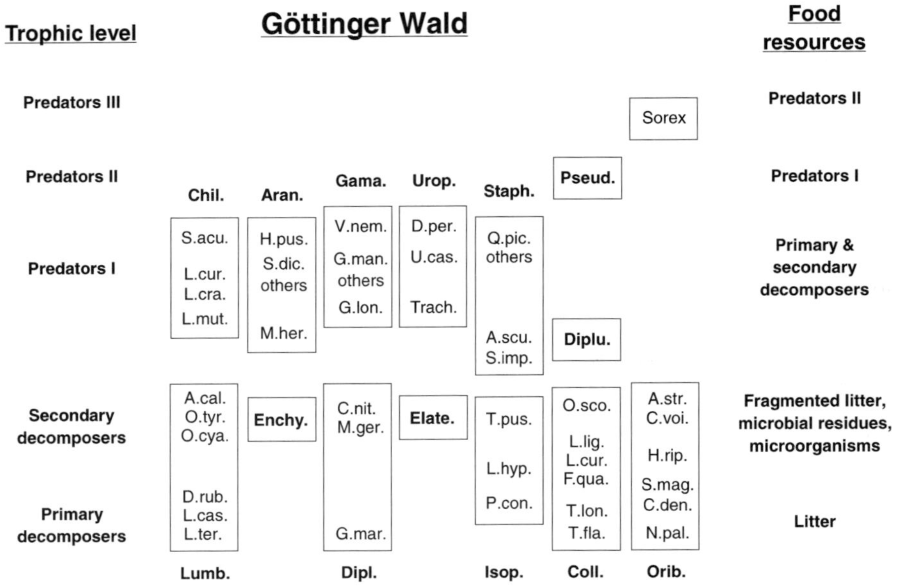

Fig. 2. Trophic structure of the decomposer community of a beechwood on limestone ("Göttinger Wald") as indicated by ^{15}N analysis (from Scheu and Falca 2000). *Aran* Araneida, *Chil* Chilopoda, *Coll* Collembola, *Dipl* Diplopoda, *Diplu* Diplura, *Elate* elaterid larvae (Coleoptera), *Enchy* Enchytraeidae, *Gama* Gamasina, *Isop* Isopoda, *Lumb* Lumbricidae, *Orib* Oribatida, *Pseud* Pseudoscorpionida, *Staph* Staphylinidae (Coleoptera), *Urop*: Uropodina; for full species names, see Scheu and Falca (2000)

Scheu and Falca 2000; Fig. 2). Both microbi-detritivores and predators span two trophic levels, suggesting that microbi-detritivores do indeed comprise a gradient from primary to secondary decomposers. Interestingly, microbi-detritivorous groups, such as Collembola, oribatid mites, diplopods and earthworms span a very similar range, indicating that species which are taxonomically unrelated exploit similar resources (Scheu and Falca 2000). A major implication of this is that higher taxonomic units are inadequate for depicting trophic levels as they reflect very broad trophic groups, such as microbi-detritivores and predators.

Similar to microbi-detritivores, predators in soil uniformly appear to be generalist feeders. This presumably is related to the high density of potential prey and the difficulties of locating specific ones in the opaque and porous soil habitat. Most important, soil microarthropods are so ubiquitous and mobile that for many predators it does not pay to search for them. This implies that soil predators almost uniformly function as trophic level omnivores, since the trophic range of their prey, microbi-detritivorous species, extends over at least two trophic levels. In addition, predators are likely to feed within their own trophic level; intraguild predation (IGP) and cannibalism are certainly widespread in decomposer food-webs (Polis 1991; Gunn and Cherrett 1993), which is supported by soil food web analyses based on stable isotopes (Ponsard and Arditi 2000; Scheu and Falca 2000). The dominance of generalist predators and the high incidence of IGP and cannibalism in soil communities add to the intriguing complexity of belowground food webs. It is likely that the exceptionally high local diversity of soil organisms at least in part results from the fact that soil predators are so closely intermingled. The very large overlap in prey (among predators) may prevent competitively superior microbi-detritivores from outcompeting inferior species, despite close resource overlap among them.

The similarity of trophic positions of species in very different taxonomic groups within microbi-detritivores and predators supports the assumption that functional redundancy among soil animals is high (Wardle et al. 1997, 1998; Setälä et al. 1998). High redundancy may help to explain why decomposer populations are extraordinary stable in time (Bengtsson 1994) and respond little to external perturbations (Scheu and Schaefer 1998; Joergensen and Scheu 1999). Decomposer diversity–ecosystem functioning experiments also support the view of high redundancy in detritus communities (Mikola et al. 2002; Wardle 2002).

2.2 The Detritus vs. Root Exudate-Based Food Web

Plants supply two quite different energy sources to the soil food web, root exudates and plant debris (Fig. 3). Plant organic matter fuels the detritus food web, which acts as two distinct, bacterial and fungal-based compartments (Wardle and Yeates 1993). Carbon is transferred to successively higher trophic levels and nutrients bound in microbial biomass are liberated for plant uptake. Root exudates build the resource for a separate short and fast energy channel. In the rhizosphere, easily degraded carbon sources become available to microorganisms leading to a biomass often several times greater than in corresponding nonassociated soil (Wardle 2002). In contrast to bacterial and fungal channels, the interactions between plants, microbes and grazers form a loop ('microbial loop'; Clarholm 1994). The dominant grazers, protozoa and nematodes may effectively control microbial species, but themselves appear to be little controlled by higher order predators (Scheu and Setälä 2002; Wardle 2002), resulting in carbon to be processed in the loop and not progress to the remainder of the soil food web.

The activity of microorganisms in soil is usually limited by carbon, except in the rhizosphere where plants steadily supply easily available carbon sources. This favors a microflora typically consisting of fast-growing

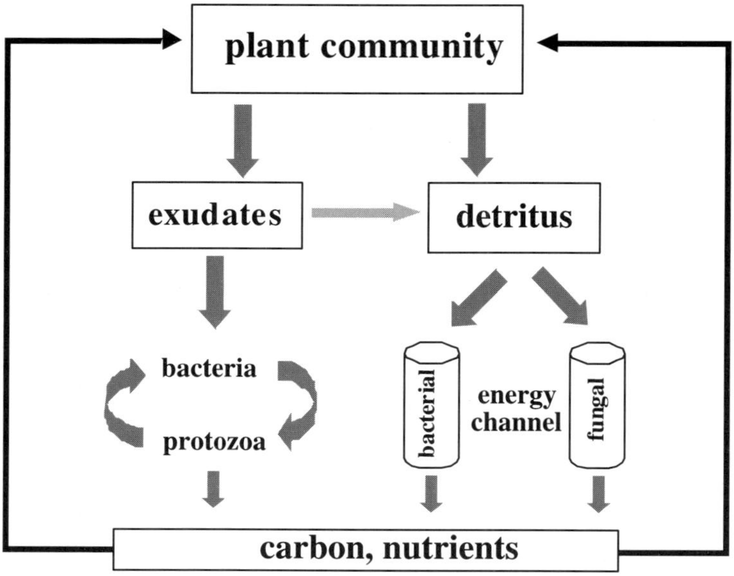

Fig. 3. Processing of matter by decomposer organisms. Differences between two basal resources and the associated decomposer system; the plant exudate and the detritus-based system. The latter constitutes two energy channels with bacteria and fungi as primary consumers

bacteria and leads to strongly increased microbial biomass and activity around roots (van Veen et al. 1989; Alphei et al. 1996). Root exudates may contain up to 40% of the net fixed carbon by plants (Lynch and Whipps 1990). In addition, microbial symbionts in the rhizosphere, such as mycorrhizae (Christensen 1989; Marschner 1992; Söderström 1992; Smith and Read 1997), or N_2-fixing microorganisms (Bezdicek and Kennedy 1979; Ryle et al. 1979), may each consume 10–20% of total net fixed carbon. Although plants direct their carbon investment to support certain microorganisms (Bonkowski et al. 2001a; Wamberg et al. 2003), in total they may release up to half of their net fixed carbon to fuel microbial interactions in the rhizosphere (Lynch and Whipps 1990).

Carbon allocated to the rhizosphere is lost in the building of light-capturing structures or defensive compounds aboveground. At first sight, therefore, it appears enigmatic why plants provide ample resources as exudates to a microbial community that is competing with roots for nutrients. The answer is provided by the loop structure of the feeding relationships in the rhizosphere. Nutrients become only temporarily locked up in the bacterial biomass near the root surface and are successively mobilized by microfaunal grazing (Bonkowski et al. 2000b).

2.3
The Bacterial vs. Fungal Food Chain

The chemical composition of soil organic matter has significant effects on long-term patterns of decomposition and detritus food chains in soils. The abundance of structural carbon compounds, such as cellulose and lignin, and the C/N ratio of organic material are important determinants (Melillo et al. 1982; Taylor et al. 1989; Verhoef and Brussaard 1990; Agren and Bosatta 1996). Pathways with a strong bacterial component mainly occur in moist soils that are rich in nitrogen and contain readily decomposable substrate (Fig. 4). These conditions promote a rapid turnover of carbon and a fast nutrient cycling to plants and, therefore, bacterially based energy channels are best described as "fast cycles". In these bacterial food chains, the major grazers of microbes are protozoa and nematodes (Moore et al. 1988; de Ruiter et al. 2002).

The fungal food chain is favored by a high C/N ratio in the soil, especially if recalcitrant materials predominate (Rosenbrock et al. 1995; Frankland 1998; Fig. 4). Fungi have higher substrate assimilation efficiency than bacteria and are able to break down complex polyaromatic compounds such as lignin and humic or phenolic acids. In contrast to bacteria, fungi have a relatively long generation time and react slowly to consumption by grazers. Fungal energy channels are therefore considered "slow cycle" systems (Moore et al. 1988; de Ruiter et al. 2002). Major grazers in the fungal food chain are fungus-feeding microarthropods, which form an abundant and species-rich community (Beare et al. 1997; Scheu and Setälä 2002). In

DECOMPOSITION PATHWAYS

Bacterial	Fungal
• moisture, N-rich, labile detritus	• high C:N, cellulose, lignin, resistant detritus
• fast cycle	• slow cylce
• major predators are protozoa and nematodes	• major predators are microarthropods
↑ **resource controlled BOTTOM UP**	↓ **top predators important TOP DOWN**

Fig. 4. Litter decomposition pathways based on either bacteria or fungi and their characteristics

the absence of microarthropods, fungus-feeding nematodes become system regulators (Parker et al. 1984). This diverse community of carnivores indicates that the functional importance of top predators, i. e., top-down control, is likely to be more significant in the fungal food chain (Scheu and Setälä 2002; Wardle 2002). In general, however, energy channels in soil run simultaneously and are not distinctly separated. There may be time lags in response or cross-exchange of energy and nutrients; they are dynamic features of the soil system.

3 The Role of Micro- and Mesofauna as Drivers of Microbial Decomposition Processes

Although decomposition is mainly a result of microbial activities, the soil fauna is an important driver of these processes by conditioning the litter and in stimulating microbial activity. The impact of the fauna on organic matter dynamics has been summarized in several reviews (Parker et al. 1984; Verhoef and Brussaard 1990; Griffiths 1994; Bengtsson et al. 1996; Beare et al. 1997; Setälä et al. 1998; Hättenschwiler et al. 2004). They indicate that the decomposer food web has a primary role in determining the mineralization of nutrients in soil, and hence plant nutrient acquisition and plant growth. Soil animals can affect the biomass, activity and diversity of the microflora directly by selectively feeding on bacteria and fungi, or indirectly by comminution of plant debris, dissemination of microbial propagules, and changes in nutrient availability (Newell 1984; Ingham et al. 1985; Visser 1985; Parkinson 1988; Beare et al. 1992; Bardgett et al. 1998; Mikola and Setälä 1998; Cragg and Bardgett 2001).

The soil microfauna affects organic matter decomposition through their grazing on and assimilation of microbial tissue and the excretion of mineral nutrients (Beare et al. 1997). About 60% of the nitrogen ingested by protozoa, and 90% by bacterial-feeding nematodes is in excess of structural needs and is excreted into the soil environment, usually as ammonia (Griffiths and Bardgett 1997). Grazing of the microfauna enhances mineralization processes in soils and liberates nutrients bound in the microbial biomass for plant uptake (Clarholm 1985; Ingham et al. 1985, 1986; Freckman 1988; Chen and Ferris 1999). Verhoef and Brussaard (1990) estimated that, depending on assimilation efficiency for nitrogen, the contribution of the soil fauna to net mineralization ranges between 10 and 49%, whereof bacteria-feeding amoebae and nematodes together accounted for over 83%. At an agriculture field site, Ferris et al. (1997) reported the bacteria-feeding nematode community to contribute between 0.2 and 1.4 kg N ha^{-1} $month^{-1}$ in bulk soil.

The soil mesofauna comprises microarthropods, the mites and springtails, and enchytraeids. The latter are assumed to be 80% microbivorous (bacteria and fungi) and 20% saprophagous (Didden et al. 1997). Mites feed either on soil microflora or on dead plant material (Luxton 1972) and springtails particularly feed on fungi, but also on bacteria or algae (Rusek 1998). As for the soil microfauna, grazing by the mesofauna may enhance mineralization of nutrients (Verhoef and Brussaard 1990; Beare et al. 1992). Moreover, microarthropods and enchytraeids affect decomposition processes largely by comminution, channeling and mixing of litter and soil. Comminution of uningested organic matter increases the surface area for microbial colonization. By fragmentation or mastication of plant debris, previously unexploited areas become available as a resource, which stimulates microbial activity and thereby enhances organic matter decomposition (Seastedt 1984; Visser 1985). Comminution followed by consumption and defecation results in a substrate with changed chemical quality and microstructure (pore volume, moisture retention, aeration) and these animal pellets can act as enrichment sites supporting microbial development (Parkinson 1988). The channeling and mixing of organic material by the mesofauna are largely done by the burrowing activity of enchytraeids. Their microtunnels in the soil system provide changed environmental conditions for microbial growth and sporulation (van Vliet et al. 1995). In addition, the mesofauna is mobile and migrates through different soil layers, passively transporting bacteria, fungi and their propagules in the gut or on the body surface to new microsites and substrates. Despite being miniscular compared to macrofauna species like earthworms and millipedes, which are the main bioturbators, the soil mesofauna may significantly contribute to forming the microbial habitat. As documented by Babel and Vogel (1989), humus material (H layer) of forest ecosystems on acid soils may almost entirely consist of fecal pellets of collembolans and enchytraeids.

Micro- and mesofauna do not affect their food source solely by harvesting; selective grazing on certain microbial species may also change the community structure of the microflora. This alters abundance and activity of bacteria and fungi and modifies the pattern of organic matter decay (Hanlon 1981; Moore et al. 1987; Klironomos et al. 1992). Feeding preferences are well known for the fungivore fauna, e.g., nematodes (Ruess et al. 2000), mites (Kaneko 1995; Maraun et al. 1998) and springtails (Visser and Whittaker 1977; Shaw 1985; Sadaka-Laulan et al. 1998). Fungivorous microarthropods with specific preferences may apply high grazing pressure on a favored species and alter the outcome of competition between fungi (Newell 1984; Klironomos et al. 1992). On the other hand, many microarthropods are general grazers and switch food sources (Chen et al. 1995; Maraun et al. 2003). Their grazing activity probably increases fungal species diversity as more species can coexist when the biomass of single individuals is kept low. This indicates that the microbial succession observed on decaying litter is influenced to a considerable extent by the different grazing activities of the soil fauna, which in turn affects decomposition processes.

4 Feedbacks of Faunal–Microbial Interactions on Plant Growth

In general, the interactions between plants and the decomposer food web are mutualistic, as they are beneficial to both. While plants add carbon sources to the soil, which promote the decomposer fauna, the latter enhances nutrient supply to plants (Scheu and Setälä 2002; Wardle 2002). Microbial and faunal activities consume carbon, whereas nutrients are temporarily immobilized in their biomass (Lavelle 1997). As documented above, the soil micro- and mesofauna affect the mobilization and immobilization of nutrients by modifying the activity of microorganisms, and this

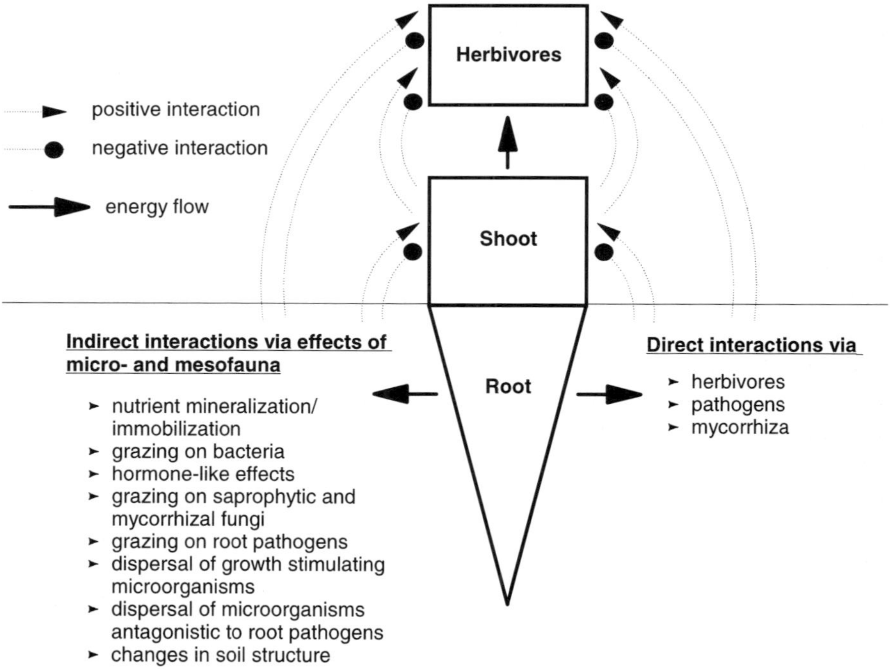

Fig. 5. Ways in which belowground interactions may affect plant growth and thus herbivore performance. Both direct and indirect interactions between roots and soil biota may stimulate or reduce herbivore development (Modified from Scheu and Setälä 2002)

has been shown to feed back to plants (Ingham 1992; Ekelund and Rønn 1994; Bonkowski et al. 2000b, 2001b; Scheu 2001; Wardle 2002). However, there is a variety of other interactions by which soil invertebrates modify plant growth, including grazing on mycorrhizal fungi (Klironomos and Kendrick 1995; Setälä 1995; Gange 2000), grazing on plant pathogens (Curl et al. 1988; Lartey et al. 1994; Pussard et al. 1994; Sabatini and Innocenti 2001; Klironomos 2002), dispersal of plant growth-stimulating microorganisms (Gange 1993; Harinikumar and Bagyaraj 1994; Lussenhop 1996), and of microorganisms antagonistic to root pathogens (Stephens and Davoren 1997). Moreover, microbial grazers indirectly modify plant growth by hormone-like effects (Jentschke et al. 1995; Muscolo et al. 1999; Bonkowski and Brandt 2002). These indirect effects of soil invertebrates add considerably to the spectrum of interactions in the rhizosphere driving plant performance (Fig. 5). Recently, it has been documented that these effects propagate further into the aboveground food web by driving the palatability of plants to herbivores (Scheu et al. 1999; Bonkowski et al. 2001a). Increasingly, it has been realized that soil invertebrate–microbial interactions are key factors, not only for plant performance, but also for the whole aboveground system (Scheu 2001; Setälä 2002; Wardle 2002). To further elaborate these interactions, we will explore feedbacks caused by bacterial and fungal grazers in more detail.

4.1 The Bacterial Food Chain

Most bacteria in bulk soil are in a dormant state, but readily burst into activity when water or easily decomposable substrates become available. This is followed by a succession of microfaunal predators, such as protozoa and bacteria-feeding nematodes (Clarholm 1981; Christensen et al. 1992; Rønn et al. 1996; de Nobili et al. 2001). Peaks of bacterial and microfaunal activity typically are found on freshly decomposing plant debris (Christensen et al. 1992; Bonkowski et al. 2000a) and in animal fecal pellets (Tajovský et al. 1992; Bonkowski and Schaefer 1997), but are also induced by root exudates (Hawes 1991; Coûteaux et al. 1988; Bonkowski and Brandt 2002), leading to a short-term increase in decomposition and mineralization of nutrients in the root vicinity, known as the 'priming effect' (Kuzyakov 2002).

Naked amoebae are the most important bacterial grazers in soil due to their high biomass and specialized feeding mode (Clarholm 1994). In contrast to suspension and filter feeders like bacterivorous nematodes or other protozoa, amoebae are grazing bacterial biofilms and colonies attached to surfaces and thus have access to the majority of bacteria in soil. With pseudopodia, amoebae reach bacterial colonies in soil pores inaccessible to other predators (Ekelund and Rønn 1994). Protozoa respond quickly to the growth of bacterial prey and enable an effective control of bacteria (Clarholm 1981). Compared to protozoa, nematodes are more mobile and may migrate to 'hot spots' of bacterial activity (Griffiths and Caul 1993), where they graze on both bacteria and amoebae (Alphei et al. 1996; Rønn et al. 1996; Bonkowski et al. 2000a).

The interplay between bacteria and bacterial predators determines the rates of nutrient cycling and strongly enhances the availability of mineral nutrients to plants (Clarholm 1984; Ingham et al. 1985; Gerhardson and Clarholm 1986; Ritz and Griffiths 1987; Kuikman et al. 1990; Jentschke et al. 1995; Alphei et al. 1996; Bonkowski et al. 2000a). The assumed mechanism, known as the 'microbial loop in soil' (Clarholm 1985), is triggered by the release of root exudates from plants which increase bacterial growth in the rhizosphere. The bacterial biomass uses carbon resources from exudates, while other nutrients, particularly nitrogen, are mobilized from soil organic matter. Microfaunal grazing in the rhizosphere is particularly important because nutrients are sequestered by bacteria (Kaye and Hart 1997; Wang and Bakken 1997), and remain locked up in bacterial biomass without remobilization by protozoa and nematodes (Christensen et al. 1992; Griffiths and Caul 1993; Griffiths et al. 1993; Bonkowski et al. 2000a). However, other indirect interactions of microfaunal grazers and plants may even be more important (Bonkowski and Brandt 2002). Protozoa have been documented to increase plant biomass without increasing nitrogen concentrations in plant tissue and plant nitrogen uptake (Alphei et al. 1996). Jentschke et al. (1995) documented that biomass of spruce seedlings increased in the presence of protozoa by >60% even though nutrients were provided in excess.

Protozoa do not indiscriminately ingest bacteria, rather they selectively feed on certain species (Boenigk and Arndt 2002). In freshwater systems significant changes in bacterial diversity due to protozoan grazing have been documented (Pernthaler et al. 1997; Jürgens et al. 1999; Posch et al. 1999), and this also appears to occur in the rhizosphere of plants (Griffiths et al. 1999; Bonkowski and Brandt 2002). Grazing-induced changes in microbial community composition and microbial functioning affect fundamental ecosystem properties because soil bacteria essentially control nutrient cycling and plant growth. For example, nitrogen fixing, nitrifying and denitrifying bacteria are in command of the nitrogen cycle (Mengel 1996). Nitrifying bacteria have been shown to be stimulated by protozoan grazers, presumably by preying on fast-growing bacterial competitors (Griffiths 1989; Alphei et al. 1996; Bonkowski et al. 2000a).

Specialized bacteria are the dominant colonizers of plant roots (Marilley and Aragno 1999), and many of the rhizosphere bacteria affect plant performance by producing hormones (Brown 1972; Costacurta and Vanderleyden 1995; Arshad and Frankenberger 1998; Lambrecht et al. 2000). Up to 80% of the bacteria isolated from plant rhizospheres are considered to produce auxins (Barea et al. 1976; Patten and Glick 1996), and Holland (1997) suggested that cytokinins in plants may originate exclusively from microorganisms. The widespread ability of both beneficial and deleterious rhizosphere microorganisms to produce plant hormones suggests that rhizosphere bacteria play an important role in manipulating root and plant growth (Shishido et al. 1996; Rolfe et al. 1997). However, the effects of rhizobacteria on root architecture seem to be directed by protozoan grazing (Bonkowski and Brandt 2002). Plants develop an extensive and highly branched root system in the presence of protozoa (Jentschke et al. 1995), corresponding to the hormonal effects on root growth by beneficial rhizobacteria (Chanway et al. 1988; Petersen et al. 1996; Rolfe et al. 1997). Thus, in addition to the stimulation of gross nutrient flows, protozoa promote a loosely mutualistic interaction between plant roots and rhizobacteria (Bonkowski and Brandt 2002). Protozoan grazing has been found to promote auxin-producing rhizobacteria. Accordingly, the growth of the root system is stimulated and allows more nutrients to be absorbed, but will also increase exudation rates, thereby further stimulating bacterial–protozoan interactions (Bonkowski and Brandt 2002).

4.2 The Fungal Food Chain

As stated above, feeding on fungi by soil invertebrates affects plant performance by fostering plant nutrition due to liberating nutrients locked in hyphae of saprophytic fungi, by changing the effectiveness of plant–mycorrhiza associations and by reducing the severity of fungal root pathogens. We concentrate here on the effects of micro- and mesofauna on saprophytic vs. mycorrhizal fungi.

Soil fungi generally dominate in bulk soils rich in organic matter. Due to their diverse enzymatic capabilities, saprophytic fungi are determinants of decomposition processes in soils (Dighton and Boddy 1989; Rosenbrock et al. 1995; Frankland 1998). Fungi degrade polymer substrates like cellulose, hemicellulose or chitin and are responsible for the breakdown of complex aromatic compounds such as lignin or humic and phenolic acids. The degradation of these recalcitrant substrates releases simple molecules that are subsequently used by other soil organisms.

Food preferences of the fauna affect fungal community structure and decomposition processes (see above). However, feeding on hyphae of saprophytic fungi has ecological implications quite different from feeding on mycorrhizal fungi. Preferential grazing on saprophytes primarily affects net mineralization, which modifies the soil nutrient pool (Trofymow et al. 1983; Faber 1991; Rihani et al. 1995; Takeda 1995; Chen and Ferris 1999). Grazing can increase the amount of nutrients in the soil that may otherwise become immobilized in the fungal biomass (Hanlon 1981). Plants may benefit from this and respond with enhanced growth and tissue nutrient content (Setälä et al. 1998; Laakso and Setälä 1999).

Mycorrhizal associations form a closer linkage where the fungi directly access nutrients in organic matter and translocate these to the plant. Thus, fungi increase plant nutrient uptake and, consequently, plant growth, and have the potential to induce important aboveground effects. The fungivorous fauna has been shown to control the growth of mycorrhizal fungi, resulting in a modified plant–fungus association. Grazing of nematodes on ectomycorrhizal fungi may restrict mycorrhizal development and limit nutrient uptake by plants, which can lead to a reduction in the yield of mycorrhizal host plants (e.g., trees; Clark 1964; Riffle 1975; Giannakis and Sanders 1987, 1989). Heavy grazing pressure on vesicular-arbuscular mycorrhizae by fungus-feeding nematodes can destroy sufficient hyphae to limit phosphorus uptake to a level inadequate for nodulation in legumes and cause reduced mycorrhizal colonization (Ingham et al. 1986; Ingham 1988). Springtails have been shown to negatively affect arbuscular mycorrhizal and ectomycorrhizal symbioses, mainly by grazing and severing the associated external fungal network from host roots (Ek et al. 1994; Hiol et al. 1994; Klironomos and Ursic 1998; Klironomos et al. 1999). However, the reduction in fungal biomass due to faunal grazing is not always harmful for the plant–fungus symbiosis; it rather depends on sufficient mobilization of nutrients in the mycorrhizosphere and the nitrogen status of the soil (Setälä 1995; Setälä et al. 1997). Choice experiments suggest that Collembola preferentially feed on saprophytic fungi in the rhizosphere, which might indirectly be beneficial for plants through an enhancement of mycorrhizal functioning (Gange 2000).

Soil food webs are regarded as donor-controlled, where the density of the resource controls the density of the consumers (Pimm 1982). However, the fungal food chain is a species-rich community of fungivores where trophic interactions and top-down control can have an important impact. Although predator–prey relationships among the fungivorous fauna probably affect soil processes, their effects on nutrient availability and plant growth are not well investigated. Some studies provide evidence that mineralization of nitrogen and carbon increases with food web complexity (Allen-Morley and Coleman 1989; Setälä et al. 1998), whereas others report primary production to be insensitive (Laasko and Setälä 1999). In general, species composition within a functional group of fungivores, i.e., the physiological attributes of the dominant species, has a greater impact than diversity within a group, indicating functional redundancy (Setälä et al. 1998; Cragg and Bardgett

2001).

Microcosm experiments that compared different subsets of microflora and microfauna (nematodes) showed that food web structure did affect bacterial, but not fungal biomass (Mikola 1998; Mikola and Setälä 1998, 1999). Supplementary energy (glucose) increased the carbon flow through the fungal channel, but decreased it through the bacterial channel. This indicates that the nature of trophic interactions may differ between the fungal and bacterial pathways. Fungi are not restricted to the same extent by grazing as bacteria. Bacteria generally have higher nutrient contents and are consumed entirely, whereas fungi are modular organisms and possess a wide range of antigrazing defences (morphological and chemical), which may hinder the attack of small faunal grazers like nematodes (Wardle and Yeates 1993). However, it should be emphasized that nematodes are only a part of the fungal food chain. Other fungal feeders, like microarthropods, caused changes in the nutrient cycle due to food web architecture. Mebes and Filser (1998) showed that soil systems containing a single springtail species had an immobilizing effect on nitrogen turnover, whereas species mixtures increased nitrogen mineralization. Higher trophic levels, such as predatory mites, can affect these fungivores, resulting in an enhanced decomposition rate due to a lower level of fungal grazing (Hedlund and Öhrn 2000). This suggests that different, but trophically equivalent groups, i. e., the fungivorous micro- and mesofauna, can have dissimilar influences on energy and nutrient flux in the fungal decomposition pathway, which results in a different impact on plant growth.

5 Conclusions

The soil micro- and mesofauna is intimately linked to microbial life in soil. Microfauna groups such as protozoa and bacterial-feeding nematodes heavily graze on soil bacteria which is of particular importance in the rhizosphere of plants. Bacterial feeders appear to be little controlled by top predators, resulting in an effective control of bacteria particularly in rhizosphere soil. Fungivorous mesofauna, most importantly collembolans, oribatid mites and enchytraeids, may also exert strong grazing pressure on their prey organisms, however, they appear to be controlled more strongly by top predators, which may limit their impact on soil fungi. In general, grazing on microorganisms stimulates decomposition processes and results in increased mineralization of nutrients with important feedbacks to plant growth. However, micro- and mesofauna-mediated changes in plant growth are complex. Grazing on bacteria significantly alters the composition of the bacterial community and has been shown, e. g., to foster nitrifiers with important implications for nutrient cycling and ecosystem functioning. Changes in the composition of the bacterial community are also likely to be responsible for hormone-like effects of protozoa on plant growth. The effects of fungal grazers on plant performance strongly depend on whether fungivorous mesofauna feeds on saprophytic, mycorrhizal or pathogenic species. Since decomposition of litter materials often is limited by nutrients such as nitrogen and phosphorus, saprophytic fungi probably compete with mycorrhiza for mineral nutrients. An important mechanism by which fungal grazers modify plant growth is, therefore, selective grazing on either saprophytic or mycorrhizal fungi. For arbuscular mycorrhizal plants, there is increasing evidence that beneficial effects of fungal grazers in part are due to selective grazing on saprophytic fungi, thereby favoring mycorrhiza. Although it is still difficult to define the explicit effects of the different faunal groups on decomposition processes, nutrient cycling and plant growth, our knowledge has advanced considerably. New methodological developments including molecular and stable isotope techniques provide the opportunity to analyse the complex interactions between microorganisms and their micro- and mesofauna grazers in unprecedented detail, promising fundamental progress in the near future.

References

Agren GI, Bosatta E (1996) Quality: a bridge between theory and experiment in soil organic matter studies. Oikos 76:522–528

Allen-Morley CR, Coleman DC (1989) Resilience of soil biota in various food webs to freezing perturbations. Ecology 70:1124–1141

Alphei J, Bonkowski M, Scheu S (1996) Protozoa, Nematoda and Lumbricidae in the rhizosphere of *Hordelymus europaeus* (Poaceae): faunal interactions, response of microorganisms and effects on plant growth. Oecologia 106:111–126

Anderson JM (1975) The enigma of soil animal species diversity. In: Vanek J (ed) Progress in soil zoology. Academia Prague, Prague, pp 51–58

Arshad M, Frankenberger WT (1998) Plant growth-regulating substances in the rhizosphere: microbial production and functions. Adv Agron 62:45–151

Babel U, Vogel H-J (1989) An evaluation of the activity of Enchytraeidae and Collembola by soil thin sections. Pedobiologia 33:167–172

Bardgett RD, Keiller S, Cook R, Gilburn AS (1998) Dynamic interactions between soil animals and microorganisms in upland grassland soils amended with sheep dung: a microcosm experiment. Soil Biol Biochem 30:531–539

Barea JM, Navarro E, Montoya E (1976) Production of plant growth regulators by rhizosphere phosphate-solubilizing bacteria. J Appl Bacteriol 40:129–134

Beare MH, Parmelle RW, Hendrix PF, Cheng W, Coleman DC, Crossley DA Jr (1992) Microbial and faunal interactions and effects on litter nitrogen and decomposition in agroecosystems. Ecol Monogr 62:569–591

Beare MH, Reddy MV, Tian G, Srivastava SC (1997) Agricultural intensification, soil biodiversity and agroecosystem function in the tropics – the role of decomposer biota. Appl Soil Ecol 6:87–108

Bengtsson J (1994) Temporal predictability in forest soil communities. J Anim Ecol 63:653–665

Bengtsson J, Setälä H, Zheng DW (1996) Food webs and nutrient cycling in soils: interactions and positive feedbacks. In: Polis GA, Winemiller KO (eds) Food webs. Integration of patterns and dynamics. Chapman and Hall, New York, pp 30–38

Bezdicek DF, Kennedy AC (1979) Economic microbial ecology: Symbiotic nitrogen fixation and nitrogen cycling in terrestrial environments. In: Lynch JM, Hobbie JE (eds) Microorganisms in action: concepts and applications in microbial ecology. Blackwell, Oxford, pp 241–260

Boenigk J, Arndt H (2002) Bacterivory by heterotrophic flagellates: community structure and feeding strategies. Antonie van Leeuwenhoek 81:465–480

Bonkowski M, Schaefer M (1997) Interactions between earthworms and soil protozoa – a trophic component in the soil food web. Soil Biol Biochem 29:499–502

Bonkowski M, Brandt F (2002) Do soil protozoa enhance plant growth by hormonal effects? Soil Biol Biochem 34:1709–1715

Bonkowski M, Griffiths B, Scrimgeour C (2000a) Substrate heterogeneity and microfauna in soil organic 'hotspots' as determinants of nitrogen capture and growth of ryegrass. Appl Soil Ecol 14:37–53

Bonkowski M, Cheng W, Griffiths BS, Alphei J, Scheu S (2000b) Microbial-faunal interactions in the rhizosphere and effects on plant growth. Eur J Soil Biol 36:135–147

Bonkowski M, Geoghegan IE, Birch ANE, Griffiths BS (2001a) Effects of soil decomposer invertebrates (protozoa and earthworms) on an above-ground phytophagous insect (cereal aphid) mediated through changes in the host plant. Oikos 95:441–450

Bonkowski M, Jentschke G, Scheu S (2001b) Contrasting effects of microbial partners in the rhizosphere: interactions between Norway spruce seedlings (*Picea abies* Karst.), mycorrhiza (*Paxillus involutus* (Batsch) Fr.) and naked amoebae (protozoa). Appl Soil Ecol 18:193–204

Brown ME (1972) Plant growth substances produced by micro-organisms of soil and rhizosphere. J Appl Bacteriol 35:443–451

Chanway CP, Nelson LM, Holl FB (1988) Cultivar specific growth promotion of spring wheat (*Triticum aestivum* L.) by co-existent *Bacillus* species. Can J Microbiol 34:925–929

Chen B, Snider RJ, Snider RM (1995) Food preference and effects of food type on the life history of some soil Collembola. Pedobiologia 39:496–505

Chen J, Ferris H (1999) The effects of nematode grazing on nitrogen mineralization during fungal decomposition of organic matter. Soil Biol Biochem 31:1265–1279

Christensen M (1989) A view of fungal ecology. Mycologia 81:1–19

Christensen S, Griffiths BS, Ekelund F, Rønn R (1992) Huge increase in bacterivores on freshly killed barley roots. FEMS Microbiol Ecol 86:303–309

Clarholm M (1981) Protozoan grazing of bacteria in soil – impact and importance. Microb Ecol 7:343–350

Clarholm M (1984) Heterotrophic, free-living protozoa: neglected microorganisms with an important task in regulating bacterial populations. In: Klug MJ, Reddy CA (eds) Current perspectives on microbial ecology. American Society for Microbiology, Washington, DC, pp 321–326

Clarholm M (1985) Interactions of bacteria, protozoa and plants leading to mineralization of soil nitrogen. Soil Biol Biochem 17:181–187

Clarholm M (1994) The microbial loop. In: Ritz K, Dighton J, Giller KE (eds) Beyond the biomass. Wiley, New York, pp 221–230

Clark WC (1964) Fungal-feeding nematodes as possible plant pathogens. NZ J Agric Res 7:441–443
Costacurta A, Vanderleyden J (1995) Synthesis of phytohormones by plant-associated bacteria. Crit Rev Microbiol 21:1–18
Coûteaux M-M, Faurie G, Palka L, Steinberg C (1988) Le relation prédateur-proie (protozoaires-bactéries) dans les sols: rôle dans la régulation des populations et conséquences sur les cycles du carbone et de l'azote. Rev Ecol Biol Sol 25:1–31
Cragg RG, Bardgett RD (2001) How changes in soil faunal diversity and composition within a trophic group influence decomposition processes. Soil Biol Biochem 33:2073–2081
Curl EA, Lartey R, Peterson CM (1988) Interactions between root pathogens and soil microarthropods. Agric Ecosyst Environ 24:249–261
De Nobili M, Contin M, Mondini C, Brookes PC (2001) Soil microbial biomass is triggered into activity by trace amounts of substrate. Soil Biol Biochem 33:1163–1170
De Ruiter PC, Griffiths B, Moore JC (2002) Biodiversity and stability in soil ecosystems: patterns, processes and the effects of disturbance. In: Loreau M, Naeem S, Inchausti P (eds) Biodiversity and ecosystem functioning – Synthesis and perspectives. Oxford Univ Press, Oxford, pp 102–113
Didden WAM, Fründ H-C, Graefe U (1997) Enchytraeids. In: Benckiser G (ed) Fauna in soil ecosystems. Recycling processes, nutrient fluxes, and agricultural production. Dekker, New York, pp 135–172
Digthon J, Boddy L (1989) Role of fungi in nitrogen, phosphorus and sulphur cycling in temperate forest ecosystems. In: Boddy L, Marchant R, Read DJ (eds) Nitrogen, phosphorus and sulphur utilization by fungi. Cambridge Univ Press, Cambridge, pp 269–298
Ek H, Sjögren M, Arnebrant K, Söderström B (1994) Extramatrical mycelial growth, biomass allocation and nitrogen uptake in ectomycorrhizal systems in response to collembolan grazing. Appl Soil Ecol 1:155–169
Ekelund F, Rønn R (1994) Notes on protozoa in agricultural soil with emphasis on heterotrophic flagellates and naked amoebae and their ecology. FEMS Microbiol Rev 15:321–353
Faber JH (1991) Functional classification of soil fauna: a new approach. Oikos 62:110–117
Ferris H, Venette RC, Lau SS (1997) Population energetics of bacterial-feeding nematodes: carbon and nitrogen budgets. Soil Biol Biochem 29:1183–1194
Frankland JC (1998) Fungal succession – unravelling the unpredictable. Mycol Res 102:1–15
Freckman DW (1988) Bacterivorous nematodes and organic-matter decomposition. Agric Ecosyst Environ 24:195–217
Gange AC (1993) Translocation of mycorrhizal fungi by earthworms during early succession. Soil Biol Biochem 25:1021–1026
Gange AC (2000) Arbuscular mycorrhizal fungi, Collembola and plant growth. Trends Ecol Evol 15:369–372
Gerhardson B, Clarholm M (1986) Microbial communities on plant roots. In: Jensen V, Kjøller A, Sørensen LH (eds) Microbial communities in soil. Elsevier, London, pp 19–34
Giannakis N, Sanders FE (1987) Interactions between nematodes and vesicular-arbuscular mycorrhizal fungi. J Sci Food Agric 40:123–125
Giannakis N, Sanders FE (1989) Interactions between mycophagous nematodes, mycorrhizal and other soil fungi. Agric Ecosyst Environ 29:163–167
Giller PS (1996) The diversity of soil communities, the poor mans tropical rainforest. Biodiv Cons 5:135–168
Griffiths BS (1989) Enhanced nitrification in the presence of bacteriophagous protozoa. Soil Biol Biochem 21:1045–1052
Griffiths BS (1994) Soil nutrient flow. In: Darbyshire JF (ed) Soil Protozoa. CAB International, Wallingford, pp 65–91

Griffiths BS, Robinson D (1992) Root-induced nitrogen mineralization: a nitrogen balance model. Plant Soil 139:253–263
Griffiths BS, Caul S (1993) Migration of bacterial-feeding nematodes, but not protozoa, to decomposing grass residues. Biol Fertil Soils 15:201–207
Griffiths BS, Bardgett RD (1997) Interactions between microbe-feeding invertebrates and soil microorganisms. In: Van Elsas JD, Trevors JT, Wellington EMH (eds) Modern soil microbiology. Dekker, New York, pp 165–182
Griffiths BS, Ekelund F, Rønn R, Christensen S (1993) Protozoa and nematodes on decomposing barley roots. Soil Biol Biochem 25:1293–1295
Griffiths BS, Bonkowski M, Dobson G, Caul S (1999) Changes in soil microbial community structure in the presence of microbial-feeding nematodes and protozoa. Pedobiologia 43:297–304
Gunn A, Cherrett JM (1993) The exploitation of food resources by soil meso-invertebrates and macro-invertebrates. Pedobiologia 37:303–320
Hanlon RDG (1981) Influence of grazing by Collembola on the activity of senescent fungal colonies grown on media of different nutrient concentration. Oikos 36:362–367
Harinikumar KM, Bagyaraj DJ (1994) Potential of earthworms, ants, millipedes, and termites for dissemination of vesicular-arbuscular mycorrhizal fungi in soil. Biol Fertil Soils 18:115–118
Hättenschwiler S, Tiunov AV, Scheu S (2004) Biodiversity and litter decomposition in terrestrial ecosystems Annu Rev Ecol Syst (in press)
Hawes MC (1991) Living plant cells released from the root cap: a regulator of microbial populations in the rhizosphere? In: Kleister DL, Cregan PB (eds) The rhizosphere and plant growth. Kluwer, Dordrecht, pp 51–59
Hedlund K, Öhrn MS (2000) Tritrophic interactions in a soil community enhance decomposition rates. Oikos 88:585–591
Hiol FH, Dixon RK, Curl EA (1994) The feeding preference of mycophagous Collembola varies with the ectomycorrhizal symbiont. Mycorrhiza 5:99–103
Holland MA (1997) Occam's razor applied to hormonology: are cytokinins produced by plants? Plant Physiol 115:865–868
Ingham ER, Trofymow JA, Ames RN, Hunt HW, Morley CR, Moore JC, Coleman DC (1986) Trophic interactions and nitrogen cycling in a semi-arid grassland soil. II. System responses to removal of different groups of soil microbes or fauna. J Appl Ecol 23:615–630
Ingham RE (1988) Interaction between nematodes and vesicular-arbuscular mycorrhizae. Agric Ecosyst Environ 24:169–182
Ingham RE (1992) Interactions between invertebrates and fungi: Effects on nutrient availability. In: Carroll GC, Wicklow DT (eds) The fungal community: its organization and role in the ecosystem, 2nd edn. Mycological series 9. Dekker, New York, pp 669–690
Ingham RE, Trofymow JA, Ingham ER, Coleman DC (1985) Interactions of bacteria, fungi and their nematode grazers: effects of nutrient cycling and plant growth. Ecol Monogr 55:119–140
Jentschke G, Bonkowski M, Godbold DL, Scheu S (1995) Soil protozoa and forest tree growth: non-nutritional effects and interaction with mycorrhizae. Biol Fertil Soils 20:263–269
Joergensen RG, Scheu S (1999) Response of soil microorganisms to the addition of carbon, nitrogen and phosphorus in a forest rendzina. Soil Biol Biochem 31:859–866
Jones CG, Lawton JH, Shachak M (1994) Organisms as ecosystem engineers. Oikos 69:373–386
Jürgens K, Pernthaler J, Schalla S, Amann R (1999) Morphological and compositional changes in a planktonic bacterial community in response to enhanced protozoan grazing. Appl Environ Microbiol 65:1241–1250

Kaneko N (1995) Composition of feeding types in oribatid mite communities in forest soils. Acta Zool Fenn 196:160–161
Kaye JP, Hart SC (1997) Competition for nitrogen between plants and soil microorganisms. Trends Ecol Evol 12:139–143
Klironomos JN (2002) Feedback with soil biota contributes to plant rarity and invasiveness in communities. Nature 417:67–70
Klironomos JN, Kendrick B (1995) Relationships among microarthropods, fungi, and their environment. Plant Soil 170:183–197
Klironomos JN, Ursic M (1998) Density-dependent grazing on the extraradical hyphal network of the arbuscular mycorrhizal fungus, *Glomus intraradices*, by the collembolan, *Folsomia candida*. Biol Fertil Soils 26:250–253
Klironomos JN, Widden P, Deslandes I (1992) Feeding preferences of the collembolan *Folsomia candida* in relation to microfungal successions on decaying litter. Soil Biol Biochem 24:685–692
Klironomos JN, Bednarczuk EM, Neville J (1999) Reproductive significance of feeding on saprobic and arbuscular mycorrhizal fungi by the collembolan, *Folsomia candida*. Funct Ecol 13:756–761
Kuikman PJ, Jansen AG, van Veen JA, Zehnder AJB (1990) Protozoan predation and the turnover of soil organic carbon and nitrogen in the presence of plants. Biol Fertil Soils 10:22–28
Kuzyakov Y (2002) Review: factors affecting rhizosphere priming effects. J Plant Nutr Soil Sci 165:382–396
Laakso J, Setälä H (1999) Sensitivity of primary production to changes in the architecture of belowground food webs. Oikos 87:57–64
Lambrecht M, Okon Y, Vande Broek A, Vanderleyden J (2000) Indole-3-acetic acid: a reciprocal signalling molecule in bacteria–plant interactions. Trends Microbiol 8:298–300
Lartey RT, Curl EA, Peterson CM (1994) Interactions of mycophagous collembola and biological control fungi in the suppression of *Rhizoctonia solani*. Soil Biol Biochem 26:81–88
Lavelle P (1997) Faunal activities and soil processes: adaptive strategies that determine ecosystem function. Adv Ecol Res 27:93–122
Lussenhop J (1996) Collembola as mediators of microbial symbiont effects upon soybean. Soil Biol Biochem 28:363–369
Luxton M (1972) Studies on the oribatid mites of a Danish beech wood soil. I. Nutritional biology. Pedobiologia 12:434–463
Lynch JM, Whipps JM (1990) Substrate flow in the rhizosphere. Plant Soil 129:1–10
Maraun M, Migge S, Schaefer M, Scheu S (1998) Selection of microfungal food by six oribatid mite species (Oribatida, Acari) from two different beech forests. Pedobiologia 42:232–240
Maraun M, Migge S, Theenhaus A, Scheu S (2003) Adding to 'the enigma of soil animal diversity': fungal feeders and saprophagous soil invertebrates prefer similar food substrates. Eur J Soil Biol 39:85–95
Marilley L, Aragno M (1999) Phylogenetic diversity of bacterial communities differing in degree of proximity of *Lolium perenne* and *Trifolium repens* roots. Appl Soil Ecol 13:127–136
Marschner H (1992) Nutrient dynamics at the soil–root interface (rhizosphere). In: Read DJ, Lewis DH, Fitter AH, Alexander IJ (eds) Mycorrhizas in ecosystems. CAB International, Wallingford, pp 3–12
Mebes H, Filser J (1998) Does the species composition of Collembola affect nitrogen turnover? Appl Soil Ecol 9:241–247

Melillo JM, Aber JD, Muratare JF (1982) Nitrogen and lignin control of hardwood leaf litter decomposition dynamics. Ecology 63:621–626

Mengel K (1996) Turnover of organic nitrogen in soils and its availability to crops. Plant Soil 181:83–93

Mikola J (1998) Effects of microbivore species composition and basal resource enrichment on trophic-level biomasses in an experimental microbial-based soil food web. Oecologia 117:396–403

Mikola J, Setälä H (1998) No evidence of trophic cascades in an experimental microbial-based soil food web. Ecology 79:153–164

Mikola J, Setälä H (1999) Interplay of omnivory, energy channels and C availability in a microbial-based soil food web. Biol Fertil Soils 28:212–218

Mikola J, Bardgett RD, Hedlund K (2002) Biodiversity, ecosystem functioning and soil decomposer food webs. In: Loreau M, Naeem S, Inchausti P (eds) Biodiversity and ecosystem functioning – synthesis and perspectives. Oxford Univ Press, Oxford, pp 169–180

Moore JC, Ingham ER, Coleman DC (1987) Interspecific and intraspecific feeding selectivity of *Folsomia candida* (Willem) (Collembola, Isotomidae) on fungi. Biol Fertil Soils 5:6–12

Moore JC, Walter DE, Hunt HW (1988) Arthropod regulation of micro- and mesobiota in below-ground detrital food webs. Annu Rev Entomol 33:419–439

Muscolo A, Bovalo F, Gionfriddo F, Nardi S (1999) Earthworm humic matter produces auxin-like effects on *Daucus carota* cell growth and nitrate metabolism. Soil Biol Biochem 31:1303–1311

Newell K (1984) Interactions between two decomposer basidiomycetes and a collembolan under Sitka spruce: grazing and its potential effects on fungal distribution and litter decomposition. Soil Biol Biochem 16:235–239

Parker LW, Santos PF, Phillips J, Whitford WG (1984) Carbon and nitrogen dynamics during the decomposition of litter and roots of a Chihuahuan desert annual, *Lepidium lasiocacarpum*. Ecol Monogr 54:339–360

Parkinson D (1988) Linkages between resource availability, microorganisms and soil invertebrates. Agric Ecosyst Environ 24:21–32

Patten CL, Glick BR (1996) Bacterial biosynthesis of indole-3-acetic acid. Can J Microbiol 42:207–220

Pernthaler J, Posch T, Simek K, Vrba J, Amann R, Psenner R (1997) Contrasting bacterial strategies to coexist with a flagellate predator in an experimental microbial assemblage. Appl Environ Microbiol 63:596–601

Petersen DJ, Srinivasan M, Chanway CP (1996) *Bacillus polymyxa* stimulates increased *Rhizobium etli* populations and nodulation when co-resident in the rhizosphere of *Phaseolus vulgaris*. FEMS Microbiol Lett 142:271–276

Pimm SL (1982) Food webs. Chapman and Hall, London

Polis GA (1991) Complex trophic interactions in deserts: an empirical critique of food-web theory. Am Nat 138:123–155

Ponsard S, Arditi R (2000) What can stable isotopes (δ^{15}N and δ^{13}C) tell about the food web of soil macro-invertebrates? Ecology 81:852–864

Posch T, Simek K, Vrba J, Pernthaler J, Nedoma J, Sattler B, Sonntag B, Psenner R (1999) Predator-induced changes of bacterial size-structure and productivity studied on an experimental microbial community. Aquat Microb Ecol 18:235–246

Pussard M, Alabouvette C, Levrat P (1994) Protozoa interactions with soil microflora and possibilities for biocontrol of plant pathogens. In: Darbyshire JF (ed) Soil protozoa. CAB International, Wallingford, pp 123–146

Riffle JW (1975) Two *Aphelenchoides* species suppress formation of *Suillus granulatus* ectomycorrhizae with *Pinus ponderosa* seedlings. Plant Dis Rep 59:951–955

Rihani M, Cancela de Fonseca JP, Kiffer E (1995) Decomposition of beech leaf litter by microflora and mesofauna. 2. Food preferences and action of oribatid mites on different substrates. Eur J Soil Biol 31:67–79

Ritz K, Griffiths BS (1987) Effects of carbon and nitrate additions to soil upon leaching of nitrate, microbial predators and nitrogen uptake by plants. Plant Soil 102:229–237

Rolfe BG, Djordjevic MA, Weinman JJ, Mathesius U, Pittock C, Gärtner E, Ride EM, Dong Z, McCully M, McIver J (1997) Root morphogenesis in legumes and cereals and the effect of bacterial inoculation on root development. Plant Soil 194:131–144

Rønn R, Griffiths BS, Ekelund F, Christensen S (1996) Spatial distribution and successional pattern of microbial activity and micro-faunal populations on decomposing barley roots. J Appl Ecol 33:662–672

Rosenbrock P, Buscot F, Munch JC (1995) Fungal succession and changes in the fungal degradation potential during the initial stage of litter decomposition in a black alder forest (*Alnus glutinosa* (L) Gaertn.). Eur J Soil Biol 31:1–11

Ruess L, García Zapata EJ, Dighton J (2000) Food preferences of a fungal-feeding *Aphelenchoides* species. Nematology 2:223–230

Rusek J (1998) Biodiversity of Collembola and their functional role in the ecosystem. Biodiv Cons 7:1207–1219

Ryle GJA, Powell CE, Gordon AJ (1979) The respiratory costs of nitrogen fixation in soybean, cowpea and white clover. J Exp Bot 30:145–153

Sabatini MA, Innocenti G (2001) Effects of Collembola on plant-pathogenic fungus interactions in simple experimental systems. Biol Fertil Soils 33:62–66

Sadaka-Laulan N, Ponge JF, Roquebert MF, Bury E, Boumezzough A (1998) Feeding preferences of the collembolan *Onychiurus sinensis* for fungi colonizing holm oak litter (*Quercus rotundifolia* Lam.). Eur J Soil Biol 34:179–188

Scheu S (2001) Plants and generalist predators as links between the below-ground and above-ground system. Basic Appl Ecol 2:3–13

Scheu S, Schaefer M (1998) Bottom-up control of the soil macrofauna community in a beechwood on limestone: manipulation of food resources. Ecology 79:1573–1585

Scheu S, Falca M (2000) The soil food web of two beech forests (*Fagus sylvatica*) of contrasting humus type: stable isotope analysis of a macro- and a mesofauna-dominated community. Oecologia 123:285–296

Scheu S, Setälä H (2002) Multitrophic interactions in decomposer communities. In: Tscharntke T, Hawkins BA (eds) Multitrophic level interactions. Cambridge Univ Press, Cambridge, pp 223–264

Scheu S, Theenhaus A, Jones TH (1999) Links between the detritivore and the herbivore system: effects of earthworms and Collembola on plant growth and aphid development. Oecologia 119:541–551

Seastedt TR (1984) The role of microarthropods in decomposition and mineralization processes. Annu Rev Entomol 29:25–46

Setälä H (1995) Growth of birch and pine seedlings in relation to grazing by soil fauna on ectomycorrhizal fungi. Ecology 76:1844–1851

Setälä H (2002) Sensitivity of ecosystem functioning to changes in trophic structure, functional group composition and species diversity in belowground food webs. Ecol Res 17:207–215

Setälä H, Rissanen J, Markkola AM (1997) Conditional outcomes in the relationship between pine and ectomycorrhizal fungi in relation to biotic and abiotic environment. Oikos 80:112–122

Setälä H, Laakso J, Mikola J, Huhta V (1998) Functional diversity of decomposer organisms in relation to primary production. Appl Soil Ecol 9:25–31

Shaw PJA (1985) Grazing preferences of *Onychiurus armatus* (Insecta: Collembola) for mycorrhizal and saprophytic fungi in pine plantations. In: Fitter AH (ed) Microbial interactions in soil. Blackwell, Oxford, pp 333–337

Shishido M, Massicotte HB, Chanway CP (1996) Effect of plant growth promoting *Bacillus* strains on pine and spruce seedling growth and mycorrhizal infection. Ann Bot 77:433–441

Smith SE, Read DJ (1997) Mycorrhizal symbiosis. Academic Press, London

Söderström B (1992) The ecological potential of the ectomycorrhizal mycelium. In: Read DJ, Lewis DH, Fitter AH, Alexander IJ (eds) Mycorrhizas in ecosystems. CAB International, Wallingford, pp 77–83

Stephens PM, Davoren CW (1997) Influence of the earthworms *Aporrectodea trapezoides* and *A. rosea* on the disease severity of *Rhizoctonia solani* on subterranean clover and ryegrass. Soil Biol Biochem 29:511–516

Tajovský K, Sautrichova H, Háněl L, Balik V, Lukesova A (1992) Decomposition of faecal pellets of the millipede *Glomeris hexasticha* (Diplopoda) in forest soil. Pedobiologia 36:146–158

Takeda H (1995) Changes in the collembolan community during the decomposition of needle litter in a coniferous forest. Pedobiologia 39:304–317

Taylor BR, Parkinson D, Parsons WFJ (1989) Nitrogen and lignin content as predictors of litter decay rates: a microcosm test. Ecology 70:97–104

Trofymow JA, Morley CR, Coleman DC, Anderson RV (1983) Mineralization of cellulose in the presence of chitin and assemblages of microflora and fauna in soil. Oecologia 60:103–110

Van Veen JA, Merckx R, van de Geiju SC (1989) Plant and soil related controls of the flow of carbon from roots through the soil microbial biomass. Plant Soil 115:179–188

Van Vliet PCJ, Beare MH, Coleman DC (1995) Population dynamics and functional roles of Enchytraeidae (Oligochaeta) in hardwood forest and agricultural ecosystems. Plant Soil 170:199–207

Verhoef HA, Brussaard L (1990) Decomposition and nitrogen mineralization in natural and agroecosystems: the contribution of soil animals. Biogeochemistry 11:175–212

Visser S (1985) Role of soil invertebrates in determining the composition of soil microbial communities. In: Fitter AH (ed) Ecological interactions in soil. Blackwell, Oxford, pp 297–317

Visser S, Whittaker JB (1977) Feeding preferences for certain litter fungi by *Onychiurus subtenuis* (Collembola). Oikos 29:320–325

Wamberg C, Christensen S, Jakobsen I (2003) Interaction between foliar-feeding insects, mycorrhizal fungi, and rhizosphere protozoa on pea plants. Pedobiologia 47:281–287

Wang JG, Bakken LR (1997) Competition for nitrogen during mineralization of plant residues in soil: microbial response to C and N. Soil Biol Biochem 29:163–170

Wardle DA (2002) Linking the aboveground and belowground components. Princeton Univ Press, Princeton

Wardle DA, Yeates GW (1993) The dual importance of competition and predation as regulatory forces in terrestrial ecosystems: evidence from decomposer food-webs. Oecologia 93:303–306

Wardle DA, Bonner KI, Nicholson KS (1997) Biodiversity and plant litter – experimental evidence which does not support the view that enhanced species richness improves ecosystem function. Oikos 79:247–258

Wardle DA, Verhoef HA, Clarholm M (1998) Trophic relationships in the soil microfood-web: predicting the responses to a changing global environment. Global Change Biol 4:713–727

Part V
Function of Microbes in Specific Soil Compartments

13 Transgenic Rhizospheres of Crop Plants: Their Impact on Indigenous Soil Fungi

Valeria Bianciotto[1], Mariangela Girlanda[1], Alexandra Lazzari[1], Gilda Cappellazzo[1], Silvia Perotto[1], Paola Bonfante[1]

1 Introduction

Durable and broad-spectrum resistance to soilborne fungal pathogens is a desirable agronomic trait in the rhizosphere of crop plants. This may be achieved by following different strategies: (1) by introducing genetically modified (GM) biocontrol microorganisms that grow and multiply in the rhizosphere (Cook et al. 1996) and (2) by modifying the plant genome to obtain constitutive expression of antimicrobial compounds (Broekaert et al. 1995; Lorito et al. 1998; Punja 2001). Several crops have been recently transformed to resist not only viruses and bacterial pathogens, but also fungal pathogens (Broglie et al. 1991; Gao et al. 2000). However, an obvious concern is the possible adverse effects on resident nontarget fungal populations, which are likely to come into contact with the microbial inoculants or the transgenic plant products during colonization of the roots and/or through phenomena such as rhizodeposition, root tissue senescence, sloughing-off and wounding (Whipps et al. 1996; Glandorf et al. 1997; Gidding 1998). Beneficial soil fungi may also be indirectly affected by transgenic antimicrobial products after incorporation in the soil of plant residues following specific cultural management procedures.

Such nontarget fungi comprise primarily arbuscular mycorrhizal (AM) fungi, ubiquitous obligate biotrophic symbionts which are well known to enhance plant nutrition and water acquisition, as well as resist biotic and nonbiotic stress factors (Harrison 1999). The tight association they establish with the plant is morphologically characterized by the formation of arbuscules, which are the main site of nutrient exchange between the two partners (Perotto and Bonfante 1995; Harrison et al. 2002).

Another, perhaps less known, beneficial fungal component that can be potentially affected by the transgenes is the assemblage of saprotrophic microfungi, which play key roles in organic matter decomposition, mineralization and solubilization, thus increasing nutrient availability to the

[1]Istituto per la Protezione delle Piante del C.N.R and Dipartimento di Biologia Vegetale dell'Università, Viale Mattioli 25, 10125 Torino, Italy, e-mail: v.bianciotto@ipp.cnr.it, Tel: +39-11-6502927, Fax: +39-11-55839

Soil Biology, Volume 3
Microorganisms in Soils: Roles in Genesis and Functions
(ed. by F. Buscot and A. Varma)

plant (Cooke and Rayner 1984; Curl and Truelove 1986; Dix and Webster 1995). These fungi also contribute to disease suppression through a variety of mechanisms. Their relevance to natural control of plant pathogens is perhaps best illustrated by soil suppressiveness, a phenomenon many

Table 1. Available studies on the impact of rhizosphere GMOs producing antifungal compounds on nontarget soil fungi

GMOs	Transgene	Organisms monitored	Processes monitored	Effect	Reference
Bacteria					
Pseudomonas fluorescens	Polyketides + pyoluteorin	Soil fungi of cucumber rhizosphere	Biodiversity	Similar to repeated cucumber monoculture	Girlanda et al. (2001)
Pseudomonas putida	Phenazine-1-carboxylic acid	Soil fungi of wheat rhizosphere	Biodiversity	Transient variation	Glandorf et al. (2001)
Plants					
Nicotiana sylvestris	Chitinase	AM fungus (*Glomus mosseae*)	Root colonization	Similar to wild-type plants	Vierheilig et al. (1993)
Nicotiana tabacum	Chitinase and/or glucanase	AM fungus (*Glomus mosseae*)	Root colonization	Similar to wild-type plants. Only with a glucanase isoform a delay in AM colonization was observed	Vierheilig et al. (1995)
Nicotiana tabacum	Chitinase	AM fungi inoculum	Root colonization and enzymatic activity	Similar to wild-type plants	Tahiri-Alaoui et al. (1994)
Nicotiana tabacum	*Rs-AFP2*	AM fungus (*Gigaspora margarita*)	Root colonization	Similar to wild-type plants	This study
Solanum tuberosum	*Hs*-AFP1 and *Ace*-AMP1	AM fungus (*Gigaspora margarita*)	Root colonization	Similar to wild-type plants	This study
Solanum tuberosum	*Hs*-AFP1 and *Ace*-AMP1	AM agricultural soil inoculum	Root colonization	Similar to wild-type plants	This study

genera (for instance *Trichoderma* and nonpathogenic *Fusarium*) have been involved in (Cook and Baker 1983; Alabouvette and Steinberg 1995; Chet et al. 1997).

Despite the crucial roles of these fungal components, only a few studies have addressed the possible effects of transgenic rhizospheres with fungal resistance traits on indigenous soil populations (Table 1). By contrast, several recent studies have investigated the effects of plants transformed with novel traits on nontarget soil fungi. For example, the impact of transgenic plants with potential for insect or nematode control was investigated by Donegan et al. (1996), Griffiths et al. (2000), and Cowgill et al. (2002), whereas other authors have studied the possible effects of plants with altered phytohormone balance or sugar metabolism on ecto- and endomycorrhizal fungi (Hampp et al. 1996; Schellenbaum et al. 1999; Gianinazzi-Pearson et al. 2000; Kaldorf et al. 2002).

The purpose of this chapter is to review current knowledge on the possible effects of GM organisms, modified for increased resistance to fungal pathogens, on indigenous beneficial soil fungi: saprotrophic and arbuscular mycorrhizal microfungi.

2
Experiments with Saprotrophic and Mycorrhizal Fungi

2.1
Saprotrophic Microfungi

Saprotrophic fungi have been largely neglected as nontarget, beneficial resident microorganisms potentially affected by transgenic rhizospheres. The few studies dealing with these fungi (Table 1) have mostly monitored the impact on total fungal counts. While this approach may allow the assessment of catastrophic effects, it does not address the possibility of specific changes in microfungal community organization, in terms of the diversity and relative abundance of fungal species. Such alterations in the composition and structure of fungal communities might have lasting effects on ecosystem functions (Kennedy and Smith 1995), as there is experimental evidence of a link between microbial diversity and maintenance and regulation of ecosystem functions (Naeem et al. 1994).

Pseudomonas spp. are the bacterial inoculants most commonly used for biocontrol and have been frequently modified genetically to increase antibiotic and/or siderophore production for enhanced biocontrol efficacy (Dunne et al. 1996), to protect a number of important crop plants.

In our laboratories, we have investigated the effects on rhizospheric fungi of two biocontrol *Pseudomonas fluorescens* inoculants: strain CHA0-Rif, which produces the antimicrobial polyketides 2,4-diacetylphloroglucinol and pyoluteorin and protects cucumber against several fungal pathogens, including *Pythium* species, and strain CHA0-Rif(pME3424), a GM derivative overproducing phloroglucinol and pyoluteorin, and thus displaying enhanced antifungal activity (Schnider et al. 1995). A cycle experiment was performed in which cucumber was grown repeatedly in the same soil, which was left uninoculated, was inoculated with CHA0-Rif or the GM derivative, or was treated with a chemical fungicide with selective action almost exclusively against oomycetes (Girlanda et al. 2001). We assessed the impact of GM bacteria on the biodiversity of rhizosphere microfungi with a morphology-based identification approach. Fungal species identification data were subjected to the classical community analyses, species abundance distribution models, calculation of a number of diversity indices weighing towards species richness and evenness, and discriminant analysis. Such analyses were performed on abundance data from over 11,000 fungal colonies which were assigned to approx. 150 species and sterile morphotypes: they did not outline significant treatment effects with either species abundance distribution models or diversity indices. Control treatments and treatments with GM bacteria were applied to the soil at the start of each of five cucumber growth cycles in soil microcosms. Differences between the control and the other treatments were found with discriminant analysis at the end of the first and the fifth cucumber growth cycle. However, the difference between the bacterial treatments and the respective controls was always smaller than the difference between the two controls, thus indicating that the effects of growing cucumber repeatedly in the same soil altered the fungal community more than the GM treatments.

We took the same approach to analyze biodiversity of saprotrophic microfungi in the rhizosphere of transgenic tomato plants, provided by Zeneca, which carried tobacco genes for two antifungal hydrolases with chitinase and β-1,3-glucanase activity (which are active against cell wall components occurring in most fungi). In this case again, some differences between the microfungal populations associated with GM and wild-type (WT) plants were found at different plant growth stages, but the effects of the transformed line were less than those linked to the plant age. Similar results were obtained when considering the phyllosphere saprotrophic microfungi, another fungal component, which may also benefit the plant through natural control of leaf pathogens (Marois and Coleman 1995), and which is perhaps less "buffered" against transgenic products than the soil populations.

2.2 Mycorrhizal Fungi

What are instead the effects on AM fungi, which live in a more intimate association with their hosts? A major, potential nontarget effect of transgenic rhizospheres may be a decrease in the activity of these symbionts, associated with reduced root colonization. Therefore, it is crucial to assess whether the ability to form arbuscular mycorrhizae is maintained in GM rhizospheres from a qualitative and quantitative point of view, by monitoring the phenotype of infection and the rate of mycorrhiza formation, as expressed with different indices.

In our laboratory, colonization by AM fungi was checked in potato (*Solanum tuberosum*) and tobacco (*Nicotiana tabacum*) plants provided by Zeneca and transformed for constitutive expression of two different defensins. Plant defensins, which bear strong homology to insect and mammalian defensins, are small cysteine-rich proteins with broad antifungal activity (Terras et al. 1995). They have been studied mainly in seeds, but are also expressed in other plant organs. Expression of defensins can be either constitutive or regulated by pathogenic attacks (Penninckx et al. 1996; Thomma and Broekaert 1998). We have analysed two potato lines (30 and 34) carrying two defensin genes: Hs-AFP1 codes for a morphogenic defensin from *Heuchera sanguinea*, causing hyphal swelling and increased branching in the fungal pathogen (Osborn et al. 1995), whereas Ace-AMP1 is derived from onion and encodes a nonmorphogenic defensin inhibiting hyphal growth of numerous pathogenic fungi as *Botritis cinerea*, *Fusarium oxysporum* and *Verticillium dahliae* (Cammue et al. 1995).

The genes Hs-AFP1 and Ace-AMP1 were driven respectively by two strong constitutive promoters, sMAS and puB3. The protein expression, as checked by immunoblotting experiments on 1-month-old plants, demonstrated the presence of the Hs-AFP1 in both leaves and roots of both high expression potato lines. In contrast, the Ace-AMP1 protein was not found to be expressed in any plant organ, thus suggesting some silencing effects of this defensin gene.

These two high expression lines were compared at 4 and 8 weeks with two control lines (line 8, wild type, and line 5, transformed with an empty vector), for arbuscular mycorrhizal infection by either a selected AM strain (*Gigaspora margarita* BEG 34), or a natural inoculum from a *Medicago* field. In addition, transgenic tobacco plants transformed for overexpression of the Rs-AFP2 defensin were also analysed. However, the protein, under the control of the cauliflower mosaic virus 35S promoter (CaMV35S), was only expressed in 1-month-old leaves. No significant differences between transgenic and control lines were found at either sampling times both in terms of the fungal structures formed or the percentage of mycorrhizal

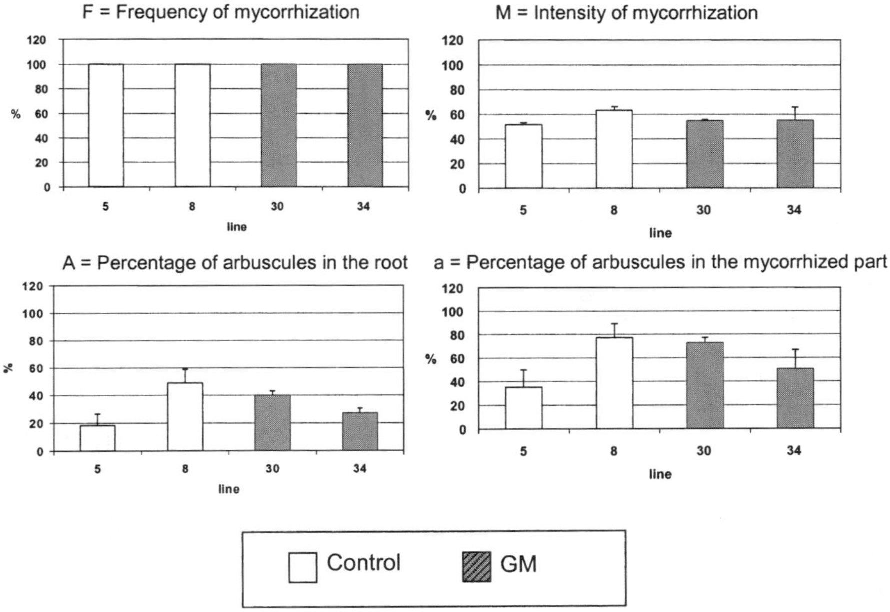

Fig. 1. Comparison between control (*lines 5* and *8*) and GM potato plants (*lines 30* and *34*) using the percentage of mycorrhizal colonization (according to Trouvelot et al. 1986). No statistical significant difference was found comparing the four lines for each indices. *Line 8* Wild type, *line 5* transformed with the plasmid without the double construct; *lines 30* and *34* transformed with two defensin genes, Hs-AFP1and Ace-AMP1

infection achieved with either kind of AM inocula (Fig. 1). It is interesting to note that a higher percentage of mycorrhizal infection was observed with the natural inoculum (~ 50%) than with the monospecific *G. margarita* inoculum (max. ~ 10%). Observations of the transgenic roots by light and electron transmission microscopy showed that, with both inocula, AMF were able to form the fungal structures typical of mycorrhizal colonization (Figs. 2 and 3).

Similar behavior was observed with mycorrhizal inocula on tomato plants transformed with chitinase and glucanase (described in the previous paragraph on saprotrophic microfungi), grown in agricultural soil.

3 Conclusions

In the situations studied to date, it appears that genetically modified microorganisms (GMM) engineered to have improved antifungal activity can exert detectable nontarget effects on the diversity of the natural sapro-

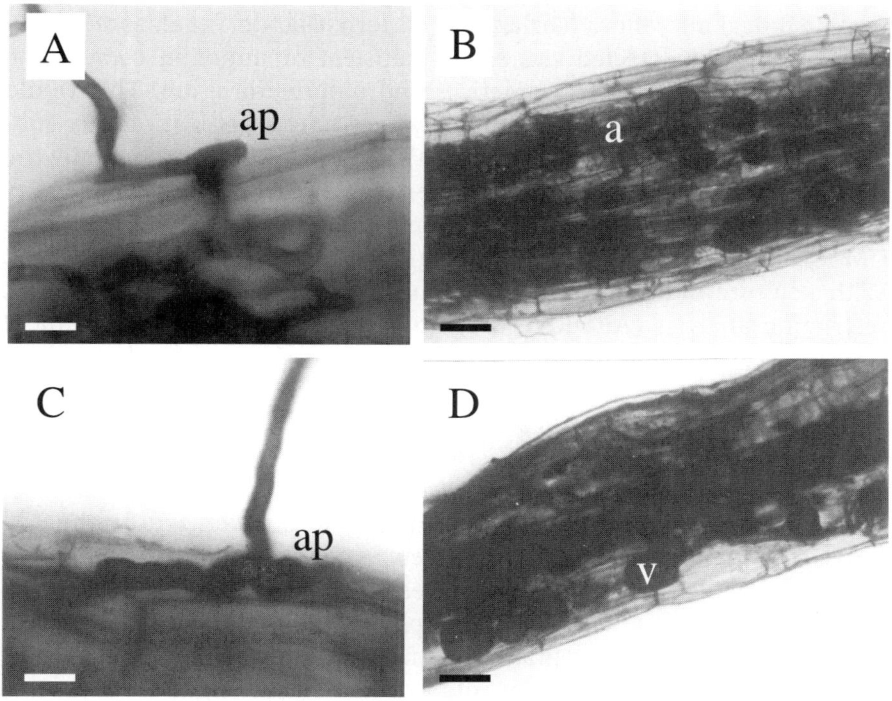

Fig. 2. Micrographs showing the colonization of control (**A, B**) and transgenic (**C, D**) potato roots. In **A** and **C**, the appressoria (*ap*) are shown; in **B** and **D**, the cortical cells are colonized by the AM fungi in a natural inoculum. In **B** and **D**, the fungal structures typical of intraradical colonization, arbuscules (*a*) and vesicles (*v*), are visible. In **A** and **C**, *bars* correspond to 15 µm, in **B** and **D** to 65 µm

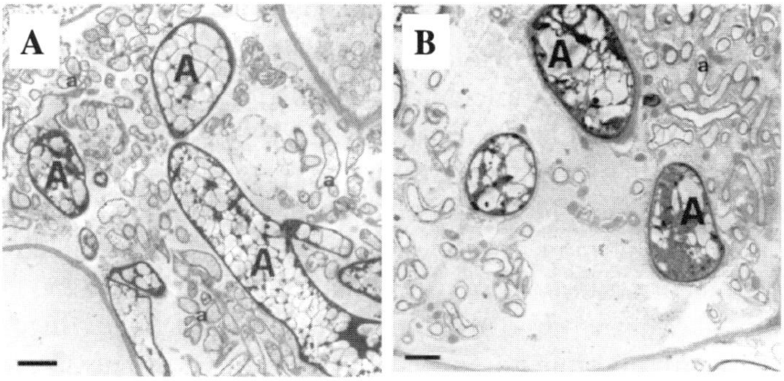

Fig. 3. Electron micrographs of tobacco (*Nicotana tabacum*) roots colonized by *Gigaspora margarita*. **A** Colonized cortical cell of wild-type tobacco; an arbuscule is visible with its large (*A*) and thin (*a*) branches. **B** A cortical cell of the transgenic plant colonized by the AM fungus. No ultrastructural alterations are visible in the arbuscule organization, comprising large (*A*) and thin (*a*) branches. *Bars* correspond to 2 µm

trophic fungal microflora (Girlanda et al. 2001; Glandorf et al. 2001). However, the impact detected was either transient or minor in comparison to the influence of cultural practices and natural variation. This would particularly apply to field conditions, where fungal populations are subjected to wide fluctuations (Gams 1992). As far as we are aware, only the study of Glandorf and colleagues (2001) has also addressed biodiversity of saprotrophic microfungi in transgenic rhizospheres. They considered GM *Pseudomonas putida* strains in the rhizosphere of field-grown wheat. With a culture-independent approach based on amplified ribosomal DNA restriction analysis (ARDRA), they observed that introduction of these GMMs can affect, although transiently, the composition of the rhizosphere fungal microflora.

A few other data are available as concerns AM fungi. Three studies by Tahiri-Alaoui et al. (1994) and Vierheilig and colleagues (1993, 1995) have shown that tobacco and *Nicotiana sylvestris* plants, constitutively expressing heterologous chitinases and glucanases, were colonized by the AM fungus *Glomus mosseae* to the same extent as control plants. No differences in the morphology of fungal structures could be outlined when mycorrhizal roots of control and transgenic plants were observed under a light microscope. The only exception was an acidic isoform of a tobacco glucanase, which caused a delay in colonization, but the significance of this finding is somewhat unclear. In general, these data, combined with our results, indicate that enhanced resistance conferred by constitutive expression of antifungal proteins in transgenic plants does not interfere with the arbuscular mycorrhizal symbiosis.

Why do root-associated fungi display such little sensitivity to plant-produced antimicrobial compounds? A possible explanation may be the long-lasting interaction of some soil fungi with the plant roots. For arbuscular mycorrhizal fungi, a long-dating co-evolution with plants has been well demonstrated, which may explain adaptation of AMF to plant defense. AMF are in fact thought to be very ancient organisms, already present at a time when the first plants colonized the land (Redecker et al. 2000).

Saprotrophic microfungal assemblages, which feature less tightly integrated association with the host plant, possibly exhibit a dynamic response to disturbance by the antimicrobial compounds, through resilience mechanisms which leave the community structure substantially unaltered.

Although permanent negative effects on the colonization by mycorrhizal fungi do not seem to occur, possible nontarget effects of GMOs on the diversity of this fungal component in the soil cannot be ruled out. In their field study on transgenic aspens for the *rol*C gene of *Agrobacterium rhizogenes*, Kaldorf et al. (2002) found that the ectomycorrhizal fungal community associated with one transgenic line differed in structure. Up to now, this aspect has not been investigated for AM fungi and would be an important issue to be addressed in the future. In the last few years, the molecular diversity and the community composition of the AM fungi have been studied in semi-natural and natural ecosystems (Daniell et al. 2001; Husband et al. 2002; Kowalchuk et al. 2002). Therefore, a similar approach could be used to compare experimental systems with control and transgenic plants. Clearly, since AM fungi are nonculturable obligate biotrophs, the analysis of biodiversity requires molecular techniques, which also hold promise for the study of saprotrophic microfungi. A topical issue is indeed the contribution to diversity studies of soil fungi of molecular techniques omitting the culturing step [such as ARDRA, denaturing gradient gel electrophoresis (DGGE) or temperature gradient gel electrophoresis (TGGE)], in comparison with conventional culture-dependent methods. Both have pros and cons, the main point being that both are selective to some extent, and thus, ideally, they should complement each other in future investigations.

Acknowledgements. The transgenic plants were supplied by Zeneca Agrochemicals. This research was supported by the Italian National Council of Research (IPP-CNR) and by Italian MURST (Cofin 2000) and by CE-BIOVEM (Center of Excellence for Plant and Microbial Biosensing)

References

Alabouvette C, Steinberg C (1995) Suppressiveness of soils to invading micro-organisms. In: Hokkanen HMT, Lynch JM (eds) Plant and microbial biotechnology research series. Biological control: benefits and risks. Cambridge Univ Press, Cambridge, pp 3–12

Broekaert WF, Terras RG, Cammue BPA, Osborn RW (1995) Plant defensins: novel antimicrobial peptides as components of the host defense system. Plant Physiol 108:1353–1358

Broglie K, Chet I, Holliday M, Cressman R, Biddle P, Knowlton S, Mauvais CJ, Broglie R (1991) Transgenic plants with enhanced resistance to the fungal pathogen *Rhizoctonia solani*. Science 254:1194–1197

Cammue BA, Thevissen K, Hendriks M, Eggermont K, Goderis I, Proost P, van Damme J, Osborn RW, Guerbette F, Kader J, Broekaert WP (1995) A potent antimicrobial protein from onion seeds showing sequence homology to plant lipid transfer proteins. Plant Physiol 109:445–455

Chet I, Inbar J, Hadar Y (1997) Fungal antagonists and mycoparasites. In: Wicklow DT, Söderström B (eds) The Mycota, vol 4. Environmental and microbial relationships. Springer, Berlin Heidelberg New York, pp 165–184

Cooke RC, Rayner ADM (1984) Ecology of saprotrophic fungi. Longman, London

Cook RJ, Baker KF (1983) The nature and practice of biological control of plant pathogens. American Phytopathological Society, St Paul

Cook RJ, Bruckart WL, Coulson JR, Goettel MS, Humber RA, Lumsden RD, Maddox JV, McManus ML, Moore L, Meyer SF, Quimby PC Jr, Stack JP, Vaughn JL (1996) Safety of microorganisms intended for pest and plant disease control: a framework for scientific evaluation. Biol Control 7:333–351

Cowgill SE, Bardgett RD, Kiezebrink DT, Atkinson HJ (2002) The effect of transgenic nematode resistance on non-target organisms in the potato rhizosphere. J Appl Ecol 39:915–923

Curl EA, Truelove B (1986) The rhizosphere. Springer, Berlin Heidelberg New York
Daniell TJ, Husband R, Fitter AH, Young JPW (2001) Molecular diversity of arbuscular mycorrhizal fungi colonising arable crops. FEMS Microbiol Ecol 36:203–209
Dix NJ, Webster J (1995) Fungal ecology. Chapman and Hall, London
Donegan KK, Schaller DL, Stone JK, Ganio LM, Reed G, Hamm PB, Seidler RJ (1996) Microbial populations, fungal species diversity and plant pathogen levels in field plots of potato plants expressing the *Bacillus thuringensis* var *tenebrionis* endotoxin. Transgenic Res 5:25–35
Dunne C, Delany I, Fenton A, Lohrke S, Moenne-Loccoz Y, O'Gara F (1996) The biotechnology and application of *Pseudomonas* inoculants for the biocontrol of phytopathogens. In: Stacey G, Mullin B, Gresshoff PM (eds) Biology of plant-microbe interactions. IS-MPMI, St Paul, pp 441–448
Gams W (1992) The analysis of communities of saprophytic microfungi with special reference to soil fungi. In: Winterhoff W (ed) Fungi in vegetation science. Kluwer, Dordrecht, pp 183–223
Gao A, Hakimi SM, Mittanck CA, Wu Y, Woerner BM, Stark DM, Shah DM, Liang J, Rommens CMT (2000) Fungal pathogen protection in potato by expression of a plant defensine peptide. Nature Biotechnol 18:1307–1310
Gianinazzi-Pearson V, Arnould C, Oufattole M, Arango M, Gianinazzi S (2000) Differential activation of H+-ATPase genes by an arbuscular mycorrhizal fungus in root cells of transgenic tobacco. Planta 211:609–613
Giddings G (1998) The release of genetically engineered micro-organisms and viruses into the environment. New Phytol 140:173–184
Girlanda M, Perotto S, Moenne-Loccoz Y, Bergero R, Lazzari A, Defago G, Bonfante P, Luppi AM (2001) Impact of biocontrol *Pseudomonas fluorescens* CHA0 and a genetically modified derivative on the diversity of culturable fungi in the cucumber rhizosphere. Appl Environ Microbiol 67:1851–1864
Glandorf DCM, Bakker PAHM, van Loon LC (1997) Influence of the production of antibacterial and antifungal proteins by transgenic plants on the saprophytic soil microflora. Acta Bot Neerl 46:85–104
Glandorf DCM, Verheggen P, Jansen T, Jorritsma J, Smit E, Leeflang P, Wernars K, Thomashow LS, Laureijs E, Thomas-Oates JE, Bakker PAHM, van Loon LC (2001) Effect of genetically modified *Pseudomonas putida* WCS358R on the fungal rhizosphere microflora of field-grown wheat. Appl Environ Microbiol 67:3371–3378
Griffiths BS, Geoghegan IE, Robertson WM (2000) Testing genetically engineered potato, producing the lectines GNA and Con A, on non-target soil organisms and processes. J Appl Ecol 37:159–170
Hampp R, Ecke M, Schaeffer C, Wallenda T, Wingler A, Kottke I, Sundberg B (1996) Axenic mycorrhization of wild type and transgenic hybrid aspen expressing T-DNA indolacetic acid-biosynthetic genes. Tree 11:59–64
Harrison M (1999) Molecular and cellular aspects of the arbuscular mycorrhizal symbiosis. Annu Rev Plant Physiol Plant Mol Biol 50:361–389
Harrison MJ, Dewbre GR, Liu J (2002) A phosphate transporter from *Medicago truncatula* involved in the acquisition of phosphate released by arbuscular mycorrhizal fungi. Plant Cell 14:2413–2429
Husband R, Herre EA, Turner SL, Gallery R, Young JPW (2002) Molecular diversity of arbuscular mycorrhizal fungi and patterns of host association over time and space in a tropical forest. Mol Ecol 11:2669–2678
Kaldorf M, Fladung M, Muhs HJ, Buscot F (2002) Mycorrhizal colonization of transgenic aspen in a field trial. Planta 214:653–660

Kennedy AC, Smith KL (1995) Soil microbial diversity and the sustainability of agricultural soils. Plant Soil 170:75–86

Kowalchuk GA, de Souza FA, van Veen JA (2002) Community analysis of arbuscular mycorrhizal fungi associated with *Ammophila arenaria* in Dutch coastal sand dunes. Mol Ecol 11:571–581

Lorito M, Woo SL, Garcia Fernandez I, Colucci G, Harman GE, Pintor-Toro JA, Filippone E, Muccifora S, Lawrence CB, Zoina A, Tuzun S, Scala F (1998) Genes from mycoparasitic fungi as a source for improving plant resistance to fungal pathogens. Proc Natl Acad Sci USA 95:7860–7865

Marois JJ, Coleman PM (1995) Ecological succession and biological control in the rhizosphere. Can J Bot 73:76–82

Naeem S, Thomposon LJ, Lawler SP, Lawton JH, Woodfin RM (1994) Declining biodiversity can alter the performance of ecosystem. Nature 368:734–737

Osborn RW, De Samblanx GW, Thevissen K, Goderis I, Torrekens S, Van Leuven F, Attenborough S, Rees SB, Broekaert WF (1995) Isolation and characterisation of plant defensins from seeds of Asteraceae, Fabaceae, Hippocastanaceae and Saxifragaceae. FEBS Lett 368:257–262

Penninckx, IAMA, Eggermont K, Terras FRG, Thomma BPHJ, de Samblanx, GW, Buchala A, Métraux J-P, Manners JM, Broekaert WF (1996) Pathogen-induced systemic activation of a plant defensin gene in *Arabidopsis* follows a salicylic acid-independent pathway. Plant Cell 8:2309–2323

Perotto S, Bonfante P (1995) Strategies of arbuscular mycorrhizal fungi when infecting host plants. New Phytol 130:3–21

Punja ZK (2001) Genetic engineering of plants to enhance resistance to fungal pathogens – a review of progress and future prospects. Can J Plant Pathol 23:216–235

Redecker K, Kodner R, Graham LE (2000) Glomalean fungi from the Ordovician. Science 289:1920–1921

Schellenbaum L, Sprenger N, Schüepp H, Wiemken A, Boller T (1999) Effects of drought, transgenic expression of a fructan synthesizing enzyme and of mycorrhizal symbiosis on growth and soluble carbohydrate pools in tobacco plants. New Phytol 142:67–77

Schnider U, Keel C, Blumer C, Troxler J, Défago G, Haas D (1995) Amplification of the housekeeping sigma factor in *Pseudomonas fluorescens* CHA0 enhances antibiotic production and improves biocontrol abilities. J Bacteriol 177:5387–5392

Tahiri-Alaoui, Grison R, Gianinazzi-Pearson V, Toppan A, Gianinazzi S (1994) The impact of the constitutive expression of chitinases in roots of transgenic tobacco on arbuscular mycorrhizal fungi. 7th International Symposium on Molecular plant-microbe interaction, Edinburgh, p 406

Terras FRG, Eggermont K, Kovaleva V, Raihkel NV, Osborn RW, Kester A, Rees SB, Torrekens S, van Leuven F, Vanderleyden J, Cammue BPA, Brokaert WF (1995) Small cysteine-rich antifungal proteins from radish: their role in host defence. Plant Cell 7:573–588

Thomma BPHJ, Broekaert WF (1998) Tissue-specific expression of plant defensin genes PDF2.1 and PDF2.2 in *Arabidopsis thaliana.* Plant Physiol Biochem 36:533–537

Trouvelot A, Kough JL, Gianinazzi-Pearson V (1986) Mesure du taux de mycorhization VA d'un système radiculaire. Recherche de méthodes d'estimation ayant une signification fonctionnelle. Les mycorhizes: physiologie et génétique, Ier sem, Dijon, 1–5 July, pp 217–222

Vierheilig H, Alt M, Neuhaus JH, Boller T, Wiemken A (1993) Colonization of transgenic *Nicotiana sylvestris* plants, expressing different forms of *Nicotiana tabacum* chitinase, by the root pathogen *Rhizoctonia solani* and by the mycorrhizal symbiont *Glomus mosseae.* Mol Plant Microbe Interact 6:261–264

Vierheilig H, Alt M, Lange J, Gut-Rella M, Wiemken A, Boller T (1995) Colonization of transgenic tobacco constitutively expressing pathogenesis-related proteins by the vesicular-arbuscular mycorrhizal fungus *Glomus mosseae.* Appl Environ Microbiol 61:3031–3034
Whipps JM, de Leij FAAM, Lynch JM, Bailey MJ (1996) Impact of genetically modified microorganisms on the terrestrial microbiota including fungi. In: Frankland JC, Magan N, Gadd GM (eds) Fungi and environmental change. Cambridge Univ Press, Cambridge, pp 299–316

14 Regulation of Microbial Activities in Functional Domains of Roots and Invertebrates

Patrick Lavelle, Corinne Rouland, Michel Diouf[1],
Françoise Binet, Anne Kersanté[2]

1 Introduction

Microorganisms perform over 90% of the chemical transformation of organic substrates in soils (Satchell 1971; Reichle et al. 1975; Ryszkowski 1975; Lamotte 1989). Their communities may comprise more than tens of thousands of different species in 100 g of soil and their biomass amounts up to several hundreds of kilograms per hectare (Torsvik et al. 1990; Lavelle and Spain 2001). Although they are globally able to digest any natural substrate in soil, microorganisms have specific temperature and moisture requirements that may severely limit their activity. They also have limited to null ability to move towards new substrates to decompose, with some notable exceptions in some taxonomic groups such as actinomycetes. Fungi that may extend their hyphae several centimetres to metres from their starting point also depend to a large extent on larger organisms or any abiotic process to disseminate their propagules in soil and litter environments. As a result, microbial activities in soils depend on a number of abiotic and biotic factors that adjust their effects on organic matter recycling and nutrient release at scales of time and space from microns to hundreds of kilometres and hours to centuries. We describe here the mechanisms that regulate their activities and discuss the implications regarding soil management.

2 Determinants of Microbial Activities: The Hierarchical Model

Soil processes like decomposition, aggregation or water storage in porosity are regulated by a hierarchy of determinants operating in nested scales of time and space (Lavelle et al. 1993). At the largest scale, abiotic determinants, i. e. climate followed by edaphic properties, constrain biological determinants that operate at smaller scales, i. e. the composition and struc-

[1]IRD, University of Paris 06, UMR 137, 93143 Bondy, France,
e-mail: Patrick.Lavelle@bondy.ird.fr
[2]CNRS, UMR ECOBIO, Université de Rennes 1, 35042 Rennes, France

Soil Biology, Volume 3
Microorganisms in Soils: Roles in Genesis and Functions
(ed. by F. Buscot and A. Varma)

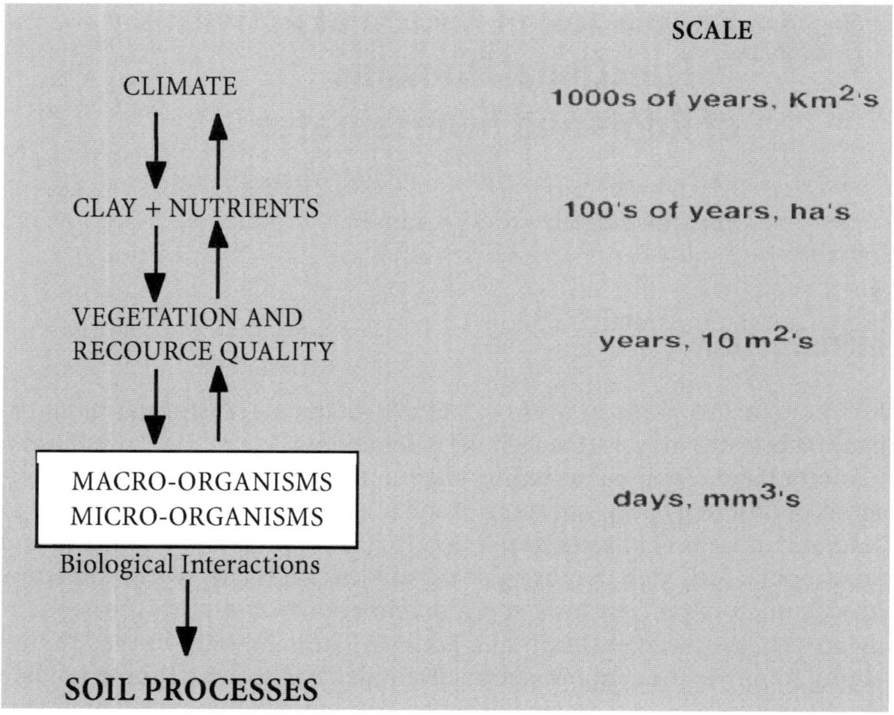

Fig. 1. A hierarchical model for determinants of decomposition and other processes in soil. (Lavelle et al. 1993)

ture of plant communities (which determine the quality and quantity of organic inputs to the soil), soil "macroorganisms" (macroinvertebrates + roots) that effect bioturbation and transport of propagules and microorganisms (Fig. 1). In addition, feedback retroactions allow determinants at lower levels of the hierarchy to influence upper levels. This hierarchy is a potential one that may not be fully operational locally. When climate is not constraining (e.g., in the humid tropics), when soils have no active clay minerals, such as smectites that strongly influence microbial activities through several mechanisms, and when the organic matter produced is homogeneous and easy to decompose, the main regulation for microbial activity may be exerted by macroinvertebrates (earthworms and termites) during the gut transit of soil and microbes and in the biogenic structures that they create.

The value of this model may be illustrated by examples from two contrasting environments. In Swedish pine forests, the combined effect of seasonal variations of temperature and moisture may explain alone 95–99% of the rate of litter decomposition (Jansson and Berg 1985). More generally, Heal et al. (1981) showed a clear decline in litter decomposition rates from tropical forests to tundra areas. Extreme climatic conditions may thus override other determinants at the scale of biomes and years. In contrast, several publications have emphasised the importance of chemical quality of litter and faunal activities on decomposition rates in tropical environments (Spain and Le Feuvre 1987; Nussbaumer et al. 1997; Lavelle and Spain 2001; Loranger et al. 2002). This illustrates the prediction that in situations where climate and soil properties are not restricting microbial activities, regulations are operated by factors at lower levels of the hierarchy, i. e. the quality of organic matter or activities of soil invertebrate engineers.

3 Microbial Adaptive Strategies: The Sleeping Beauty Paradox

Microorganisms may decompose any kind of natural substrate found in soils and those that can be cultivated in the laboratory multiply and greatly increase their biomass in short periods of time (in the order of days). The turnover time of microbial biomass under field conditions, however, generally varies between 6 and 18 months (Voroney 1983; Andren et al. 1990), which indicates that they are inactive most of the time. A likely interpretation of this inactivity is starvation that results from the inability of microorganisms to move towards new substrates once their immediate surroundings are exhausted. The apparent contradiction between laboratory observations on cultivable species and field measurements has been named the "sleeping beauty paradox" (Lavelle et al. 1995). The "Prince Charming" of the story is a macroorganism or any physical process that may bring microorganisms in contact with new substrates to decompose.

In turn, invertebrates and roots are known to have limited digestive abilities, and they largely rely on the faculty of microorganisms to digest a wide range of substrates for them (see Bignell 1984; Coleman et al. 1984; Clarholm 1985; Barois and Lavelle 1986; Higashi et al. 1992; Trigo et al. 1999; Rouland-Lefevre 2000). They have developed three different mechanisms to take advantage of microbial digestive capabilities: predation, the external rumen strategy and internal mutualisms.

4 Predation in Micro-Food Webs

A number of soil invertebrates take advantage of microbial digestive capabilities by simply predating on them. They comprise a number of microinvertebrates such as protozoa and 50–80% of nematodes, several tens to hundreds of microns in size that live in the water-filled soil pores. Bacterial feeders are active predators that can ingest up to 5000 cells/min and 800 kg bacteria ha^{-1} $year^{-1}$ (Coleman et al. 1984). Collembola and some Acari are other active microbial consumers that mainly feed on fungi (Vannier 1985).

Predation on microorganisms generally stimulates mineralisation of C and N, as observed in a large number of microcosm studies (Ingham et al. 1986; Setälä et al. 1991; de Ruiter et al. 1993). A similar observation was made by Tiunov et al. (2001) in the burrow walls of the large anecic earthworm *Lumbricus terrestris* where the grazing pressure of protozoa and nematodes on microorganisms may control the dynamics of the microbial succession, strongly affecting nutrient cycling processes in these microhabitats. The rhizosphere is another environment where microbial activity is stimulated by the provision of easily assimilable exudates and further regulated as microorganisms are preyed upon by the micropredator food web that colonises this environment (Bonkowski et al. 2000).

Large invertebrates can also feed at least partly on microbial biomass. This is the case for fungus-growing termites and ants. Microbial consumption by earthworms, although suggested in a few studies, is not really supported by results (Brown et al. 2000).

5 The External Rumen Strategy

Over time, a large number of organisms in soil and litter have evolved different types of mutualist relationships with microorganisms. In the external rumen strategy (Swift et al. 1979), organisms stimulate microbial activities outside their body in organic structures that they create: the faecal pellets of litter invertebrates, fungus combs made by fungus-growing termites or leaf-cutting ants, and decomposing leaves accumulated by some earthworms at the entrance of their burrows named middens (Hamilton and Sillman 1989; Maraun et al. 1999).

Litter invertebrates fragment and ingest litter, thus impregnating it with saliva before releasing their faecal pellets. Microbial activity is significantly enhanced in these structures and invertebrates that re-ingest faecal pellets assimilate the products of the external digestions of microbes (Swift et al. 1979; Hanlon and Anderson 1980). This interaction results in a three-phase response of microbial communities as shown by Hanlon and Anderson in a microcosm experiment. After the introduction of invertebrates (the diplopod *Glomeris marginata*), a preliminary decrease in CO_2 production is observed during the first 2 days, probably due to grazing on hyphae and ingestion of decomposing litter; a strong activation is further observed after day 2, culminating at day 10 during the active phase of deposition of fresh faecal pellets. After this peak, microbial activity decreases and may become lower than in the control after 20 days with a high density of *Glomeris* as microbial activity drops down in dry faecal pellets that become microsites for C sequestration due to their compact structure and low moisture content.

Roots also develop external rumen strategies with soil microflora. Part of the inactive soil microorganisms is selectively awakened by the production of exudates in the immediate surroundings of root tips (Bowen and Rovira 1976; Zak et al. 1994; Grayston et al. 1998; Yang and Crowley 2000; Baudoin et al. 2003). They significantly increase their biomass by digesting organic matter at their contact. The release of nutrients from their biomass follows their predation by Protozoa in the root hair region, as suggested by Coleman et al. (1984) and Clarholm (1985). In this case, a priming effect (sensu Jenkinson 1966) is triggered by the addition of an easily assimilable substrate, i. e. root exudates that act as an ecological mediator (Lavelle et al. 1995).

Other examples of external rumen strategies are fungus cultures by leaf-cutting ants and fungus-growing termites. Anecic earthworms also develop similar systems when they accumulate decomposing leaves at the mouth of their burrows and deposit a fungicide mucus on them to favour bacterial growth before they finally ingest the leaves (Cooke 1983; Cortez et al. 1989; Hamilton and Silman 1989).

6
Internal Mutualisms in Earthworms and Termites

Large invertebrates have developed inhabitational mutualist systems that allow them to regulate more efficiently their interaction with the ingested microflora. Earthworms ingest soil and litter and mix them thoroughly while adding significant amounts of water (1vol of water for 1vol of soil) and intestinal mucus that acts as an ecological mediator similar to root exudates. Mucus is a highly assimilable substrate mainly composed of low molecular weight glycoproteins (Martin et al. 1987). Earthworms add large amounts of this compound that may represent 5–40% of the dry weight of soil in the anterior part of the gut (Trigo et al. 1999), depending on general climate conditions and the quality of the ingested substrate. In the median and posterior parts of the gut, mucus is no longer found. It is hypothesised that the part that has not been used by microorganisms is recycled by the worm, a very likely hypothesis given the huge energetic investment represented by the production of this mucus.

In the gut content of several temperate and tropical earthworms, Lattaud et al. (1998) found enzymes that were not produced by sterile earthworm gut tissue cultures. Gut tissues of the tropical earthworms *Pontoscolex corethrurus* and *Millsonia anomala* did not produce cellulase and mannanase (an enzyme that degrades mannan, which is an essential component of seed material). However, another tropical earthworm, *Polypheretima elongata*, was able to produce these enzymes showing that inside a general pattern specific situations may occur.

Termites have highly sophisticated and efficient digestion systems that associate external and internal rumen systems in rather diverse manners. Rouland-Lefèvre (2000) indicates that mutualist relationships between termites and fungi may take one of the three possible modes: (1) exploitation of the cellulolytic capacity of protozoan symbionts residing in the hindgut; (2) exploitation of the cellulolytic capacity of bacteria residing in the hindgut; (3) reliance upon fungal cellulase, originating in the food.

7
Selection of Microflora in the Functional Domains of Soil Ecosystem Engineers

Microbial communities seem to be highly variable in soils, depending on the functional domain they belong to. Diverse techniques have allowed us to compare their communities based on nucleic acid separations or fatty acid analyses. Enzymatic activities assessed, for example, by the BIOLOG set of reactions also allowed us to approach their functional diversity. The following examples detail some of the results obtained recently.

Rhizospheres generally have much larger numbers of non-symbiotic microorganisms than the outer rhizosphere soil, although their diversity is often lower. Densities in the rhizoplane are commonly 2–20 times, and up to 100 times greater than in the non-rhizosphere soil (Curl and Truelove 1986). Microbial communities in the rhizosphere seem to have a rather strict spatial pattern with a suite of different microorganisms depending on the distance to the root (Fig. 2; Bazin et al. 1990). Rhizosphere communities seem to be rather different from the surrounding soil (Vancura 1980; Grayston et al. 1998; Kozdroj and van Elsas 2000; Baudoin et al. 2001).

Blackwood and Paul (2003), for example, found significant differences in communities of, respectively, the rhizosphere, the light and heavy fractions of soils in agricultural systems of Michigan (USA). Rhizosphere soil had a significantly higher number of microbial cells per gram of the fraction than the heavy soil fraction, and a lower density than the light fraction. When expressed as a function of fraction C content, bacterial numbers were generally higher in the heavy fraction as opposed to the light fraction and rhizosphere. The patterns, however, very much depended on the type of soil or agroecosystem as also shown in a number of other studies (Lup-

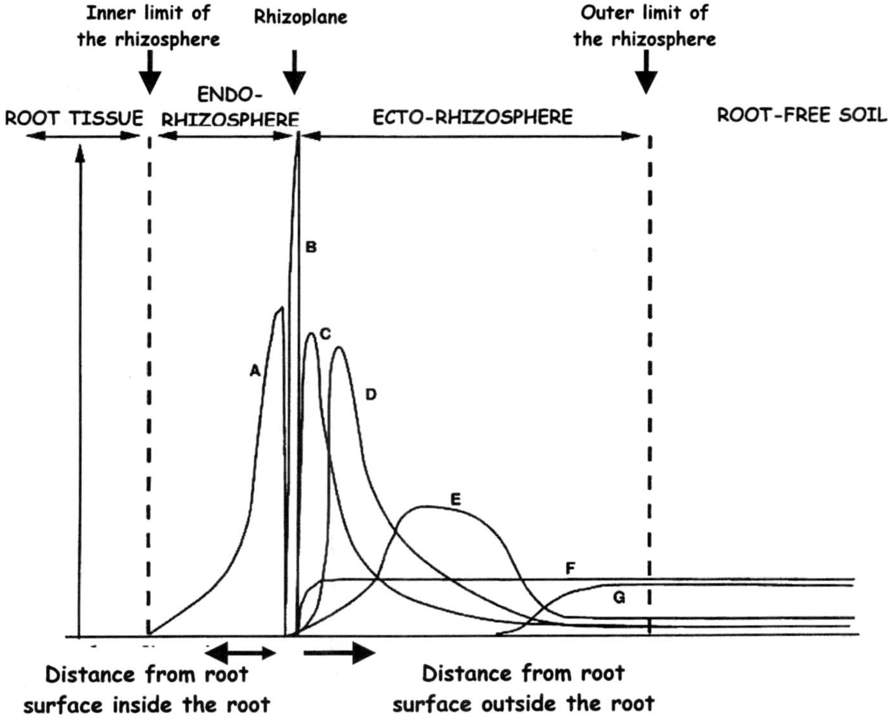

Fig. 2. Distribution of different types of microorganisms in the rhizosphere. *A* Root-penetrating species growing in the intracellular space; *B* species growing exclusively in the rhizoplane, but not root penetrating; *C* species physically excluded from the rhizoplane by type B organisms, but which can utilise organic materials from both the root surface and type B microorganisms; *D* organisms physically and biochemically excluded from the root surface by type B and C organisms, but living on their metabolites or utilising organic material that type B and C organisms cannot use; *E* organisms which utilise secondary products originating from the rhizosphere: *F* organisms excluded from niches on or close to the rhizoplane, but not sensitive to the inhibitory products of organisms types *A–E*; *G* organisms excluded from the rhizosphere by the activities and inhibitory products of rhizosphere organisms

wayli et al. 1998; Marschner et al. 2001), indicating the large-scale effect of a higher order determinant as predicted by the hierarchical model (Lavelle et al. 1993). The size of cells, however, was not significantly different in the rhizosphere, heavy fraction and bulk soil; the light fraction had the largest average cell size. Molecular analysis of 16S ribosomal DNA by the T-RFLP technique opposed the microbial community of the heavy fraction to the one living in the light fraction and rhizosphere. The last two microhabitats had the most variable communities and also smaller numbers of T-RF genetic units. Blackwood and Paul (2003) note that microbial communities were more diverse and stable in the soil heavy fraction. This fraction would constitute the original community from which a rather randomly assembled community formed when rhizosphere activities awoke dormant bacteria. Only those having the life history traits enabling them to respond to high amounts of labile C would multiply. The authors also reported that when plant communities were in place for short periods (i. e. 1 year as compared to 6–10 years in their experiment), the effect of plants on microbial communities in their rhizosphere may be predominant (Latour et al. 1996; Maloney et al. 1997).

In this case, therefore, the authors stress that the light fraction of soil that represented a pool of active organic matter greater than five times as large as the rhizosphere (including fresh root mass) was the main site for microbial activities. This finding actually draws special attention to the dead rhizosphere effects since a large part of the residues of the light soil fraction is composed of actually dead root debris, the decomposition of which is often greatly influenced by live root activities (Billès and Bottner 1981).

Several studies have shown clear differences in the biochemical environment of rhizospheres of different plants using different techniques. Priming effects of root exudates on soil organic matter have been shown (Sallih and Bottner 1988; Mary et al. 1993) and clear differences in the biochemical environments of rhizospheres of different plant species. Bachman and Kinzel (1992) showed significant differences in enzyme and metabolite concentrations in the rhizosphere of six different annual plants from Germany and a control soil with no plants. Rhizosphere characteristics also greatly differed between soil types.

In an analysis of the rhizosphere community structure of five different temperate grass species, Fang et al. (2001) found marked differences in fatty acid methyl ester (FAME) profiles and substrate utilisation patterns as assessed by the BIOLOG technology. These authors, however, and Miethling et al. (2003) failed to find a clear relationship between the community structures assessed in this way and their abilities to degrade atrazine and phenanthrene.

Soil invertebrates may also significantly influence microbial communities and activities in their respective functional domains (Lavelle and Spain 2001). Mutualistic digestive processes involving microorganisms are reasonably well understood, at least in termites (Bignell 1994; Rouland-Lefèvre 2000). The occurrence of mutualist digestive systems in earthworms is supported by a number of studies on temperate and tropical earthworm species (Barois and Lavelle 1986; Lattaud et al. 1998; Trigo et al. 1999; Garvin et al. 2000), although evidence based on observations and direct assessments of microbial communities is still largely lacking.

Molecular analyses of microbial DNA in invertebrate biogenic structures show some specific patterns the origin of which is still not comprehen-

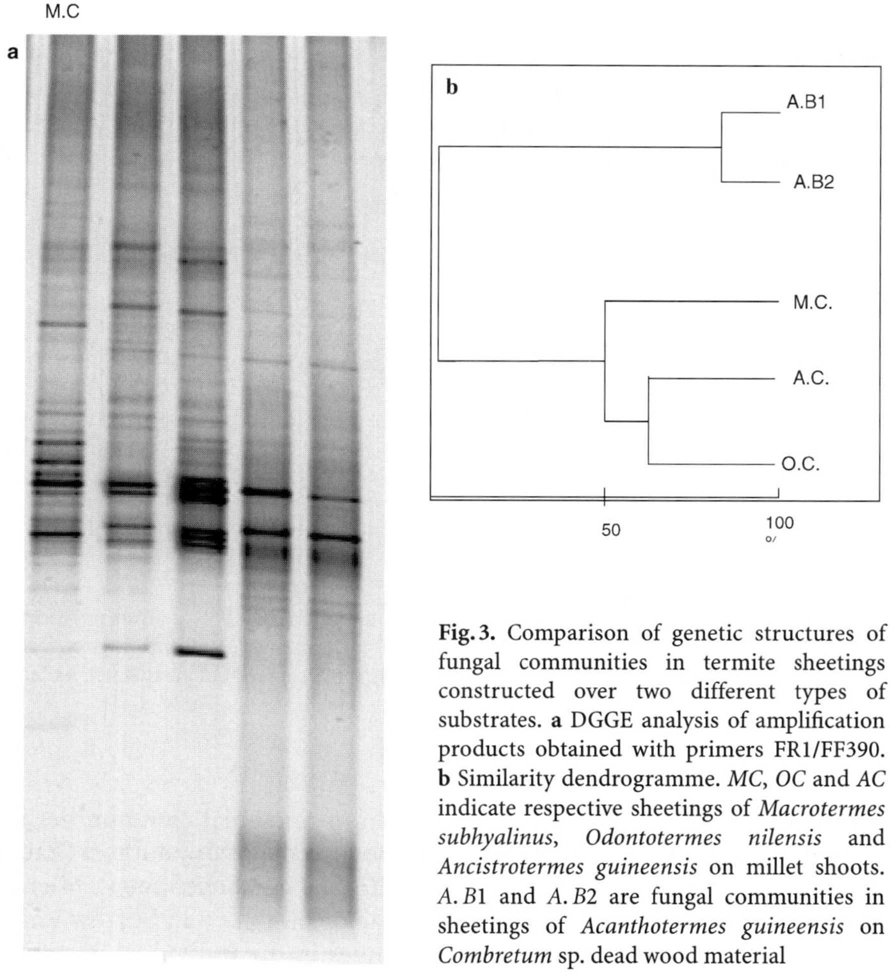

Fig. 3. Comparison of genetic structures of fungal communities in termite sheetings constructed over two different types of substrates. **a** DGGE analysis of amplification products obtained with primers FR1/FF390. **b** Similarity dendrogramme. *MC*, *OC* and *AC* indicate respective sheetings of *Macrotermes subhyalinus*, *Odontotermes nilensis* and *Ancistrotermes guineensis* on millet shoots. *A. B1* and *A. B2* are fungal communities in sheetings of *Acanthotermes guineensis* on *Combretum* sp. dead wood material

sively understood (Fall 2002; Diouf 2003; Kersanté 2003; Mora et al. 2003). Diouf (2003) analysed fungal communities in sheetings of three different termite species from Sénégal (*Ancistrotermes guineensis*, *Odontotermes nilensis*, *Macrotermes subhyalinus*). He found that fresh structures had larger numbers of fungal propagules than the adjacent soil of the 0–10 cm layer. Older ones, aged 8 days or more, no longer sustained these densities. Communities assessed by DGGE analysis of PCR-amplified extracted DNA showed significant differences in their structures; fresh sheetings of the three species were closer to each other than to the bulk soil. However, as they aged, these communities became closer to those in soil. It is assumed that the community in fresh structures is directly reflecting the fungal community colonising plant material that was covered with the sheetings (Fig. 3).

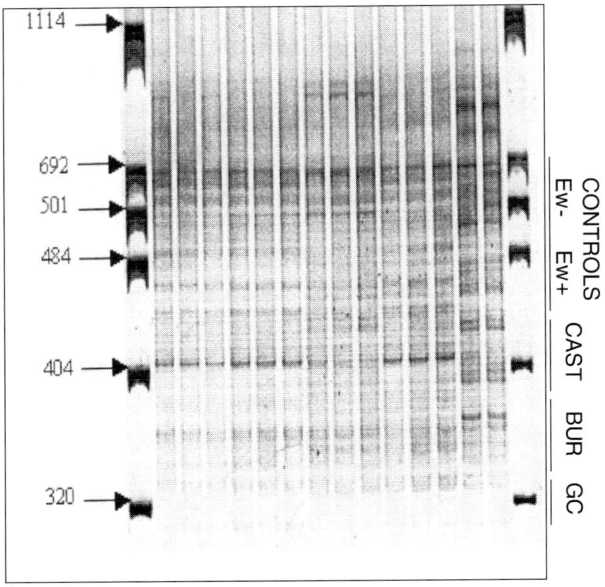

Fig. 4. Comparison of genetic structures of bacteria based on PCR IGS analysis of 16S and 23S DNA, in soil microsites: *Ew–* control with no earthworms; *Ew +* non-ingested soil treated with earthworms; *CAST* earthworm casts; *BUR* burrow wall of earthworm gallery; *GC* earthworm gut content

Changes in the activity and composition of microbial communities in earthworm biogenic structures have been observed by many authors (Parle 1963; Scheu 1987; Tiwari and Mishra 1993; Tiunov and Scheu 1999). Microbial numbers and biomass increase by a factor of 2.3–4.7 in burrow walls of *Lumbricus terrestris* mainly due to changes in the individual bacterial cell volume (Tiunov and Scheu 1999), with a 3.7–9.1 increase in respiration rate. These results confirm earlier estimates by Bhatnagar (1975) who claimed that up to 5–25% of total soil microflora was included in this microenvironment in temperate pastures of central France. Recent studies using molecular techniques have shown evidence of the occurrence of selective microbial communities in Lumbricidae earthworm casts and burrows (Fig. 4; Kersanté 2003). The bacterial community in the gut contents differed the most from the original soil, casts and to a lesser extent, burrow walls, having intermediate communities. This suggests that digestive processes would select a rather specific microbial community through the Sleeping Beauty effect, and progressive changes would occur in ageing structured communities that would slowly converge towards a common "bulk soil" type of community. An alternative hypothesis, although difficult to support, could be that earthworms harbour in their guts a specific permanent microflora that could be important enough to determine the pattern of DNA analyses observed, and would manage not to drift away in the casts. An explanation for this situation might be also found in the separation proposed by Hattori and Hattori (1976) between inter- and intra-aggregate microbial communities. Intra-aggregate communities, probably dominant in the heavy fraction community of Blackwood and Paul (2003), would respond slowly to the huge disturbance of the environment produced by ecosystem engineer activities. The more opportunistic inter-aggregate community would first respond and produce the dramatic changes of microbial communities observed in gut contents and fresh structures shown by molecular techniques. At a later stage, in ageing structures, opportunist microbes would develop first, and then decrease in importance, while inter-aggregate microbes would increase, taking advantage of substrates made available to them by the disturbance and earlier microbial transformations. An alternative hypothesis might be that once past the peak of activity and abundance of microorganisms able to respond to the stimulation, the more stable microbial community linked to the soil heavy fraction that had not really disappeared would become dominant again and be detectable by molecular techniques.

These results clearly indicate that microbial communities should be studied as much as possible at scales that make sense according to their temporal dynamics and spatial distributions. The sum of biogenic structures that comprise the functional domain of a soil ecosystem engineer seems to be the right scale at which to study these phenomena as suggested by the examples detailed in this section. These results clearly question the common practice of having a "control" or "bulk soil" sample for comparison since the "bulk" soil belongs to one or several discrete functional domains of ecosystem engineers and any sample taken haphazardly will comprise one or several of these, with no real control of their nature.

8 Conclusion and Implications for Soil Management

Large organisms living in soils closely interact with soil microflora and most studies show dramatic differences in microbial communities of engineered microsites as compared to non-disturbed microsites. However, much remains to be understood about the exact processes that are involved, and caution has to be taken regarding the significance of results obtained by a diversity of methods that may create artefacts or may be misinterpreted. However, there is no doubt that a significant part of microbial activity is attributable to stimulations at small scales of time and space in microhabitats created by the bioturbation effects of invertebrates and roots.

This has major implications on the management options necessary to sustain intensive microbial activity. Ecosystem services that depend on microbial activities, such as nutrient cycling, detoxification or C sequestration, are certainly more intensive and sustainable when biological regulators such as roots and larger invertebrates are present. This emphasises the value of feeding these organisms with appropriate organic residues and focusing specific attention on the maintenance of their diversity (Lavelle et al. 2001).

References

Andren O, Lindberg T, Paustian K, Rosswall T (1990) Ecology of arable land. Organisms, carbon and nitrogen cycling. Ecological bulletins, Munksgaard, Copenhagen, 221 pp

Bachman G, Kinzel (1992) Biochemical screening of the soil-root system of six different plants. Soil Biol Biochem 24:543–552

Barois I, Lavelle P (1986) Changes in respiration rate and some physicochemical properties of a tropical soil during transit through *Pontoscolex corethrurus* (Glossoscolecidæ, Oligochæta). Soil Biol Biochem 18(5):539–541

Baudoin E, Benizri E, Guckert A (2001) Metabolic fingerprint of microbial communities from distinct maize rhizosphere compartments. Eur J Soil Biol 37:85–93

Baudoin E, Benizri E, Guckert A (2003) Impact of artificial root exudates on the bacterial community structure in bulk soil and maize rhizosphere. Soil Biol Biochem 35:1183–1192

Bazin MJ, Markham P, Scott EM, Lynch JM (1990) Population dynamics and rhizosphere interactions. In: Lynch JM (ed) The rhizosphere. Wiley, Chichester, pp 99–127

Bhatnagar T (1975) Lumbricids and humification: a new aspect of the microbial incorporation of nitrogen induced by earthworm. In: Kilbertus G, Reisinger O, Mourey A, Cancela da Fonseca JA (eds) Biodégradation et humification. Pierron, Nancy, France, pp 168–182

Bignell DE (1994) Soil-feeding and gut morphology in higher termites. In: Hunt JH, Nalepa CA (eds) Nourishment and evolution in insect societies. Westview Press, Boulder, Colorado, pp 131–158

Billès G, Bottner P (1981) Effet des racines vivantes sur la décomposition d'une litière racinaire marquée au 14C. Plant Soil 62:193–208

Blackwood CB, Paul EA (2003) Eubacterial community structure and population size within the soil light fraction, rhizosphere, and heavy fraction of several agricultural systems. Soil Biol Biochem 35:1245–1255

Bonkowski M, Cheng WX, Griffiths BS, Alphei G, Scheu S (2000) Microbial-faunal interactions in the rhizosphere and effects on plant growth. Eur J Soil Biol 36:135–147

Bowen GD, Rovira AD (1976) Microbial colonization of plant roots. Annu Rev Phytopathol 14:121–144

Brown GG, Barois I, Lavelle P (2000) Regulation of soil organic matter dynamics and microbial activity in the drilosphere and the role of interactions with other edaphic functional domains. Eur J Soil Biol 36:177–198

Clarholm M (1985) Interactions of bacteria, protozoa and plant leading to mineralization of soil nitrogen. Soil Biol Biochem 17:181–187

Coleman DC, Ingham RE, McClellan JF, Trofymow JA (1984) Soil nutrient transformations in the rhizosphere via animal–microbial interactions. In: Rayner ADM, Anderson JM, Walton DWH (eds) Invertebrates – microbial interactions. Cambridge Univ Press, Cambridge, pp 35–58

Cooke A (1983) The effect of fungi on food selection by *Lumbricus terrestris* L. In: Satchell JE (ed) Earthworm ecology. Chapman and Hall, London, pp 365–373
Cortez J, Hameed R, Bouché MB (1989) C and N transfer in soil with or without earthworms fed with ^{14}C- and ^{15}N-labelled wheat straw. Soil Biol Biochem 21:491–497
Curl EA, Truelove B (1986) The Rhizosphere. Springer, Berlin Heidelberg New York
De Ruiter PC, Vanveen JA, Moore JC, Brussaard L, Hunt HW (1993) Calculation of nitrogen mineralization in soil food webs. Plant Soil 157:263–273
Diouf MN (2003) Les communautés fongiques des structures de récolte de plusieurs espèces de termites Macrotermitinae: origine et évolution. Thèse de l'Université, Paris XII, 165 pp
Fall S (2002) Réponse des communautés microbiennes d'un sol Sahélien à la modification de ses habitats: cas de l'activité de construction d'un termite humivore *Cubitermes niokolensis*. Thèse de l'Université Claude Bernard (Lyon 1), 176 pp
Fang C, Radosevich M, Fuhrmann JJ (2001) Characterisation of rhizosphere microbial community structure in five similar grass species using FAME and BIOLOG analyses. Soil Biol Biochem 33:679–682
Garvin MH, Lattaud C, Trigo D, Lavelle P (2000) Activity of glycolytic enzymes in the gut of *Hormogaster elisae* (Oligochaeta, Hormogastridae). Soil Biol Biochem 32:929–934
Grayston SJ, Wang S, Campbelle CD, Edwards AC (1998) Selective influence of plant species on microbial diversity in the rhizosphere. Soil Biol Biochem 30:369–378
Hamilton WE, Sillman DY (1989) Influence of earthworm middens on the distribution of soil microarthropods. Biol Fertil Soils 8:279–284
Hanlon RDG, Anderson JM (1980) Influence of Macroarthropod feeding activities on microflora in decomposing oak leaves. Soil Biol Biochem 12:255–261
Hattori T, Hattori R (1976) The physical environment in soil microbiology. An attempt to extend principles of microbiology to soil micro-organisms. Crit Rev Microbiol 26:423–461
Heal OW, Flannagan PW, French DD, MacLean SF (1981) Decomposition and accumulation of organic matter in tundra. In: Bliss LC, Heal OW, Moore JJ (eds) Tundra ecosystems: a comparative analysis. Cambridge Univ Press, Cambridge, pp 587–633
Higashi M, Abe T, Burns TP (1992) Carbon-nitrogen balance and termite ecology. Proc R Soc Lond B 249:303–308
Ingham ER, Trofymow JA, Ames RN, Hunt HW, Morley CR, Moore JC, Coleman DC (1986) Trophic interactions and nitrogen cycling in a semi-arid grassland soil. I. Seasonal dynamics of the natural populations, their interactions and effects on nitrogen cycling. J Appl Ecol 23:597–612
Jansson PE, Berg B (1985) Temporal variation of litter decomposition in relation to simulated soil climate. Long-term decomposition in a Scots pine forest. V. Can J Bot 63:1008–1016
Jenkinson DS (1966) The priming action. J Appl Radiat Isotopes [Suppl]:199–208
Kersanté A (2003) Rôle régulateur de la macrofaune lombricienne dans la dynamique de l'herbicide atrazine en sol cultivé tempéré. Thèse de l'Université Rennes, 166 pp
Kozdroj J, van Elsas JD (2000) Response of bacterial community to root exudates in soil polluted with heavy metals assessed by molecular and cultural approaches. Soil Biol Biochem 32:1405–1417
Lamotte M (1989) Place des animaux détritivores et des microorganismes décomposeurs dans les flux d'énergie des savanes africaines. Pedobiologia 33:17–35
Latour X, Corberand T, Laguerre G, Allard F, Lemanceau P (1996) The composition of fluorescent Pseudomonad populations associated with roots as influenced by plant and soil type. Appl Environ Microbiol 62:2449–2456
Lattaud C, Locati S, Mora P, Rouland C, Lavelle P (1998) The diversity of digestive systems in tropical geophagous earthworms. Appl Soil Ecol 9:189–195

Lavelle P, Spain AV (2001) Soil ecology. Kluwer, Dordrecht, 654 pp
Lavelle P, Blanchart E, Martin A, Martin S, Spain A, Toutain F, Barois I, Schaefer R (1993) A hierarchical model for decomposition in terrestrial ecosystems: application to soils of the humid tropics. Biotropica 25:130–150
Lavelle P, Lattaud C, Trigo D, Barois I (1995) Mutualism and biodiversity in soils. In: Collins HP, Robertson GP, Klug MJ (eds) The significance and regulation of soil biodiversity. Kluwer, Dordrecht, pp 23–33
Lavelle P, Barros E, Blanchart E, Brown G, Desjardins T, Mariani L, Rossi J (2001) Soil organic matter management in the tropics: why feed the soil macrofauna? Nutr Cycl Agroecosyst 61:53–61
Loranger G, Ponge JF, Imbert D, Lavelle P (2002) Leaf decomposition in two semi-evergreen tropical forests: influence of litter quality. Biol Fertil Soil 35:247–252
Lupwali NZ, Rice WA, Clayton GW (1998) Soil microbial diversity and community structure under wheat as influenced by tillage and crop rotation. Soil Biol Biochem 30:1733–1741
Maloney PE, van Bruggen AHC, Hu S (1997) bacterial community structure in relation to the carbon environment in lettuce and tomato rhizosphere and in bulk soil. Microb Ecol 34:109–117
Maraun M, Alphei J, Bonkowski M, Buryn R, Migge S, Peter M, Schaefer M, Scheu S (1999) Middens of the earthworm *Lumbricus terrestris* (Lumbricidae) microhabitats for micro- and mesofauna in forest soil. Pedobiologia 43:276–287
Marschner P, Yang CH, Lieberei R, Crowley DE (2001) Soil and plant specific effects on bacterial community composition in the rhizosphere. Soil Biol Biochem 33:1437–1445
Martin A, Cortez J, Barois I, Lavelle P (1987) Les mucus intestinaux de Ver de Terre, moteur de leurs interactions avec la microflore. Rev Ecol Biol Sol 24:549–558
Mary B, Fresneau C, Morel JL, Mariotti A (1993) C and N cycling during decomposition of root mucilage, roots and glucose in soil. Soil Biol Biochem 25:1005–1014
Miethling R, Ahrends K, Tebbe CC (2003) Structural differences in the rhizosphere communities of legumes are not equally reflected in community-level physiological profiles. Soil Biol Biochem 35:1405–1410
Mora P, Seugé C, Chotte JL, Rouland C (2003) Physico-chemical typology of the biogenic structures of termites and earthworms: a comparative analysis. Biol Fertil Soil 37:245–249
Nussbaumer U, Ascher J, Kraft A, Insam H (1997) Litter decomposition of a tropical understory species (*Ctenanthe lubbersiana*) grown under ambient and elevated CO_2. Acta Oecol 18:377–381
Parle JN (1963) Micro-organisms in the intestines of earthworms. J Gen Microbiol 31:1–11
Reichle DE, McBrayer JF, Ausmus S (eds) (1975) Ecological energetics of decomposer invertebrates in a deciduous forest and total respiration budget. Academia, Prague, pp 283–293
Rouland-Lefèvre C (2000) Symbiosis with fungi. In: Abe T, Bignell DE, Higashi M (eds) Termites: evolution, society, symbioses, ecology. Kluwer, Dordrecht, pp 289–306
Ryszkowski L (1975) Energy and matter economy of ecosystems. In: Van Dobben WH, Lowe-MacConnell RH (eds) Unifying concepts in ecology. Junk, The Hague, pp 109–126
Sallih Z, Bottner P (1988) Effect of wheat (*Triticum aestivum*) roots on mineralization rates of soil organic matter. Biol Fertil Soils 7:67–70
Satchell JE (1971) Feasibility study of an energy budget for Meathop Wood. In: Duvigneaud P (ed) Productivité des écosystèmes forestiers. UNESCO, Paris
Scheu S (1987) Microbial activity and nutrient dynamics in earthworm casts (Lumbricidae). Biol Fertil Soil 5:230–234
Setälä H, Tyynismaa M, Martikainen E, Huhta V (1991) Mineralisation of C, N, P in relation to decomposition community structure in coniferous forest soil. Pedobiologia 35:285–296

Spain AV, Le Feuvre RP (1987) Breakdown of four litters of contrasting quality in a tropical Australian rainforest. J Appl Ecol 24:279–288
Swift MJ, Heal OW, Anderson JM (1979) Decomposition in terrestrial ecosystems. Blackwell, Oxford
Tiunov AV, Scheu S (1999) Microbial respiration, biomass, biovolume and nutrient status in burrow walls of *Lumbricus terrestris* L. (Lumbricidae). Soil Biol Biochem 31:2039–2048
Tiunov AV, Bonkowski M, Alphei J, Scheu S (2001) Microflora, Protozoa and Nematoda in *Lumbricus terrestris* burrow walls: a laboratory experiment. Pedobiologia 45:46-60
Tiwari SC, Mishra RR (1993) Fungal abundance and diversity in earthworm casts and in uningested soil. Biol Fertil Soils 16:131–134
Torsvik V, Salte K, Sorheim R, Goksoyr J (1990) Comparison of phenotypic diversity and DNA heterogeneity in a population of soil bacteria. Appl Environ Microbiol 56:776–781
Trigo D, Barois I, Garvin MH, Huerta E, Irisson S, Lavelle P (1999) Mutualism between earthworms and soil microflora. Pedobiologia 43:866–873
Vancura V (1980) Fluorescent pseudomonads in the rhizosphere of plants and their relation to root exudates. Folia Microbiol 25:168–173
Vannier G (1985) Modes d'exploitation et partage des ressources alimentaires dans le système saprophage par les microarthropodes du sol. Bull Ecol 16:19–34
Voroney RP (1983) decomposition of crop residues. PhD Thesis, University of Saskatchewan, Canada
Yang CH, Crowley DE (2000) Rhizosphere microbial community structure in relation to root location and plant iron nutritional status. Appl Environ Microbiol 66:345–351
Zak JC, Willig MR, Moorhead DL, Wildman HG (1994) Functional diversity of microbial communities: a quantitative approach. Soil Biol Biochem 26:1101–1108

15 Microorganisms of Biological Crusts on Soil Surfaces

Burkhard Büdel[1]

1 Introduction

Biological soil crusts are an intimate association between soil particles and cyanobacteria, algae, microfungi, lichens, and bryophytes in different proportions which live within, or immediately on top of, the uppermost millimeters of soils. As a result, soil particles are glued together through the presence and activity of these biota and the resultant living crust covers the surface of the ground as a coherent layer (Belnap et al. 2001). This definition clearly excludes microbial mats and thick bryophyte or lichen mats, where soil particles are not aggregated by organisms.

Two main types of biological soil crusts can be distinguished according to their appearance in time and space: (1) the pioneer crusts as an initial succession phase, developing after disturbances and (2) the zonal (climax) crusts as a permanent state of vegetation development. Pioneer crusts are mainly composed of cyanobacteria in arid and semiarid regions or wherever an arid/semiarid microclimate is present, and of green algae in temperate regions. Permanent biological soil crusts are by far more complex, as they are formed by a mixture of cyanobacteria, algae, microfungi, lichens and bryophytes in different combinations and abundances. While pioneer biological soil crusts occur in many biomes of the world, e.g., savannas, temperate deciduous forests, semideserts and deserts, permanent biological soils crusts are restricted to biomes with harsh conditions for vascular plants, e.g., hot and cold deserts, xerothermic steppe formations and arctic-alpine and subarctic tundra (Büdel 2001a–d; Eldridge 2001; Galun and Garty 2001; Green and Broady 2001; Hansen 2001; Patova and Sivkov 2001; Rosentreter and Belnap 2001; Türk and Gärtner 2001; Ullmann and Büdel 2001a).

The microorganisms of biological crusts on soils have been studied under various aspects. In most cases, however, these investigations focused on one or several groups of organisms (e.g., cyanobacteria, algae, microfungi,

[1] University of Kaiserslautern, Department of Biology/Botany, P.O. Box 3049, 67653 Kaiserslautern, Germany, e-mail: buedel@rhrk.uni-kl.de

Soil Biology, Volume 3
Microorganisms in Soils: Roles in Genesis and Functions
(ed. by F. Buscot and A. Varma)

lichens, bryophytes) to the exclusion of other groups. Almost nothing is known about decomposition processes related to biological soil crusts.

2 Oxygenic Phototrophs

By far the largest and most important group of organisms in biological soil crusts includes the primary producers, performing oxygenic photosynthesis. Among them, the cyanobacteria are present in almost all types and developmental phases of soil crusts (Schwabe 1963) and are the pioneer group after disturbances. In many cases, cyanobacteria are accompanied by unicellular green algae and sometimes also lichens and mosses in different proportions. So far, about 46 genera of cyanobacteria, 70 genera of eukaryotic algae, 14 cyanolichen genera, 70 chlorolichen genera, and 65 moss and liverwort genera have been reported from soil crusts of the Earth (Büdel 2001a–d; Eldridge 2001; Galun and Garty 2001; Green and Broady 2001; Hansen 2001; Patova and Sivkov 2001; Rosentreter and Belnap 2001; Türk and Gärtner 2001; Ullmann and Büdel 2001a).

2.1 Cyanobacteria

The cyanobacteria provide much of the crust biomass and stability, and are represented by unicellular and filamentous growth forms. Cyanobacteria are generally more widely distributed (genus and species level) than eukaryotic algae which might be due to their long presence on Earth ($\sim$3.5 billion years). Filamentous species and in particular the genus *Microcoleus* (Fig. 1a) provide most of the cohesive quality of soil crusts (Fig. 2), due to their sticky sheath. The sheath tube is normally left behind when trichomes move, thus still keeping soil particles together. Once the trichome movement has settled, a new sheath tube is secreted. While the species *Microcoleus paludosus* (Kütz.) Gom., *M. sociatus* W. et G.S. West (Fig. 1a), *M. steenstrupii* Boye-Petersen and *M. vaginatus* (Vauch.) Gom. are typical for soil crusts with no or only slight salinity, *M. chtonoplastes* Thuret is a characteristic cyanobacterium of crusts with higher salinity and in the marine environment. The main biomass in most desert soil crusts is contributed by the species *M. vaginatus* (Belnap at al. 2001). However, the taxonomy of the genus is not sufficiently solved yet and data from molecular studies (16S rRNA) suggest heterogeneity of the genus (F. Garcia-Pichel, A. Lòpez-Cortéz, and U. Nübel, pers. comm.). In a molecular characterization using 16S rRNA and associated ITS region of the North American

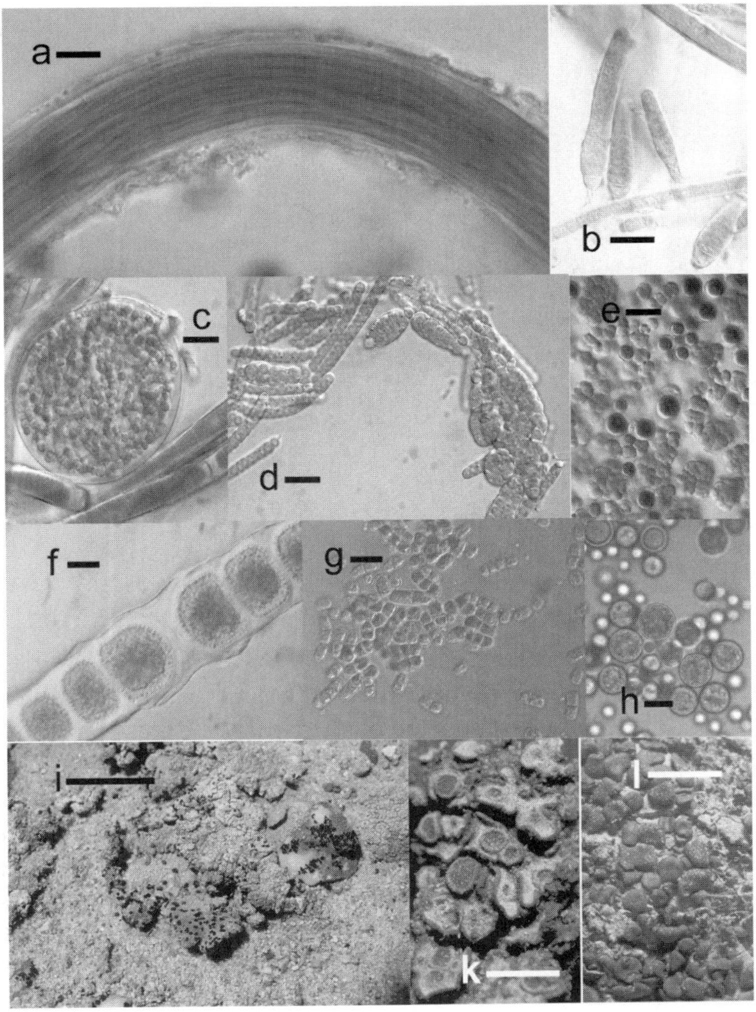

Fig. 1. Microorganisms of biological soil crusts. **a** *Microcoleus sociatus* (Cyanobacteria), soil crust, Negev Desert, Israel. *Bar* 10 µm. **b** *Calothrix* sp. (Cyanobacteria), pre-cultured material, dry savanna soil crust, Namibia. *Bar* 10 µm. **c** *Nostoc* sp. (Cyanobacteria), fully developed colony (thalli); cultured material, dry savanna soil crust, Namibia. *Bar* 20 µm. **d** *Nostoc* sp. (Cyanobacteria), as in **c**, young thallus primordia with primary heterocysts. *Bar* 20 µm. **e** *Chroococcidiopsis* sp. (Cyanobacteria), pre-cultured material in the late colonial phase, dry savanna soil crust, Namibia; large brown cells not yet identified green algae. *Bar* 10 µm. **f** *Zygogonium ericetorum* (green algae), soil crust from a forest margin, Germany. *Bar* 10 µm. **g** *Klebsormidium* sp. (green algae), cultured material, semidesert soil crust, South Africa. *Bar* 10 µm. **h** *Neochloris* sp. (green algae), cultured material, semidesert soil crust, South Africa. *Bar* 20 µm. **i** *Lecidella crystallina* (lichenized Ascomycotina), desert soil crust, Namib Desert, Namibia. *Bar* 10 cm. **k** *Peltula patellata* (lichenized Ascomacotina), desert soil crust, central Australia. *Bar* 0.5 cm. **l** *Peltula radicata* (lichenized Ascomycotina), desert soil crust, Gobi Desert, Mongolia. *Bar* 1 cm

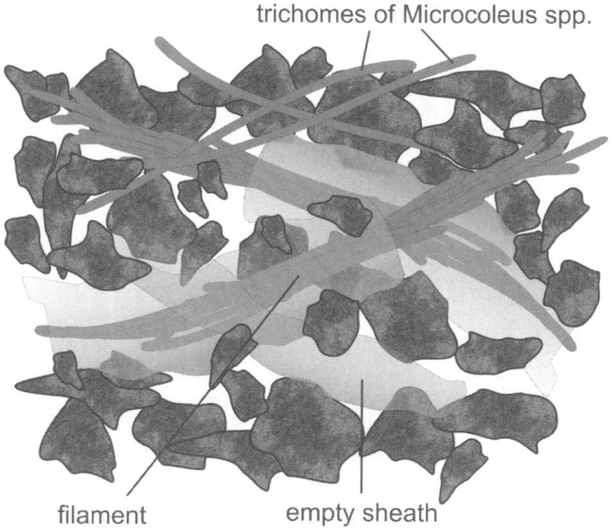

Fig. 2. Scheme of a *Microcoleus*-dominated soil crust, gelatinous sheaths gluing soil particles together; not drawn to scale

Microcoleus species, Boyer et al. (2002) found evidence that the identity of *Microcoleus* specimens from desert soils, normally attributed to the species *M. chtonoplastes*, might better belong to what they call the morphospecies *M. steenstrupii*. This finding is supported by the clearly different ecophysiology of the two species.

Another common genus is *Nostoc*, occurring in almost all soil crusts exposed to extended arid phases in the warmer climates, while in cold climates, *Nostoc* is restricted to soil crusts with an occasional surplus (e.g., flooding) of water (Fig. 1c,d). Macroscopic colonies of *Nostoc* species (e.g., *N. commune* (L.) Vaucher ex Bornet et Flahault, *N. flagelliforme* Berkely et Curtis ex Bornet et Flahault) spend a substantial period of their life cycle lying loosely on soil surfaces, while taxa with microscopic colonies (*N. punctiforme* (Kützing) Hariot) might spend most of their lives inside soils, directly underneath the surface (Fig. 3). The motile hormogonia and most of the primordia stay inside the soil (Fig. 3). In many cases, soil inhabiting *Nostoc* is commonly referred to as *Nostoc commune*. However, this might be problematic from two points of view: (1) the epithet often is applied schematically to terrestrial *Nostoc* material; (2) at present, taxonomy of Nostocacean organisms is rather confused and inhomogeneous and the very wide range of growth conditions suggests that *N. commune* might be an aggregate species (Mollenhauer et al. 1994, 1999). The horse hair-like, vagrant threads of *Nostoc flagelliforme*, from the central Asian desert, and the cosmopolitan *N. commune* might be the only cyanobacteria from soil crusts used for human food.

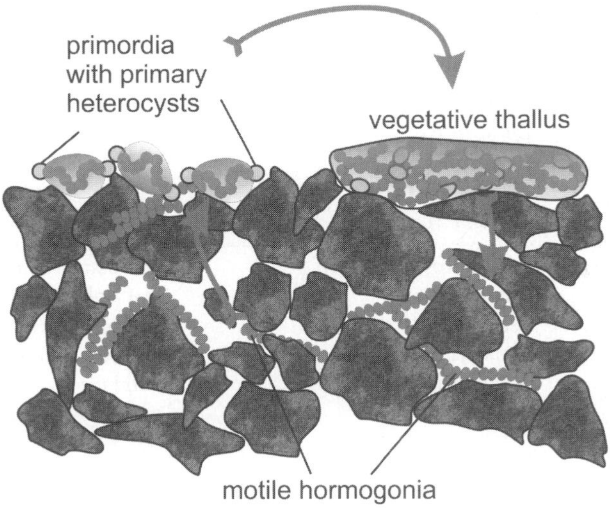

Fig. 3. Scheme of a *Nostoc*-dominated soil crust. While adult *Nostoc* colonies are on the soil surface, the motile phase (hormogonia) lives inside the soil; not drawn to scale

Further frequent filamentous cyanobacteria in soil crusts are the false branched genera *Scytonema* and *Calothrix* (Fig. 1b), both characterized by a brown to bright yellow sheath, with the color being created by the UV-A protecting indol-alkaloid scytonemin embedded within the gelatinous sheath (Garcia-Pichel and Castenholz 1991; Proteau et al. 1993). The species of both genera form small, cushion- or turf-like colonies emerging from the uppermost part of the soil and are visible with a hand lens as small brown or black patches when wet. In humid savanna soil crusts, the genus *Schizothrix*, together with other cyanobacteria of the Oscillatoriales, are the major constituents in terms of biomass and crust formation (San José and Bravo 1991; Büdel et al. 1994). Further filamentous cyanobacteria occurring in soil crusts are listed in Table 1.

Among unicellular cyanobacteria, the coenobia-forming genera *Gloeocapsa*, *Gloeothece*, and *Chroococcus* occur frequently in soil crusts, while the genera *Chroococcidiopsis* (Fig. 1e) and *Pleurocapsa* dominate in hypolithic crusts (Fig. 4; Büdel 2001d; Ullmann and Büdel 2001a). Species of *Chroococcidiopsis* and the species *Pleurocapsa minor* Hansgirg are very sensitive to high irradiance and the hypolithic habitat seems to provide just the appropriate light climate in addition to a less extreme water regime (Vogel 1955). The genus *Chroococcidiopsis* not only occurs inside soils and underneath translucent rocks, but also can be found in lichens of soil crusts. From a recent study on their molecular phylogeny and systematics, we know that they do not belong to the pleurocapsalean group of cyanobacteria, but are the closest living relatives to the filamentous, heterocyst-differentiating cyanobacteria (Nostocales; Fewer et al. 2002). Other unicellular genera found in biological soil crusts are listed in Table 1.

Table 1. Genera of cyanobacteria in biological soil crusts. (According to Büdel 2001d, with additions from Patova and Sivkov 2001)

Chroococcales	*Aphanocapsa*	*Aphanothece*
Chroococcidiopsis	*Chroococcus*	*Coccochloris*
Cyanothece	*Gloeocapsa*	*Gloeothece*
Katagnymene	*Myxosarcina*	*Pleurocapsa*[a]
Rhabdogloea	*Synechococcus*	*Xenotholos*
Oscillatoriales	*Crinalium*	*Heteroleibleinia*
Leptolyngbya	*Lyngbya*	*Komvophoron*
Microcoleus[a]	*Oscillatoria*[a]	*Phormidium*
Plectonema	*Porphyrosiphon*	*Pseudanabaena*
Pseudophormidium	*Schizothrix*	*Symploca*
Tychonema	–	–
Nostocales	*Anabaena*	*Calothrix*[a]
Cylindrospermum	*Microchaete*	*Nodularia*
Nostoc[a]	*Rivularia*	*Scytonema*[a]
Tolypothrix	*Trichormus*	–
Stigonematales	*Chlorogloea*	*Chlorogloeopsis*
Fischerella	*Hapalosiphon*	*Mastigocladus*
Stigonema[a]	–	–

[a]Directly involved in crust formation

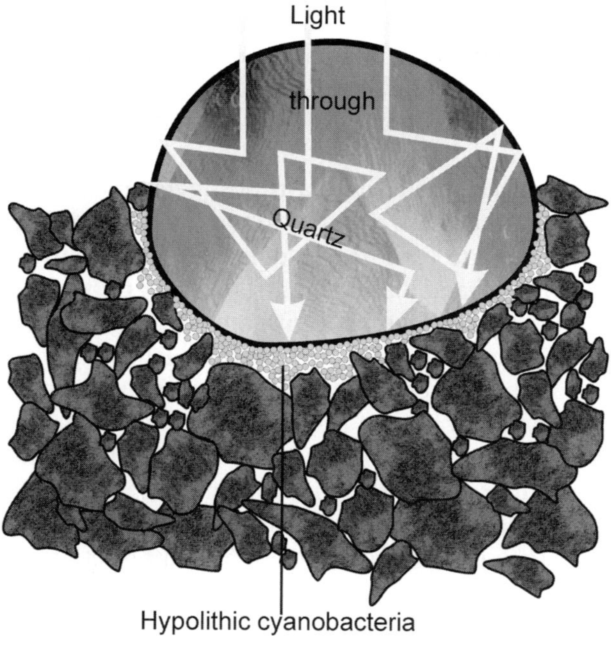

Fig. 4. Scheme of crust formation by hypolithic growth; cyanobacteria-dominated crusts develop underneath translucent rocks of all sizes where enough light reaches them, thus stabilizing the soil; not drawn to scale

2.2 Algae

So far, some 70 genera of eukaryotic algae have been identified from biological soil crusts belonging to the green algae and diatoms (Büdel 2001d). For a complete list of genera, see Table 2. Green algae occur in high species numbers in biological soil crusts with cyanobacteria, only very rarely does this correspond with abundance and the main contribution to biomass comes from the cyanobacteria (Belnap et al. 2001; Büdel 2001d). Soil crusts that are dominated or even exclusively formed by green algae are rare. The filamentous green alga *Zygogonium ericetorum* Kützing (Fig. 1f) occurs in many cyanobacteria-dominated soil and rock crusts from the tropics to the boreal region. So far, it is known from temperate regions in Europe only

Table 2. Genera of eukaryotic algae from biological soil crusts. (Büdel 2001d)

Euglenophyta	*Astasia*	–
Green algae	*Bracteacoccus*[a]	*Chlamydocapsa*
Chlamydomonas[a]	*Chlorella*[a]	*Chlorococcum*[a]
Chlorosarcinopsis	*Coccomyxa*[a]	*Cystococcus*
Deasonia	*Desmococcus**[a]	*Dictyococcus*
Ellipsoidion	*Elliptochloris*	*Fottea*
Friedmannia	*Fritschiella*	*Geminella*
Gloeocystis	*Interfilum*	*Leptosira*
Lobococcus	*Macrochloris*	*Myrmecia*
Neochloris	*Oedogonium*	*Palmellopsis*
Palmogloea	*Pleurastrum*	*Prasiococcus*
Protosiphon	*Pseudococcomyxa*	*Radiosphaera*
Rhizoclonium	*Stichococcus*[a]	*Tetracystis*
Trebouxia	*Trochiscia*	*Ulothrix*
Zygnematophyceae	*Cosmarium*	*Cylindrocystis*
Euastrum	*Penium*	*Zygogonium*[a]
Klebsormidiophyceae	*Klebsormidium*[a]	–
Xanthophyceae	*Botrydiopsis*	*Botrydium*
Chloridella	*Gloeobotrys*	*Heterococcus*
Monocilia	*Pleurochloris*	*Tribonema*
Vaucheria	*Xanthonema*	–
Bacillariophyceae	*Achnanthes*	*Amphora*
Caloneis	*Cymbella*	*Diploneis*
Eunotia	*Fragilaria*	*Gomphonema*
Hantzschia	*Navicula*[a]	*Nitzschia*
Pinnularia	*Stauroneis*	–

[a]Common in soil crusts

that this species forms a pioneer crust on disturbed and exposed soils of the central European *Fagus* forests (Büdel 2001b; Paus 1997), later on followed by mosses (e.g., *Polytrichum piliferum* Schreb. ex Brid.). Further green alga-dominated soil crusts, formed mainly by *Klebsormidium*, were found on disturbed soils of forest areas and sand dunes in Europe, often together with the xanthophyte genus *Tribonema* and a few other unicellular green algae (e.g., Komáromy 1976; Lukešová and Komárek 1987; Büdel 2001b).

The most common eukaryotic algal genera are *Bracteacoccus*, *Chlamydomonas*, *Chlorella*, *Chlorococcum*, *Coccomyxa*, *Stichococcus*, and *Klebsormidium* (Fig. 1g). The unicellular green alga *Neochloris* regularly occurs in soil crusts of dry, thornbush savannas in southern Africa (Fig. 1h, unpubl. results).

Green algae occurring in hypolithic crusts of the Knersvlakte are *Chlorella vulgaris* Beyerinck, *Cystococcus humicola* Näg., and *Coccomyxa hypolithica* Vogel (1955). In addition to cyanobacteria and green algae, diatoms can also be found in the hypolithic habitat. Rumrich et al. (1989) enumerated 51 diatom species associated with the hypolithic environment in the Namib Desert and adjacent arid regions. The species reported by them, e.g., *Achnanthes* (4 species), *Amphora* (2), *Cymbella* (5), *Eunotia* (2), *Fragilaria* (5), *Gomphonema* (3) *Navicula* (9), *Nitzschia* (8), and *Pinnularia* (2), were all generalists, and confined to neither hypolithic habitats nor arid areas.

2.3 Microlichens

Microlichens, possessing either cyanobacteria (cyanolichens) or green algae as photobionts (chlorolichens), regularly occur in biological soil crusts of the zonal vegetation type, but also, to a certain extent, in pioneer crusts. Lichen crusts (Fig. 5) are well developed in the xerothermic steppe formations of temperate and Mediterranean biomes, as well as in semideserts and some desert regions. While some 70 chlorolichen genera have been found so far, only about 15 genera of cyanolichens have been reported from soil crusts (Table 3). As a more or less regular feature, lichen-dominated crusts combine with cyanobacteria, with the proportion of both groups to each other dependent on the soil properties given (Ullmann and Büdel 2001b).

Probably the most widely spread lichen-dominated soil crust type is the so-called *Bunte Erdflechtengesellschaft* (community of colored lichens; Reimers 1950) that can be found in xerothermic steppe formations all over the world in only slightly differing species composition (Büdel 2001a–c; Eldridge 2001; Galun and Garty 2001; Green and Broady 2001; Hansen 2001; Rosentreter and Belnap 2001; Türk and Gärtner 2001; Ullmann and

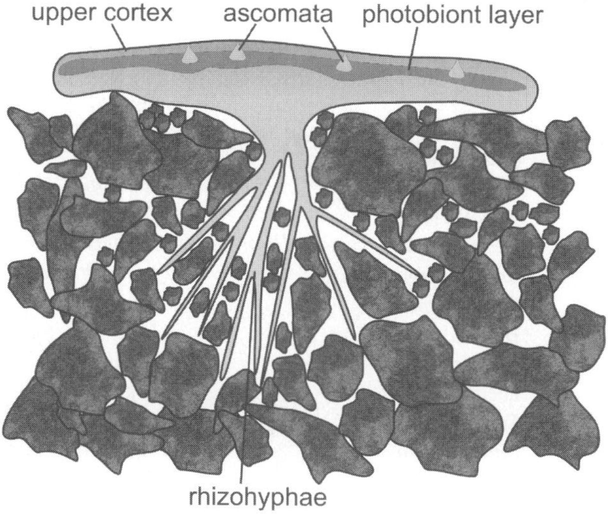

Fig. 5. Scheme of a lichen-dominated soil crust, the soil particles are kept together by rhizines and rhizohyphae; not drawn to scale

Büdel 2001a). Their main constituents are the species *Fulgensia fulgens* (Sw.) Elenk., *F. bracteata* (Hoffmann) Räsenen, *Psora decipiens* (Hedw.) Ach., *Squamarina lentigera* (GH Weber) Poelt, *Toninia sedifolia* (Scop.) Timdal, *Catapyrenium lachneum* (Ach.) R. Sant., *Collema tenax* (Sw.) Ach. and often *Diploschistes diacapsis* (Ach.) Lumbsch.

Common lichens of biological soil crusts in xerothermic habitats worldwide are *Catapyrenium* (e.g., *C. cinerascens* (Nyl.) Breuss, *C. squamulosum* (Ach.) Breuss, *C. pilosellum* Breuss), *Collema* (e.g., *C. cristatum* (L.) Wigg., *C. coccophorum* Tuck.), *Baeomyces* (e.g., *B. placophilus* Ach., *B. rufescens* (Huds.) Rebent.), *Dibaeis* (e.g., *D. baeomyces* (L.f.) Rambold et Hertel), *Diploschistes* (e.g., *D. diacapsis*, *D. muscorum* (Her.) Zahlbr.), *Endocarpon* (e.g., *E. pusillum* Hedwig), *Fulgensia* (e.g., *F. bracteata*, *F. fulgens*), *Heppia* (e.g., *H. despreauxii* (Mont.) Tuck., *H. lutosa* (Ach.) Nyl., *H. solorinoides* (Nyl.) Nyl., *Peltula* (*P. patellata* (Bagl.) Swinscow et Krog [Fig. 1k], *P. radicata* Nyl. [Fig. 1l], *P. richardssii* (Herre) Wetm.), *Squamarina* (e.g., *S. cartilaginea* (With.) P.W. James, *S. lentigera*, *S. concrescens* (Müll.Arg.) Poelt) and *Toninia* (*T. sedifolia*, *T. toepferi* (B. Stein) Navas). The crustose chlorolichen *Lecidella crystallina* Vézda et Wirth (Fig. 1i), together with hypolithic cyanobacteria, forms a soil crust that covers tens of square kilometers in the Namib Desert (Ullmann and Büdel 2001a). As can be seen from numerous publications, the state of research concerning the diversity of lichen species from biological soil crusts seems to be fairly good, but, considering the scattered investigation pattern of biological soil crusts on the Earth, further species living in this peculiar association might well be found.

Table 3. Lichen genera found in biological soil crusts. (Büdel 2001d)

Chlorolichens	*Acarospora*[a]	*Arthrorhaphis*
Aspicilia[a]	*Bacidia*	*Baeomyces*[a]
Buelliaa	*Caloplaca*[a]	*Candelariella*
Catapyrenium[a]	*Catolechia*	*Cetraria*
Chondropsis	*Cladia*	*Cladonia*
Coelocaulon	*Dacampia*	*Dactylina*
Dermatocarpon	*Desmazieria*	*Dibaeis*[a]
Diploschistes[a]	*Endocarpon*[a]	*Eremastrella*[a]
Fulgensiaa	*Fuscopannaria*	*Gyalecta*
Gypsoplaca	*Heterodea*	*Heteroplacidium*
Involucropyrenium	*Lecanora*[a]	*Lecidea*[a]
Lecidella[a]	*Lecidoma*	*Leprocaulon*
Leproloma	*Leptochidium*	*Massalongia*
Megaspora	*Multiclavula*	*Mycobilimbia*
Neofuscelia	*Ochrolechia*	*Pannaria*
Paraporpidia	*Parmelia*	*Pertusaria*
Phaeophyscia	*Phaeorrhiza*	*Physconia*
Placynthiella	*Polyblastia*	*Protoblastenia*
Psora[a]	*Psoroma*	*Rinodina*
Siphula	*Solorina*[a]	*Squamarina*[a]
Stereocaulon	*Teloschistes*	*Texosporium*
Toninia[a]	*Trapelia*	*Trapeliopsis*
Xanthoparmelia	*Xanthoria*	–
Cyanolichens	*Collema*[a]	*Gloeoheppia*[a]
Gonohymenia	*Heppia*[a]	*Leciophysma*
Leptogium	*Lichinella*	*Peccania*
Peltigera[a]	*Peltula*[a]	*Pseudopeltula*
Psorotichia	*Synalissa*	–

[a]Important crust-forming genera

3 Heterotrophic Organisms

Most studies on biological soil crusts almost exclusively focused on the primary production side with cyanobacteria, eukaryotic algae, lichens, and mosses. Bacteria (including actinomycetes), fungi, protists and microscopic invertebrate animals have received less attention (e.g., Wheeler et al. 1993; Belnap 2001), while decomposition is widely neglected (Greenfield 1997). As a consequence, the state of our knowledge of the heterotrophic part of biological soil crusts is comparatively poor (Evans and Johansen 1999).

3.1 Bacteria

Bacteria including actinomycetes are the numerically (not biomass!) dominant microorganisms of soils (Kieft 1991) with gram-positive bacteria usually outnumbering the gram-negative ones (Brock et al. 1994). While bacterial biomass and activity in soils have been determined for many soil types (e.g., Bölter 1989, 1996), this has been done only very rarely in combination with biological soil crusts. In a study by Wheeler et al. (1993), investigating the bacterial flora of biological soil crusts in the Colorado National Monument, the density and composition of bacteria were determined. The bacterial densities showed a range differing by several orders of magnitude, values varying between 4.0×10^5 and 3.8×10^7 colony forming units/g dry weight from 40 topsoil samples. This range of density was consistent with values from similar desert soils (Cameron 1969; Skujins 1984). The bacterial composition included members of the genera *Bacillus*, *Micrococcus*, and *Arthrobacter*. Actinomycetes were also present, but the genus was not determined. In his review of bacterial heterotrophs of Antarctic soils, not clearly related to biological soil crusts, however, Vincent (1988) summarized the detection of bacterial genera by various authors. In it, the genera *Pseudomonas*, *Achromobacter*, and *Alcaligenes* were reported from the Antarctic Peninsula, *Corynebacterium*, *Arthrobacter*, and *Micrococcus* from Dry Valley soils, and *Bacillus*, *Achromobacter*, *Arthrobacter*, *Aerobacter*, *Pseudomonas*, and the actinomycete *Streptomyces* from a moss site at Cape Bird. Killian and Fehér (1939) investigated the microbiology of soils at the northern margin of the Sahara, and listed more than 40 species of bacteria, however, it was not mentioned in the text whether soil crust formation was present or not.

During the preparation of this review, the great lack in our knowledge about the composition and ecological role of bacteria in biological soil crusts became obvious. There is an urgent need for research in the microbiology of biological soil crusts with respect to bacteria.

3.2 Microfungi

Although recognized as an important element of soil crusts in earlier studies, fungi are often neglected or only incidentally mentioned in recent investigations about biological soil crusts. In the most recent review on fungi in soil crusts, States et al. (2001) pointed out our limited knowledge of the diversity and the role of fungi in biological soil crusts. They distinguish between five different types of fungi/crust associations:

1. Physical crusts; formed by rain and desiccation cycles; moisture conditions promote spore germination and growth of bacteria and fungi (Fletcher and Martin 1948; Bond and Harris 1964; Cameron 1971).
2. Cyanobacterial/algal crusts; actinomycetes and filamentous fungi seem to play an important, but almost unknown role.
3. Lichen crusts; with a variety of saprotrophic, fungicolous, and lichenicolous fungi associated.
4. Bryophyte crusts; with an almost unknown and unstudied fungal flora.
5. Mixed crusts; normally a much higher diversity of fungi is associated with these crusts.

A list of microfungal taxa associated with biological soil crust formation is given in Table 4. Kieft (1991) published an excellent review of desert soil fungal communities, which the interested reader could consult. Mostly, soil fungi are classified according to their biotrophic interactions with other components of the soil crust and their ecological niche (Hawksworth 1988; States et al. 2001). According to these authors, crust-inhabiting fungi belong to the following groups: (1) terricolous/humicolous, saprotrophic on

Table 4. Microfungal taxa from biological soil crusts from Wyoming and Utah (States et al. 2001), from Death Valley, California (Hunt and Durrell 1966) and from the northern Sahara region. (Killian and Fehér 1939)

Zygomycetes	*Mortierella* sp.	*Rhizopus nigricans*
Syncephalastrum sp.	–	–
Ascomycetes	*Chaetomium perlucidum*	*Graphyllium permundum*
Kalmusia utahensis	*Macroventuria wentii*	*Pleospora richtophensis*
Basidiomycetes	*Podaxis pistilaris*	–
Pale Hyphomycetes	*Aspergillus leporis*	*Aspergillus ustus*
Aspergillus wentii	*Chrysosporium pannorum*	*Fusarium flocciferum*
Fusarium spp.	*Penicillium oxalicum*	*Penicillium* spp.
Pseudozyma sp.	*Stemphylium iliois*	Yeast II
Coelomycetes and dark Hyphomycetes		
Alternaria tenuis	*Alternaria tenuissima*	*Bipolaris ravenelii*
Cladosporium herbarum	*Cladosporium macrocarpum*	*Embellisia clamydospora*
Embellisia telluster	*Epicoccum purpurascens*	*Phoma anserina*
Phoma fimeti	*Phoma leveillei*	*Phoma nebulosa*
Phoma sp.	*Sphaeronaema* spp.	*Trichoderma* sp.
Ulocladium chartarum	*Ulocladium multiforme*	*Mycelia sterilia*, light
Mycelia sterilia, dark	–	–

organic matter in soil particles; (2) bryophillous/herbicolous, saprotrophic on mosses and plants; (3) fungicolous/lichenicolous, parasites, parasymbionts or saprobes on lichens and on other fungi and (4) mycorrhiza-forming, with vascular plant roots. However, what really becomes clear from the literature on the diversity and role of fungi in biological soil crusts, is our rather limited knowledge. Future systematic and more standardized studies are badly needed in order to understand the true diversity and function of fungi in soil crust dynamics.

3.3 Heterotrophic Protists and Invertebrate Animals

During decomposition, early-colonizing yeast and bacteria are grazed by protozoans and nematodes, while mites control nematode number. Later stages of decomposition are dominated by fungi, which are grazed by nematodes, collembola, and mites (Ingham et al. 1985). In Bamforth's study (1984), a total of 49 protozoan taxa were found in the soil crusts of deserts and semiarid woodlands of Arizona. This number included 13 flagellates, 23 ciliates and 13 testate amoebae. Feeding of protists on cyanobacteria and algae has been observed several times (e.g., Tchan and Whitehouse 1953; Whitford 1996). The amoebae feeding on unicellular cyanobacteria from rocks and soils have been regularly observed in crust formations of southern Africa and Arizona (own observations, unpubl.).

Grazing of nematodes on coccoid algae has been observed in moistened biological soil crusts (Evans and Johansen 1999), but is also known from soils of Antarctica (Powers et al. 1998). Arthropod taxa, feeding on

Table 5. Soil arthropods feeding on cyanobacteria, algae, and fungi from the Chihuahuan Desert. (Adapted from Belnap 2001)

Coleoptera	*Anotylus* sp.	–
Collembola	*Folsomia elongata*	*Hypogastura scotti*
Oribatida	*Ceratozetes* sp.	*Haplozetes* sp.
Oppia sp.	*Peloribates* sp.	*Pergalumna* sp.
Pilogalumna sp.	–	–
Astigmata	*Tryophagus similis*	*Tryophagus zachvakini*
Endeostigmata	*Alicorhagia fragilis*	*Alicorhagia usitata*
Prostigmata	*Eupodes* spp.	*Linopodes* sp.
Paratydeus sp.	–	–
Tardigrada	Not specified	–

cyanobacteria, algae, and fungi of biological soil crusts from the Chihuahua Desert have been reported by Walter (1988) and States in Belnap (2001). See Table 5 for a list of arthropod organisms. However, although a few organisms are known to feed on the primary producers of biological soil crusts, the diversity and trophic interaction of them is still very poorly understood.

4 Conclusions

Our knowledge of the biodiversity of soil crust biota from different geographical regions is rather dissimilar. This, on the one hand, is based on different methods applied by most floristic studies (e.g., determination is only rarely based on cultured material in the case of cyanobacteria, algae and fungi). On the other hand, the species concept, especially of cyanobacteria, is currently in a state of flux. For future research, the application of molecular tools should be of great help (e.g., fluorescence in situ hybridization (FISH), DNA chips for the most common taxa). More attention must also be drawn to the heterotrophic part of soil crusts in order to fully understand turnover rates.

In the ecological context, it would be important to focus on their role as primary producers, as C and N sinks or as soil stabilizers against erosion. Moreover, their influence on higher plant diversity and succession needs more investigation before we really understand the role of biological soil crusts in the ecosystem context.

Acknowledgements. The author wishes to thank the DFG (Deutsche Forschungsgemeinschaft) for several research grants and the BMBF (Federal Ministry for Education and Research, Germany) for funding. Stephanie Dojani is thanked for providing Figs. 1b–e, and Mario Salisch for Fig. 1h.

References

Bamforth SS (1984) Microbial distributions in Arizona deserts and woodlands. Soil Biol Biochem 16:133–137

Belnap J (2001) Microbes and microfauna associated with biological soil crusts. In: Belnap J, Lange OL (eds) Biological soil crusts: structure, function, and management. Ecological studies, vol 150. Springer, Berlin Heidelberg New York, pp 167–174

Belnap J, Büdel B, Lange OL (2001) Biological soil crusts: characteristics and distribution. In: Belnap J, Lange OL (eds) Biological soil crusts: structure, function, and management. Ecological studies, vol 150. Springer, Berlin Heidelberg New York, pp 3–30

Bölter M (1989) Microbial activity in soils from Antarctica (Casey Station, Wilkes Land). Proc NIPR Symp Polar Biol 2:146–153

Bölter M (1996) Analysis of soil microbial communities (autotrophs and heterotrophs) from King George Island (Arctowski Station). Proc NIPR Symp Polar Biol 9:283–298
Bond RD, Harris RD (1964) The influence of the microflora on physical properties of soils. I. Effects associated with filamentous algae and fungi. Aust J Soil Res 2:111–122
Boyer SL, Johansen JR, Flechtner VR (2002) Phylogeny and genetic variance in terrestrial *Microcoleus* (Cyanophyceae) species based on sequence analysis of the 16S rRNA gene and associated 16S–23S ITS region. J Phycol 38:1222–1235
Brock TD, Madigan MT, Martinko JM, Parker J (1994) Biology of microorganisms. Prentice Hall, Englewood Cliffs
Büdel B (2001a) Biological soil crusts of South America. In: Belnap J, Lange OL (eds) Biological soil crusts: structure, function, and management. Ecological studies, vol 150. Springer, Berlin Heidelberg New York, pp 51–55
Büdel B (2001b) Biological soils crusts in European temperate and Mediterranean regions. In: Belnap J, Lange OL (eds) Biological soil crusts: structure, function, and management. Ecological studies, vol 150. Springer, Berlin Heidelberg New York, pp 75–86
Büdel B (2001c) Biological soil crusts of Asia including the Don and Volga region. In: Belnap J, Lange OL (eds) Biological soil crusts: structure, function, and management. Ecological studies, vol 150. Springer, Berlin Heidelberg New York, pp 87–94
Büdel B (2001d) Synopsis: Biogeography of soil-crust biota. In: Belnap J, Lange OL (eds) Biological soil crusts: structure, function, and management. Ecological studies, vol 150. Springer, Berlin Heidelberg New York, pp 141–152
Büdel B, Lüttge U, Stelzer R, Huber O, Medina E (1994) Cyanobacteria of rocks and soils of the Orinoco lowlands and the Guyana uplands, Venezuela. Bot Acta 107:422–431
Cameron RE (1969) Abundance of microflora in soils of desert regions. JPL Space Prog Techn Rep no 32(1378):1–16
Cameron RE (1971) Antarctic soil microbial and ecological investigations. In: Quam LO, Porter HD (eds) Research in the Antarctic American Association for the Advancement of Science. Washington, DC, pp 137–189
Eldridge DJ (2001) Biological soil crusts of Australia. In: Belnap J, Lange OL (eds) Biological soil crusts: structure, function, and management. Ecological studies, vol 150. Springer, Berlin Heidelberg New York, pp 119–131
Evans RD, Johansen JR (1999) Microbiotic crusts and ecosystem processes. Crit Rev Plant Sci 18:183–225
Fewer D, Friedl T, Büdel B (2002) *Chroococcidiopsis* and heterocyst-differentiating cyanobacteria are each other's closest living relatives. Mol Phylogenet Evol 23:82–90
Fletcher JE, Martin WP (1948) Some effects of algae and molds in the rain crust of desert soils. Ecology 29:95–100
Galun M, Garty J (2001) Biological soil crusts of the Middle East. In: Belnap J, Lange OL (eds) Biological soil crusts: structure, function, and management. Ecological studies, vol 150. Springer, Berlin Heidelberg New York, pp 95–106
Garcia-Pichel F, Castenholz RW (1991) Characterization and biological implications of scytonemin, a cyanobacterial sheath pigment. J Phycol 27:395–409
Green TGA, Broady PA (2001) Biological soil crusts of Antarctica. In: Belnap J, Lange OL (eds) Biological soil crusts: structure, function, and management. Ecological studies, vol 150. Springer, Berlin Heidelberg New York, pp 133–139
Greenfield LG (1997) Nitrogen in soil *Nostoc* mats: forms, release and implications for nutrient cycling in Antarctica. NZ Nat Sci 23: 101–107
Hansen ES (2001) Lichen-rich soil crusts of arctic Greenland. In: Belnap J, Lange OL (eds) Biological soil crusts: structure, function, and management. Ecological studies, vol 150. Springer, Berlin Heidelberg New York, pp 57–65
Hawksworth DL (1988) The variety of fungal/algal symbioses, their evolutionary significance, and the nature of lichens. Bot J Linn Soc 96:3–20

Hunt CD, Durrell LW (1966) Distribution of fungi and algae. US Geol Surv Profess Pap 509:55–66
Ingham RE, Trofymow JA, Ingham ER, Coleman DC (1985) Interactions of bacteria, fungi, and their nematode grazers: effects on nutrient cycling, and plant growth. Ecol Monogr 55:119–140
Kieft TL (1991) Soil microbiology in reclamation of arid and semi-arid lands. In: Skujins J (ed) Semiarid lands and deserts: soil resource and reclamation. Decker, New York, pp 209–256
Killian C, Fehér D (1939) Recherches sur la phénomène microbiologie des sols désertiques. Encycl Biol 21:1–127
Komáromy ZP (1976) Soil algal growth types as edaphic adaptations in Hungarian forest and grass steppe ecosystems. Acta Bot Acad Sci Hung 22:373–379
Lukešová A, Komárek J (1987) Succession of soil algae on dumps from strip coal-mining in the Most region (Czechoslovakia). Folia Geobot Phytotax, Praha 22:355–362
Mollenhauer D, Büdel B, Mollenhauer R (1994) Approaches to species delimitations in the genus *Nostoc* Vaucher 1803 ex Bornet et Flahault 1888. Arch Hydrobiol Suppl 105 (Algol Stud 75):189–209
Mollenhauer D, Bengtson R, Lindstrøm E-A (1999) Macroscopic cyanobacteria of the genus *Nostoc*: a neglected and endangered constituent of European inland aquatic biodiversity. Eur J Phycol 34:349–360
Patova E, Sivkov M (2001) Diversity of soil cyanophyta, CO_2-gas exchange and acetylene reduction of the soil crust in the cryogenic soils (East-European tundra). Nova Hedwigia, Beih 123:387–395
Paus SM (1997) Die Erdflechtenvegetation Nordwestdeutschlands und einiger Randgebiete. Bibl Lichenol 66:1–222
Powers LE, Ho M, Freckman DW, Virginia RA (1998) Distribution, community structure, and microhabitats of soil invertebrates along an elevational gradient in Taylor Valley, Antarctica. Arctic Alpine Res 30:133–141
Proteau PJ, Gerwick WH, Garcia-Pichel F, Castenholz RW (1993) The structure of scytonemin, an ultraviolet sunscreen pigment from the sheaths of cyanobacteria. Experientia 49:825–829
Reimers H (1950) Beiträge zur Kenntnis der Bunten Erdflechten-Gesellschaft. I. Zur Systematik und Verbreitung der Charakterflechten der Gesellschaft besonders im Harzvorland. Ber Dtsch Bot Ges 63:148–156
Rosentreter R, Belnap J (2001) Biological soil crusts of North America. In: Belnap J, Lange OL (eds) Biological soil crusts: structure, function, and management. Ecological studies, vol 150. Springer, Berlin Heidelberg New York, pp 31–50
Rumrich U, Rumrich M, Lange-Bertalot H (1989) Diatomeen als "Fensteralgen" in der Namib-Wüste und anderen ariden Gebieten von SWA/Namibia. Dinteria 20:23–29
San José JJ, Bravo CR (1991) CO_2 exchange in soil algal crusts occurring in the *Trachypogon* savannas of the Orinoco Llanos, Venzuela. Plant Soil 135:233–244
Schwabe GH (1963) Blaualgen der phototrophen Grenzschicht. Pedobiol 2:132–152
Skujins J (1984) Microbial ecology of desert soils. Adv Microb Ecol 7:49–92
States JS, Christensen M, Kinter CL (2001) Soil fungi as components of biological soil crusts. In: Belnap J, Lange OL (eds) Biological soil crusts: structure, function, and management. Ecological studies, vol 150. Springer, Berlin Heidelberg New York, pp 155–166
Tchan YT, Whitehouse JA (1953) Study of soil algae II. The variation of the algal populations in sandy soil. Proc Linn Soc NSW 78:160–170
Türk R, Gärtner G (2001) Biological soil crusts of the subalpine, alpine and nival areas in the Alps. In: Belnap J, Lange OL (eds) Biological soil crusts: structure, function, and management. Ecological studies, vol 150. Springer, Berlin Heidelberg New York, pp 67–73

Ullmann I, Büdel B (2001a) Biological soil crusts of Africa. In: Belnap J, Lange OL (eds) Biological soil crusts: structure, function, and management. Ecological studies, vol 150. Springer, Berlin Heidelberg New York, pp 107–118

Ullmann I, Büdel B (2001b) Ecological determinants of species composition of biological soil crusts on a landscape scale. In: Belnap J, Lange OL (eds) Biological soil crusts: structure, function, and management. Ecological studies, vol 150. Springer, Berlin Heidelberg New York, pp 203–213

Vincent WF (1988) Microbial ecosystems of Antarctica. Studies in polar research. Cambridge Univ Press, Cambridge

Vogel S (1955) Niedere "Fensterpflanzen" in der südafrikanischen Wüste. Beitr Biol Pflanz 31:45–135

Walter DE (1988) Nematophagy by soil arthropods from the shortgrass steppe, Chihuahuan Desert and Rocky Mountains of the Central United States. Agric Eco Syst Environ 24:307–316

Wheeler CC, Flechtner VR, Johansen JR (1993) Microbial spatial heterogeneity in microbiotic crusts in Colorado National Monument. II. Bacteria. Great Basin Nat 53:31–39

Whitford WG (1996) The importance of the biodiversity of soil biota in arid ecosystems. Biodiv Conserv 5:185–195

16 Microorganisms in Toxic Metal-Polluted Soils

Geoffrey M. Gadd[1]

1 Introduction

Metals are, of course, significant natural components of all soils where their presence in the mineral fraction comprises a pool of potentially mobile metal species, many essential nutrients for plants and microorganisms, and important solid components that can have a fundamental effect on soil biogeochemical processes, e.g. clays, minerals, iron and manganese oxides. Metals are also present in organic fractions, frequently in bound forms, with some metal recycling occurring as a result of biomass and organic matter degradation and decay. The aqueous phase provides a mobile medium for metal transfer and circulation through the soil and to organisms, but also to the aquatic environment. Microorganisms are intimately associated with the biogeochemical cycling of metals, and associated elements, where their activities can result in mobilization and immobilization depending on the process involved and the microenvironment where the organism is located (Ehrlich 1997). Root-inhabiting rhizosphere microorganisms, including mycorrhizal fungi, have a major influence on plant nutrition by means of effects on phosphate availability for example, but also concomitant metal circulation. Indeed, during the early phases of soil formation the contribution of microbial activities to rock weathering, mineral dissolution and element cycling is significant and also intimately related to metal movements and microbial strategies for metal transformations.

Metals exhibit a range of toxicities towards microorganisms, depending on physico-chemical factors, speciation etc., and while toxic effects can arise from natural processes in the soil, toxic effects on microbial communities are more commonly associated with anthropogenic contamination or redistribution of toxic metals in terrestrial ecosystems. Such contamination can arise from aerial and aquatic sources, as well as agricultural practices, industrial activity, and domestic and industrial wastes. In some

[1]Division of Environmental and Applied Biology, Biological Sciences Institute, School of Life Sciences, University of Dundee, Dundee, DD1 4HN, Scotland, UK, e-mail: g.m.gadd@dundee.ac.uk, Tel: +44-(0)1382-344765, Fax: +44-(0)1382-348216

Soil Biology, Volume 3
Microorganisms in Soils: Roles in Genesis and Functions
(ed. by F. Buscot and A. Varma)

cases, microbial activities can result in remobilization of metals from other wastes and transfer into aquatic systems. It is commonly accepted that toxic metals, and their chemical derivatives, metalloids, and organometals can have significant effects on microbial populations and almost every index of microbial activity can be affected, depending on the particular situation. However, metal toxicity is greatly affected by the physico-chemical nature of the environment and the chemical behaviour of the particular metal species in question (Gadd and Griffiths 1978). The degree of metal toxicity reduction can be quantitatively related to adsorption capacity and selectivity of adsorbents for example (Malakul et al. 1998). Further, unlike certain xenobiotics perhaps, not always is there a link between bioavailability and toxicity, a good example being the frequent amelioration of metal fungitoxicity by acidic pH (Gadd 1993a). Despite apparent toxicity, many microorganisms survive, grow and even flourish in apparently metal-polluted locations and a variety of mechanisms, both active and incidental, contribute to tolerance. All mechanisms depend on some change in metal speciation leading to decreased or increased mobility. Such metal transformations between soluble and insoluble phases are also at the heart of metal biogeochemistry, thus providing a direct link between microbial responses and element cycles. Thus, interactions of microorganisms with metals are extremely important and underpin many aspects of soil biology. The objective of this chapter is to detail major interactions of microorganisms with metals with particular reference to effects on speciation, toxicity, and biogeochemical cycling. In addition, attention is drawn to their significance in plant nutrition and in environmental biotechnology where certain microbial processes have application in metal recovery or detoxification.

2 Metals in Soils

All soils contain metals, the concentrations in uncontaminated soil being related to the geology of the parental material. 'Normal' total concentrations may be high with speciation and bioavailability determining any effects on microbial populations. Metal release by environmental disturbance, weathering processes and biological activities may result in toxic effects on indigenous microbial populations, as well as metal transfer to aquatic habitats. However, effects on microbial populations are more commonly associated with pollution events, where concentrations of introduced toxic metals may greatly exceed indigenous levels. To appreciate and understand possible effects on microbial populations, it is necessary to consider some aspects of metal chemistry in soil since this determines speciation, bioavailability and toxicity, and is itself greatly influenced by environmental factors (McLean and Bledsoe 1992).

In soil, metals may be dissolved in the soil solution, associated with organic and inorganic soil constituents and precipitated as pure or mixed compounds, including secondary minerals. These pools are the major locations of introduced metals, although natural metals will also show a distribution within these pools, as well as being found in the primary and secondary mineral soil components. Concentrations of metals in the soil solution is governed by many chemical processes including inorganic and organic complexation, oxidation–reduction reactions, precipitation/dissolution, adsorption/desorption etc., some of course mediated by microbial activities. Metals can exist in the soil solution as free cations (e.g. Cu^{2+}, Cd^{2+}, Zn^{2+}), as soluble complexes with inorganic or organic ligands (e.g. $ZnCl^{+}$, $CdCl_3^{-}$, metal citrates) and associated with colloidal material. Common inorganic ligands that can complex metals include SO_4^{2-}, Cl^{-}, OH^{-}, PO_4^{3-}, NO_3^{-} and CO_3^{2-}. Organic complexing agents include humic and fulvic acids, aromatic and aliphatic compounds and carboxylic acids (McLean and Bledsoe 1992). This is obviously a complex phenomenon in view of the enormous variety of organic compounds that may be present and their physical, chemical and biological origins. Solubility will influence transport through the soil environment as well as interactions with the biota. The oxidation state of several metals also determines solubility with, e.g. Cr(VI) being soluble and toxic with Cr(III) being immobile and less toxic. Such reductive transformations may be mediated by microbes, with accompanying consequences for survival. Colloidal materials of significance in affecting metal bioavailability and transport include iron and manganese oxides, clay minerals and organic matter. It is also possible to envisage microorganisms as agents of metal immobilization and transport if mobile/motile in the soil solution.

Metals can precipitate as solid phases in soils, e.g. $CdCO_3$, $Pb(OH)_2$, ZnS, CuS, as well as mixed compounds. Toxic metals may also substitute for other metals in indigenous minerals via solid solution formation, e.g. Cd may substitute for Ca in $CaCO_3$. In addition, toxic metals may sorb onto pre-existing minerals. Precipitation may be a significant process in highly contaminated soils. Metal sorption to soil organic matter (including organisms) is also highly important. Since organic matter decreases with depth in soils, mineral components will become more significant for sorption with depth. Clay minerals, e.g. montmorillonite and kaolinite, and iron and manganese oxides are major inorganic determinants of metal availability in soil. It should be noted that metals associated with exchange sites are a significant reserve of potentially mobile metals, depending on environmental changes/biological activity. Anionic contaminants, such as arsenic, selenium and chromium oxyanions like AsO_4^{3-}, AsO_2^{-}, SeO_4^{2-}, SeO_3^{2-}

and CrO_4^{2-}, can sorb to positive charges on insoluble organic matter as well as iron, manganese and aluminium oxides, carbonates, the density of positive charges increasing with decreasing pH values (McLean and Bledsoe 1992). The affinity sequence for anion absorption onto iron oxide was phosphate = silicate = arsenate > bicarbonate/carbonate > citrate = selenate > molybdate > oxalate > fluoride = selenate > sulphate, though capacities are very much smaller than those for metal cations (Balistrieri and Chao 1987). Metal cation adsorption is correlated with pH, redox potential, clay minerals, organic and inorganic matter, Fe and Mn oxides, and calcium carbonate content, and alterations in these may affect metal mobility: microbial activity can contribute greatly to changes in pH, redox potential and soil organic matter. Thus, complex arrays of physical, chemical and biological processes govern metal mobility in soil and interactions with microorganisms. Microbial activity may also have significant effects on metal mobility and chemical changes in soil, those effects which predominate often being site-specific. Clearly, soils vary widely in composition, with the nature of the contamination often adding a further highly specific dimension. In addition, there may be clear temporal changes in metal mobility and microbial populations due to many factors such as environmental effects, natural weathering, and plant growth. In a microbiological or molecular context, the nano- and microenvironments are also highly significant with important processes often being obscured by macroscopic methods for assessment of physico-chemical attributes, metal concentrations and microbial processes. Subsequent sections in this chapter will attempt to illustrate how diverse microbial groups may affect metal speciation in soil, and conversely, how metals may affect microbial populations.

3 Effects of Toxic Metals on Microbial Communities

Toxic metals may have a considerable impact on populations of soil microorganisms and their activities (Brookes and McGrath 1984). Adverse effects may be observed as a reduction in biomass as well as activity: almost every index of microbial activity has been shown to be potentially affected by toxic metal exposure (Brookes and McGrath 1984; Aoyama and Nagumo 1997a; Kuperman and Carreiro 1997). Such changes may limit other important soil processes such as organic matter decomposition (Chander and Brookes 1991; Aoyama and Nagumo 1997b). In addition, changes in microbial community structure in response to toxic metal contamination may result in reduced activity, such as altered abilities to degrade organic pollutants (Doelman et al. 1994). Soil containing toxic metals, e.g. Cu, Ni, Zn,

Cd, from long-term inputs of contaminated sewage sludge may contain less biomass and altered microbial functionality. A general order of inhibition was found to be Zn > Cu > Cd for single metals while for metal combinations, it was (Cd + Cu + Zn) > [(Cd + Zn) and (Cu + Zn)] > (Cd + Cu). In both cases, grassland >arable (Renella et al. 2002). However, short-term incubations are a poor model of changes in microbial biomass or activity due to chronic metal exposure (Renella et al. 2002). In general, experimentation based around metal additions to soils is unpredictable and may have little relation to field results (Giller et al. 1998). Undoubtedly, metal effects on natural soil communities are complex and difficult to characterize because of the complex array of contributing factors. Most knowledge about toxic metal effects on soil microorganisms is derived from data relating to only a few toxic metals, e.g. Cu and Zn, or from studies on sewage sludge applications that contain metal mixtures at relatively low concentrations (Shi et al. 2002). Many contaminated sites contain mixtures of metals as well as organic pollutants: each may have reciprocal physical and chemical effects on the other with complexation and other phenomena affecting toxicity, bioavailability and degradation for example. Toxic metals are often associated with organic pollutants in polluted sites. Such mixtures may provide special problems for a functional soil community. In one study, it was found that the soil microbial community was predominantly affected by hydrocarbons rather than associated Pb and Cr contamination (Shi et al. 2002). Therefore, although some gross generalizations are possible regarding toxic metal influence on microbial communities, individual cases are likely to be site-specific and extremely complex.

Numerous studies have shown that microbial population responses to toxic metals are characterized by a population shift from bacteria, including streptomycetes, to fungi (Mineev et al. 1999; Chander et al. 2001a, b; Kostov and van Cleemput 2001; Olayinka and Babalola 2001; Khan and Scullion 2002). However, other studies have shown a higher metal sensitivity of the fungal component of the microbial biomass (Pennanen et al. 1996). The time needed for development of community tolerance may be highly variable, some studies detecting increased tolerance of bacterial communities after a few days of metal exposure (Diaz-Ravina and Baath 1996), other studies demonstrating several years (Doelman and Hanstra 1979). An increase in metal tolerance of a bacterial community after metal addition may be attributed to the immediate death of sensitive species followed by differing competitive abilities and adaptation of bacterial survivors (Diaz-Ravina and Baath 1996). For soil denitrifying communities, an immediate metal effect (1 day) was a reduction in the denitrification rate, but also a decrease in N_2O reduction more than the N_2O production rate, i.e. nitrate reductase activity was a sensitive metal target. Metal exposure resulted in an increased Cd-, Cu- and Zn tolerance of N_2O reductase activity (Holtan-Hartwig et al. 2002). Using phospholipid fatty acid (PLFA) to reveal changes in species composition, bacterial community tolerance increased in all metal treatments compared to an unpolluted control, with tolerance to specific metals increasing the most when the same metal was added to the soil, i. e. tolerance to Cu increased most in Cu-polluted treatments (Baath et al. 1998). For β-subgroup ammonia oxidizers, metal additions caused specific changes in the community, which did not occur in the presence of inoculated metal-resistant bacteria (Stephen et al. 1999). Archaea from metal-contaminated soils cluster within the terrestrial non-thermophilic group, *Crenarchaeota* (Sandaa et al. 1999). In anaerobic bacterial consortia, dehalogenation, aromatic degradation and methanogenesis showed differential sensitivities to different toxic metals, indicating that metals may affect anaerobic degradation of aromatic pollutants (Kuo and Genthner 1996).

All nutritional groups of fungi (saprotrophs, biotrophs and necrotrophs) can be affected by toxic metals. A relative decrease in an indicator fatty acid for arbuscular mycorrhizal fungi and an increase for other fungi has been reported for zinc-polluted soil (Kelly et al. 1999). Toxic metals (Cd, Cr, Cu, Ni, Pb and Zn) led to a decrease in the number of arbuscular mycorrhizal fungi and low colonization of plant roots, and, as a result, changes in mycorrhizal species diversity (del Val et al. 1999; Moynahan et al. 2002; Mozafar et al. 2002). Toxic metals also reduce plant root colonization by ectomycorrhizal fungi and ectomycorrhizal species composition (Fay and Mitchell 1999; Hartley et al. 1999; Markkola et al. 2002). It should be noted that phytoextraction practices, e. g. choice of plant species and soil amendments, may have significant effects on the indigenous community of arbuscular mycorrhizas as well as their function during long-term remediation treatments (Pawlowska et al. 2000). However, a major limitation to predicting the consequences of pollution-mediated changes in mycorrhizal fungal communities is a limited understanding of the functional significance of mycorrhizal biodiversity (Cairney and Meharg 1999).

The most frequent soil saprotrophic microfungi isolated from heavily metal-polluted habitats in Argentina, Czech Republic and Ukraine were reported to be species of *Penicillium*, *Aspergillus*, *Trichoderma*, *Fusarium*, *Rhizopus*, *Mucor*, as well as *Paecilomyces lilacinus*, *Nectria invertum*, *Cladosporium cladosporioides*, *Alternaria alternata* and *Phoma fimeti* (Kubatova et al. 2002; Massaccesi et al. 2002; Fomina, Manichev, Kadoshnikov and Nakonechnaya, unpubl. data). Melanized fungi, such as *Cladosporium* sp., *Alternaria alternata* and *Aureobasidium pullulans*, were often isolated from soil containing high concentrations of copper and mercury (Zhdanova et al. 1986) and may also be dominant members of the mycoflora of metal-contaminated phylloplanes (Mowll and Gadd 1985). Dark septate endophytes were found to be dominant fungi among isolates from roots of *Erica herbacea* L. in Pb, Cd, and Zn-polluted soil (Cevnik et al. 2000). *Penicillium* species are often reported to be dominant in copper-contaminated environments (Ribeiro et al. 1972).

Investigations on metal toxicity in mycorrhizal fungi have revealed wide inter- and intraspecific variation in metal sensitivity (Jones and Muehlchen 1994; Hartley et al. 1997a, b; Vodnik et al. 1998; Blaudez et al. 2000; Meharg and Cairney 2000). Some studies have suggested that selection for resistant ecotypes occurs where the degree of toxic metal contamination and selection pressure is high (Colpaert et al. 2000; Sharples et al. 2000, 2001). For example, isolates of the ericoid mycorrhizal fungus *Oidiodendron maius*, recovered from *Vaccinium myrtillus* growing in metal-polluted areas, were in general less sensitive to metals than strains from non-polluted sites (Lacourt et al. 2000). In vitro Zn tolerance of isolates of the ectomycorrhizal fungus *Suillus luteus* from a zinc-polluted habitat was significantly higher than isolates from a non-polluted site (Colpaert et al. 2000).

Another metal-rich niche inhabited by fungi is the rock environment which often contains toxic metal minerals. Fungi have been reported from a wide range of rock types and building stone including limestone, soapstone, marble, granite, sandstone, andesite, basalt, gneiss, dolerite, amphibolite, quartz and cement even from the most harsh environments, e.g. hot and cold deserts (Staley et al. 1982; Gorbushina et al. 1993; Gerrath et al. 1995; Bogomolova et al. 1998; Kumar and Kumar 1999; Sterflinger 2000; Verrecchia 2000; Burford et al. 2003a,b). Fungi form a major component of microbial biofilms (an assemblage of surface-associated microbial cells enclosed in an extracellular polymeric substance matrix) in rocks. They can exist as free-living fungi or as lichens (symbiosis between ascomycete or basidiomycete fungi and green algae (mainly *Trebouxia*) or less often, cyanobacteria. Among the most commonly reported fungal groups inhabiting exposed rock surfaces are the microcolonial fungi (MCF) described by Staley et al. (1982). These are also known as the black yeasts or yeast-like black meristematic fungi (Gorbushina et al. 1993; Wollenzien et al. 1995; Sterflinger 2000; de Leo et al. 2003). Meristematic growth is characterized by the production of swollen cells with thick melanin-containing cell walls. In addition, many of the black fungi exhibit yeast-like stages (Gadd 1980; Gorbushina et al. 1993; Wollenzien et al. 1995; Bogomolova et al. 1998; Sterflinger 2000; de Leo et al. 2003).

4 Metal Resistance and Tolerance Mechanisms

4.1 Bacteria

Bacterial plasmids have resistance genes to many toxic metals and metalloids, e.g. Ag^{+}, AsO_2^{-}, AsO_4^{3-}, Cd^{2+}, Co^{2+}, CrO_4^{2-}, Cu^{2+}, Hg^{2+}, Ni^{2+}, Sb^{3+}, TeO_3^{2-}, Tl^{+} and Zn^{2+}. Related systems are also frequently located on bacterial chromosomes, e.g. Hg^{2+} resistance in *Bacillus*, Cd^{2+} efflux in *Bacillus*, arsenic efflux in *E. coli* (Silver and Phung 1996). Copper tolerance genes are generally genome-located. Generalizations regarding bacterial metal resistance may include that (1) plasmid-determined resistances are highly specific, (2) resistance systems have been found on plasmids in all bacterial groups tested and (3) resistance mechanisms generally involve efflux from the cells or enzymatic detoxification (Silver and Phung 1996; Nies 1992a, 1995, 1999, 2003). However, other less-specific interactions, e.g. sorption, may contribute to the overall response. Efflux pumps, determined by plasmid and chromosomal systems, are either ATPases or chemiosmotic systems, with mechanisms often showing similarity in different types of bacteria. Cd^{2+} resistance may involve (1) an efflux ATPase in Gram-positive bacteria, (2) cation-H^{+} antiport in Gram-negative bacteria, and (3) intracellular metallothionein in cyanobacteria (Nies 1992b; Silver and Phung 1996). Arsenic-resistant Gram-negative bacteria have an arsenite efflux ATPase and an arsenate reductase (which reduces arsenate [As(V)] to arsenite [As(III)] which comprise the underlying biochemical mechanism (Ji and Silver 1992a, b; Ji et al. 1994). Similar systems for Hg^{2+} resistance occur on plasmids from Gram-positive and Gram-negative bacteria with component genes being involved in transport of Hg^{2+} to the detoxifying enzyme, mercuric reductase, which reduces Hg^{2+} to elemental Hg^{o} (Silver and Phung 1996). The enzyme organomercurial lyase can break the C–Hg bond in organomercurials (Schottel et al. 1974). The plant *Arabidopsis thaliana* has been transformed to express bacterial mercuric reductase, providing potential for phytoremediation (Rugh et al. 1996). It appears that contaminating mercury selects for higher frequencies of Hg^{2+}-resistant bacteria in polluted habitats (Silver and Phung 1996).

Plasmid-determined chromate resistance appears unconnected with chromate [Cr(VI)] reduction to Cr(III) (Ohtake et al. 1987), resistance depending on reduced CrO_4^{2-} uptake (Ohtake et al. 1987). Similarly, plasmid-mediated Ag^{+} resistance appears not to involve Ag^{+} reduction to Ag^{o} (Silver and Phung 1996). A Cd^{2+} efflux ATPase is widely found in Gram-positive bacteria including soil *Bacillus* sp. (Ivey et al. 1992). The large plasmids of

Alcaligenes eutrophus have numerous toxic metal resistance determinants, e.g. three for Hg^{2+}, one for Cr^{6+}, and two for divalent cations, *czc* (Cd^{2+}, Zn^{2+} and Co^{2+} resistance) and *cnr* (Co^{2+} and Ni^{2+} resistance; Nies et al. 1990; Nies 1992b; Silver and Phung 1996). Czc functions as a chemiosmotic divalent cation/H^{+} antiporter (Nies et al. 1989a,b; Nies 1995; Nies and Silver 1995). In *Enterococcus hirae* (previously *Streptococcus faecalis*), copper resistance is determined by two genes, *copA* and *copB* which respectively determine uptake and efflux P-type ATPases (Solioz et al. 1994; Solioz and Odermatt 1995). Plasmid-determined Cu^{2+} resistance has been described in *Pseudomonas* (Cooksey 1993, 1994), *Xanthomonas* (Lee et al. 1994) and *Escherichia coli* (Brown et al. 1994, 1995). Chromosomal genes also affect Cu^{2+} transport and resistance by determining functions such as uptake, efflux and intracellular Cu^{2+} binding (Brown et al. 1995). Bacterial arsenic resistance is plasmid-mediated in Gram-positive bacteria and several mechanisms of plasmid-mediated tellurite resistance have been suggested, including reduction, reduced uptake and enhanced efflux, although as with Ag^{+}, resistance does not appear to depend on reduction to the elemental form (Te^{o}; Walter and Taylor 1992; Turner et al. 1995). Thus, there are still many unknown interactions to be characterized, and even for the well-known bacterial mercury, arsenic, cadmium and copper resistance mechanisms, the fundamental mechanisms of resistance or gene regulation at the molecular level are not yet fully understood (Silver and Phung 1996).

4.2 Fungi

As with bacteria, intracellular metal concentrations may be regulated by transport, including efflux mechanisms (Gadd 1993b; Macreadie et al. 1994; Blaudez et al. 2000). Such mechanisms are involved in normal metal homeostasis, but also have a role in the detoxification of potentially toxic metals. The fungal vacuole also has an important role in the regulation of cytosolic metal ion concentrations and the detoxification of potentially toxic metal ions (White and Gadd 1986; Gadd 1993b, 1996; Gharieb and Gadd 1998; Liu and Culotta 1999). Metals preferentially sequestered by the vacuole include Mn^{2+} (Okorokov et al. 1985; Gadd and Lawrence 1996), Fe^{2+} (Bode et al. 1995), Zn^{2+} (White and Gadd 1987), Co^{2+} (White and Gadd 1986), Ca^{2+} and Sr^{2+} (Okorokov et al. 1985; Borst-Pauwels 1989; Gadd 1993b; Okorokov 1994), Ni^{2+} (Joho et al. 1995) and the monovalent cations K^{+}, Li^{+} and Cs^{+} (Okorokov et al. 1980; Perkins and Gadd 1993a, b). The absence of a vacuole or a functional vacuolar H^{+}-ATPase in *S. cerevisiae* is associated with increased sensitivity and a largely decreased capacity of the cells to accumulate Zn, Mn, Co and Ni (Ramsay and Gadd 1997), metals known to be mainly detoxified in the vacuole (Gadd 1993b; Joho et al. 1995).

For Cu and Cd, intracellular detoxification in fungi appears to predominantly depend on sequestration in the cytosol by induced metal-binding molecules (Hayashi and Mutoh 1994; Macreadie et al. 1994; Ow et al. 1994; Rauser 1995). These include low molecular weight cysteine-rich proteins (metallothioneins) and peptides derived from glutathione (phytochelatins; Mehra and Winge 1991; Macreadie et al. 1994; Ow et al. 1994; Rauser 1995; Wu et al. 1995; Inouhe et al. 1996). The latter peptides have the general structure of $(\gamma\text{Glu-Cys})_n$-Gly where the γGlu-Cys repeating unit may extend up to 11 (Ow et al.1994). In *Schizosaccharomyces pombe* the value of n ranges from 2–5, while in *S. cerevisiae*, only an n_2 isopeptide has been observed (Macreadie et al. 1994). As well as being termed phytochelatins, such peptides are also known as cadystins and metal γ-glutamyl peptides, although the chemical structure, $(\gamma EC)_nG$, is a more useful description. Although $(\gamma EC)_nG$ can be induced by a wide variety of metal ions, including Ag, Au, Hg, Ni, Pb, Sn and Zn, metal binding has only been shown for a few, primarily Cd and Cu (Ow et al. 1994). For Cd, two types of complexes exist in *S. pombe* and *Candida glabrata*. A low molecular weight complex consists of $(\gamma EC)_nG$ and Cd, whereas a higher molecular weight complex also contains acid-labile sulphide (Murasugi et al. 1983; Ow et al. 1994). The $(\gamma EC)_nG$-Cd-S^{2-} complex has a greater stability and higher Cd binding capacity than the low molecular weight complex, and consists of a CdS crystallite core and an outer layer of $(\gamma EC)_nG$ peptides (Dameron et al. 1989). The higher binding capacity of the sulphide-containing complex confers tolerance to Cd (Ow et al. 1994). In *S. pombe*, evidence has also been presented for subsequent vacuolar localization of $(\gamma EC)_nG$-Cd-S^{2-} complexes (Ortiz et al. 1992, 1995; Ow 1993), illustrating a link between cytosolic sequestration and vacuolar compartmentation. Although the main function of *S. cerevisiae* metallothionein (yeast MT) is cellular copper homeostasis, induction and synthesis of MT as well as amplification of MT genes leads to enhanced copper resistance in both *S. cerevisiae* and *C. glabrata* (Macreadie et al. 1994; Howe et al. 1997). Production of MT has been detected in both Cu- and Cd-resistant strains of *S. cerevisiae* (Tohoyama et al. 1995; Inouhe et al. 1996). However, it should be noted that other determinants of tolerance also occur in these and other organisms, e.g. transport phenomena and extracellular precipitation (Gadd and White 1989; Inouhe et al. 1996; Yu et al. 1996), while some organisms, e.g. *Kluyveromyces lactis*, are not capable of MT or $(\gamma EC)_nG$ synthesis (Macreadie et al. 1994). In *S. cerevisiae*, it has been shown that changes in amino acid pools can occur in response to nickel exposure with the formation of vacuolar nickel–histidine complexes being proposed as a survival mechanism (Joho et al. 1995). Little work has been carried out on MT or $(\gamma EC)_nG$ peptides in filamentous fungi (see Gadd 1993b; Galli et al. 1994; Howe et al. 1997; Kameo et al. 2000).

5 Microbial Transformations of Toxic Metals

5.1 Mobilization

Microorganisms can mobilize metals through autotrophic and heterotrophic leaching, chelation by microbial metabolites and siderophores, and methylation, which can result in volatilization. Such processes can lead to dissolution of insoluble metal compounds and minerals, including oxides, phosphates, sulphides and more complex mineral ores, and desorption of metal species from exchange sites on, e.g. clay minerals or organic matter.

Microorganisms can acidify their environment by proton efflux via plasma membrane H^+-ATPases, maintenance of charge balance or as a result of respiratory carbon dioxide accumulation. Acidification can lead to metal release via a number of obvious routes, e.g. competition between protons and the metal in a metal–anion complex or in a sorbed form, resulting in the release of free metal cations. Organic acids can supply both protons and metal complexing anions (Burgstaller and Schinner 1993; Gadd 1999; Gadd and Sayer 2000). For example, citrate and oxalate can form stable complexes with a large number of metals. Many metal citrates are highly mobile and not readily degraded (Francis et al. 1992). Oxalic acid can also act as a leaching agent for those metals that form soluble oxalate complexes, including Al and Fe (Strasser et al. 1994). Organic acid production is also an important agent of mineral deterioration, playing a role in both biogenic chemical weathering and soil formation (Gadd 1999). Such solubilization phenomena can also have consequences for mobilization of metals from toxic metal-containing minerals, e.g. pyromorphite ($Pb_5(PO_4)_3Cl$) which can form in urban and industrially contaminated soils. Pyromorphite can be solubilized by phosphate-solubilizing fungi, with concomitant production of lead oxalate (Sayer et al. 1999).

Most chemolithotrophic (autotrophic) leaching is carried out by acidophilic bacteria which fix CO_2 and obtain energy from the oxidation of Fe(II) or reduced sulphur compounds which causes the solubilization of metals because of the resulting production of Fe(III) and H_2SO_4 (Rawlings 1997; Schippers and Sand 1999). The microorganisms involved include sulphur-oxidizing bacteria, e.g. *Thiobacillus thiooxidans*, iron- and sulphur-oxidizing bacteria, e.g. *Thiobacillus ferrooxidans* and iron-oxidizing bacteria, e.g. *Leptospirillum ferrooxidans* (Ewart and Hughes

1991; Bosecker 1997). As a result of sulphur- and iron-oxidation, metal sulphides are solubilized concomitant with the pH of their immediate environment being decreased, therefore resulting in solubilization of other metal compounds including metals sorbed to soil and mineral constituents (Ewart and Hughes 1991; Rawlings and Silver 1995; Bosecker 1997; Rawlings 1997).

Siderophores are highly specific Fe(III) ligands (formation constants often $> 10^{30}$) which are excreted by microorganisms to aid iron assimilation. Such assimilation may be improved by attachment to solid Fe oxides in soil. Although primarily produced as a means of obtaining iron, siderophores are also able to bind other metals such as magnesium, manganese, chromium(III), gallium(III) and radionuclides such as plutonium(IV) (Birch and Bachofen 1990).

Microorganisms can also mobilize metals, metalloids and organometallic compounds by reduction and oxidation processes (Gadd 1993a; Gharieb et al. 1999; Lovley 2000). For example, solubilities of Fe and Mn increase on reduction of Fe(III) to Fe(II) and Mn(IV) to Mn(II) (Lovley and Coates 1997; Lovley 2000; McLean et al. 2002). Most iron reduction is carried out by specialized anaerobic bacteria that use Fe(III) as a terminal electron acceptor. Dissimilatory metal-reducing bacteria can use a variety of metal(loid)s with an appropriate redox couple, including Fe(III), Mn(IV), Se(IV), Cr(VI) and U(VI) (Oremland et al. 1991; Stolz and Oremland 1999). While Fe and Mn increase their solubility upon reduction, the solubility of other metals such as U(VI) to U(IV) and Cr(VI) to Cr(III) decreases, resulting in immobilization (Phillips et al. 1995; Smith and Gadd 2000; see later). Reduction of Hg(II) to Hg(0) by bacteria and fungi results in diffusion of elemental Hg out of cells (Silver 1996, 1998; Hobman et al. 2000). *Bacillus* and *Streptomyces* sp. can oxidize Hg(0) to Hg(II) and, therefore participate in the oxidative phase of the global Hg cycle (Smith et al. 1998).

Fe(III) and Mn(IV) oxides absorb metals strongly and this may hinder metal mobilization. Microbial reduction of Fe(III) and Mn(IV) may be one way for releasing such metals and this process may be enhanced with the addition of humic materials, or related compounds. Such compounds may also act as electron shuttles for, e.g. U(VI) and Cr(VI), converting them to less soluble forms, especially if located in tight pore spaces where microorganisms cannot enter (Lovley and Coates 1997). Bacterial Fe(III) reduction resulted in release of, e.g. Mn and Co, from goethite where 5% of the iron was substituted by these metals (Bousserrhine et al. 1999). Iron-reducing bacterial strains solubilized 40% of the Pu present in contaminated soils within 6–7 days (Rusin et al. 1993) and both iron- and sulphate-reducing bacteria were able to solubilize Ra from uranium mine tailings, with solubilization occurring largely by disruption of reducible host minerals (Landa and Gray 1995).

5.2 Immobilization

A number of processes lead to immobilization of metals. Although immobilization reduces the external free metal species, it may also promote solubilization in some circumstances by shifting the equilibrium to release more metal into solution.

Biosorption can be defined as the microbial uptake of organic and inorganic metal species, both soluble and insoluble, by physico-chemical mechanisms such as adsorption. In living cells, metabolic activity may also influence this process because of changes in pH, E_h, organic and inorganic nutrients and metabolites. Biosorption can also provide nucleation sites for the formation of stable minerals (Beveridge and Doyle 1989; Southam 2000; McLean et al. 2002; see later). As well as sorption to cellular surfaces, cationic metal species can be accumulated within cells via membrane transport systems of varying affinity and specificity. Once inside cells, metal species may be bound (e.g. to metallothioneins, phytochelatins), precipitated (e.g. as reduced forms), localized within intracellular structures or organelles (e.g. fungal vacuoles), or translocated to specific structures (e.g. fungal fruiting bodies) depending on the element concerned and the organism (Gadd 1996, 1997; White et al. 1997; Gadd and Sayer 2000).

Peptidoglycan carboxyl groups are the main binding site for cations in Gram-positive bacterial cell walls with phosphate groups contributing significantly in Gram-negative species (Beveridge and Doyle 1989; McLean et al. 2002). Chitin is an important structural component of fungal cell walls and this is an effective biosorbent for radionuclides (Tsezos and Volesky 1982a, b; Tobin et al. 1994). Fungal phenolic polymers and melanins possess many potential metal-binding sites with oxygen-containing groups including carboxyl, phenolic and alcoholic hydroxyl, carbonyl and methoxyl groups being particularly important (Gadd 1993b). Fungi can be efficient sorbents of metal ions over a wide range of pH values and although they may take up less metal per unit dry weight than clay minerals (the most important metal-sorbing component in the soil), they are more efficient sorbents per unit surface area (Morley and Gadd 1995). It seems likely that microbial binding and biomineralization (mineral formation) reactions have a more significant role in metal speciation and mobility in the terrestrial environment than has previously been supposed (Krantz-Rulcker et al. 1993, 1996; Ledin et al. 1996; Morley et al. 1996; McLean et al. 2002). It should also be appreciated that the morphology of filamentous fungi allows interconnection of hyphae and continuity of cytoplasm. The ability of fungi to translocate nutrients (and organelles, oxygen, and metabolites) may be important in the translocation of metal species and concentration in specific regions such as fruiting bodies (Haselwandter and Berreck 1994;

Dighton and Terry 1996; Gray et al. 1996; Connolly and Jellison 1997). Elevated metal and radionuclide concentrations, particularly radiocaesium, commonly occur in fruiting bodies of basidiomycetes during growth in polluted environments (Gadd 1993a, 1997; Haselwandter and Berreck 1994; Anderson et al. 1997; Wainwright and Gadd 1997). In fact, it has been concluded that the fungal component of soil can immobilize the total Chernobyl radiocaesium fallout received in upland grasslands (Dighton et al. 1991) though grazing of fruit bodies by animals may lead to radiocaesium transfer along the food chain (Bakken and Olsen 1990).

A range of specific and non-specific metal-binding compounds are produced by microorganisms. Non-specific metal-binding compounds range from simple organic acids and alcohols to macromolecules such as polysaccharides, humic and fulvic acids (Birch and Bachofen 1990; Beech and Cheung 1995; Bridge et al. 1999; Sayer and Gadd 2001). Extracellular polymeric substances (EPS), a mixture of polysaccharides, mucopolysaccarides and proteins (Zinkevich et al. 1996) are produced by bacteria and fungi and also bind metals (Beech and Cheung 1995; White and Gadd 1998a). Extracellular polysaccharides can also adsorb or entrap particulate matter such as precipitated metal sulphides and oxides (Flemming 1995; Vieira and Melo 1995) and these processes may be particularly important in microbial biofilms (White and Gadd 1998a, 2000).

Where microbial reduction of a metal to a lower redox state occurs, mobility and toxicity may be reduced (Lovley 2001; Finneran et al. 2002). Such processes may also accompany other indirect reductive metal precipitation mechanisms, e.g. in sulphate-reducing bacterial systems where reduction of Cr(VI) can be a result of indirect reduction by Fe^{2+} and the produced sulphide. Aerobic or anaerobic reduction of Cr(VI) to Cr(III) is widespread in microorganisms (Smith and Gadd 2000; McLean and Beveridge 2001). U(VI) can be reduced to U(IV) by certain Fe(III)-dissimilatory microorganisms and this reduction in solubility can be the basis of U removal from contaminated waters and leachates (Lovley and Coates 1997; Lovley 2001; Finneran et al. 2002). Sulphur and sulphate-reducing bacteria are geochemically important in reductive precipitation of toxic metals, e.g. U(VI), Cr(VI), Tc(VII), Pd(II) (Aubert et al. 1998; Lloyd and Macaskie 1998; Lloyd et al. 1999a, b). Some sulphate-reducing bacteria like *Desulfotomaculum reducens* share physiological properties of both sulphate- and metal-reducing groups of bacteria, and can use Cr(VI), Mn(IV), Fe(III) and U(IV) as sole electron acceptors (Tebo and Obraztsova 1998).

Sulphate-reducing bacteria (SRB) oxidize organic compounds or hydrogen coupled with the reduction of sulphate, producing sulphide (White and Gadd 1997, 1998a, b). The solubility products of most heavy metal sulphides are very low, so that even a moderate output of sulphide can remove metals (White and Gadd 1998a, b). Sulphate-reducing bacteria can also create extremely reducing conditions which can chemically reduce species such as U(VI) (White and Gadd 1998b).

Bacterial Fe oxidation is ubiquitous in environments with sufficient Fe^{2+} and conditions to support bacterial growth such as drainage waters and tailings piles in mined areas, pyritic and hydric soils (bogs and sediments), drain pipes and irrigation ditches, and plant rhizospheres. Iron-oxidizers found in acidic soil environments are acidophilic chemolithotrophs, such as *Thiobacillus ferrooxidans*, significant for its role in generating acid mine drainage (Ewart and Hughes 1991). Fungi also oxidize metals in their environment. Desert varnish is an oxidized metal layer (patina) a few millimetres thick found on rocks and in soils of arid and semi-arid regions, and is believed to be of fungal and bacterial origin.

6 Metalloid Transformations

Main microbial transformations carried out in the soil are reduction and methylation which can lead to alterations in bioavailability and toxicity. For selenium, some bacteria can use SeO_4^{2-} as a terminal e^- acceptor in dissimilatory reduction and also reduce and incorporate Se into organic components, e. g. selenoproteins (assimilatory reduction). Methylation and subsequent volatilization of methylated selenium derivatives is also a widely found property of soil bacteria and fungi and may be an important process in Se transport from terrestrial to aquatic environments (Dungan and Frankenberger 1999). Selenate (SeO_4^{2-}) and selenite (SeO_3^{2-}) can be reduced to Se^o, with SeO_3^{2-} reduction appearing more ubiquitous than SeO_4^{2-} reduction. However, only SeO_4^{2-} can support bacterial growth under anaerobic conditions: SeO_4^{2-} reduction to Se^o is a major sink for Se oxyanions in anoxic sediments (Oremland et al. 1989; Stolz and Oremland 1999; Oremland and Stolz 2000). Anaerobic sulphate-reducing bacteria like *Desulfovibrio desulfuricans* can reduce selenate/selenite to Se^o, but neither oxyanion could be used for respiratory growth (Tomei et al. 1995). Reduction to Se^o can be considered a detoxification mechanism (Dungan and Frankenberger 1999).

The opposite process of Se^o oxidation can occur in soils and sediments (Dowdle and Oremland 1998; Losi and Frankenberger 1998). It is possible that Se^o oxidation is a similar process to S oxidation, and may be mediated by heterotrophs and autotrophs (Losi and Frankenberger 1998). In aerobic soil slurries, Se^{4+} was the main product with lower amounts of Se^{6+} being produced: heterotrophic and autotrophic thiobacilli were believed to be the active organisms (Dowdle and Oremland 1998).

Methylation of Se is a ubiquitous property of microorganisms and can occur in soils, sediments and water (Gadd 1993b). Bacteria and fungi are the most important Se-methylaters in soil (Karlson and Frankenberger 1988) with the most frequently produced volatile being dimethyl selenide (DMSe; Karlson and Frankenberger 1988, 1989; Thompson-Eagle et al. 1989). Other volatiles produced in smaller amounts include dimethyl diselenide (DMDSe; Dungan and Frankenberger 1999). Like reduction, volatilization can be considered a detoxification mechanism since volatile Se derivatives are lost from the soil system. Those environmental factors that affect microbial activity can markedly affect Se methylation, e. g. pH, temperature, organic amendments, Se speciation etc., addition of organic amendments can stimulate methylation (Dungan and Frankenberger 1999). The opposite process of demethylation can also occur in soil and water systems. Anaerobic demethylation may be mediated by methylotrophic bacteria (Oremland et al. 1989).

Tellurium may also be transformed by analogous mechanisms as selenium, i. e. reduction and methylation (Chasteen and Bentley 2003). Reduction of tellurite to Te^{o} results in a grey to black colouration of microbial colonies with extracellular and intracellular precipitation being observed (Gharieb et al. 1999). Dimethyl telluride (DMTe) is the main product of Te methylation (Chasteen and Bentley 2003).

Arsenic methylation can be mediated by many organisms with compounds having the general structure $(CH_3)_nAsH_{3-n}$ and mono-, di- and trimethylarsine (n=1, 2, 3 respectively) being major volatile compounds (Bentley and Chasteen 2002). Reduction of arsenic oxyanions by reductase enzymes is also frequent and a determinant of As resistance. However, there appears to be no involvement of such reductases in biomethylation (Bentley and Chasteen 2002).

7 Biomineralogy of Metal–Microbe Interactions

Biomineralization refers to biologically induced mineralization where an organism modifies the local microenvironment creating conditions that promote chemical precipitation of extracellular mineral phases (Hamilton 2003). Commonly, this results from microbial oxidation or reduction of a metal species, and as well as the plethora of metabolism-dependent microbial transformations of metal species, microbial surfaces provide chemically reactive sites for adsorption and complexation (biosorption) as detailed previously. In a mineralogical context, the latter processes can lead to the nucleation and formation of mineral precipitates around biomass (Beveridge 1989; Fortin et al. 1997; McLean et al. 2002). The nature of the resultant mineral may depend on the nature of the cell surface, the cellular microenvironment (which may be influenced by metabolic activity) and the presence of reactive anions, e. g. sulphide, carbonate, and phosphate, some of these arising from microbial activity. However, such biomineralization can also occur independent of microbial activity and on dead biomass. Mineral phases may undergo further changes in, e. g. crystallinity, with time and in relation to physico-chemical characteristics of the environment; some minerals may incorporate other metals into their structure (Watson et al. 1995, 2000; Brown et al. 1999; McLean et al. 2002). It should also be appreciated that microbial modification of their microenvironment may result in physico-chemical conditions that promote spontaneous metal precipitation (McLean et al. 2002).

The formation of solid mineral phases by bacterial processes may result in mineral deposition over geological time scales by diagenesis of sediments (Beveridge et al. 1983). While much work on microbial biomineralization has been carried out using bacteria, it should be stressed that all major microbial groups have roles in metal immobilization and mineral formation, e. g. cyanobacteria, microalgae, and fungi. While cyanobacterial and microalgal processes are generally of greater significance in aquatic environments and in early stages of soil formation, e. g. rock colonization, fungi also have important roles in the terrestrial environment regarding mineral dissolution, metal and anion cycling, and also mineral formation.

Calcium oxalate is the most common form of oxalate encountered in the environment, mostly occurring as the dihydrate (weddellite) or the more stable monohydrate (whewellite; Gadd 1999). Calcium oxalate crystals are commonly associated with free-living, pathogenic and plant symbiotic fungi and are formed by the precipitation of solubilized calcium as the oxalate (Gharieb et al. 1998; Gadd 1999). This has an important influence on biogeochemical processes in soils, acting as a reservoir for calcium, and also influencing phosphate availability. Fungi can also produce other metal oxalates with a variety of different metals and metal-bearing minerals, e. g. Cd, Co, Cu, Mn, Sr and Zn (White et al. 1997; Gadd 1999; Sayer et al. 1999).

In many arid and semi-arid regions, calcareous soils and near surface limestones (calcretes) are often secondarily cemented with calcite ($CaCO_3$). This phenomenon has been partly attributed to physicochemical processes; however, the abundance of calcified fungal filaments in weathered profiles of chalky limestone and Quaternary calcretes indicates fungal activity (Verrecchia and Dumont 1996; Gadd 1999). Mineralized carbonate precipitates are also found in association with bacterial biofilms (Glasauer et al. 2003).

8
Mycorrhizas

Plant symbiotic mycorrhizal fungi can accumulate metals from soil components and this may have consequences for both essential metal nutrition of the symbiosis as well as increased or decreased toxicity. Since plants growing on metalliferous soils are generally mycorrhizal, an important ecological role for the fungus has frequently been postulated although such a role, e.g. phytoprotection, is often difficult to establish experimentally. Mycorrhizal fungi exhibit "constitutive and adaptive resistance" to metals (Meharg and Cairney 2000) although the relative contributions of passive and active processes in overall response is seldom elucidated (Gadd 1993a, b). Ericaceous plants appear to be entirely dependent on the presence of ericoid mycorrhizas for protection against copper, the fungus preventing metal translocation to aerial plant shoots (Bradley et al. 1981, 1982). Arbuscular mycorrhizas (AM) from metal-contaminated sites are often more metal-tolerant to, e.g. Cd and Zn, than other isolates, suggesting a possible benefit to the plant via increased metal resistance, nutrient uptake etc., though in some instances, AM plants do not necessarily require fungal colonization for survival (Griffioen 1994). For ectomycorrhizal fungi (EcM), it appears that these organisms possess constitutive as well as adaptive properties of resistance (Colpaert and van Assche 1987, 1992; Hartley et al. 1997a, b). It is often postulated that mycorrhizas provide a barrier to the uptake of potentially toxic metals (Wilkins 1991; Hetrick et al. 1994; Wilkinson and Dickinson 1995; Leyval et al. 1997; Meharg and Cairney 2000) though this has not been confirmed in every case. Further, in some instances, AM may mediate enhanced accumulation of essential metals, which unless regulated, may lead to phytotoxicity (Killham and Firestone 1983). It is generally concluded that local conditions in metal-contaminated sites may determine the cost–benefit relationship between the plant and the AM fungus, since detrimental, neutral or beneficial interactions have all been documented (Meharg and Cairney 2000). For ericaceous mycorrhizas, clear host protection is observed for ericaceous plants, e.g. *Calluna* sp., *Erica* sp., *Vaccinium* sp. growing on polluted and/or naturally metalliferous soils particularly regarding Cu and Zn (Bradley et al. 1981, 1982). Further, ericaceous plants are generally found on nutrient-deficient soils and it is likely the mycorrhiza could additionally benefit the plants by enhanced nutrient uptake (Smith and Read 1997). A protective metal-binding effect of ectomycorrhizal fungi has been postulated frequently (e.g. Leyval et al. 1997) though other workers point out the lack of clear evidence for this (Dixon and Buschena 1988; Colpaert and van Assche 1993). However, EcM plants possessed higher tissue P concentrations, indicating some benefit from the association (Meharg and Cairney 2000).

9 Bioremediation

Some of the processes outlined above have potential for application for treating contaminated land. In a bioremediation context, production of sulphuric acid by *Thiobacillus* species has been used to solubilize metals from sewage sludge, thus enabling separation from the sludge which can then be used as a fertilizer (Sreekrishnan and Tyagi 1994). Autotrophic leaching has been used to remediate other metal-contaminated solid materials including soil and red mud, the main waste product of Al extraction from bauxite (Vachon et al. 1994). Although some processes could be used *in situ* (e.g. leaching using S-oxidising bacteria), many are probably most suitable for *ex situ* use in bioreactors, where the mobilized or immobilized metal can be separated from soil components (White et al. 1998). Living or dead fungal and bacterial biomass and metabolites have been used to remove metals and metalloids from solution by biosorption or chelation (Macaskie 1991; Gadd 2001).

Microbial activities in anaerobic, subsurface environments also offer possibilities for the bioremediation of metal contaminants. Metal(loid)s that form insoluble precipitates when reduced may be of particular interest for *in situ* treatment, such as Se(0), Cr(III), Tc(IV) and U(IV) (Thompson-Eagle and Frankenberger 1992; Lovley and Coates 1997; Stolz and Oremland 1999). The sulphide produced from sulphate reduction plays a major role in metal sulphide immobilization in sediments, but has also been applied to bioremediation of metals in waters and leachates. A process integrating bacterial sulphate-reduction with bioleaching by sulphur-oxidizing bacteria has also been developed to remove contaminating toxic metals from soils. Sulphur- and iron-oxidising bacteria liberated metals from soils in the form of an acid sulphate solution that enabled almost all the metals to be removed by bacterial sulphate reduction (White et al. 1998). Large-scale bioreactors have been developed using bacterial sulphate-reduction for treating metal-contaminated waters (Barnes et al. 1992; Gadd 1992)

10 Phytoremediation

Phytoremediation is the use of plants to remove or detoxify environmental pollutants (Baker and Brooks 1989; Salt et al. 1998). Although free-living and symbiotic microorganisms influence plant productivity, metal bioavailability and interactions, there are few integrated studies, and many phytoremediation studies are carried out without reference to contributory microbial processes. Phytoremediation can be divided into phytoextraction (pollutant removal from soil into shoots and leaves), phytodegradation (pollutant degradation by plant–microbe systems), rhizofiltration (absorption of pollutants by plant roots), phytostabilization (plant-mediated reduction of pollutant bioavailability), phytovolatilization (plant-mediated volatilization of pollutants) and phytoscrubbing (plant removal of atmospheric pollutants). Most attention has focused on metals to date. Two basic strategies are chelate-assisted and continuous phytoextraction. Chelate-assisted phytoextraction has been invoked since plants do not naturally accumulate important toxic elements, e. g. Pb, Cd, As, and many radionuclides, to levels that would be significant in a remediative context. Application of various synthetic chelates can enhance plant metal accumulation (Huang et al. 1997; Salt et al. 1998). Continuous phytoextraction of metals relies on intrinsic properties of plants that lead to accumulation in aerial plant tissues. However, many natural 'hyperaccumulators' often exhibit low biomass, slow growth rates and none are known for important elements like Pb, Cd, As and U (Salt et al. 1998). Ni, Zn and Se appear to be the elements accumulated to the highest levels (Salt et al. 1998).

Plants possess analogous metal resistance mechanisms as microorganisms, i. e. chelation, intracellular compartmentation, transformations etc., although plants may be relatively metal-sensitive compared to microorganisms. Manipulation of metal tolerance may provide a means for phytoremediation: bacterial Hg^{2+}-reductase has been expressed in *Arabidopsis thaliana* (Rugh et al. 1996).

Biodegradation of organic pollutants is generally accelerated in vegetated soils confirming plant roles and that of rhizosphere-inhabiting microorganisms (Schnoor et al. 1995; Cunningham and Ow 1996). There is often uncertainty in the relative roles of plants and microorganisms in the overall process. It seems clear that microorganisms will often be most important in this context, their soil populations being markedly stimulated by the rhizosphere effect (Salt et al. 1998). Organic compounds in root exudates can also serve as co-metabolites for bacterial xenobiotic degradation (Donnelly et al. 1994).

11 Conclusions

This chapter has outlined the importance of metal–microbe interactions in several soil contexts, not least the fundamental position in the biogeochemical cycling of metals and associated elements and nutrients, and in plant productivity. The real application and unrealised potential of many natural microbial and microbe/plant processes are also growing topics in the area of bioremediation. However, analysis and understanding of the effects of toxic metals on microbial communities is relatively limited despite extensive research. Clearly, the complexity of interactions between metal species and soil components, as well as between metal species and microorganisms, are major factors as are the limitations of chemical and biological techniques that can be meaningfully applied. Laboratory and short-term field experimentation may provide extensive information although this often has limited relevance to natural and contaminated natural environments. The lack of meaningful baseline data and ignorance of kinetic changes in metal speciation, microbial communities and soil structure and composition can also hamper studies of long-term and historically contaminated sites. Pure culture studies may be open to the usual predictable criticisms, although in the area of metal–microbe interactions, fundamental data obtained have been of paramount importance, having wide relevance to our understanding of prokaryotic and eukaryotic biology in a wider context, including molecular and cell biology, genetics, physiology and biochemistry, as well as providing insights into natural roles and potential in environmental biotechnology. It is likely that novel and developing molecular and genomic techniques for characterising microbial communities and their functions will catalyse important steps forward in our understanding of metals and microbial communities as well in the soil in general. As genomic information on environmentally relevant microorganisms becomes available along with reliable techniques for, e.g. DNA extraction, phylogenetic analysis, cell imaging, biolabelling and use of marker molecules, improvements in understanding should be marked and will serve to reconcile important areas of laboratory and field-based research. Future studies are also dependent on an interdisciplinary approach involving all aspects of biology, chemistry, mathematical modelling, mineralogy, and geochemistry, for example: the complexity of the soil environment demands this. There is a lot of work to be carried out before daylight appears over the (soil) horizon!

Acknowledgements. The author gratefully acknowledges research support from the Biotechnology and Biological Sciences Research Council, the Natural Environment Research Council, the Royal Societies of London and Edinburgh, and British Nuclear Fuels plc.

References

Anderson P, Davidson CM, Littlejohn D, Ure AM, Shand CA, Cheshire MV (1997) The translocation of caesium and silver by fungi in some Scottish soils. Comm Soil Sci Plant Anal 28:635–650

Aoyama M, Nagumo T (1997a) Effects of heavy metal accumulation in apple orchard soils on microbial biomass and microbial activities. Soil Sci Plant Nutrition 43:601–612

Aoyama M, Nagumo T (1997b) Comparison of the effects of Cu, Pb, and As on plant residue decomposition, microbial biomass, and soil respiration. Soil Sci Plant Nutr 43:613–622
Aubert C, Lojou E, Bianco P, Rousset M, Durand M-C, Bruschi M, Dolla A (1998) The *Desulfuromonas acetoxidans* triheme cytochrome c_7 produced in *Desulfovibrio desulfuricans* retains its metal reductase activity. Appl Environ Microbiol 64:1308–1312
Baath E, Diaz-Ravina M, Frostegard A, Campbell CD (1998) Effect of metal-rich sludge amendments on the soil microbial community. Appl Environ Microbiol 64:238–245
Baker AJM, Brooks RR (1989) Terrestrial higher plants which accumulate metallic elements – a review of their distribution, ecology and phytochemistry. Biorecovery 1:81–126
Bakken LR, Olsen RA (1990) Accumulation of radiocaesium in fungi. Can J Microbiol 36:704–710
Balistrieri LS, Chao TT (1987) Selenium adsorption by goethite. Soil Sci Soc Am J 51:1145–1151
Barnes LJ, Janssen FJ, Sherren J, Versteegh JH, Koch RO, Scheeren PJH (1992) Simultaneous removal of microbial sulphate and heavy metals from wastewater. Trans Inst Mining Metall 101:183–190
Beech IB, Cheung CWS (1995) Interactions of exopolymers produced by sulphate-reducing bacteria with metal ions. Int Biodeter Biodeg 35:59–72
Bentley R, Chasteen TG (2002) Microbial methylation of metalloids: arsenic, antimony and bismuth. Micro Mol Bio Rev 66:250–271
Beveridge TJ (1989) Role of cellular design in bacterial metal accumulation and mineralization. Annu Rev Microbiol 43:147–171
Beveridge TJ, Doyle RJ (1989) Metal ions and bacteria. Wiley, New York
Beveridge TJ, Meloche JD, Fyfe WS, Murray RGE (1983) Diagenesis of metals chemically complexed to bacteria: laboratory formation of metal phosphates, sulfides and organic condensates in artificial sediments. Appl Environ Microbiol 45:1094–1108
Birch L, Bachofen R (1990) Complexing agents from microorganisms. Experientia 46:827–834
Blaudez D, Jacob C, Turnau K, Colpaert JV, Ahonen-Jonnarth U, Finlay R, Botton B, Chalot M (2000) Differential responses of ectomycorrizal fungi to heavy metals in vitro. Mycol Res 104:1366–1371
Bode H-P, Dumschat M, Garotti S, Fuhrmann GF (1995) Iron sequestration by the yeast vacuole. A study with vacuolar mutants of *Saccharomyces cerevisiae*. Eur J Biochem 228:337–342
Bogomolova EV, Vlasov Yu D, Panina LK (1998) On the nature of the microcolonial morphology of epilithic black yeasts *Phaeococcomyces* de Hoog. Doklady Russian Acad Sci 363:707–709
Borst-Pauwels GWFH (1989) Ion transport in yeast including lipophilic ions. Methods Enzymol 174:603–616
Bosecker K (1997) Bioleaching: metal solubilization by microorganisms. FEMS Microbiol Rev 20:591–604
Bousserrhine N, Gasser UG, Jeanroy E, Berthelin J (1999) Bacterial and chemical reductive dissolution of Mn-, Co- Cr-, and Al-substituted geothites. Geomicrobiol J 16:245–258
Bradley R, Burt AJ, Read DJ (1981) Mycorrhizal infection and resistance to heavy metals. Nature 292:335–337
Bradley B, Burt AJ, Read DJ (1982) The biology of mycorrhiza in the Ericaceae. VIII. The role of mycorrhizal infection in heavy metal resistance. New Phytol 91:197–209
Bridge TAM, White C, Gadd GM (1999) Extracellular metal-binding activity of the sulphate-reducing bacterium *Desulfococcus multivorans*. Microbiol 145:2987–2995
Brookes PC, McGrath SP (1984) Effects of metal toxicity on the size of the soil microbial biomass. J Soil Sci 35:341–346
Brown NL, Lee BTO, Silver S (1994) Bacterial transport of and resistance to copper. In: Sigel

H, Sigel A (eds) Metal ions in biological systems, vol 30. Dekker, New York, pp 405–434
Brown NL, Barrett SR, Camakaris J, Lee BTO, Rouch DA (1995) Molecular genetics and transport analysis of the copper-resistance determinant (*pco*) from *Escherichia coli* plasmid pRJ1004. Mol Microbiol 17:1153–1166
Brown GE, Foster AL, Ostergren JD (1999) Mineral surfaces and bioavailability of heavy metals: a molecular-scale perspective. Proc Natl Acad Sci USA 96:3388–3395
Burford EP, Kierans M, Gadd GM (2003a) Geomycology: fungi in mineral substrata. Mycologist 17:98–107
Burford EP, Fomina M, Gadd GM (2003b) Fungal involvement in bioweathering and iotransformation of rock aggregates and minerals. Mineral Mag 67:1127–1155
Burgstaller W, Schinner F (1993) Leaching of metals with fungi. J Biotechnol 27:91–116
Cairney JWG, Meharg AA (1999) Influences of anthropogenic pollution on mycorrhizal fungal communities. Environ Poll 106:169–182
Cevnik M, Jurc M, Vodnik D (2000) Filamentous fungi associated with the fine roots of *Erica herbacea* L. from the area influenced by the Zerjav lead smelter (Slovenia). Phyton Ann Rei Bot 40:61–64
Chander K, Brookes PC (1991) Effects of heavy metals from past applications of sewage sludge on microbial biomass and organic matter accumulation in a sandy loam and silty loam UK soil. Soil Biol Biochem 23:927–932
Chander K, Dyckmans J, Hoeper H, Joergensen RG, Raubuch M (2001a) Long-term effects on soil microbial properties of heavy metals from industrial exhaust deposition. J Plant Nutr Soil Sci 164:657–663
Chander K, Dyckmans J, Joergensen RG, Meyer B, Raubuch M (2001b) Different sources of heavy metals and their long-term effects on soil microbial properties. Biol Fertil Soil 34:241–247
Chasteen TG, Bentley R (2003) Biomethylation of selenium and tellurium: microorganisms and plants. Chem Rev 103:1–26
Colpaert JV, van Assche JA (1987) Heavy metal resistance in some ectomycorrhizal fungi. Funct Ecol 1:415–421
Colpaert JV, van Assche JA (1992) Zinc toxicity in ectomycorrhizal *Pinus sylvestris*. Plant Soil 143:201–211
Colpaert JV, van Assche JA (1993) The effect of cadmium on ectomycorrhizal *Pinus sylvestris* L. New Phytol 123:325–333
Colpaert JV, Vandenkoornhuyse P, Adriaensen K, Vangronsveld J (2000) Genetic variation and heavy metal tolerance in the ectomycorrhizal basidiomycete *Suillus luteus*. New Phytol 147:367–379
Connolly JH, Jellison J (1997) Two-way translocation of cations by the brown rot fungus *Gloeophyllum trabeum*. Int Biodet Biodeg 39:181–188
Cooksey DA (1993) Copper uptake and resistance in bacteria. Mol Microbiol 7:1–5
Cooksey DA (1994) Molecular mechanisms of copper resistance and accumulation in bacteria. FEMS Microbiol Rev 14:381–386
Cunningham SD, Ow DW (1996) Promises and prospects of phytoremediation – update on biotechnology. Plant Physiol 110:715–719
Dameron CT, Reese RN, Mehra RK, Kortan AR, Carrol PJ, Steigerwald ML, Brus LE, Winge DR (1989) Biosynthesis of cadmium sulfide quantum semiconductor crystallites. Nature 338:596–597
De Leo F, Urzi C, de Hoog GS (2003) A new meristematic fungus, *Pseudotaeniolina globosa*. Ant van Leeuwenhoek 83:351–360
Del Val C, Barea JM, Azcon-Aguilar C (1999) Diversity of arbuscular mycorrhizal fungus populations in heavy-metal-contaminated soils. Appl Environ Microbiol 65:718–723
Diaz-Ravina M, Baath E (1996) Development of metal tolerance in soil bacterial com-

munities exposed to experimentally increased metal levels. Appl Environ Microbiol 62:2970–2977
Dighton J, Terry G (1996) Uptake and immobilization of caesium in UK grassland and forest soils by fungi, following the Chernobyl accident. In: Frankland JC, Magan N, Gadd GM (eds) Fungi and environmental change. Cambridge Univ Press, Cambridge, pp 184–200
Dighton J, Clint GM, Poskitt J (1991) Uptake and accumulation of ^{137}Cs by upland grassland soil fungi: a potential pool of Cs immobilization. Mycol Res 95:1052–1056
Dixon RK, Buschena CA (1988) Response of ectomycorrhizal *Pinus banksia* and *Picea glauka* to heavy metals in soil. Plant Soil 105:265–271
Doelman P, Hanstra L (1979) Effects of lead on the soil bacterial microflora. Soil Biol Biochem 11:487–491
Doelman P, Jansen E, Michels M, van Til M (1994) Effects of heavy metals in soil on microbial diversity and activity as shown by the sensitivity-resistance index, an ecologically relevant parameter. Biol Fertil Soil 17:177–184
Donnelly PK, Hegde RS, Fletcher JS (1994) Growth of PCB-degrading bacteria on compounds from photosynthetic plants. Chemosphere 28:981–988
Dowdle PR, Oremland RS (1998) Microbial oxidation of elemental selenium in soil slurries and bacterial cultures. Environ Sci Technol 32:3749–3755
Dungan RS, Frankenberger WT (1999) Microbial transformations of selenium and the bioremediation of seleniferous environments. Bioremed 3:171–188
Ehrlich HL (1997) Microbes and metals. Appl Microbiol Biotechnol 48:687–692
Ewart DK, Hughes MN (1991) The extraction of metals from ores using bacteria. Adv Inorg Chem 36:103–135
Fay DA, Mitchell DT (1999) A preliminary study of the mycorrhizal associations of the tree seedlings growing on mine spoil at Avoca, Co. Wicklow. Biol Environ Proc R Irish Acad 99B:19–26
Finneran KT, Anderson RT, Nevin KP, Lovley DR (2002) Bioremediation of uranium-contaminated aquifers with microbial U(VI) reduction. Soil Sediment Contam 11:339–357
Flemming H-K (1995) Sorption sites in biofilms. Water Sci Technol 32:27–33
Fortin D, Ferris FG, Beveridge TJ (1997) Surface-mediated mineral development by bacteria. In: Banfield J, Nealson KH (eds) Reviews in mineralogy, vol 35. Mineralogical Society of America, Washington, DC, pp 161–180
Francis AJ, Dodge CJ, Gillow JB (1992) Biodegradation of metal citrate complexes and implications for toxic metal mobility. Nature 356:140–142
Gadd GM (1980) Melanin production and differentiation in batch cultures of the polymorphic fungus *Aureobasidium pullulans*. FEMS Microbiol Lett 9:237–240
Gadd GM (1992) Microbial control of heavy metal pollution. In: Fry JC, Gadd GM, Herbert RA, Jones CW, Watson-Craik I (eds) Microbial control of pollution. Cambridge Univ Press, Cambridge, pp 59–88
Gadd GM (1993a) Interactions of fungi with toxic metals. New Phytol 124:25–60
Gadd GM (1993b) Microbial formation and transformation of organometallic and organometalloid compounds. FEMS Microbiol Rev 11:297–316
Gadd GM (1996) Influence of microorganisms on the environmental fate of radionuclides. Endeavour 20:150–156
Gadd GM (1997) Roles of microorganisms in the environmental fate of radionuclides. In: Lake JV, Bock GR, Cardew G (eds) CIBA foundation symposia 203. Health impacts of large releases of radionuclides. Wiley, Chichester, pp 94–108
Gadd GM (1999) Fungal production of citric and oxalic acid: importance in metal speciation, physiology and biogeochemical processes. Adv Microb Physiol 41:47–92
Gadd GM (2001) Accumulation and transformation of metals by microorganisms. In: Rehm H-J, Reed G, Puhler A, Stadler P (eds) Biotechnology, a multi-volume comprehensive

treatise, vol 10. Special processes. Wiley-VCH, Weinheim, pp 225–264
Gadd GM, Griffiths AJ (1978) Microorganisms and heavy metal toxicity. Microb Ecol 4:303–317
Gadd GM, White C (1989) Heavy metal and radionuclide accumulation and toxicity in fungi and yeasts. In: Poole RK, Gadd GM (eds) Metal-microbe interactions. IRL Press, Oxford, pp 19–38
Gadd GM, Lawrence OS (1996) Demonstration of high-affinity Mn^{2+} uptake in *Saccharomyces cerevisiae* – specificity and kinetics. Microbiol 142:1159–1167
Gadd GM, Sayer JA (2000) Fungal transformations of metals and metalloids. In: Lovley DR (ed) Environmental microbe-metal interactions. American Society of Microbiology, Washington, DC, pp 237–256
Galli U, Schuepp H, Brunold C (1994) Heavy metal binding by mycorrhizal fungi. Physiol Plant 92:364–368
Gerrath JF, Gerrath JA, Larson DW (1995) A preliminary account of endolithic algae of limestone cliffs of the Niagara Escarpment. Can J Bot 73:788–793
Gharieb MM, Gadd GM (1998) Evidence for the involvement of vacuolar activity in metal(loid) tolerance: vacuolar-lacking and -defective mutants of *Saccharomyces cerevisiae* display higher sensitivity to chromate, tellurite and selenite. BioMetals 11:101–106
Gharieb MM, Sayer JA, Gadd GM (1998) Solubilization of natural gypsum ($CaSO_4.2H_2O$) and the formation of calcium oxalate by *Aspergillus niger* and *Serpula himantiodes*. Mycol Res 102:825–830
Gharieb MM, Kierans M, Gadd GM (1999) Transformation and tolerance of tellurite by filamentous fungi: accumulation, reduction and volatilization. Mycol Res 103:299–305
Giller K, Witter E, McGrath SP (1998) Toxicity of heavy metals to microorganisms and microbial processes in agricultural soils: a review. Soil Biol Biochem 30:1389–1414
Glasauer S, Burford EP, Harper FA, Gadd GM, Beveridge TJ (2004) Transformation of metals and metalloids by bacteria and fungi. In: Hillel D, Rosenzweig C, Powlson D, Scow K, Singer M, Sparks D (eds) Encyclopedia of soils in the environment. Academic Press, London (in press)
Gorbushina AA, Krumbein WE, Hamann R, Panina L, Soucharjevsky S, Wollenzien U (1993) On the role of black fungi in colour change and biodeterioration of antique marbles. Geomicrobiol J 11:205–221
Gray SN, Dighton J, Jennings DH (1996) The physiology of basidiomycete linear organs. 3. Uptake and translocation of radiocaesium within differentiated mycelia of *Armillaria* spp growing in microcosms and in the field. New Phytol 132:471–482
Griffioen WAJ (1994) Characterization of a heavy metal-tolerant endomycorrhizal fungus from the surroundings of a zinc refinery. Mycorrhiza 4:197–200
Hamilton WA (2003) Microbially influenced corrosion as a model system for the study of metal microbe interactions: a unifying electron transfer hypothesis. Biofouling 19:65–76
Hartley JW, Cairney G, Meharg AA (1997a) Do ectomycorrhizal fungi exhibit adaptive tolerance to potentially toxic metals in the environment? Plant Soil 189:303–319
Hartley C, Cairney JWG, Sanders FE, Meharg AA (1997b) Toxic interactions of metal ions (Cd^{2+}, Pb^{2+}, Zn^{2+} and Sb^{3-}) on in vitro biomass production of ectomycorrhizal fungi. New Phytol 137:551–562
Hartley J, Cairney JWG, Freestone P, Woods C, Meharg AA (1999) The effects of multiple metal contamination on ectomycorrhizal Scots pine (*Pinus sylvestris*) seedlings. Environ Poll 106:413–424
Haselwandter K, Berreck M (1994) Accumulation of radionuclides in fungi. In: Winkelmann G, Winge DR (eds) Metal ions in fungi. Dekker, New York, pp 259–277
Hayashi Y, Mutoh N (1994) Cadystin (phytochelatin) in fungi. In: Winkelmann G, Winge DR (eds) Metal ions in fungi. Dekker, New York, pp 311–337

Hetrick BAD, Wilson GWT, Figge DAH (1994) The influence of mycorrhizal symbiosis and fertilizer amendments on establishment of vegetation in heavy metal mine spoil. Environ Pollut 86:171–179

Hobman JL, Wilson JR, Brown NL (2000) Microbial mercury reduction. In: Lovley DR (ed) Environmental microbe–metal interactions. ASM Press, Washington, DC, pp 177–197

Holtan-Hartwig L, Bechmann M, Hoyas TR, Linjordet R, Bakken LR (2002) Heavy metals tolerance of soil denitrifying communities: N_2O dynamics. Soil Biol Biochem 34:1181–1190

Howe R, Evans RL, Ketteridge SW (1997) Copper binding proteins in ectomycorrhizal fungi. New Phytol 135:123–131

Huang JWW, Chen JJ, Berti WR, Cunningham SD (1997) Phytoremediation of lead-contaminated soils: role of synthetic chelates in lead phytoextraction. Environ Sci Technol 31:800–805

Inouhe M, Sumiyoshi M, Tohoyama H, Joho M (1996) Resistance to cadmium ions and formation of a cadmium-binding complex in various wild-type yeasts. Plant Cell Physiol 37:341–346

Ivey DM, Guffanti AA, Shen Z, Kudyan N, Krulwich TA (1992) The CadC gene product of alkaliphilic *Bacillus firmus* OF4 partially restores Na^+ resistance to an *Escherichia coli* strain lacking an Na^+/H^+ antiporter (NhaA). J Bacteriol 174:4878–4884

Ji G, Silver S (1992a) Regulation and expression of the arsenic resistance operon from *Staphylococcus aureus* plasmid pI258. J Bacteriol 174:3684–3694

Ji G, Silver S (1992b) Reduction of arsenate to arsenite by the arsC protein of the arsenic resistance operon of *Staphylococcus aureus* plasmid pI258. Proc Natl Acad Sci USA 89:9474–9478

Ji G, Garber EA, Armes LG, Chen C-M, Fuchs JA, Silver S (1994) Arsenate reductase of *Staphylococcus aureus* plasmid p1258. Biochemistry 33:7294–7299

Joho M, Inouhe M, Tohoyama H, Murayama T (1995) Nickel resistance mechanisms in yeasts and other fungi. J Indust Microbiol 14:164–168

Jones D, Muehlchen A (1994) Effects of the potentially toxic metals, aluminium, zinc and copper on ectomycorrhizal fungi. J Environ Sci Health A Environ Sci Eng 29:949–966

Kameo S, Iwahashi H, Kojima Y, Satoh H (2000) Induction of metallothioneins in the heavy metal resistant fungus *Beauveria bassiana* exposed to copper or cadmium. Analusis 28:382–385

Karlson U, Frankenberger WT (1988) Effects of carbon and trace element addition on alkylselenide production by soil. Soil Sci Soc Am J 52:1640–1644

Karlson U, Frankenberger WT (1989) Accelerated rates of selenium volatilization from California soils. Soil Sci Soc Am J 53:749–753

Kelly JJ, Haggblom M, Tate RL (1999) Changes in soil microbial communities over time resulting from one time application of zinc: a laboratory microcosm study. Soil Biol Biochem 31:1455–1465

Khan M, Scullion J (2002) Effects of metal (Cd, Cu, Ni, Pb or Zn) enrichment of sewage-sludge on soil micro-organisms and their activities. Appl Soil Ecol 20:145–155

Killham K, Firestone MK (1983) Vesicular arbuscular mycorrhizal mediation of grass response to acidic and heavy metal depositions. Plant Soil 72:39–48

Kostov O, van Cleemput O (2001) Microbial activity of Cu contaminated soils and effect of lime and compost on soil resiliency. Compost Sci Util 9:336–351

Krantz-Rulcker C, Allard B, Schnurer J (1993) Interactions between a soil fungus, *Trichoderma harzianum* and IIB metals – adsorption to mycelium and production of complexing metabolites. Biometals 6:223–230

Krantz-Rulcker C, Allard B, Schnurer J (1996) Adsorption of IIB metals by 3 common soil fungi – comparison and assessment of importance for metal distribution in natural soil

systems. Soil Biol Biochem 28:967–975
Kubatova A, Prasil K, Vanova M (2002) Diversity of soil microscopic fungi on abandoned industrial deposits. Crypt Mycol 23:205–219
Kumar R, Kumar AV (1999) Biodeterioration of stone in tropical environments: an overview. John Paul Getty Trust, Los Angeles
Kuo C-W, Genthner BRS (1996) Effect of added heavy metal ions on biotransformation and biodegradation of 2-chlorophenol and 3-chlorobenzoate in anaerobic bacterial consortia. Appl Environ Microbiol 62:2317–2323
Kuperman RG, Carreiro MM (1997) Soil heavy metal concentrations, microbial biomass and enzyme activities in a contaminated grassland ecosystem. Soil Biol Biochem 29:179–190
Lacourt I, D'Angelo S, Girlanda M, Turnau K, Bonfante P, Perotto S (2000) Genetic polymorphism and metal sensitivity of *Oidiodendron maius* strains isolated form polluted soil. Ann Microbiol 50:157–166
Landa ER, Gray JR (1995) US Geological Survey – results on the environmental fate of uranium mining and milling wastes. J Ind Microbiol 26:19–31
Ledin M, Krantz-Rulcker C, Allard B (1996) Zn, Cd and Hg accumulation by microorganisms, organic and inorganic soil components in multicompartment systems. Soil Biol Biochem 28:791–799
Lee Y-A, Hendson M, Panopoulus NJ, Schrott MN (1994) Molecular cloning, chromosomal mapping, and sequence analysis of copper resistance genes from *Xanthomonas campestris* pv. juglandis: homology with blue copper proteins and multicopper oxidase. J Bacteriol 176:173–188
Leyval C, Turnau K, Haselwandter K (1997) Effect of heavy metal pollution on mycorrhizal colonization and function: physiological, ecological and applied aspects. Mycorrhiza 7:139–153
Liu XF, Culotta VC (1999) Mutational analysis of *Saccharomyces cerevisiae* Smf1p, a member of the Nramp family of metal transporters. J Mol Biol 289:885–891
Lloyd JR, Macaskie LE (1998) Enzymatic recovery of elemental palladium using sulfate-reducing bacteria. Appl Environ Microbiol 64:4607–4609
Lloyd JR, Ridley J, Khizniak T, Lyalikova NN, Macaskie LE (1999a) Reduction of technetium by *Desulfovibrio desulfuricans*: biocatalyst characterization and use in a flow-through bioreactor. Appl Environ Microbiol 65:2691–2696
Lloyd JR, Thomas GH, Finlay JA, Cole JA, Macaskie LE (1999b) Microbial reduction of technetium by *Escherichia coli* and *Desulfovibrio desulfuricans*: enhancement via the use of high activity strains and effect of process parameters. Biotechnol Bioeng 66:122–130
Losi ME, Frankenberger WT (1998) Microbial oxidation and solubilization of precipitated elemental selenium in soil. J Environ Qual 27:836–843
Lovley DR (ed) (2000) Environmental microbe–metal interactions. ASM Press, Washington, DC
Lovley DR (2001) Anaerobes to the rescue. Science 293:1444–1446
Lovely DR, Coates JD (1997) Bioremediation of metal contamination. Curr Opin Biotechnol 8:285–289
Macaskie LE (1991) The application of biotechnology to the treatment of wastes produced by the nuclear fuel cycle – biodegradation and bioaccumulation as a means of treating radionuclide-containing streams. Crit Rev Biotechnol 11:41–112
Macreadie IG, Sewell AK, Winge DR (1994) Metal ion resistance and the role of metallothionein in yeast. In: Winkelmann G, Winge DR (eds) Metal ions in fungi. Dekker, New York, pp 279–310
Malakul P, Srinivasan KR, Wang HY (1998) Metal toxicity reduction in naphthalene biodegradation by use of metal-chelating adsorbents. Appl Environ Microbiol 64:4610–4613

Markkola AM, Ahonen-Jonnarth U, Roitto M, Strommer R, Hyvarinen M (2002) Shift in ectomycorrhizal community composition in Scots pine (*Pinus sylvestris* L.) seedling roots as a response to nickel deposition and removal of lichen cover. Environ Pollut 120:797–803
Massaccesi G, Romero MC, Cazau MC, Bucsinszky AM (2002) Cadmium removal capacities of filamentous soil fungi isolated from industrially polluted sediments, in La Plata (Argentina). World J Microbiol Biotechnol 18:817–820
McLean JE, Bledsoe BE (1992) Behavior of metals in soils. EPA/540/S-92/018, US EPA, Washington, DC
McLean J, Beveridge TJ (2001) Chromate reduction by a pseudomonad isolated from a site contaminated with chromated copper arsenate. Appl Environ Microbiol 67:1076–1084
McLean JS, Lee J-U, Beveridge TJ (2002) Interactions of bacteria and environmental metals, fine-grained mineral development, and bioremediation strategies. In: Huang PM, Bollag J-M, Senesi N (eds) Interactions between soil particles and microorganisms. Wiley, New York, pp 228–261
Meharg AA, Cairney JWG (2000) Co-evolution of mycorrhizal symbionts and their hosts to metal-contaminated environments. Adv Ecol Res 30:69–112
Mehra RK, Winge DR (1991) Metal ion resistance in fungi: molecular mechanisms and their related expression. J Cell Biochem 45:30–40
Mineev VG, Gomonova NF, Zenova GM, Skvortsova IN (1999) Changes in the properties of soddy-podzolic soil and its microbocenosis under intensive anthropogenic impact. Eurasian Soil Sci 32:413–417
Morley GF, Gadd GM (1995) Sorption of toxic metals by fungi and clay minerals. Mycol Res 99:1429–1438
Morley GF, Sayer JA, Wilkinson SC, Gharieb MM, Gadd GM (1996) Fungal sequestration, solubilization and transformation of toxic metals. In: Frankland JC, Magan N, Gadd GM (eds) Fungi and environmental change. Cambridge Univ Press, Cambridge, pp 235–256
Mowll JL, Gadd GM (1985) The effect of vehicular lead pollution on phylloplane mycoflora. Trans Br Mycol Soc 84:685–689
Moynahan OS, Zabinski CA, Gannon JE (2002) Microbial community structure and carbon-utilization diversity in a mine tailings revegetation study. Restoration Ecol 10:77–87
Mozafar A, Ruh R, Klingel P, Gamper H, Egli S, Frossard E (2002) Effect of heavy metal contaminated shooting range soils on mycorrhizal colonization of roots and metal uptake by leek. Environ Monit Assess 79:177–191
Murasugi A, Wada C, Hayashi Y (1983) Occurrence of acid labile sulfide in cadmium binding peptide 1 from fission yeast. J Biochem 93:661–664
Nies DH (1992a) Resistance to cadmium, cobalt, zinc, and nickel in microbes. Plasmid 27:17–28
Nies DH (1992b) Czcr and Czcd, gene-products affecting regulation of resistance to cobalt, zinc, and cadmium (*czc* system) in *Alcaligenes eutrophus.* J Bacteriol 174:8102–8110
Nies DH (1995) The cobalt, zinc, and cadmium efflux system *czcabc* from *Alcaligenes eutrophus* functions as a cation-proton antiporter in *Escherichia coli.* J Bacteriol 177:2707–2712
Nies DH (1999) Microbial heavy-metal resistance. Appl Microbiol Biotechnol 51:730–750
Nies DH (2003) Efflux-mediated heavy metal resistance in prokaryotes. FEMS Microbiol Rev 27:313–339
Nies DH, Silver S (1995) Ion efflux systems involved in bacterial metal resistances. J Ind Microbiol 14:186–199
Nies DH, Nies A, Chu L, Silver S (1989a) Expression and nucleotide sequence of a plasmid-determined divalent-cation efflux system from *Alcaligenes eutrophus.* Proc Natl Acad Sci USA 86:7351–7355
Nies A, Nies DH, Silver S (1989b) Cloning and expression of plasmid genes en-

coding resistances to chromate and cobalt in *Alcaligenes eutrophus*. J Bacteriol 171:5065–5070
Nies A, Nies DH, Silver S (1990) Nucleotide sequence and expression of plasmid-encoded chromate resistance determinant from *Alcaligenes eutrophus*. J Biol Chem 265:5648–5653
Ohtake H, Cervantes C, Silver S (1987) Decreased chromate uptake in *Pseudomonas fluorescens* carrying a chromate resistance plasmid. J Bacteriol 169:3853–3856
Okorokov LA (1994) Several compartments of *Saccharomyces cerevisiae* are equipped with Ca^{2+} ATPase(s). FEMS Microbiol Lett 117:311–318
Okorokov LA, Lichko LP, Kulaev IS (1980) Vacuoles: main compartments of potassium, magnesium and phosphate in *Saccharomyces carlsbergensis* cells. J Bacteriol 144:661–665
Okorokov LA, Kulakovskaya TV, Lichko LP, Polorotova EV (1985) H^+/ion antiport as the principal mechanism of transport systems in the vacuolar membrane of the yeast *Saccharomyces carlsbergensis*. FEBS Lett 192:303–306
Olayinka A, Babalola GO (2001) Effects of copper sulphate application on microbial numbers and respiration, nitrifier and urease activities, and nitrogen and phosphorus mineralization in an alfisol. Biol Agric Hort 19:1–8
Oremland R, Stolz J (2000) Dissimilatory reduction of selenate and arsenate in nature. In: Lovley DR (ed) Environmental microbe–metal interactions. ASM Press, Washington, DC, pp 199–224
Oremland RS, Hollibaugh JT, Maest AS, Presser TS, Miller LG, Culbertson CW (1989) Selenate reduction to elemental selenium by anaerobic bacteria in sediments and culture: biogeochemical significance of a novel sulfate-independent respiration. Appl Environ Microbiol 55:2333–2343
Oremland RS, Steinberg NA, Presser TS, Miller LG (1991) *In situ* bacterial selenate reduction in the agricultural drainage systems of Western Nevada. Appl Environ Microbiol 57:615–617
Ortiz DF, Kreppel DF, Speiser DM, Scheel G, McDonald G, Ow DW (1992) Heavy metal tolerance in the fission yeast requires an ATP-binding cassette-type vacuolar membrane transporter. EMBO J 11:3491–3499
Ortiz DF, Ruscitti T, McCue KF, Ow DW (1995) Transport of metal-binding peptides by HMT1, a fission yeast ABC-type vacuolar membrane protein. J Biol Chem 270:4721–4728
Ow DW (1993) Phytochelatin-mediated cadmium tolerance in *Schizosaccharomyces pombe*. In Vitro Cell Dev Biol Plant 29P:213–219
Ow DW, Ortiz DF, Speiser DM, McCue KF (1994) Molecular genetic analysis of cadmium tolerance in *Schizosaccharomyces pombe*. In: Winkelmann G, Winge DR (eds) Metal ions in fungi. Dekker, New York, pp 339–359
Pawlowska TE, Chaney RL, Chin M, Charvat I (2000) Effects of metal phytoextraction practices on the indigenous community of arbuscular mycorrhizal fungi at a metal-contaminated landfill. Appl Environ Microbiol 66:2526–2530
Pennanen T, Frostegard A, Fritze, Baath E (1996) Phospholipid fatty acid composition and heavy metal tolerance of soil microbial communities along two heavy metal-polluted gradients in coniferous forests. Appl Environ Microbiol 62:420–428
Perkins J, Gadd GM (1993a) Accumulation and intracellular compartmentation of lithium ions in *Saccharomyces cerevisiae*. FEMS Microbiol Lett 107:255–260
Perkins J, Gadd GM (1993b) Caesium toxicity, accumulation and intracellular-localization in yeasts. Mycol Res 97:717–724
Phillips EJP, Landa ER, Lovley DR (1995) Remediation of uranium contaminated soils with bicarbonate extraction and microbial U(VI) reduction. J Ind Microbiol 14:203–207
Ramsay LM, Gadd GM (1997) Mutants of *Saccharomyces cerevisiae* defective in vacuolar function confirm a role for the vacuole in toxic metal ion detoxification. FEMS Microbiol

Lett 152:293–298
Rauser WE (1995) Phytochelatins and related peptides. Plant Physiol 109:1141–1149
Rawlings DE (1997) Mesophilic, autotrophic bioleaching bacteria: description, physiology and role. In: Rawlings DE (ed) Biomining: theory, microbes and industrial processes. Springer, Berlin Heidelberg New York, pp 229–245
Rawlings DE, Silver S (1995) Mining with microbes. Biotechnology 13:773–338
Renella G, Chaudri AM, Brookes PC (2002) Fresh additions of heavy metals do not model long-term effects on microbial biomass and activity. Soil Biol Biochem 34:121–124
Ribeiro RM, Moureaux C, Mussi Santos A (1972) Essai de mise en évidence sur milieu électif d'une microflore fongique adaptée aux sols à teneur élevée en cuivre. Cahiers ORSTOM Série Pédol 10:305–308
Rugh CL, Wilde HD, Stack NM, Thompson DM, Summers AO, Meagher RB (1996) Mercuric ion reduction and resistance in transgenic *Arabidopsis thaliana* plants expressing a modified bacterial *merA* gene. Proc Natl Acad Sci USA 93:3182–3187
Rusin PA, Sharp JE, Oden KL, Arnold RG, Sinclair NA (1993) Isolation and physiology of a manganese-reducing *Bacillus polymyxa* from an Oligocene silver-bearing ore and sediment with reference to Precambrian biogeochemistry. Precamb Res 61:231–240
Salt DE, Smith RD, Raskin I (1998) Phytoremediation . Annu Rev Plant Physiol Plant Mol Biol 49:643–668
Sandaa R-A, Enger O, Torsvik V (1999) Abundance and diversity of *Archaea* in heavy-metal-contaminated soils. Appl Environ Microbiol 65:3293–3297
Sayer JA, Gadd GM (2001) Binding of cobalt and zinc by organic acids and culture filtrates of *Aspergillus niger* grown in the absence or presence of insoluble cobalt or zinc phosphate. Mycol Res 105:1261–1267
Sayer JA, Cotter-Howells JD, Watson C, Hillier S, Gadd GM (1999) Lead mineral transformation by fungi. Curr Biol 9:691–694
Schippers A, Sand W (1999) Bacterial leaching of metal sulfides proceeds by two indirect mechanisms via thiosulfate or via polysulfides and sulphur. Appl Environ Microbiol 65:319–321
Schnoor JL, Licht LA, McCutcheon SC, Wolfe NL, Carreira LH(1995) Phytoremediation of organic and nutrient contaminants. Environ Sci Technol 29:318A–323A
Schottel J, Mandal A, Clark D, Silver S, Hedges RW (1974) Volatilization of mercury and organomercurials determined by inducible R-factor systems in enteric bacteria. Nature 251:335–337
Sharples JM, Chambers SM, Meharg AA, Cairney JWG (2000) Genetic diversity of root-associated fungal endophytes from *Calluna vulgaris* at contrasting field sites. New Phytol 148:153–162
Sharples JM, Meharg AA, Chambers SM, Cairney JWG (2001) Arsenate resistance in the ericoid mycorrhizal fungus *Hymenoscyphus ericae*. New Phytol 151:265–270
Shi W, Becker J, Bischoff M, Turco RF, Konopka AE (2002) Association of microbial community composition and activity with lead, chromium, and hydrocarbon contamination. Appl Environ Microbiol 68:3859–3866
Silver S (1996) Bacterial resistances to toxic metals – a review. Gene 179:9–19
Silver S (1998) Genes for all metals – a bacterial view of the Periodic Table. J Indust Microbiol Biotechnol 20:1–12
Silver S, Phung Le T (1996) Bacterial heavy metal resistance: new surprises. Annu Rev Microbiol 50:753–789
Smith SE, Read DJ (1997) Mycorrhizal symbiosis, 2nd edn. Academic Press, San Diego
Smith T, Pitts K, McGarvey JA, Summers AO (1998) Bacterial oxidation of mercury metal vapour, Hg(0). Appl Environ Microbiol 64:1328–1332
Smith WL, Gadd GM (2000) Reduction and precipitation of chromate by mixed culture

sulphate-reducing bacterial biofilms. J Appl Microbiol 88:983–991
Solioz M, Odermatt A (1995) Copper and silver transport by CopB-ATPase in membrane vesicles of *Enterococcus hirae*. J Biol Chem 270:9217–9221
Solioz M, Odermatt A, Krapf R (1994) Copper pumping ATPases: common concepts in bacteria and man. FEBS Lett 346:44–47
Southam G (2000) Bacterial surface-mediated mineral formation. In: Lovley DR (ed) Environmental microbe–metal interactions. ASM Press, Washington, DC, pp 257–276
Sreekrishnan TR, Tyagi RD (1994) Heavy metal leaching from sewage sludges: a techno-economic evaluation of the process options. Environ Technol 15:531–543
Staley JT, Palmer F, Adams JB (1982) Microcolonial fungi: common inhabitants on desert rocks. Science 215:1093–1095
Stephen JR, Chang Y-J, Macnaughton SJ, Kowalchuk GA, Leung KT, Flemming CA, White DC (1999) Effect of toxic metals on indigenous soil β-subgroup proteobacterium ammonia oxidizer community: structure and protection against toxicity by inoculated metal-resistant bacteria. Appl Environ Microbiol 65:95–101
Sterflinger K (2000) Fungi as geologic agents. Geomicrobiol J 17:97–124
Stolz JF, Oremland RS (1999) Bacterial respiration of arsenic and selenium. FEMS Microbiol Rev 23:615–627
Strasser H, Burgstaller W, Schinner F (1994) High yield production of oxalic acid for metal leaching purposes by *Aspergillus niger*. FEMS Microbiol Lett 119:365–370
Tebo BM, Obraztsova AY (1998) Sulfate-reducing bacterium grows with Cr(VI), U(VI), Mn(IV), and Fe(III) as electron acceptors. FEMS Microbiol Lett 162:193–198
Thompson-Eagle ET, Frankenberger WT (1992) Bioremediation of soils contaminated with selenium. In: Lal R, Stewart BA (eds) Advances in soil science. Springer, Berlin Heidelberg New York, pp 261–309
Thompson-Eagle ET, Frankenberger WT, Karlson U (1989) Volatilization of selenium by *Alternaria alternata*. Appl Environ Microbiol 55:1406–1413
Tobin JM, White C, Gadd GM (1994) Metal accumulation by fungi – applications in environmental biotechnology. J Ind Microbiol 13:126–130
Tohoyama H, Inouhe M, Joho M, Murayama T (1995) Production of metallothionein in copper-resistant and cadmium-resistant strains of *Saccharomyces cerevisiae*. J Ind Microbiol 14:126–131
Tomei FA, Barton LL, Lemanski CL, Zocco TG, Fink NH, Sillerud LO (1995) Transformation of selenate and selenite to elemental selenium by *Desulfovibrio desulfuricans*. J Ind Microbiol 14:329–336
Tsezos M, Volesky B (1982a) The mechanism of uranium biosorption by *Rhizopus arrhizus*. Biotechnol Bioeng 24:385–401
Tsezos M, Volesky B (1982b) The mechanism of thorium biosorption by *Rhizopus arrhizus*. Biotechnol Bioeng 24:955–969
Turner RJ, Weiner JH, Taylor DE (1995) Neither reduced uptake nor increased efflux is encoded by tellurite resistance determinants expressed in *Escherichia coli*. Can J Microbiol 41:92–98
Vachon RPD, Tyagi J, Auclair C, Wilkinson KJ (1994) Chemical and biological leaching of aluminium from red mud. Environ Sci Technol 28:26–30
Verrecchia EP (2000) Fungi and sediments. In: Riding RE, Awramik SM (eds) Microbial sediments. Springer, Berlin Heidelberg New York, pp 69–75
Verrecchia EP, Dumont J-L (1996) A biogeochemical model for chalk alteration by fungi in semiarid environments. Biogeochemistry 35:447–470
Vieira MJ, Melo LF (1995) Effect of clay particles on the behaviour of biofilms formed by *Pseudomomonas fluorescens*. Wat Sci Technol 32:45–52
Vodnik D, Byrne AR, Gogala N (1998) The uptake and transport of lead in some ectomycorrhizal fungi in culture. Mycol Res 102:953–958

Wainwright M, Gadd GM (1997) Fungi and industrial pollutants. In: Wicklow DT, Soderstrom BE (eds) The Mycota V environmental and microbial relationships. Springer, Berlin Heidelberg New York, pp 85–97

Walter EG, Taylor DE (1992) Plasmid-mediated resistance to tellurite: expressed and cryptic. Plasmid 27:52–64

Watson JHP, Ellwood DC, Deng QX, Mikhalovsky S, Hayter CE, Evans J (1995) Heavy metal adsorption on bacterially-produced FeS. Min Eng 8:1097–1108

Watson JHP, Cressey BA, Roberts AP, Ellwood DC, Charnock JM, Soper AK (2000) Structural and magnetic studies on heavy-metal-adsorbing iron sulphide nanoparticles produced by sulphate-reducing bacteria. J Magnet Magn Mat 214:13–30

White C, Gadd GM (1986) Uptake and cellular distribution of copper, cobalt and cadmium in strains of *Saccharomyces cerevisiae* cultured on elevated concentrations of these metals. FEMS Microbiol Ecol 38:277–283

White C, Gadd GM (1987) The uptake and cellular distribution of zinc in *Saccharomyces cerevisiae.* J Gen Microbiol 133:727–737

White C, Gadd GM (1997) An internal sedimentation bioreactor for laboratory-scale removal of toxic metals from soil leachates using biogenic sulphide precipitation. J Ind Microbiol 18:414–421

White C, Gadd GM (1998a) Accumulation and effects of cadmium on sulphate-reducing bacterial biofilms. Microbiol 144:1407–1415

White C, Gadd GM (1998b) Reduction of metal cations and oxyanions by anaerobic and metal-resistant organisms: chemistry, physiology and potential for the control and bioremediation of toxic metal pollution. In: Grant WD, Horikoshi T (eds) Extremophiles: physiology and biotechnology. Wiley, New York, pp 233–254

White C, Gadd GM (2000) Copper accumulation by sulphate-reducing bacterial biofilms and effects on growth. FEMS Microbiol Lett 183:313–318

White C, Sayer JA, Gadd GM (1997) Microbial solubilization and immobilization of toxic metals: key biogeochemical processes for treatment of contamination. FEMS Microbiol Rev 20:503–516

White C, Sharman K, Gadd GM (1998) An integrated microbial process for the bioremediation of soil contaminated with toxic metals. Nature Biotechnol 16:572–575

Wilkins DA (1991) The influence of sheathing (ecto-)mycorrhizas of trees on the uptake and toxicity of metals. Agric Ecosyst Environ 35: 245–260

Wilkinson DM, Dickinson NM (1995) Metal resistance in trees – the role of mycorrhizae. Oikos 72:298–300

Wollenzien U, de Hoog GS, Krumbein WE, Urzi C (1995) On the isolation of microcolonial fungi occurring on and in marble and other calcareous rocks. Sci Total Environ 167:287–294

Wu JS, Sung HY, Juang RJ (1995) Transformation of cadmium-binding complexes during cadmium sequestration in fission yeast. Biochem Mol Biol Int 36:1169–1175

Yu W, Farrell RA, Stillman DJ, Winge DR (1996) Identification of SLF1 as a new copper homeostasis gene involved in copper sulfide mineralization in *Saccharomyces cerevisiae.* Mol Cell Biol 16:2464–2472

Zhdanova NN, Redchitz TI, Vasilevskaya AI (1986) Species composition and sorption properties of Deuteromycetes in soils polluted by industrial wastewater (in Russian). Mikrobiol Zh 48:44–50

Zinkevich V, Bogdarina I, Kang H, Hill MAW, Tapper R, Beech IB (1996) Characterization of exopolymers produced by different isolates of marine sulphate-reducing bacteria. Int Biodet Biodeg 37:163–172

Part VI
Techniques to Investigate Soil Microorganisms

17 Marker Genes in Soil Microbiology

Christoph C. Tebbe[1]

1 Introduction

The central issue of soil microbiological research is to characterize the microorganisms that live in the soil, to find out how they are regulated and to understand their interactions with each other and with their environment. Since the beginning of soil microbiology with the groundbreaking work of Sergey Winodgradsky (1859–1953), many ecologically important processes in soil have been linked to microbial activity. We know today that soil microorganisms represent an enormous pool of biomass on our planet. In fact, it has been estimated that the global amount of organic carbon stored in microbial biomass in surface and subsurface soils is comparable to the amount of carbon in plant material (Whitman et al. 1998). Soil microbial research has revealed that soil microorganisms carry out key functions in the global cycling of carbon, nitrogen, sulfur and other elements. In this context, soil microorganisms also contribute to the sustainability of ecosystems, e. g., by mineralizing pesticides on agricultural fields. On the other hand, soil microbial activities can have detrimental ecological effects, as indicated by the production of greenhouse gases such as methane or nitrous oxides.

Although soil microbiology has developed dramatically since its outset, the field is still hampered by many technical problems in its limited ability to answer very basic questions. Certain activities of soil samples, like the degradation of a specific compound, can be measured, but it is not yet possible to clearly identify all of the microorganisms that might be involved in such a process. The "black box" metaphor is often used to describe this situation (Tiedje et al. 1999). The question of whether ecologically important processes in soils are carried out by microbial communities that are structurally highly similar or very different, depending on the type of soil or other factors, is still unresolved. Consequently, it remains difficult to understand the importance and value of microbial diversity for soil function.

[1]Institut für Agrarökologie, Bundesforschungsanstalt für Landwirtschaft (FAL), Bundesallee 50, 38116 Braunschweig, Germany, e-mail: christoph.tebbe@fal.de

Soil Biology, Volume 3
Microorganisms in Soils: Roles in Genesis and Functions
(ed. by F. Buscot and A. Varma)

The huge gap between viable microorganisms that can be observed under a microscope and those that can be cultivated under standard laboratory conditions clearly indicates that classical cultivation-dependent techniques are not sufficient for describing the actual microbial diversity of a soil sample. There is no doubt that laboratory cultivation and pure culture work are of fundamental importance for understanding microbiological activities and their regulations. However, in many cases, the direct transfer of such data to the more complex level of a microbial community or to specific habitats in the heterogeneous soil structure is not possible. In addition, pure culture work is time-consuming and can only be done on a very small proportion of microorganisms. Moreover, often it is not clear how representative a selected pure culture is for an actual activity found in soil.

Soil microbiology today has arrived at the critical point at which the increasing knowledge gained by molecular biology and pure culture work must be connected to a larger scale. These scales refer to microbial communities, ecological niches or even to landscape-determined factors and ecosystems (Tunlid 1999; Manefield and Turner 2002; Torsvik and Ovreas 2002; Paerl and Steppe 2003).

At this critical point, marker genes are one of the most important tools to help connect the molecular to the environmental level. Marker genes are used to characterize the diversity of microorganisms in a soil sample without isolating them by cultivation. In another context, they also allow one to follow the fate of single microbial cells in microcosms or in the field. Furthermore, marker genes can be applied to study the expression and regulation of specific genes in complex ecological matrices such as soil. In addition, they can work as biosensors, e.g., to detect biologically active compounds found in an environmental sample.

2 Definition of Marker Genes and Their First Applications in Soil Microbiology

Marker genes are nucleic acid sequences that tag an organism with a specific property. This property can either be within a DNA molecule itself or it can be encoded by DNA and become detectable after gene expression. In general, two types of marker genes can be differentiated: (1) intrinsic marker genes, which are natural constituents of the genome of an organism (de Lorenzo et al. 1998; Jansson et al. 2000; Prosser 2002), and (2) recombinant marker genes which are inserted into an organism by genetic engineering (Jansson 2003; Table 1). In the latter context, the term "reporter genes" is used for recombinant marker genes that are used to study gene expres-

Table 1. Definitions for marker genes and potential applications

Term	Definition	Potential applications (examples)
Intrinsic marker genes	Natural occurrence in organisms	Study of structural diversity Study of functional diversity Analysis of community structure, or biofilm architecture
Recombinant marker genes	Insertion by genetic engineering	Fate (survival, growth, biological activity) of a selected organism introduced into a complex environmental substrate (soil, rhizosphere, etc.) Study of in situ gene transfer
Reporter gene	Regulated recombinant marker gene	Study of in situ gene expression, detection of inducers and inducible promoters Biosensors

sion, e. g., when a marker gene is under the control of a specific promoter (Burlage et al. 1994; Killham and Yeomans 2001; Leveau and Lindow 2002). When microorganisms with reporter genes are used to detect compounds or signals from the environment, they are often referred to as biosensors (Simpson et al. 1998; Alexander et al. 2000; Casavant et al. 2002; Hakkila et al. 2002).

The first applications of marker genes to study natural microbial communities date back to the mid-1980s, when gene probes were used on environmental samples. Gene probes are single-stranded DNA molecules, radioactively or otherwise labeled that are complementary to specific genomic sequences. Based on the method developed earlier for colony hybridization with *Escherichia coli* (Grunstein and Hogness 1975), this DNA–DNA hybridization technique was used to detect specific genotypes. The precondition was the cultivation, growth and colony formation of the targeted organisms on agar-plates. The first use of the colony hybridization technique for studying environmental samples was to detect pathogens in food (Fitts et al. 1983; Hill et al. 1983). Soon thereafter, the technique was successfully applied to detect genetic potentials for toluene degradation (Sayler et al. 1985) and mercury resistance in bacteria (Barkay et al. 1985; Barkay and Olson 1986) isolated from contaminated soil and sediments. The transposons Tn5 and Tn7 were used as vehicles for the first recombinant marker genes to tag a bacteria with an antibiotic resistance or a metabolic activity (lactose utilization) and follow their fate after inoculation into a soil microcosms (Frederickson et al. 1988) or after a deliberate field release (Drahos et al. 1986; Kluepfel et al. 1991).

3 Ribosomal RNA as an Intrinsic Marker

The most widely applied and important marker genes in environmental microbiology to date, however, are genes encoding for the RNA of the small subunit of ribosomes (SSU rRNA; Head et al. 1998). These intrinsic marker genes and their corresponding gene product, the SSU rRNA molecules, occur in all living organisms, as they all possess ribosomes. Ribosomes are an essential part of the protein biosynthesis machinery and thus functionally highly conserved. However, despite this functional conservation, certain parts of the rRNA molecule and its corresponding gene are variable as they do not seem that important for function. These variable sites can serve as "molecular clocks" to trace evolution and phylogenetic relationships (Tourasse and Gouy 1997; Otsuka et al. 1999).

The outstanding importance of the rRNA genes for investigating the phylogeny of all forms of life, i. e., for building up the "universal tree of life," was recognized early by Woese and Fox in 1977 (Woese and Fox 1977). Norman Pace and his coworkers were able to use fluorescence-labeled rRNA targeted oligonucleotide probes to detect specific genotypes, independently of cultivation, in environmental samples by means of in situ hybridization (Olsen et al. 1986; Pace et al. 1986). Due to the high copy numbers of ribosomes in each cell, the SSU rRNA can directly and specifically be visualized under the microscope with complementary fluorescence-labeled gene probes (Amann et al. 1996). Depending on the specificity of the gene probe, different phylogenetic groups or different species can be detected and quantified without cultivation (Amann and Ludwig 2000). In addition, assemblages of microorganisms as they occur in biofilms (Christensen et al. 1998), microbial aggregates (Daims et al. 2001a), or in symbiotic relationships (Czarnetzki and Tebbe 2003) can be analyzed for their structure and thus give information about their functional interactions. In this context, fluorescence in situ hybridization (FISH) is especially powerful if it is combined with confocal laser scanning microscopy and flow cytometry (Amann et al. 1990; Assmus et al. 1995; Behrens et al. 2003). As many so far undescribed SSU rRNA genes are being discovered today in environmental samples (see following paragraph), FISH with SSU rRNA-specific gene probes offers a key technique for confirming the existence, abundance, and biological activity of microorganisms with such genes in an environmental sample (Amann and Kühl 1998; Daims et al. 2001b). Combined with cell sorting devices or optical tweezers, noncultivated FISH tagged bacterial cells can also be mechanically enriched for further analyses, thus circumventing the need for cultivation (Huber et al. 1995).

The methodology available in the early 1980s for isolating and analyzing intrinsic marker genes directly from the environment was very limited. In

1984, David Stahl and coworkers published their success in characterizing a symbiotic bacterium from a marine invertebrate without cultivation by directly extracting and sequencing the 5S rRNA molecule from the large ribosomal subunit (Stahl et al. 1984). This work was followed by the detection of three other bacteria from a hot spring habitat (Stahl et al. 1985). However, the size of the 5S rRNA molecule, 109–114 nucleotides, was rather small, which limited the potential for using the nucleotide sequence to differentiate and characterize environmental isolates. The SSU rRNA gene (16S for bacteria and archaea, 18S for eucarya) with a size of approx. 1540 nucleotides was obviously more suitable for ecological studies. The first 16S rRNA genes from noncultivated bacteria were obtained after direct extraction of 16S rRNA and cloning in *E. coli*. Weller and Ward isolated 16S rRNA genes from hot spring microbial communities after pre-enrichment of these sequences with three different enzymatic steps, i. e., reverse transcriptase, DNA polymerase I, and a terminal desoxynucleotidyl-transferase (Weller and Ward 1989). At about the same time, Giovannoni and coworkers used a technique which was less time and labor consuming than the Weller and Ward approach (Giovannoni et al. 1990) and which is still one of the key techniques contributing to many outstanding developments and results in microbial ecology and molecular soil microbiology: the polymerase chain reaction (PCR).

4 Polymerase Chain Reaction and Soil-Extracted Nucleic Acids

Polymerase chain reaction (PCR) is the process that allows the amplification of specific genes from the background of any DNA with an in vitro reaction. The first application of PCR was reported in 1985 (Saiki et al. 1985). In order to specifically amplify genes, short nucleotide sequences at the beginning and end of that gene need to be known. Primers with complementary nucleotides hybridize to these regions when the DNA of the target molecule is single stranded. These single strands are generated by melting the double helix with heat. A heat-stable DNA polymerase recognizes the short double-stranded molecules of target DNA and primers and starts synthesizing complementary DNA strands and thus replicating the DNA molecule. After replication, the double-stranded products are again melted into single strands and then the process can start again by decreasing the temperature to allow the DNA polymerase to synthesize another double-stranded molecule. Theoretically, 25 rounds of temperature cycles lead to an amplification of one gene copy to 2^{25} (33,554,432) copies.

Extraction of DNA from soils is a precondition for detecting intrinsic marker genes by PCR without cultivation. Two strategies have been applied to obtain PCR-amplifiable material: cell extraction and direct DNA extraction. Cell extraction means that the bacterial cells are first separated from the soil matrix and collected as intact cells. These cells are subsequently washed and then common protocols for extracting DNA from cultivated microorganisms can be applied (Holben et al. 1988; Steffan et al. 1988). The technique requires centrifugation steps and is time-consuming. The yield of DNA can be relatively low, but the quality, i.e., the purity and size of DNA fragments of the DNA, is generally high. The alternative approach is the direct lysis of the microbial cells in the soil matrix followed by the extraction and purification of DNA (Ogram et al. 1987; Steffan et al. 1988; Tebbe and Vahjen 1993). The method generates DNA that is initially highly contaminated with humic material. This DNA must be purified before PCR can be applied. The key enzyme for PCR, the *Taq*-polymerase is relatively sensitive to contamination by humic acids (Tebbe and Vahjen 1993). Direct extraction of DNA is straightforward and generally well suited for amplifying DNA for most applications. A large number of different protocols have been developed to obtain PCR-amplifiable DNA from soil, but to date, commercially available kits can also be used for this purpose.

The threshold of detecting single genes in soil DNA can be as low as approx. 10–100 copies/g of soil. In particular, recombinant genes can be detected efficiently because the specific primers are less likely to cross-react with DNA of indigenous microorganisms. The most limiting factors for detecting smaller numbers of genes relate to problems associated with sample compression and the dimension of a PCR tube. Sample compression simply means that the total DNA extracted from 1 g of soil, e.g., 80 μg, can hardly be packed into a single microliter which then would serve as a template for PCR. In addition, high concentrations of nonspecific DNA can interfere with the specificity of the primers designed to amplify a low copy number gene. The dilution of template DNA also alleviates problems associated with co-extracted compounds such as humic acids. Nevertheless, for many studies with PCR and soil DNA, it is desirable to use relatively concentrated DNA instead of dilutions.

5 Cloning, Sequencing and Profiling Marker Genes from Soil

For the detection of more abundant intrinsic marker genes, primer design becomes a very critical point. Most studies on PCR amplification of intrinsic marker genes try to retrieve such genes from unknown organisms. This can be achieved by using primers, i.e., primers which bind to phylogenetically conserved regions within a selected gene. A bioinformatic approach based on aligning the known sequences of a gene is used to select such primers. Most knowledge in this regard has been collected from work with SSU rRNA genes, and a large number of primers with specificities for different phylogenetic lineages, groups or clusters is now available. A PCR reaction with primers hybridizing to evolutionary conserved regions of the SSU rRNA genes and soil DNA normally yields PCR products having the same size, but different nucleotide sequences (due to the variable and hypervariable regions). A general strategy to sequence different SSU rRNA genes from soil is to clone the PCR products in *E. coli*, screen the cloned products for differences, e. g., with restriction endonucleases, and sequence the different products.

The first primers allowing the amplification of bacterial SSU rRNA genes from noncultivated bacteria were described in 1990 (Edwards et al. 1990; Giovannoni et al. 1990) and their first applications on environmental DNA already indicated that the majority of sequences were derived from unknown or not-yet cultured microorganisms. The first SSU rRNA genes that were amplified from soil DNA were reported by Liesack and Stackebrandt in 1992 (Liesack and Stackebrandt 1992). Since then, many publications have reported on the diversity of SSU rRNA genes from soil and still, even with public databases containing more than 35,000 rRNA gene sequences, the majority of sequences that are isolated from soil are typically not identical to these sequences. In addition, the closest relatives that can be found for newly recovered sequences are often SSU rRNA genes of other uncultured microorganisms. We know today that many soil bacteria and archaea belong to phylogenetic groups with only a few cultivated organisms, e. g., *Acidobacteria*, *Verrucomicrobia* or *Planctomycetes*. Other sequences that are frequently recovered from soil samples fall into phylogenetic groups like TM7, WS6, or OP11, which are to date exclusively represented by sequences of noncultivated organisms (Borneman et al. 1996; Bintrim et al. 1997; Dojka et al. 1998; Hugenholtz et al. 1998). Indeed, one of the major objectives for soil microbiological research today is to succeed in cultivating representative organisms from these groups or to study the structure of their genomes without cultivation.

The cloning and sequencing of PCR products amplified from soil DNA are excellent strategies for assessing the total microbial diversity in a soil sample. However, for many ecological studies, different samples need to be compared with each other, e. g., to characterize an effect of a certain treatment or environmental change. In such situations, genetic profiling techniques are very useful, as they generate genetic fingerprints from the diversity of PCR products amplified with universal primers. Genetic profiling techniques allow the inclusion of larger numbers of replicates, which is important in differentiating actual effects from the natural variability of a microbial community. The most commonly applied techniques for generating genetic profiles are DGGE (denaturing gradient gel elec-

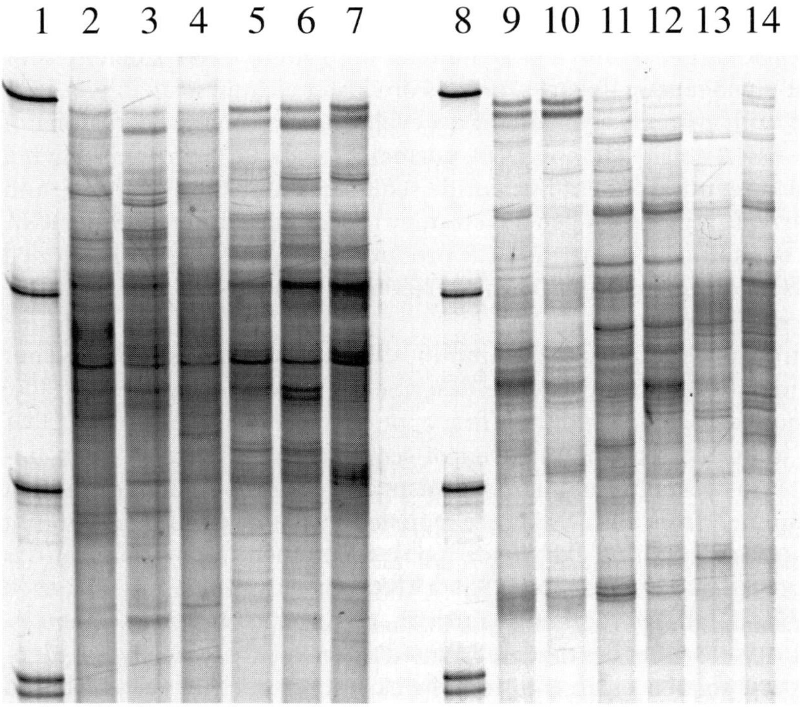

Fig. 1. SSCP profiles of partial SSU rRNA genes amplified by PCR from DNA extracted from different horizons of a forest soil (parabrown earth pseudogley). Each horizon is represented by three independent replicates taken at a distance of 5 m from each other. Horizon A_h (8 cm depth; *lanes 2–4*), horizon S_wB_v (25 cm depth; *lanes 5–7*), horizon S_w (40 cm depth; *lanes –9 to 11*) and horizon IIS_d (78 cm depth; *lanes 12–14*). *Lanes 1* and *8*, DNA-SSCP markers. The patterns indicate similarities and differences between bacterial communities inhabiting the different horizons. (Rabold and Tebbe 2001, unpubl. results)

trophoresis; Muyzer et al. 1993, 1998), SSCP (single-strand conformation polymorphism; Schwieger and Tebbe 1998; Tebbe et al. 2001), TRFLP (terminal restriction fragment length polymorphism; Liu et al. 1997) or RISA (ribosomal intergeneric spacer analysis; Ranjard et al. 1997). Each of these techniques has its own advantages and disadvantages in regard to technical handling, required equipment, sources of artifacts, sensitivity of detection and potentials for identifying members of the microbial community. Figure 1 provides an example demonstrating the similarity of SSCP patterns obtained from soil DNA which was extracted from replicate soil samples and compared to differences which were found in different horizons of that soil. The intrinsic marker gene that was used to generate these profiles was a partial, PCR-amplified sequence of the bacterial SSU rRNA.

6 Structural and Functional Diversity of Soil Microbial Communities as Seen with Intrinsic Marker Genes

A limitation of using SSU rRNA genes for the characterization of microbial communities independent of cultivation is the fact that those genes are not perfect markers for "function". For ecological studies, it may often be more meaningful to investigate whether the genetic potential for a specific function exists, and if yes, how genetically variable this function is represented. Thus, instead of a "structural diversity" that would ideally be assessed by analyzing the dominant SSU rRNA genes, "functional diversity" would be the focus. Depending on the selected function, however, SSU rRNA genes may or may not be useful for studying functional diversity. They are useful for characterizing the diversity of autotrophic ammonium-oxidizing bacteria in soil, since all known soil bacteria with this property belong only to two genera, *Nitrospira* and *Nitrosomonas*. Both genera fall into the beta-subgroup of the *Proteobacteria* and therefore primers with specificities for these two groups could be designed which allow the study of the diversity of ammonium oxidizers based on SSU rRNA genes. For other functional groups such as sulfate-reducing bacteria, the situation is more complicated, since six different phylogenetic groups are already known to be capable of using sulfate in anaerobic environments as an electron acceptor. However, primer sets targeting these different groups have been developed and they can serve as a valuable tool for understanding the importance of those groups in different environmental niches (Daly et al. 2000).

Many functional potentials, however, cannot be tagged with SSU rRNA genes as markers because they are not strictly linked to the phylogenetic lineages of their hosts. Examples include key activities in carbon or nitrogen cycling, such as cellulose degradation, nitrogen fixation or denitrification. The genetic potentials for these properties can be found in very different phylogenetic groups, and within these groups, not all members may have the respective potential. As an alternative to SSU rRNA genes, the functional genes themselves may serve as intrinsic markers. A precondition for using such genes as markers for cultivation-independent analyses is that common primer binding sites can be identified. In several cases this is possible with the use of degenerate (less specific) nucleotides with primers. Functional marker genes that have been successfully detected in soil DNA now include many enzymatic activities that are considered to be important in the cycling of carbon or nitrogen, such as the bacterial ammonium monooxygenase (Kowalchuk et al. 2000; Norton et al. 2002), particulate and soluble methane monooxygenase (Baker et al. 2001; Auman and Lid-

Table 2. Examples of intrinsic marker genes and the objective of using them in soil microbiological studies

Intrinsic marker gene	Encoded function	Targeted group of organisms or phenotype
rrna	RNA of the small subunit of ribosomes	Assess the structural diversity of the total or a selected subfraction of the microbial communities in soil, rhizosphere or the gut of soil animals
amoA	Ammonium monooxygenase	Study the functional diversity of autotrophic, ammonium oxidizing bacteria
pmoA	Particulate methane monooxygenase	Study the diversity of methanotrophic bacteria
mmoX	Soluble methane monooxygenase	Detection of the potential to oxidize halogenated compounds, i. e., tricholoroethylene (TCE)
nifH	Nitrogenase	Study the genetic potential to assimilate molecular nitrogen from the atmosphere
narG	Membrane-bound nitrate reductase	Detection of the genetic potential and diversity of soil bacteria capable of denitrification
mcrA	Methyl-coenzyme M reductase	Detection and diversity of methanogenic bacteria in soil
chiA	Chitinase	Characterization of the diversity of group A bacterial chitinases in soil
merA	Mercury reductase	Genetic potential for the detoxification of mercury (Hg^{2+})

strom 2002), nitrate reductase (Philippot et al. 2002), nitrogenase (Widmer et al. 1999), or the methyl-coenzyme M reductase of methanogenic Archaea (Lueders et al. 2001; Luton et al. 2002; Table 2). In addition, genes encoding for catabolic potentials in soils have also been detected and characterized for their diversity, e. g., chitinases (Williamson et al. 2000; Metcalfe et al. 2002), dehalogenases (Marchesi and Weightman 2003), catechol dioxygenases (Okuta et al. 1998), or phenol hydroxylases (Watanabe et al. 1998).

7 Expression of Intrinsic Marker Genes and Detection of Gene Transfer Potentials

Marker genes themselves cannot directly indicate activity, but rather genetic potentials. For many ecological studies, however, it is more meaningful to analyze general or specific activities within a microbial community. A certain indicator for activity is the cell concentration of ribosomes and consequently SSU rRNA, which can specifically be detected by FISH, and studies have tried to compare SSU rRNA diversity to that of SSU rRNA genes in order to identify active members of a microbial community. To detect a specific activity of an intrinsic gene it would be ideal to quantify the transcripts of the gene, the mRNA, instead of the gene. For eukaryotic mRNA, e. g., fungal genes in soil, this approach is feasible (Bogan et al. 1996), but for bacterial mRNA it is problematic, as the latter molecules only have an average half-life value of some minutes (Wellington et al. 2003). Nevertheless, with an improvement of RNA extraction methods, there is recent evidence in the literature that in selected soil environments transcripts of bacterial genes can be isolated and characterized. Alfreider and coworkers (2003) demonstrated the presence of mRNA of two catabolic chlorocatechol dioxygenases genes which indicated active bioremediation in an aquifer.

The detection of intrinsic marker genes with less conservation than the SSU rRNA genes will often be a trade-off between choosing primer specificities that are high enough to avoid the amplification of false-positives (DNA fragments that have nothing to do with the functional gene that was targeted), and still low enough to hybridize to conserved regions that may be somewhat different from the predicted sequence and have the risk of being false-negative if primer hybridizations should fail. An interesting exception, however, are genes that are frequently transferred between different phylogenetic lineages, such as genes encoding for replication and transfer functions of broad-host range plasmids (mobile genetic elements). On such plasmids, these genes build the backbone structure which is often complemented with different functional genes, e. g., encoding for antibiotic or heavy metal resistances, toxin production or biodegradative pathways. Plasmids can be differentiated according to the compatibility of their replication genes, either allowing coexistence of plasmids or incompatibilities. Using primers specific for different incompatibility groups, it was possible to study the diversity and prevalence of plasmids in a variety of environmental substrates after direct extraction of DNA, PCR and DNA–DNA hybridization (Smalla et al. 2000).

8 Recombinant Marker Genes

Recombinant marker genes have become important tools in both basic and applied soil microbiology. There is a choice of different marker genes available, some confer an enzymatic activity that can be detected by a color reaction, others produce light or fluorescence, and others confer resistance while growing in the presence of an antibiotic or heavy metal (Table 3). In order to follow the survival of an organism introduced into soil, a good recombinant marker gene would confer a unique and easily detectable property to a selected organism and thus allow the study of the behavior of that organism within the background of an indigenous microbiological community.

The methods for detecting a marker gene-tagged organism can either be direct or indirect. Direct methods measure the recombinant property directly in an environmental sample, e.g., by quantifying the intensity of a light signal produced by a luciferase activity or by visualizing cells with the green fluorescent protein (GFP) in a confocal laser scanning microscope. An example of direct detection of GFP-tagged bacterial cells, even without a microscope, is shown in Fig. 2.

Indirect methods require the recombinant cells first to be isolated by cultivation and then analyzed for the presence of the marker. The indirect detection has been widely used to follow the fate of genetically modified bacteria after inoculation into soil microcosms or after field release (Tebbe

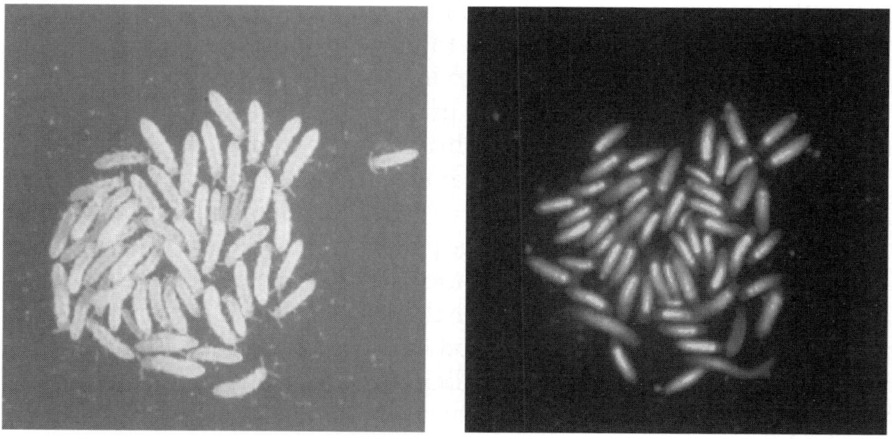

Fig. 2. Use of GFP to study the fate of *Pseudomonas putida* in the gut of the soil microarthropod *Folsomia candida* (Collembola). The pictures show a group of *F. candida* specimens as seen in regular daylight (*left*) and in UV-light (*right*). The *gfp*-tagged *P. putida* cells are clearly visible in the gut of the nonpigmented *F. candida* and in some fecal depositions at the periphery of the group. (Thimm and Tebbe 1998, unpubl. results)

Table 3. Examples of recombinant marker genes used in soil microbiology and microbial ecology

Type of marker gene (origin)	**Name, encoded function (examples)**	**Properties**
Resistance to antibiotics or heavy metals (bacteria)	*nptII*, resistance to kanamycin *cam*, resistance to chloramphenicol *merA*, resistance to mercury, Hg^{2+} *telA/telB*, resistance to tellurite	Selectable marker for enrichment by cultivation Possible use of antibiotics or metals in the laboratory
Chromogenic marker, i.e., color reaction by enzymatic activity (bacteria)	*lacZY*, galactosidase *gusA*, galacturonidase *xylE*, catechol 2, 3, catechol dioxygenase	Simple assays for detection Possible indigenous background activity, e.g., for *xylE* in soil
Fluorescence marker (jellyfish)	*gfp*, green fluorescent protein	Single cell detection in the microscope Unstable variants for gene expression measurements Possible decrease in host fitness Possible nonfluorescence as a consequence of non-proper molecule folding
Luciferase (marine bacteria)	*luxAB* (luminescence with decanal as a substrate) *luxCDABE* (autoluminescence)	Specific luminometric detection Suitable for in situ measurements Suitable to measure metabolic activity Autoluminescence may be a metabolic burden
Luciferase (firefly)	*luc* (luminescence with luciferin as a substrate)	Easily detectable No indigenous background activity Highly stable and low metabolic burden due to recombinant gene expression Expensive substrate
Ice crystallization (bacteria)	*inaZ* (protein, inducing the formation of ice crystals)	Simple detection Possible indigenous background activity

2000; Amarger 2002; Choi et al. 2003). For the indirect method, however, the tagged organisms should also carry a resistance marker to allow the enrichment of the selected strain on selective media. This resistance marker can be intrinsic, like a chromosomal mutation to streptomycin resistance, or it can be recombinant, e.g., by adding an *nptII* gene for kanamycin resistance to another nonselectable marker such as luciferase or GFP (Selbitschka et al. 1992; Weilbo and Skorupska 2001). As a consequence of using a selectable marker, the threshold of detection of a single bacterial colony within the background of co-cultivated soil bacteria can change from, e.g., 10,000–100 colony-forming units of that organism per gram of soil. Heavy metal resistance genes have also been used as selectable markers, e.g., to study the survival of bacteria after a field release (Corich et al. 2001). However, it should be kept in mind that heavy metals are also toxic to humans and that used growth media may need a special waste disposal.

9 Detection of In Situ Gene Transfer and Gene Expression with Recombinant Marker Genes

Recombinant marker genes have been very useful for studying bacterial gene transfers in natural microbial communities. By using plasmids which carried marker genes such as *lacZ*, *luc*, or *gfp*, it was possible to demonstrate the importance of the growth conditions of bacteria for conducting conjugation (Christensen et al. 1996). In other studies, stimulated conjugative gene transfer could be detected in rhizospheres (van Elsas and Trevors 1990) and in the gut and feces of soil invertebrates (Hoffmann et al. 1998; Thimm et al. 2001). In order to achieve sensitive rates for detecting gene transfer, it is necessary to exclude (counter-select) the donor cells which carry the marker, as they would overgrow the relatively rare transconjugants. The counter-selection can be achieved by different techniques, such as killing the donor cells with specific phages (Smit et al. 1991), or by selecting transconjugants on carbon sources that cannot be used by the donor (Hoffmann et al. 1998). An interesting application of a recombinant marker to detect gene transfer without cultivation was demonstrated in a marine model ecosystem: the donor bacteria carried the GFP-encoding gene on a plasmid, but the expression of the gene was inhibited by a repressor protein with the responsible gene located in the chromosome. As a consequence of plasmid transfer to indigenous bacteria (without the repressor protein), the GFP-encoding gene was expressed and the transconjugants became detectable by their green fluorescence (Dahlberg et al. 1998). The strategy of marker gene repression in donors was also applied to study conjugative gene transfer in laboratory-made biofilms (Christensen et al. 1998). Since soil microorganisms mainly grow attached to surfaces rather than as plankton, such biofilm studies have a high relevance for understanding gene transfer processes in soil.

Reporter genes can be defined as inducible marker genes. With reporter genes it is possible to study bacterial and fungal gene expression in complex matrices, such as soils or rhizospheres (Bae and Knudsen 2000; Bergero et al. 2003; Jansson 2003). A typical reporter gene is a marker gene that is inserted without its own promoter into the genome of a "host" organism. Depending on the delivery and recombination technique, the marker gene can be inserted at random or at specific positions. Random insertions can be used to screen for inducible promoters or to report general metabolic activity (van Overbeek and van Elsas 1995; Suarez et al. 1997; Porteous et al. 2000). On the other hand, marker genes can be inserted into selected sites of the genome to report on the impact of environmental conditions on specific activities of the tagged organism. In order to elucidate the physiological conditions that bacteria face in the rhizosphere, selected rhizosphere-colonizing bacteria were equipped with different reporter genes, e. g., those which indicated growth (Brennerova and Crowley 1994), available carbon sources (Jaeger et al. 1999; Miller et al. 2001), or limitations of carbon, nitrogen, phosphorus, iron or oxygen (Loper and Lindow 1994; Sorensen et al. 2001; Dollard and Billard 2003). Other reporter strains were useful in characterizing the effect that the soil matrix has on the expression of biodegradative genes (Holden et al. 2002), or on the biological availability of heavy metals (Tom-Petersen et al. 2001; Ivask et al. 2002). Reporter genes, however, can be used not only to characterize how certain environmental conditions affect the physiology of an organism, but also to identify new, environmentally regulated promoters in bacteria. Examples are the detection of heavy metal responsive promoters (Rossbach et al. 2000) or promoters that are induced in the rhizosphere (van Overbeek and van Elsas 1995). It is obvious that such promoters, once they are identified, can be very helpful in agricultural biotechnology, e. g., for the improvement of bacterial inoculants, and for environmental analysis, by using them to construct biosensors.

10 Recombinant Marker Genes as Biosensors

Biosensors are living cells that are used to detect toxic or other biologically active compounds. Genetic engineering techniques have the potential to generate biosensors according to the specific needs of detection, e. g., by connecting marker genes to specifically regulated promoters. In contrast to nonengineered biosensors, such as the marine bacterium *Photobacterium phosphoreum* (Britz et al. 1997), biosensors are more specific and versatile. As an advantage over chemical measurements, biosensors can measure the biological availability of a compound in a complex environmental substrate, and they detect combined effects of different compounds of biological activities. Biosensors can be designed to be specific for a single compound (Willardson et al. 1998; Prachayasittikul et al. 2001), or they can be designed to measure a general effect on a biological system, e.g., the induction or inhibition of protein biosynthesis or the detection of stress or genotoxicity (Kostrzynska et al. 2002). Regulatory elements from heavy metal-resistant bacteria can be used to construct specific biosensors for measuring the bioavailability of a heavy metal such as mercury (Hansen and Sorensen 2000; Rasmussen et al. 2000). The threshold of detection can be as low as 1 fM Hg^{2+}. Other biosensors have been developed to detect different hydrocarbons in polluted soils (Jacob et al. 2001) by connecting the luxCDABE operon to substrate-specific promoters. With such biosensors it was possible quantify bioavailable BTEX compounds in soil (benzene, toluene, ethylbenzene, xylene) with an on-line system (Applegate et al. 1998). On the other hand, reporter genes can be inserted downstream of promotors that respond to stress, e.g., by detecting the induction of DNA repair (*recA*, *uvrA*, *alkA*). Recently, biosensors became available that can detect homoserine lactones and thereby identify bacteria capable of quorum sensing (Flavier et al. 1997; Shaw et al. 1997).

Most biosensors are designed to work under laboratory conditions, but since they provide an easy means of detection, like luminescence, they may also find some applications in the field, e.g., by indicating active bioremediation (Ripp et al. 2000; Sayler and Ripp 2000), or the presence of unwanted compounds. A striking example is the potential use of luminescent bacteria responsive to the explosive TNT (trinitrotoluene) for the detection of land mines (Yinon 2002). In considering future field applications of biosensors with recombinant marker genes, it should be noted that none of the common marker genes (Table 3) is known to cause any risk to human health or the environment, either theoretically or practically, the latter demonstrated by several field studies with genetically engineered bacteria that have been conducted in the US and Europe during the last 15 years.

11 Conclusions and the Future of Marker Genes

Marker genes have become a highly valuable and versatile tool in soil microbiology. Current activities in genome sequencing of culturable and uncultured soil microorganisms continuously improve our capacity to design specific primers for the detection and study of genetic potentials in soil by the PCR-amplification of intrinsic marker genes, e.g., those that encode for ecological key functions in soil. The expression of functional genes can be studied by detecting transcripts of these genes in nucleic acids, directly extracted from soil. Despite the relatively high instability of the bacterial mRNA, in contrast to the fungal (eukaryotic) mRNA, recent protocols make it now more feasible to address the question of bacterial gene expression in soil. A new perspective for using directly extracted nucleic acids from soil for ecological studies is given by recent developments in micro-array-technology (Cho and Tiedje 2001; Small et al. 2001; Bertilsson et al. 2002). It may not be long before the abundance of several hundred or thousand species or the expression of many different genes can be studied from one nucleic acid sample extracted from soil. However, to reach this goal, problems connected with the selection of appropriate conditions for nucleic acid hybridizations and with the detection threshold of hybridized gene probes need to be solved (Peplies et al. 2003). Ultimately, these technologies may replace current gel-based genetic profiling techniques and PCR itself, as the latter technique, despite its enormous importance in soil microbiology, as outlined in this chapter, also has its limitations due to the fact that a single PCR does not amplify each target sequence from a soil DNA at the same rate and thus generates biases (Suzuki and Giovannoni 1996; Lueders and Friedrich 2003). Recent studies with stable isotopes indicate how useful the connection between marker gene and isotope technology is for linking structural to functional community analyses (Radajewski et al. 2000; Boschker and Middelburg 2002; Manefield et al. 2002). However, isotope technology will not completely replace the use of recombinant marker genes to study gene expression, as the latter technology provides models that can easily and directly be studied in complex systems. In addition, reporter genes are very safe to use in the laboratory and, as shown by experimental evidence, also in the field. Recombinant marker genes have already been very helpful in studying the fate of microorganisms in soil, the expression of selected activities in complex microbial communities, and to detect biologically active compounds by biosensor technology. It is therefore likely that both recombinant and intrinsic marker gene technology will further develop and continue to contribute to our understanding of the microbiologically mediated processes in terrestrial ecosystems.

References

Alexander DC, Costanzo MA, Guzzo J, Cai J, Charoensri N, Diorio C, Dubow MS (2000) Blazing towards the next millennium: luciferase fusions to identify genes responsive to environmental stress. Water Air Soil Pollut 123:81–94

Alfreider A, Vogt C, Babel W (2003) Expression of chlorocatechol 1,2-dioxygenase and chlorocatechol 2,3-dioxygenase genes in chlorobenzene-contaminated subsurface samples. Appl Environ Microbiol 69:1372–1376

Amann R, Kühl M (1998) In situ methods for assessment of microorganisms and their activities. Curr Opin Microbiol 1:352–358

Amann R, Ludwig W (2000) Ribosomal RNA-targeted nucleic acid probes for studies in microbial ecology. FEMS Microbiol Rev 24:555–565

Amann R, Snaidr J, Wagner M, Ludwig W, Schleifer KH (1996) In situ visualization of high genetic diversity in a natural microbial community. J Bacteriol 178:3496–3500

Amann R, Binder BJ, Olson RJ, Chisholm SW, Devereux R, Stahl DA (1990) Combination of 16S ribosomal-RNA-targeted oligonucleotide probes with flow-cytometry for analyzing mixed microbial populations. Appl Environ Microbiol 56:1919–1925

Amarger N (2002) Genetically modified bacteria in agriculture. Biochimie 84:1061–1072

Applegate BM, Kehrmeyer SR, Sayler GS (1998) A chromosomally based *tod-luxCDABE* whole-cell reporter for benzene, toluene, ethylbenzene, and xylene (BTEX) sensing. Appl Environ Microbiol 64:2730–2735

Assmus B, Hutzler P, Kirchhof G, Amann R, Lawrence JR, Hartmann A (1995) In-Situ localization of *Azospirillum brasilense* in the rhizosphere of wheat with fluorescently labeled, ribosomal-RNA-targeted oligonucleotide probes and scanning confocal laser microscopy. Appl Environ Microbiol 61:1013–1019

Auman AJ, Lidstrom ME (2002) Analysis of sMMO-containing Type I methanotrophs in Lake Washington sediment. Environ Microbiol 4:517–524

Bae YS, Knudsen GR (2000) Cotransformation of *Trichoderma harzianum* with beta-glucuronidase and green fluorescent protein genes provides a useful tool for monitoring fungal growth and activity in natural soils. Appl Environ Microbiol 66:810–815

Baker PW, Futamata H, Harayama S, Watanabe K (2001) Molecular diversity of pMMO and sMMO in a TCE-contaminated aquifer during bioremediation. FEMS Microbiol Ecol 38:161–167

Barkay T, Olson BH (1986) Phenotypic and genotypic adaptation of aerobic heterotrophic sediment bacterial communities to mercury stress. Appl Environ Microbiol 52:403–406

Barkay T, Fouts DL, Olson BH (1985) Preparation of a DNA gene probe for detection of mercury resistance genes in gram-negative bacterial communities. Appl Environ Microbiol 49:686–692

Behrens S, Ruhland C, Inacio J, Huber H, Fonseca A, Spencer-Martins I, Fuchs BM, Amann R (2003) In situ accessibility of small-subunit rRNA of members of the domains Bacteria, Archaea, and Eucarya to Cy3-labeled oligonucleotide probes. Appl Environ Microbiol 69:1748–1758

Bergero R, Harrier LA, Franken P (2003) Reporter genes: applications to the study of arbuscular mycorrhizal (AM) fungi and their symbiotic interactions with plant. Plant Soil 255:143–155

Bertilsson S, Cavanaugh CM, Polz MF (2002) Sequencing-independent method to generate oligonucleotide probes targeting a variable region in bacterial 16S rRNA by PCR with detachable primers. Appl Environ Microbiol 68:6077–6086

Bintrim SB, Donohue TJ, Handelsman J, Roberts GP, Goodman RM (1997) Molecular phylogeny of Archaea from soil. Proc Natl Acad Sci USA 94:277–282

Bogan BW, Schoenike B, Lamar RT, Cullen D (1996) Expression of *lip* genes during growth in soil and oxidation of anthracene by *Phanerochaete chrysosporium*. Appl Environ Microbiol 62:3697–3703

Borneman J, Skroch PW, Osullivan KM, Palus JA, Rumjanek NG, Jansen JL, Nienhuis J, Triplett EW (1996) Molecular microbial diversity of an agricultural soil in Wisconsin. Appl Environ Microbiol 62:1935–1943

Boschker HTS, Middelburg JJ (2002) Stable isotopes and biomarkers in microbial ecology. FEMS Microbiol Ecol 40:85–95

Brennerova MV, Crowley DE (1994) Direct detection of rhizosphere-colonizing *Pseudomonas* sp. using an *Escherichia coli* ribosomal RNA promoter in a Tn7-*lux* system. FEMS Microbiol Ecol 14:319–330

Britz ML, Simonov N, Chun UK (1997) Stabilization of bioluminescence of immobilized *Photobacterium phosphoreum* and monitoring of environmental pollutants. J Microbiol Biotechnol 7:242–249

Burlage RS, Palumbo AV, Heitzer A, Sayler G (1994) Bioluminescent reporter bacteria detect contaminants in soil samples. Appl Biochem Biotechnol 45/46:731–740

Casavant NC, Beattie GA, Phillips GJ, Halverson LJ (2002) Site-specific recombination-based genetic system for reporting transient or low-level gene expression. Appl Environ Microbiol 68:3588–3596

Cho JC, Tiedje JM (2001) Bacterial species determination from DNA–DNA hybridization by using genome fragments and DNA microarrays. Appl Environ Microbiol 67:3677–3682

Choi HY, Ryder MH, Gillings MR, Stokes HW, Ophel-Keller KM, Veal DA (2003) Survival of a *lacZY*-marked strain of *Pseudomonas corrugata* following a field release. FEMS Microbiol Ecol 43:367–374

Christensen BB, Sternberg C, Molin S (1996) Bacterial plasmid conjugation on semi-solid surfaces monitored with the green fluorescent protein (GFP) from *Aequorea victoria* as a marker. Gene 173:59–65

Christensen BB, Sternberg C, Andersen JB, Eberl L, Moller S, Givskov M, Molin S (1998) Establishment of new genetic traits in a microbial biofilm community. Appl Environ Microbiol 64:2247–2255

Corich V, Giacomini A, Vian P, Vendramin E, Carlot M, Basaglia M, Squartini A, Casella S, Nuti MP (2001) Aspects of marker/reporter stability and selectivity in soil microbiology. Microbial Ecol 41:333–340

Czarnetzki AB, Tebbe CC (2004) Detection and phylogenetic analysis of *Wolbachia* in Collembola. Environ Microbiol 6:35–44

Dahlberg C, Bergstrom M, Hermansson M (1998) In situ detection of high levels of horizontal plasmid transfer in marine bacterial communities. Appl Environ Microbiol 64:2670–2675

Daims H, Nielsen JL, Nielsen PH, Schleifer KH, Wagner M (2001a) In situ characterization of *Nitrospira*-like nitrite-oxidizing bacteria active in wastewater treatment plants. Appl Environ Microbiol 67:5273–5284

Daims H, Ramsing NB, Schleifer KH, Wagner M (2001b) Cultivation-independent, semiautomatic determination of absolute bacterial cell numbers in environmental samples by fluorescence in situ hybridization. Appl Environ Microbiol 67:5810–5818

Daly K, Sharp RJ, McCarthy AJ (2000) Development of nucleotide probes and PCR primers for detecting phylogenetic subgroups of sulfate-reducing bacteria. Microbiology UK 146:1693–1705

De Lorenzo V, Glover A, Hill P, Jansson JK, Jorgenson K, Lindström K, Mergeay M, Molin S, Morgan A, Nybroe O, Oliver J, Prosser J, Romantshuk M, Selbitschka W, Tebbe C, Top E (1998) Marker genes as tags for monitoring microorganisms in nature – an opinion. Centraltryckeriet AB, Boras, Sweden

Dojka MA, Hugenholtz P, Haack SK, Pace NR (1998) Microbial diversity in a hydrocarbon- and chlorinated-solvent-contaminated aquifer undergoing intrinsic bioremediation. Appl Environ Microbiol 64:3869–3877

Dollard MA, Billard P (2003) Whole-cell bacterial sensors for the monitoring of phosphate bioavailability. J Microbiol Methods 55:221–229

Drahos SJ, Hemming BC, McPherson S (1986) Tracking recombinant organisms in the environment: β-galactosidase as a selectable non-antibiotic marker for fluoresecent pseudomonads. Bio/Technology 4:439–444

Edwards U, Rogall T, Blöcker H, Emde M, Böttger EC (1990) Isolation and direct sequencing of entire genes. Characterization of a gene coding for 16S ribosomal RNA. Nucleic Acids Res 17:7843–7853

Fitts R, Diamond M, Hamilton C, Neri M (1983) DNA–DNA hybridization assay for detection of *Salmonella* spp. in foods. Appl Environ Microbiol 46:1146–1151

Flavier AB, Ganova Raeva LM, Schell MA, Denny TP (1997) Hierarchical autoinduction in *Ralstonia solanacearum*: control of acyl-homoserine lactone production by a novel autoregulatory system responsive to 3-hydroxypalmitic acid methyl ester. J Bacteriol 179:7089–7097

Fredrickson JK, Bezdicek DF, Brockman FJ, Li SW (1988) Enumeration of Tn5 mutant bacteria in soil by using a most-probable-number-DNA hybridization procedure and antibiotic resistance. Appl Environ Microbiol 54:446–453

Giovannoni SJ, Britschgi TB, Moyer CL, Field KG (1990) Genetic diversity in Sargasso Sea bacterioplankton. Nature 345:60–63

Grunstein M, Hogness DS (1975) Colony hybridization: a method for the isolation of cloned DNAs that contain a specific gene. Proc Natl Acad Sci USA 72:3961–3965

Hakkila K, Maksimow M, Karp M, Virta M (2002) Reporter genes *lucFF*, *luxCDABE*, *gfp*, and *dsred* have different characteristics in whole-cell bacterial sensors. Anal Biochem 301:235–242

Hansen LH, Sorensen SJ (2000) Versatile biosensor vectors for detection and quantification of mercury. FEMS Microbiol Lett 193:123–127

Head IM, Saunders JR, Pickup RW (1998) Microbial evolution, diversity, and ecology: a decade of ribosomal RNA analysis of uncultivated microorganisms. Microb Ecol 35:1–21

Hill WE, Payne WL, Aulisio CCG (1983) Detection and enumeration of virulent *Yersinia enterolitica* in food by DNA colony hybridization. Appl Environ Microbiol 46:636–641

Hoffmann A, Thimm T, Dröge M, Moore ERB, Munch JC, Tebbe CC (1998) Intergeneric transfer of conjugative and mobilizable plasmids harbored by *Escherichia coli* in the gut of the soil microarthropod *Folsomia candida* (Collembola). Appl Environ Microbiol 64:2652–2659

Holben WE, Jansson JK, Chelm BK, Tiedje JM (1988) DNA probe method for the detection of specific microorganisms in the soil bacterial community. Appl Environ Microbiol 54:703–711

Holden PA, LaMontagne MG, Bruce AK, Miller WG, Lindow SE (2002) Assessing the role of *Pseudomonas aeruginosa* surface-active gene expression in hexadecane biodegradation in sand. Appl Environ Microbiol 68:2509–2518

Huber R, Burggraf S, Mayer T, Barns SM, Rossnagel P, Stetter KO (1995) Isolation of a hyperthermophilic archeum predicted by in situ RNA analysis. Nature 376:57–58

Hugenholtz P, Goebel BM, Pace NR (1998) Impact of culture-independent studies on the emerging phylogenetic view of bacterial diversity. J Bacteriol 180:4765–4774

Ivask A, Virta M, Kahru A (2002) Construction and use of specific luminescent recombinant bacterial sensors for the assessment of bioavailable fraction of cadmium, zinc, mercury and chromium in the soil. Soil Biol Biochem 34:1439–1447

Jacob GBA, Colin DCB, Paton GI (2001) Comparison of response of six different luminescent bacterial bioassays to bioremediation of five contrasting oils. J Environ Monit 3:404–410

Jaeger CH, Lindow SE, Miller S, Clark E, Firestone MK (1999) Mapping of sugar and amino acid availability in soil around roots with bacterial sensors of sucrose and tryptophan. Appl Environ Microbiol 65:2685–2690

Jansson JK (2003) Marker and reporter genes: illuminating tools for environmental microbiologists. Curr Opin Microbiol 6:310–316

Jansson JK, van Elsas JD, Bailey MJ (2000) Tracking genetically engineered microorganisms. Landes Bioscience, Georgetown, Texas

Killham K, Yeomans C (2001) Rhizosphere carbon flow measurement and implications: from isotopes to reporter genes. Plant Soil 232:91–96

Kostrzynska M, Leung KT, Lee H, Trevors JT (2002) Green fluorescent protein-based biosensor for detecting SOS-inducing activity of genotoxic compounds. J Microbiol Methods 48:43–51
Kowalchuk GA, Stienstra AW, Heilig GHJ, Stephen JR, Woldendorp JW (2000) Changes in the community structure of ammonia-oxidizing bacteria during secondary succession of calcareous grasslands. Environ Microbiol 2:99–110
Kluepfel DA, Kline EL, Skipper HD, Hughes TA, Gooden DT, Drahos DJ, Barry GF, Hemming BC, Brandt EJ (1991) The release and tracking of genetically engineered bacteria in the environment. Phytopathol 81:348–352
Leveau JHJ, Lindow SE (2002) Bioreporters in microbial ecology. Curr Opin Microbiol 5:259–265
Liesack W, Stackebrandt E (1992) Occurrence of novel groups of the domain Bacteria as revealed by analysis of genetic material isolated from an Australian terrestrial environment. J Bacteriol 174:5072–5078
Liu WT, Marsh TL, Cheng H, Forney LJ (1997) Characterization of microbial diversity by determining terminal restriction fragment length polymorphisms of genes encoding 16S rRNA. Appl Environ Microbiol 63:4516–4522
Loper JE, Lindow SE (1994) A biological sensor for iron available to bacteria in their habitats on plant surfaces. Appl Environ Microbiol 60:1934–1941
Lueders T, Friedrich MW (2003) Evaluation of PCR amplification bias by terminal restriction fragment length polymorphism analysis of small-subunit rRNA and *mcrA* genes by using defined template mixtures of methanogenic pure cultures and soil DNA extracts. Appl Environ Microbiol 69:320–326
Lueders T, Chin KJ, Conrad R, Friedrich M (2001) Molecular analyses of methyl-coenzyme M reductase alpha-subunit (*McrA*) genes in rice field soil and enrichment cultures reveal the methanogenic phenotype of a novel archaeal lineage. Environ Microbiol 3:194–204
Luton PE, Wayne JM, Sharp RJ, Riley PW (2002) The *mcrA* gene as an alternative to 16S rRNA in the phylogenetic analysis of methanogen populations in landfill. Microbiol SGM 148:3521–3530
Manefield M, Turner SL (2002) Quorum sensing in context: out of molecular biology and into microbial ecology. Microbiol SGM 148:3762–3764
Manefield M, Whiteley AS, Griffiths RI, Bailey MJ (2002) RNA stable isotope probing, a novel means of linking microbial community function to Phylogeny. Appl Environ Microbiol 68:5367–5373
Marchesi JR, Weightman AJ (2003) Comparing the dehalogenase gene pool in cultivated alpha-halocarboxylic acid-degrading bacteria with the environmental metagene pool. Appl Environ Microbiol 69:4375–4382
Metcalfe AC, Krsek M, Gooday GW, Prosser JI, Wellington EMH (2002) Molecular analysis of a bacterial chitinolytic community in an upland pasture. Appl Environ Microbiol 68:5042–5050
Miller WG, Brandl MT, Quinones B, Lindow SE (2001) Biological sensor for sucrose availability: relative sensitivities of various reporter genes. Appl Environ Microbiol 67:1308–1317
Muyzer G, de Waal EC, Uitterlinden AG (1993) Profiling of complex microbial populations by denaturing gradient gel electrophoresis analysis of polymerase chain reaction-amplified genes coding for 16S rRNA. Appl Environ Microbiol 59:695–700
Muyzer G, Brinkhoff T, Nübel U, Santegoeds C, Schäfer H, Wawer C (1998) Denaturing gradient gel electrophoresis (DGGE) in microbial ecology. In: Akkermans ADL, van Elsas JD, de Bruijn FJ (eds) Molecular microbial ecology manual. Kluwer, Dordrecht, pp 1–27
Norton JM, Alzerreca JJ, Suwa Y, Klotz MG (2002) Diversity of ammonia monooxygenase operon in autotrophic ammonia-oxidizing bacteria. Arch Microbiol 177:139–149

Ogram A, Sayler GS, Barkay T (1987) DNA extraction and purification from sediments. J Microbiol Methods 7:57–66
Okuta A, Ohnishi K, Harayama S (1998) PCR isolation of catechol 2,3-dioxygenase gene fragments from environmental samples and their assembly into functional genes. Gene 212:221–228
Olsen GJ, Lane DJ, Giovannoni SJ, Pace NR, Stahl DA (1986) Microbial ecology and evolution: a ribosomal RNA approach. Annu Rev Microbiol 40:337–365
Otsuka J, Terai G, Nakano T (1999) Phylogeny of organisms investigated by the base-pair changes in the stem regions of small and large ribosomal subunit RNAs. J Mol Evol 48:218–235
Pace N, Stahl DA, Lane DJ, Olsen GJ (1986) The analysis of natural microbial populations by rRNA sequences. Adv Microb Ecol 9:1–55
Paerl HW, Steppe TF (2003) Scaling up: the next challenge in environmental microbiology. Environ Microbiol 5:1025–1038
Peplies J, Glökner FO, Amann R (2003) Optimization strategies for DNA microarray-based detection of bacteria with 16S rRNA-targeting oligonucleotide probes. Appl Environ Microbiol 69:1397–1407
Philippot L, Piutti S, Martin-Laurent F, Hallet S, Germon JC (2002) Molecular analysis of the nitrate-reducing community from unplanted and maize-planted soils. Appl Environ Microbiol 68:6121–6128
Porteous F, Killham K, Meharg A (2000) Use of a *lux*-marked rhizobacterium as a biosensor to assess changes in rhizosphere C flow due to pollutant stress. Chemosphere 41:1549–1554
Prachayasittikul V, Ayudhya CIN, Bulow L (2001) Lighting *E. coli* cells as biological sensors for $Cd^{2(+)}$. Biotechnol Lett 23:1285–1291
Prosser JI (2002) Molecular and functional diversity in soil micro-organisms. Plant Soil 244:9–17
Radajewski S, Ineson P, Parekh NR, Murrell JC (2000) Stable-isotope probing as a tool in microbial ecology. Nature 403:646–649
Ranjard L, Richaume A, Jocteur-Monrozier L, Nazaret S (1997) Response of soil bacteria to Hg(II) in relation to soil characteristics and cell location. FEMS Microbiol Ecol 24:321–331
Rasmussen LD, Sorensen SJ, Turner RR, Barkay T (2000) Application of a *mer-lux* biosensor for estimating bioavailable mercury in soil. Soil Biol Biochem 32:639–646
Ripp S, Nivens DE, Ahn Y, Werner C, Jarrell J, Easter JP, Cox CD, Burlage RS, Sayler GS (2000) Controlled field release of a bioluminescent genetically engineered microorganism for bioremediation process monitoring and control. Environ Sci Technol 34:846–853
Rossbach S, Kukuk ML, Wilson TL, Feng SF, Pearson MM, Fisher MA (2000) Cadmium-regulated gene fusions in *Pseudomonas fluorescens*. Environ Microbiol 2:373–382
Saiki RK, Scharf S, Faloona F, Mullis KB, Horn GT, Erlich HA, Arnheim N (1985) Enzymatic amplification of β-globin genomic sequences and restriction site analysis for the diagnosis of sickle-cell anemia. Science 230:1350–1354
Sayler GS, Ripp S (2000) Field applications of genetically engineered microorganisms for bioremediation processes. Curr Opin Biotechnol 11:286–289
Sayler GS, Shields MS, Tedford E, Breen A, Hooper S, Sirotkin K, Davis JW (1985) Application of DNA–DNA colony hybridization to the detection of catabolic genotypes in environmental samples. Appl Environ Microbiol 49:1295–1303
Schwieger F, Tebbe CC (1998) A new approach to utilize PCR-single-strand-conformation polymorphism for 16S rRNA gene-based microbial community analysis. Appl Environ Microbiol 64:4870–4876
Selbitschka W, Pühler A, Simon R (1992) The construction of *recA* deficient containment. *Rhizobium meliloti* and *R. leguminosarum* strains marked with *gusA* or *luc* cassettes for use in risk assessment studies. Mol Ecol 1:9–19

Shaw PD, Ping G, Daly SL, Cha C, Cronan JE, Rinehart KL, Farrand SK (1997) Detecting and characterizing *N*-acyl-homoserine lactone signal molecules by thin-layer chromatography. Proc Natl Acad Sci USA 94:6036–6041

Simpson ML, Sayler GS, Applegate BM, Ripp S, Nivens DE, Paulus MJ, Jellison GE (1998) Bioluminescent-bioreporter integrated circuits form novel whole-cell biosensors. Trends Biotechnol 16:332–338

Small J, Call DR, Brockman FJ, Straub TM, Chandler DP (2001) Direct detection of 16S rRNA in soil extracts by using oligonucleotide microarrays. Appl Environ Microbiol 67:4708–4716

Smalla K, Krögerrecklenfort E, Heuer H, Dejonghe W, Top E, Osborn M, Niewint J, Tebbe C, Barr M, Bailey M, Greated A, Thomas C, Turner S, Young P, Nikolakopoulou D, Karagouni A, Wolters A, van Elsas JD, Dronen K, Sandaa R, Borin S, Brabhu J, Grohmann E, Sobecky P (2000) PCR-based detection of mobile genetic elements in total community DNA. Microbiology UK 146:1256–1257

Smit E, van Elsas JD, van Veen JA, Devos WM (1991) Detection of plasmid transfer from *Pseudomonas fluorescens* to indigenous bacteria in soil by using bacteriophage Phi-R2f for donor counterselection. Appl Environ Microbiol 57:3482–3488

Sorensen J, Jensen LE, Nybroe O (2001) Soil and rhizosphere as habitats for *Pseudomonas* inoculants: new knowledge on distribution, activity and physiological state derived from micro-scale and single-cell studies. Plant Soil 232:97–108

Stahl DA, Lane DJ, Olsen GJ, Pace NR (1984) Analysis of hydrothermal vent-associated symbionts by ribosomal RNA sequences. Science 224:409–411

Stahl DA, Lane DJ, Olsen GJ, Pace NR (1985) Characterization of a Yellowstone hot spring microbial community by 5S rRNA sequences. Appl Environ Microbiol 49:1379–1384

Steffan RJ, Goksoyr J, Bej AK, Atlas RM (1988) Recovery of DNA from soils and sediments. Appl Environ Microbiol 54:2908–2915

Suarez A, Güttler A, Strätz M, Staendner LH, Timmis KN, Guzman CA (1997) Green fluorescent protein-based reporter systems for genetic analysis of bacteria including monocopy applications. Gene 196:69–74

Suzuki MT, Giovannoni SJ (1996) Bias caused by template annealing in the amplification of mixtures of 16S rRNA genes by PCR. Appl Environ Microbiol 62:625–630

Tebbe CC (2000) Use of *luc*-tagged genetically modified microorganisms (GMMs) to study rhizobial ecology in soil columns, field lysimeters and field plots. In: Jansson JK, van Elsas JD, Bailey MJ (eds) Tracking genetically engineered microorganisms. Landes Bioscience, Georgetown, Texas, pp 127–137

Tebbe CC, Vahjen W (1993) Interference of humic acids and DNA extracted directly from soil in detection and transformation of recombinant-DNA from bacteria and a yeast. Appl Environ Microbiol 59:2657–2665

Tebbe CC, Schmalenberger A, Peters S, Schwieger F (2001) Single-strand conformation polymorphism (SSCP) for microbial community analysis. In: Rochelle PA (ed) Environmental molecular microbiology: protocols and applications. Horizon Scientific, Wymondham, UK, pp 161–175

Thimm T, Hoffmann A, Fritz I, Tebbe CC (2001) Contribution of the earthworm *Lumbricus rubellus* (Annelida, Oligochaeta) to the establishment of plasmids in soil bacterial communities. Microbial Ecol 41:341–351

Tiedje JM, Asuming-Brempong S, Nusslein K, Marsh TL, Flynn SJ (1999) Opening the black box of soil microbial diversity. Appl Soil Ecol 13:109–122

Tom-Petersen A, Hosbond C, Nybroe O (2001) Identification of copper-induced genes in *Pseudomonas fluorescens* and use of a reporter strain to monitor bioavailable copper in soil. FEMS Microbiol Ecol 38:59–67

Torsvik V, Ovreas L (2002) Microbial diversity and function in soil: from genes to ecosystems. Curr Opin Microbiol 5:240–245

Tourasse NJ, Gouy M (1997) Evolutionary distances between nucleotide sequences based on the distribution of substitution rates among sites as estimated by parsimony. Mol Biol Evol 14:287–298

Tunlid A (1999) Molecular biology: a linkage between microbial ecology, general ecology and organismal biology. Oikos 85:177–189

Van Elsas JD, Trevors JT (1990) Plasmid transfer to indigenous bacteria in soil and rhizospheres: problems and perspectives. In: Fry JC, Day MJ (eds) Bacterial genetics in natural environments. Chapman and Hall, London, pp 188–199

Van Overbeek LS, van Elsas JD (1995) Root exudate-induced promoter activity in *Pseudomonas fluorescens* mutants in the wheat rhizosphere. Appl Environ Microbiol 61:890–898

Watanabe K, Teramoto M, Futamata H, Harayama S (1998) Molecular detection, isolation, and physiological characterization of functionally dominant phenol-degrading bacteria in activated sludge. Appl Environ Microbiol 64:4396–4402

Weilbo J, Skorupska A (2001) Construction of improved vectors and cassettes containing *gusA* and antibiotic resistance genes for studies of transcriptional activity and bacterial localization. J Microbiol Methods 45:197–205

Weller R, Ward DM (1989) Selective recovery of 16S rRNA sequences from natural microbial communities in the form of cDNA. Appl Environ Microbiol 55:1818–1822

Wellington EMH, Berry A, Krsek M (2003) Resolving functional diversity in relation to microbial community structure in soil: exploiting genomics and stable isotope probing. Curr Opin Microbiol 6:295–301

Whitman WB, Coleman DC, Wiebe WJ (1998) Prokaryotes: the unseen majority. Proc Natl Acad Sci USA 95:6578–6583

Widmer F, Shaffer BT, Porteous LA, Seidler RJ (1999) Analysis of *nifH* gene pool complexity in soil and litter at a Douglas fir forest site in the Oregon Cascade Mountain Range. Appl Environ Microbiol 65:374–380

Willardson BM, Wilkins JF, Rand TA, Schupp JM, Hill KK, Keim P, Jackson PJ (1998) Development and testing of a bacterial biosensor for toluene-based environmental contaminants. Appl Environ Microbiol 64:1006–1012

Williamson N, Brian P, Wellington EMH (2000) Molecular detection of bacterial and streptomycete chitinases in the environment. Antonie Van Leeuwenhoek 78:315–321

Woese C, Fox GE (1977) Phylogenetic structure of the prokaryotic domain: the primary kingdoms. Proc Natl Acad Sci USA 74:5088–5090

Yinon J (2002) Field detection and monitoring of explosives. Trac Trends Anal Chem 21:292–301

18 Assessing Functions of Soil Microbes with Isotopic Measurements

Erik A. Hobbie[1]

1 Introduction

The difficulty of experimental manipulations, observations, and measurements in the cryptic world of the soil has long hindered studies of microbial functioning in this environment. One promising tool to explore soil microbial functioning, stable isotope measurements, is increasingly used to study links between microbes and biogeochemical processes. Such measurements rely on calculating ratios of heavy to light isotopes of many key biological elements (e.g., ^{13}C:^{12}C, ^{15}N:^{14}N, ^{18}O:^{16}O, ^{2}H:^{1}H, and ^{34}S:^{32}S). Studies can be generally classified into two main types, those using natural abundance levels of isotopes in microbes or microbially produced compounds, and those in which compounds or substances artificially enriched in one or more of the heavy isotopes are applied (tracer studies), and the subsequent fate of the isotopic label followed into different ecosystem components, including microbes. Natural abundance and tracer studies can be used at different levels of resolution, ranging from the whole organism down to specific compound classes (e.g., lipids), specific compounds, or even DNA or RNA unique to a single species. Analyses using gas chromatography on mixtures of volatile compounds linked to isotope ratio mass spectrometry have been particularly widely applied (reviewed in Boschker and Middleburg 2002). The use of isotopic techniques has increased dramatically in the past 15 years, and has been supported by a variety of technological advances, including faster analyses with the widespread adoption of continuous flow isotope ratio mass spectrometry, improvements in our ability to measure hydrogen and oxygen isotope ratios, and steady increases in the quality and variety of compound-specific measurements possible. In the following sections, I focus on the use of natural abundance measurements of carbon and nitrogen isotopes to address various research topics in soil microbial ecology, with some discussion of tracer experiments. Useful reviews on natural abundance measurements include Högberg (1997), Dawson et al. (2002), Hayes (2002) and Werner and Schmidt (2002).

[1]Complex Systems Research Center, University of New Hampshire, Durham, New Hampshire 03824, USA, e-mail: Erik.Hobbie@unh.edu, Tel: +1-603-8623581; Fax: +1-603-8620188

Soil Biology, Volume 3
Microorganisms in Soils: Roles in Genesis and Functions
(ed. by F. Buscot and A. Varma)

2
Natural Abundance Measurements

Most chemical and biochemical processes favor the initial incorporation of the lighter isotope in the product, leaving the remaining substrate enriched in the heavy isotope. Such kinetic isotope effects can be contrasted with equilibrium isotope effects, or the isotopic partitioning of elements between components at equilibrium. Equilibrium isotope effects generally result in the heavy isotope accumulating in the compound with the stronger chemical bonds (Bigeleisen 1965). The partitioning of isotopes in reactions is termed "isotopic fractionation"; the magnitude of isotopic fractionation depends on the elements involved and the specific reaction mechanism. Isotopically fractionating processes, therefore, result in variation in the isotopic ratios between the substrate and the product. These ratios depend on the isotopic ratio of the substrate, the proportion of substrate transformed to product, and whether the system is open or closed (Fig. 1). Natural abundance studies use small isotopic differences among different ecosystem pools and compounds to understand the sources and fluxes of many of the most biologically important elements. Because differences in isotopic ratios are small, they are measured using the "δ" notation, in deviations in parts per mille (‰) from a standard ratio, according to (18.1).

$$\delta^{n}\mathrm{X}(‰) = (R_{\mathrm{sample}}/R_{\mathrm{standard}} - 1) \times 1000 \tag{18.1}$$

In (18.1), n equals the atomic mass of the heavy isotope, X is the symbol for the element of interest, and R equals the molar abundance of the heavy isotope divided by the light isotope (e.g., $^{13}C/^{12}C$). The isotopic standard for carbon is Vienna PeeDee Belemnite ($^{13}C/^{12}C = 0.0112372$), for nitrogen,

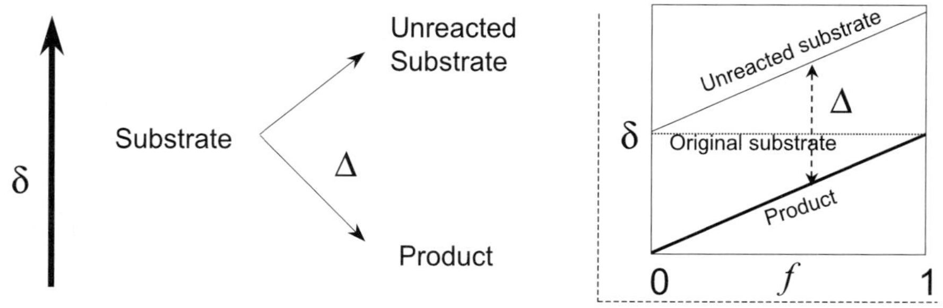

Fig. 1. Isotopic composition of substrate and product in an open system depends on the fraction (f) of substrate transformed to the product and the isotopic fractionation (Δ) of the reaction. For discussion of the somewhat more complicated case of a closed system, see Hayes (2002)

atmospheric N_2 ($^{15}N/^{14}N = 0.0036765$), for sulfur, Canyon Diablo troilite ($^{34}S/^{32}S = 0.0450045$), and for hydrogen and oxygen, Vienna Standard Mean Ocean Water (V-SMOW, $D/H = 0.00015576$, $^{18}O/^{16}O = 0.00200520$) (Hoefs 1997). Isotopic values for carbon generally range from 0‰ (carbonates) to −50‰ (methane in some systems), with most $\delta^{13}C$ values in systems dominated by plants of the C_3 photosynthetic pathway ranging from −20 to −35‰. In systems dominated by plants of the C_4 photosynthetic pathway, values range from −10 to −20‰. For $\delta^{15}N$, pools in terrestrial ecosystems are usually between 20 and −10‰. In comparisons among samples, samples with more of the heavy isotope are commonly referred to as isotopically enriched, or heavy, and samples with less of the heavy isotope are referred to as isotopically depleted, or light. Isotopic fractionation (Δ) for reactions in open systems can be calculated based on the isotopic signature of the source and product according to (18.2).

$$\Delta(\delta^n X_{source} - \delta^n X_{product})/(1 + \delta^n X_{product}) \tag{18.2}$$

2.1 Fungi

Fungi can be divided into two main life history strategies depending on their carbon source. Saprotrophic fungi obtain their carbon from the decay of dead organic matter, whereas mycorrhizal fungi form symbiotic associations with plants in which plant-supplied sugars are exchanged for nutrients obtained by fungi from the soil. Mycorrhizal fungi can be further divided into several types, of which the two most important are arbuscular mycorrhizal (AM) fungi (belonging to the Glomerales), symbiotic on most herbaceous plants and some trees, and ectomycorrhizal fungi, associated with many of the dominant tree families of temperate and boreal regions (e.g., Pinaceae, Betulaceae, Salicaceae, and Fagaceae), but also some families distributed in the tropics (e.g., Dipterocarpaceae and Myrtaceae). Higher fungi in the phyla Basidiomycota and Ascomycota have been tempting targets for isotopic measurements in recent years because many produce large, conspicuous fruiting bodies that often can be identified to species. The production of large fruiting bodies has allowed researchers to analyze isotopes at the species level in situ, without resorting to techniques such as DNA- or RNA-specific analyses. Such isotopic analyses are providing interesting insights into patterns of carbon and nitrogen cycling by both ectomycorrhizal and saprotrophic fungi in these phyla.

2.1.1 Carbon Isotopes

One of the clearest patterns emerging from field studies is that fungi are enriched in ^{13}C relative to their substrates. Saprotrophic fungi are 3–4‰ enriched in ^{13}C relative to woody or litter substrates (Kohzu et al. 1999; Hobbie et al. 2001), and ectomycorrhizal fungi are enriched by 1–5‰ in ^{13}C relative to current-year foliage (Hobbie et al. 1999; Högberg et al. 1999a; Kohzu et al. 1999). Because specific plant tissues or plant species often differ in ^{13}C content (Brugnoli and Farquhar 2000), $\delta^{13}C$ measurements could be used to determine the probable carbon sources of fungi. For example, in a study in a mixed birch-pine forest, Högberg et al. (1999a) used the fidelity between the $\delta^{13}C$ of ectomycorrhizal fungi and their putative hosts to determine that mycorrhizal fungi specific to conifer or birch hosts were 1–3‰ enriched in ^{13}C relative to plant host foliage, and that mycorrhizal fungi of broad host specificity primarily obtained carbon from the dominant, overstory Scots pine (Fig. 2).

A second clear pattern is that saprotrophic fungi are several per mille (‰) enriched in ^{13}C relative to co-occurring ectomycorrhizal fungi (Fig. 2;

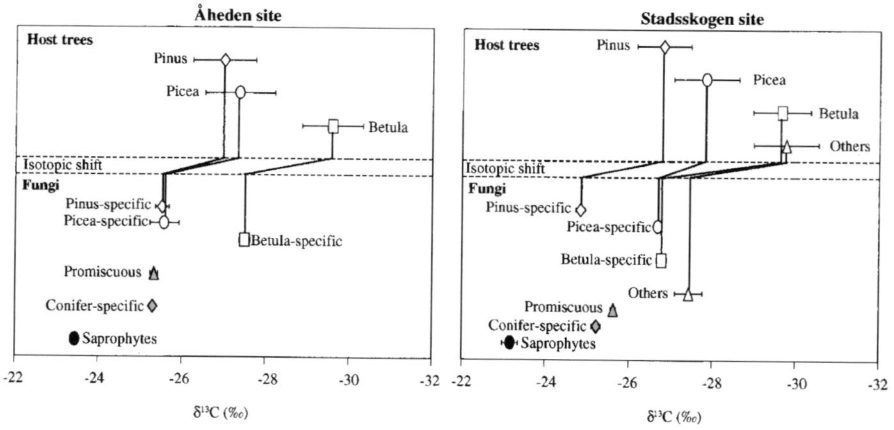

Fig. 2. Natural abundance of ^{13}C ($\delta^{13}C$) of fungal fruit bodies and of host trees of ectomycorrhizal fungi in mixed temperate forests at two sites in Sweden, Åheden and Stadsskogen. Promiscuous fungi are nonhost-specific ectomycorrhizal fungi, whereas conifer-specific fungi could associate with pine (*Pinus sylvestris*) or spruce (*Picea abies*). Pine is the dominant tree at both sites, followed in abundance by spruce and birch (*Betula pendula*). Other tree species at Stadsskogen included aspen, willow, and alder (grouped as *others*). Ectomycorrhizal fungi are arranged in groups according to their host specificity and connected to their hosts by *lines*. Values for saprotrophic fungi (*saprophytes*) are also shown. *Bars* show standard deviations of single observations. Replicates are individuals in the case of host trees; in the case of fungi, replicates are species. (Modified from Högberg et al. 1999a, with permission)

Hobbie et al. 1999; Kohzu et al. 1999; Henn and Chapela 2001). Such measurements are, therefore, useful in determining whether fungi of unknown life history status are mycorrhizal when measured along with other ecosystem pools (Högberg et al. 1999a; Hobbie et al. 2001).

Recently, independent verification of mycorrhizal or saprotrophic status in fungi has been obtained using radiocarbon (^{14}C) measurements (Hobbie et al. 2002). The ^{14}C content of atmospheric CO_2 started to increase in the early 1950s as a result of ^{14}C created during thermonuclear testing. In this study, the declining ^{14}C content of atmospheric CO_2 since the Nuclear Test Ban Treaty of 1963 among the Soviet Union, Great Britain, and the United States was used to estimate the age of carbon incorporated by mycorrhizal or saprotrophic fungi. Mycorrhizal fungi closely tracked current-year foliage or atmospheric CO_2 in ^{14}C content, whereas saprotrophic fungi ranged in apparent age from 3 to 50+ years. It appeared possible to assign several taxa of unknown life history strategy to mycorrhizal or saprotrophic status based solely on ^{14}C measurements (Fig. 3). Although radiocarbon mea-

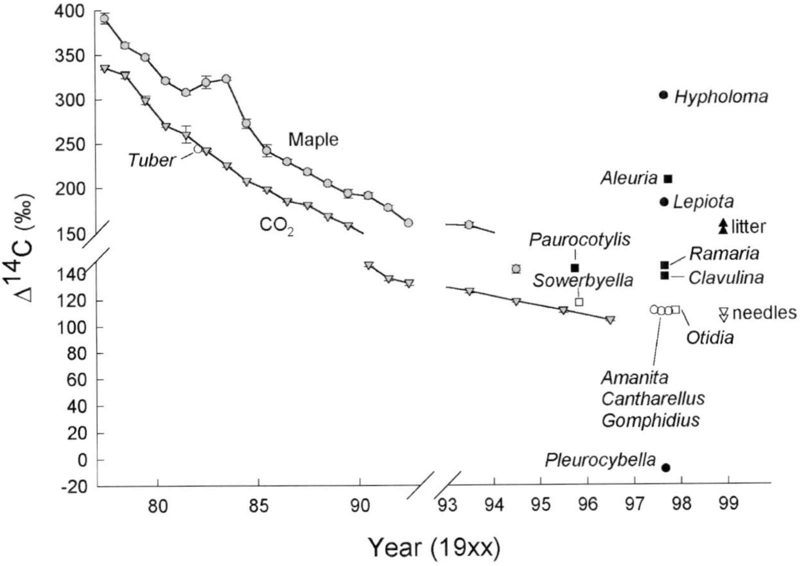

Fig. 3. Measured values for radiocarbon $\Delta^{14}C$ in samples of fungi, needles, and litter. Growing-season mean $\Delta^{14}C$ for CO_2 at Schauinsland, Germany (1977–1996), and for maple leaves in Quebec (1977–1993) are also shown +SE. Note *x*- and *y-axis* scale changes of breaks at 1993 and 150‰, respectively. *Open triangles* needles, *filled triangles* litter, *open circles* mycorrhizal fungi, *open squares* suspected mycorrhizal fungi, *filled circles* saprotrophic fungi, *filled squares* suspected saprotrophic fungi. Plotted dates for the mycorrhizal fungi *Amanita, Cantharellus*, and *Gomphidius* and the suspected mycorrhizal fungus *Otidia* collected in September 1997 from Woods Creek, Oregon, USA (four total) were slightly altered for clarity. (Modified from Hobbie et al. 2002)

surements are expensive compared to stable isotope measurements (about US$ 300 versus US$ 15 per sample), the technique could potentially address several issues about carbon cycling in fungi, such as incorporation of carbon derived from organic nitrogen uptake that previously required tracer isotope experiments. Uptake of organic nitrogen should theoretically result in a protein pool that reflects the ^{14}C age of soil amino acids rather than the ^{14}C age of atmospheric CO_2.

Culture studies are used to examine fungal $\delta^{13}C$ patterns under controlled conditions. When grown without water stress in an inert culture media (perlite), mycorrhizal hyphae of *Suillus luteus* and *Thelephora terrestris* were 3.2 ± 0.2‰ enriched relative to host foliage of Scots pine and $1.6 + 0.1$‰ enriched in ^{13}C relative to roots (Hobbie and Colpaert, in press). The enrichment of fungi relative to roots is similar to a $\sim$ 2‰ enrichment of cellulose relative to bulk tissue in needles and wood of *Picea abies* (Gleixner et al. 1993), and similar to a 2‰ enrichment in ^{13}C of ectomycorrhizal fungi relative to fine roots in a field study (Hobbie et al. 1999). Taken together, these studies suggest that at constant levels of water stress, ectomycorrhizal fungi should have similar ^{13}C enrichments to root cellulose or transferred sugars. The $\delta^{13}C$ of fixed carbon increases with water stress (Brugnoli and Farquhar 2000), and is, therefore, generally higher in carbon fixed in summer than in carbon fixed in spring or fall. Whether mycorrhizal fungi that fruit at different seasons have different $\delta^{13}C$ values has not yet been examined, but could provide information about the residence time of fungal carbon stores.

In contrast to ^{13}C enrichment patterns in ectomycorrhizal fungi relative to plant hosts, spores and hyphae of AM fungi are 1–4‰ depleted in ^{13}C relative to plant hosts (Nakano et al. 1999; Staddon et al. 1999). This was attributed to the high content of ^{13}C-depleted lipids in AM fungi, particularly in AM spores, as lipids can be up to 10‰ depleted in ^{13}C relative to microbial biomass (Hayes 2002). An additional factor that could explain ^{13}C patterns in AM fungi is the extensive transport of lipids from intracellular hyphae to extracellular hyphae and spores (Pfeffer et al. 1999). Using heterotrophically grown seedlings, Luo and Sternberg (1994) demonstrated that gluconeogenesis from lipids results in cellulose 1‰ depleted in ^{13}C relative to the substrate, whereas gluconeogenesis from starch results in cellulose 1‰ enriched in ^{13}C. The potential effects on isotopic patterns of growth on lipids versus growth on carbohydrates could also be inferred from a culture study with the yeast *Candida lipolytica*. Growth on glucose resulted in no depletion in ^{13}C of biomass or carbohydrates relative to the substrate, whereas growth on alkanes resulted in a ^{13}C depletion relative to the substrate of 2.2‰ for biomass and 4.7‰ for carbohydrates (Zyakun 1996). Metabolic processes should be similar for growth on alkanes as for growth on lipids. If these results also apply to AM fungi, then transport of lipids in AM fungi followed by subsequent gluconeogenesis (Pfeffer et al. 1999) could result in ^{13}C-depleted fungal carbohydrates in AM fungi relative to plant-supplied sugars, and could even result in ^{13}C-depleted fungal carbohydrates in AM fungi relative to transported lipids.

Because plant lignin is about 4‰ depleted in ^{13}C relative to cellulose and hemicellulose (Benner et al. 1987), it is probable that part of the enrichment of saprotrophic fungi relative to substrates arises from preferential incorporation of carbon derived from the decomposition of relatively labile, ^{13}C-enriched carbohydrates, such as cellulose and hemicellulose. In a detailed study of compound-specific $\delta^{13}C$ patterns, Gleixner et al. (1993) determined that both white-rot fungi (with known ligninolytic abilities) and soft-rot fungi (without ligninolytic abilities) were equally (about 1.8‰) enriched relative to wood cellulose, implying little incorporation of lignin-derived carbon by either fungal type. This interpretation agrees with prior studies of growth of saprotrophic fungi on synthetic, ^{14}C-labeled lignins, in which fungal metabolites such as veratryl alcohol and CO_2 were ^{14}C-labeled, but fungal biomass was not (discussed in Jennings 1995). These studies, therefore, imply that white-rot fungi do not incorporate lignin-derived carbon, but that instead their extensive ligninolytic capabilities are used solely to improve access to cellulose and other carbohydrates. We conclude that saprotrophic fungi are about 2‰ enriched in ^{13}C relative to carbohydrates of their colonized substrates. Since carbohydrates are the probable carbon source for most fungi, the ^{13}C enrichment during metabolism by wood decay fungi is also probably about 2‰.

$\delta^{13}C$ signatures of fungi appear primarily controlled by source $\delta^{13}C$ plus variable ^{13}C enrichments during metabolism. This enrichment appears high in saprotrophic fungi feeding on wood, low in ectomycorrhizal fungi, and negative in AM fungi. Examining ^{13}C patterns in microbes grown on simple carbohydrates versus complex substrates provides some insight into the causes of these differing enrichments (Table 1). In culture studies on simple substrates, microbial biomass is less than 1‰ enriched in ^{13}C relative to carbon sources, independent of microbe type. Microbial biomass is most enriched in ^{13}C relative to sources when complex sources are used, such as with colonization on wood or growth on litter-derived leachates (Table 1). Bacteria grown on glucose are generally 0–2‰ depleted in ^{13}C relative to glucose, with the notable exceptions of a 3‰ enrichment of *Pseudomonas aeruginosa* and a 1.5‰ enrichment of *Nostoc* sp. (Table 1).

Microbial efficiency, or the proportion of supplied carbon that is transformed into biomass, seems linked to ^{13}C enrichments during metabolism (Table 1). Fungi growing on glucose or sucrose, with an efficiency of $\sim$ 60% (Lundberg et al. 2001; also calculated from Henn and Chapela 2000), appear less enriched in ^{13}C relative to substrate carbon than fungi feeding on

Table 1. Isotopic fractionation between microbes and substrates versus microbial efficiency. *Trametes* and *Candida* are fungi, *Chlorella* is an alga, other organisms are bacteria. Standard errors are given when available

Substrate	Organism	Δ_{S-B} (‰)	Efficiency (%)	Reference
Fagus wood	*Trametes versicolor*	−3.5±0.5	15–21	Kohzu et al. (1999)
Wood	Decay fungi (five taxa)	−3.5±0.3		Hobbie et al. (2001)
Wood	Decay fungi	−3.1, −1.8[a]		Gleixner et al. (1993)
Quercus[b]	Bacteria[c]	−1.9		Coffin et al. (1989)
Spartina[b]	Bacteria[c]	−1.7		Coffin et al. (1989)
Marsh plants[b]	Bacteria[c]	−0.2±0.5		Créach et al. (1997)
Marsh plants[b]	*Vibrio alginolyticus*	−2.3±0.5		Créach et al. (1997)
Sucrose[d]	Soil fungi (12 taxa)	−0.3±0.4		Hobbie et al. (2003)
Sucrose	Fungi (three taxa)	0.0±0.2	54–62	Henn and Chapela (2000)
n-Alkanes	*Candida lipolytica*	2.2		Zyakun (1996)
Glucose	*Candida lipolytica*	0.1		Zyakun (1996)
Glucose	Fungi (four taxa)	1.0±0.7		Hobbie et al.
Glucose	*Chlorella pyrenoidosa*	0.1		Abelson and Hoering (1961)
Glucose	Bacteria[b]	0.8		Coffin et al. (1989)
Glucose	*Escherichia coli*	1.9		Abelson and Hoering (1961)
Glucose	*E. coli*	−0.3		Monson and Hayes (1982)
Glucose	*E. coli*	0.6	45	Blair et al. (1985)
Glucose	*E. coli*	1.7		Zhang et al. (2002)
Glucose	*E. coli*	2.3		MacGregor et al. (2002)
Glucose	*E. coli*	1.3		Créach et al. (1997)
Glucose	*Vibrio parahaemolyticus*	1.3		Créach et al. (1997)
Glucose	*V. alginolyticus*	1.4		Créach et al. (1997)
Glucose	*Pseudomonas aeruginosa*	−3.0		Coffin et al. (1990)
Glucose	*Nostoc* sp. (strain Mac)	−1.5		Ingram et al. (1973)
N-acetyl-glucosamine	*V. harveyi*	4.1		Macko and Estep (1984)

$\Delta_{S-B} = \delta^{13}C_{Substrate} - \delta^{13}C_{Biomass}$
[a]Relative to wood cellulose
[b]Leachate
[c]Mixed populations
[d]Some incorporation of malt extract

polymeric carbohydrates or complex substrates, with efficiencies of 10–25% (Lekkerkerk et al. 1990; also calculated from Kohzu et al. 1999). Metabolic processes during growth of microbes on simple sugars such as glucose and sucrose are probably similar to metabolic processes of mycorrhizal fungi growing symbiotically, and therefore, similar isotopic fractionations during metabolism should also be expected.

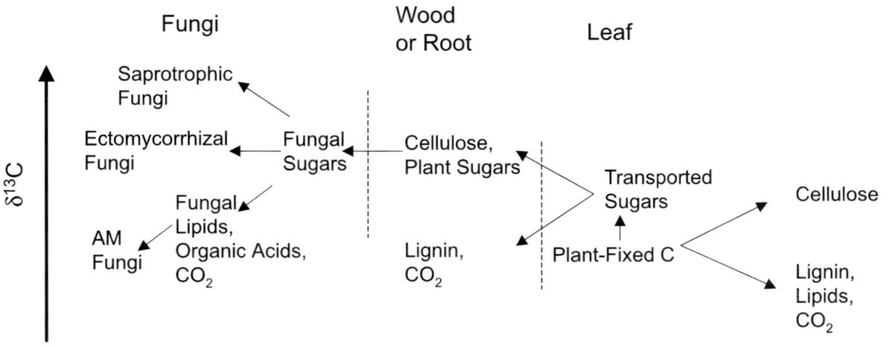

Fig. 4. Movement and isotopic fractionation of carbon isotopes in different compounds and components of plants, mycorrhizal fungi, and saprotrophic fungi. Separate pathways are indicated for arbuscular mycorrhizal (*AM*) fungi and ectomycorrhizal fungi. Isotopic fractionation along *nonhorizontal arrows* is 1–2‰

In Fig. 4, I propose a scheme of carbon isotope movement among plants, mycorrhizal fungi, and saprotrophic fungi that accounts for current observations from field and culture studies. An important contributing factor for many isotopic patterns is the enrichment in ^{13}C of carbohydrates relative to other compound classes such as lignins and lipids. Because carbohydrate polymers are abundant in plants and possess a regular structure that is relatively amenable to enzymatic attack, they are the main carbon sources for most saprotrophic fungi, ensuring that saprotrophic fungi will be enriched in ^{13}C relative to their substrates. In addition, since both plants and fungi transport carbon primarily as ^{13}C-enriched sugars, tissues such as roots, wood, and fungal fruiting bodies apparently become increasingly enriched in ^{13}C as the stream of labile carbohydrates becomes increasingly metabolized. Many of the biochemical mechanisms causing isotopic fractionation among different compounds are still unknown. Gleixner et al. (1993) proposed that isotopic fractionation during triose interconversions or by aldolase during cleavage of hexose to triose controlled the enrichment in ^{13}C of fungal carbohydrates relative to cellulose. Because a "futile" cycle of hexose to triose interconversions operates in plants (Hill et al. 1995), this mechanism could conceivably contribute to increases in ^{13}C content of carbohydrates with increasing distance from the "source" photosynthate. Since lipids are important transport compounds in AM fungi (Pfeffer et al. 1999), ^{13}C enrichment along transport pathways of these fungi appears less likely than in ectomycorrhizal fungi. Patterns of δ^{13}C in AM or ectomycorrhizal fungi relative to host-supplied sugars may accordingly reflect whether carbohydrates or lipids are the primary storage compounds used for later biosynthesis, as already shown for heterotrophically produced leaves derived from lipid- versus starch-storing seeds (Luo and Sternberg 1994).

2.1.2 Nitrogen Isotopes

The close coupling between carbon and nitrogen cycling in terrestrial ecosystems (Ågren and Bosatta 1996), the key role of mycorrhizal fungi in plant N supply, and the key role of saprotrophic fungi in decomposition and nutrient mineralization have prompted several studies of nitrogen isotope patterns in fungi and other ecosystem components in the last 10 years (Gebauer and Dietrich 1993). In many studies, fruiting bodies of saprotrophic and mycorrhizal fungi differ in $\delta^{15}N$ as well as $\delta^{13}C$ (Hobbie et al. 1999; Kohzu et al. 1999; Henn and Chapela 2001), with mycorrhizal fungi usually enriched in ^{15}N relative to saprotrophic fungi. However, $\delta^{15}N$ patterns exhibit less fidelity within and between life history strategies than $\delta^{13}C$, with both differences in source $\delta^{15}N$ and differences in internal processing of N probably controlling the $\delta^{15}N$ of fungal N.

Mycorrhizal fungi differ greatly in $\delta^{15}N$ values depending on taxa. Lilleskov et al. (2002) compared patterns of $\delta^{15}N$ and protein use in different mycorrhizal genera and concluded that those with greater proteolytic capabilities generally were higher in $\delta^{15}N$ than those without such capabilities. To date, no studies have measured the $\delta^{15}N$ of free amino acids in soil. Because peptide hydrolysis fractionates against ^{15}N about 4‰ (Silfer et al. 1992), it appears probable that free amino acids may be somewhat depleted in ^{15}N relative to bulk proteinaceous material in soils. In addition, mineral N is generally depleted in ^{15}N relative to bulk soil (Högberg 1997), so source $\delta^{15}N$ probably influences differences among mycorrhizal taxa in $\delta^{15}N$.

Differences in $\delta^{15}N$ of saprotrophic fungi may also be linked to source $\delta^{15}N$. Wood decay fungi, particularly those of high C:N ratios, often have $\delta^{15}N$ values similar to their substrate (Hobbie et al. 2001). Isotopic fractionation against ^{15}N during decomposition progressively enriches the remaining N in ^{15}N content, with humus often 5–10‰ higher in $\delta^{15}N$ than undegraded plant tissues. The $\delta^{15}N$ of saprotrophic fungi may accordingly directly reflect the enzymatic capabilities of fungi to access N in different plant and soil pools. This point must be considered with several caveats. First, metabolic processes show large isotopic fractionations in fungi, as evidenced by the consistent ^{15}N depletion in fungal chitin relative to protein of up to 10‰ (Taylor et al. 1997). Therefore, differences in chitin or protein concentrations in fruiting bodies could easily affect bulk $\delta^{15}N$. Second, tracer studies have not been done to show unequivocally which soil N pools saprotrophic or mycorrhizal fungi are using. And third, soil N pools of potentially different availabilities have in general not been isotopically characterized, making it difficult to link fungal $\delta^{15}N$ patterns to source $\delta^{15}N$.

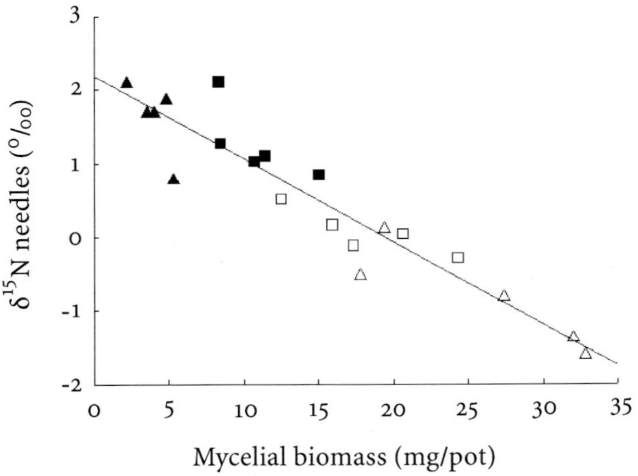

Fig. 5. Mycelial biomass correlates with foliar $\delta^{15}N$ in mycorrhizal *Pinus sylvestris*. Fungal biomass in perlite is calculated from ergosterol measurements and appropriate conversion factors for *Thelephora* or *Suillus*. $r^2 = 0.90$, $P < 0.001$. *Filled symbols* High N, *open symbols* low N; *triangles Suillus, squares Thelephora*. (Hobbie and Colpaert 2003)

Because the dominant plants in most terrestrial ecosystems are mycorrhizal, efforts to understand the controls over $\delta^{15}N$ patterns in plant cultures have recently focused on the potential role of mycorrhizal fungi. Ectomycorrhizal fungi are generally enriched in ^{15}N relative to host plants, indicating that fungi may alter the isotopic composition of nitrogen that they take up from soil and subsequently pass on to host plants (Högberg 1990; Schmidt and Stewart 1997; Hobbie et al. 1999). Several recent culture studies have confirmed that ectomycorrhizal fungi are enriched in ^{15}N relative to host pines (Högberg et al. 1999b; Kohzu et al. 2000; Hobbie and Colpaert 2003). Allocation of photosynthate to mycorrhizal fungi and foliar $\delta^{15}N$ were highly and negatively correlated (Hobbie and Colpaert 2003; Fig. 5), suggesting the potential use of foliar $\delta^{15}N$ measurements to indicate carbon allocation to mycorrhizal fungi. The form in which isotopically light N is transferred to plants is unclear in ectomycorrhizal fungi, but ^{15}N-depleted amino acids are probably created during transamination reactions and subsequently exported to host plants. This would result in mycorrhizal plants being depleted in ^{15}N relative to available N and mycorrhizal fungi being enriched in ^{15}N relative to available N.

Fractionation against ^{15}N associated with creation of transfer compounds in symbiotic microbes should influence plant and microbial $\delta^{15}N$ patterns. For example, the central role of glutamine in fungal amino acid metabolism suggests that isotopic fractionations associated with glutamine creation and transformation may control the large ^{15}N depletion between ectomycorrhizal fungi and plants. Glutamine is commonly invoked as a probable transfer compound of nitrogen between mycorrhizal fungi and plants (Smith and Smith 1990). The amido group of fungally derived glutamine may be the source N assimilated by ectomycorrhizal plants, and is the source for nitrogen in N-acetylglucosamine (Carlile et al. 2001), the monomer of the important fungal carbohydrate chitin. Chitin in fungi is depleted in ^{15}N relative to proteins and amino acids by about 9‰ (Taylor et al. 1997). Nodules in N_2-fixing *Rhizobium* are often enriched in ^{15}N relative to host legumes (discussed in Werner and Schmidt 2002), despite consistent evidence from enzymatic studies that fixation of atmospheric N_2 by nitrogenase has little isotopic fractionation (Werner and Schmidt 2002). Ammonia is known to be the transfer compound in N_2-fixing bacteria (Day et al. 2001), and is a suspected transfer compound in AM fungi (Bago et al. 2001). Although microbial symbionts of mycorrhizal and rhizobial plants use different transfer compounds, it appears probable that kinetic isotopic effects associated with movement of ammonia or amino groups can lead to similar ^{15}N depletions of host plants relative to symbionts in these two plant–microbial associations. Such a process would also lead to mycorrhizal fungi being ^{15}N-enriched relative to saprotrophic fungi.

2.2 Methane Cycling

The important trace gas methane is produced by certain bacteria under anaerobic conditions by two main microbial processes, acetate fermentation and carbon dioxide reduction in the presence of hydrogen. It is consumed during methane oxidation. Natural sources of methane include wetlands, rice paddies, termite mounds, and ruminant animals, whereas sinks include phototrophic methane oxidation in the atmosphere and biogenic methane oxidation in soils (Bréas et al. 2001). Because many estimates of the global budget of methane rely on isotopic measurements to partition fluxes between sources and sinks, a better understanding of isotopic fractionation during methane production and consumption could lead to improved estimates of methane budgets at both global and local scales. Methane production via acetate fermentation discriminates against ^{13}C by 24–27‰, whereas production through carbon dioxide reduction discriminates even more, 55–58‰ (Whiticar 1999). Methane oxidation in soils is accompanied by a smaller, but still relatively large, isotopic fractionation against ^{13}C of 4–27‰. Since these three processes also differ in their effects on the hydrogen isotope composition (δD) of methane (Whiticar 1999), comparison of methane $\delta^{13}C$ and δD values should allow researchers to partition methane cycling among these three processes.

2.3 Using Isotopic Differences Between C_3 and C_4 Photosynthetic Pathways to Probe Microbial Carbon Sources

Many studies of soil organic matter have used the isotopic difference between C_3 and C_4 photosynthetic pathways to study soil carbon cycling. This is essentially a low-resolution tracer study, with an isotopic difference of about 14‰ between C_3 and C_4 inputs. In soils planted with C_3 and C_4 grasses, Nakano et al. (1999) measured the $\delta^{13}C$ of spores of the AM fungus *Gigaspora gigantea* and showed that in mixed cultures, C_4 grasses contributed a greater proportion of fungal carbon than C_3 grasses. Ryan and Aravena (1994) studied the $\delta^{13}C$ of soil microbial biomass in a C_3 soil that had been planted with maize (a C_4 crop) for 5 years. They determined that 30% of soil microbial biomass in no-till treatments was derived from maize, against 55% in conventional till treatments. Presumably, tillage leads to faster incorporation belowground of recent carbon. A short-term labeling study with C_4-derived sucrose or ^{13}C-enriched glucose applied to soil was used to separate respiration by free-living microbes from respiration by plants and mycorrhizal fungi (Ekblad et al. 2002). The authors determined that microbial respiration was primarily limited by C supply, not nutrients, and that fractionation against ^{13}C during respiration was small (although no value was given). The $\delta^{13}C$ of soil microbial biomass has also been studied along gradients of C_3 to C_4 grasslands (Šantrùcková et al. 2000). Both soil organic matter and soil-respired CO_2 were about 2‰ depleted relative to soil microbial biomass. The authors suggested that loss of ^{13}C-depleted CO_2 in respiration and retention of ^{13}C-enriched microbial products contributed to the well-known progressive enrichment in ^{13}C with increasing depth in soil profiles. At this point it is unclear whether specific compound classes of microbial products differing in $\delta^{13}C$ are preferentially retained in soil organic matter.

2.4 Problems with Extrapolating Natural Abundance Cultures to the Field

When determining isotopic fractionation between microbial cultures and a single, simple carbon source (e.g., glucose or sucrose), microbial assimilation of complex mixtures added to the media must also be considered. For example, to ensure adequate growth it is common to add small amounts ($\sim$ 10%) of yeast extract or malt extract. These added compounds serve as potential carbon sources that are often unaccounted for in calculations of isotopic fractionation (e.g., see Will et al. 1986, 1989; Abraham et al.

1998; Henn and Chapela 2000; Zhang et al. 2002). Even supposedly inert substrates such as agar, a structural polymer of some algae consisting of galactose subunits linked by β-1,3- and α-1,4 bonds, can serve as carbon sources for some microbes (Rees et al. 1994), and may accordingly be unsuitable as substrates of precise studies of carbon nutrition (Carlile et al. 2001). In a study on agar using ^{13}C-enriched glucose to study metabolism in eight fungal taxa, several saprotrophic fungi were significantly depleted in ^{13}C relative to expected values, indicating that some fungi could obtain up to 45% of their carbon directly from agar itself (Hobbie et al. 2004). In studies of isotopic fractionation during growth on simple substrates, the degree of carbon incorporation from yeast extract, malt extract or agar can be checked through parallel culture experiments with ^{13}C-enriched substrates.

A related difficulty accompanies culture studies of nitrogen nutrition and δ^{15}N patterns. Because culture studies generally apply N at levels higher than in the field, substantial fractionation against ^{15}N on uptake and assimilation is possible (Handley and Raven 1992; Emmerton et al. 2001). Therefore, unless nitrogen supply is carefully limited or matched to uptake rates, δ^{15}N patterns in culture have uncertain relevance to field situations, where N supply rates are generally low. Fractionation against ^{15}N on uptake could also possibly affect microbial δ^{15}N in the field if nitrogen no longer limits microbial growth. For example, such fractionation was estimated at 9‰ from dramatic declines in algal δ^{15}N following nutrient addition in an Arctic stream (Peterson et al. 1993), and similar processes may partially account for wide variations in δ^{15}N of ectomycorrhizal fungi after anthropogenic nitrogen fertilization in an Alaskan forest (Lilleskov et al. 2002).

3
Compound-Specific Measurements and Isotopic Tracers

Few microorganisms, other than the subset of fungi that produce large fruiting bodies, can be directly sampled for bulk analyses of isotopic ratios. In addition, those fungi that do fruit tend to do so only during certain seasons in response to specific environmental conditions, such as sufficient soil moisture. Soil microbial ecologists, therefore, need techniques that can provide isotopic information about specific taxa or functional groups through in situ sampling of soil. Compound-specific measurements that can be linked to broad taxonomic groups are one possibility, including analyses of phospholipid fatty acids (Cifuentes and Salata 2001), fungal cell wall components such as N-acetylglucosamine (S. Frey, pers. comm.), or bacterial cell wall components such as muramic acid (S. Frey, pers. comm.),

D-alanine, and diaminopimelic acid (Pelz et al. 1998). Methods for the isotopic analyses of DNA or RNA have also been developed (Coffin et al. 1990), and could be applied at different levels of taxonomic resolution (MacGregor et al. 2002). Such analyses have been used in studies of microbial processes in sediments and aquatic systems, but have been used little in soils. RNA could be particularly useful since it degrades rapidly in the environment and reflects the component of microbial biomass that is actively synthesizing proteins. To date, these techniques have not supplied great insight into microbial processes because uncertainties in what controls isotopic fractionation at the microbial level have hindered the interpretation of isotopic patterns.

Several techniques using compound-specific measurements coupled to tracer studies have considerable promise for understanding the microbial impact on a variety of biogeochemical processes, such as methane oxidation (Bull et al. 2000; Radajewski et al. 2000), degradation of aromatic compounds (Johnsen et al. 2002), formation of fatty acids in soils (Lichtfouse et al. 1995), or in examining relative activities of fungal versus bacterial communities on ^{13}C-enriched substrates (Arao 1999). This general approach is reviewed in Jones and Bradford (2001). Nuclear magnetic resonance (NMR) studies also have great potential for new insights into metabolic processes, are nondestructive, and can provide position-specific information on labeling patterns within molecules that is extremely difficult to obtain using isotope ratio mass spectrometry. Such NMR studies have used ^{13}C-applied substrates in culture studies of soil microorganisms (Gaines et al. 1996), or directly in soil (Lundberg et al. 2001). In the latter study, ^{13}C-labeled glucose was traced for 28 days into solid-state components (NMR-invisible components of microbial biomass), respired CO_2, and triacylglycerols, with the triacylglycerols probably located in oil droplets within fungi.

4
Conclusions and Future Research

This review will, I hope, stimulate further applications of isotopic techniques to the roles of microbes in belowground processes. Several challenges remain. Our ability to interpret results from culture studies at natural abundance must be improved through better knowledge of the main metabolic pathways of microbes. Once that is accomplished, we can then extrapolate from culture studies to field studies with greater confidence than now possible. Ongoing work, much of it driven by the potential for microbes in industrial production of specific compounds, has demonstrated that the main metabolic fluxes within bacteria and fungi can be determined by using ^{13}C-labeled substrates and tracking ^{13}C through various metabolites using NMR techniques (Portais and Delort 2002). Such approaches are also proving fruitful in studies of carbon metabolism in symbioses between plants and AM fungi (Pfeffer et al. 2001). Similar experiments at natural abundance levels using mass spectrometry could firmly link specific enzymatic reactions to natural abundance patterns in key microbial metabolites, such as amino acids or lipids, and thereby allow researchers to move away from the correlative approaches used to date in field studies. Field studies are likely to increasingly rely on compound-specific measurements that can provide some taxonomic resolution, such as analyses of phospholipid fatty acids. Our ability to interpret isotopic patterns from such studies should also improve through paired natural abundance and tracer studies in cultures. Finally, promising efforts to combine isotopic tracers and computer modeling in field studies (Currie and Nadelhoffer 1999) could be expanded to computer models of ecosystem function that include key microbial processes (such as PnET-N-DNDC; Li et al. 2000), or expanded to models of soil food web functioning (Moore et al. 1996). Such modeling could be further extended to isotopic predictions at natural abundances. For example, by using the NESIS model (http://ecosystems.mbl.edu/Research/Models/nesis/welcome.html) developed by Ed Rastetter, isotopic predictions can be created for any model that includes elemental fluxes. Such isotopic predictions could allow researchers to test many hypotheses about the movement of isotopes through soil ecosystems, and thereby improve our understanding of fundamental soil processes driven by microbes.

Acknowledgements. This article was supported in part by the US National Science Foundation, award DEB-0235727. Earlier versions were greatly improved by insightful comments from Steve Trudell, John Hobbie, and Bruce Peterson. I thank Peter Högberg and Agneta Planboeck for permission to reproduce Fig. 2.

References

Abelson PH, Hoering TC (1961) Carbon isotope fractionation in formation of amino acids by photosynthetic organisms. Proc Natl Acad Sci USA 47:623–632

Abraham W-R, Hesse C, Pelz O (1998) Ratios of carbon isotopes in microbial lipids as an indicator of substrate usage. Appl Environ Microbiol 61:4202–4209

Ågren GI, Bosatta E (1996) Theoretical ecosystem ecology. Cambridge Univ Press, Cambridge

Arao T (1999) In situ detection of changes in soil bacterial and fungal activities by measuring ^{13}C incorporation into soil phospholipid fatty acids from ^{13}C acetate. Soil Biol Biochem 31:1015–1020

Bago B, Pfeffer P, Shachar-Hill Y (2001) Could the urea cycle be translocating nitrogen in the arbuscular mycorrhizal symbiosis? New Phytol 149:4–8

Benner R, Fogel ML, Sprague EK, Hodson RE (1987) Depletion of ^{13}C in lignin and its implications for stable carbon isotope studies. Nature 329:708–710

Bigeleisen J (1965) Chemistry of isotopes. Science 147:463–471

Blair N, Leu A, Munoz E, Olsen J, Kwong E, Des Marais D (1985) Carbon isotopic fractionation in heterotrophic microbial metabolism. Appl Environ Microbiol 50:996–1001

Boschker HTS, Middelburg JJ (2002) Stable isotopes and biomarkers in microbial ecology. FEMS Microbiol Ecol 40:85–95

Bréas O, Guillou C, Reniero F, Wada E (2001) The global methane cycle: isotopes and mixing ratios, sources and sinks. Isotopes Environ Health Stud 37:257–379

Brugnoli E, Farquhar GD (2000) Photosynthetic fractionation of carbon isotopes. In: Leegood RC, Sharkey TD, von Caemmerer S (eds) Photosynthesis: physiology and metabolism. Kluwer, Dordrecht, pp 399–434

Bull ID, Parekh NR, Hall GH, Ineson P, Evershed RP (2000) Detection and classification of atmospheric methane oxidising bacteria in soil. Nature 405:175–178

Carlile MJ, Watkinson SC, Gooday GW (2001) The fungi, 2nd edn. Academic Press, Boston

Cifuentes LA, Salata GG (2001) Significance of carbon isotope discrimination between bulk carbon and extracted phospholipid fatty acids in selected terrestrial and marine environments. Org Geochem 32:613–621

Coffin RB, Fry B, Peterson BJ, Wright RT (1989) Carbon isotopic compositions of estuarine bacteria. Limnol Oceanogr 34:1305–1310

Coffin R, Devereux R, Price W, Cifuentes L (1990) Analysis of stable isotopes of nucleic acids to trace sources of dissolved substrates used by estuarine bacteria. Appl Environ Microbiol 66:2012–2020

Créach V, Bertru G, Mariotti A (1997) Natural isotopic composition of heterotrophic bacteria to determine the origin of bioavailable dissolved organic carbon. CR Acad Sci 320:339–347

Currie WS, Nadelhoffer KJ (1999) Dynamic redistribution of isotopically labeled cohorts of nitrogen inputs in two temperate forests. Ecosystems 2:4–18

Dawson TE, Mambelli S, Plamboeck AH, Templer PH, Tu KP (2002) Stable isotopes in plant ecology. Annu Rev Ecol Syst 33:507–559

Day DA, Poole PS, Tyerman SD, Rosendahl L (2001) Ammonia and amino acid transport across symbiotic membranes in nitrogen-fixing legume nodules. Cell Mol Life Sci 58:61–71

Ekblad A, Nyberg G, Högberg P (2002) ^{13}C-discrimination during microbial respiration of added C_3-, C_4- and ^{13}C-labelled sugars to a C_3-forest soil. Oecologia 131: 245–249

Emmerton KS, Callaghan TV, Jones HE, Leake JR, Michelsen A, Read DJ (2001) Assimilation and isotopic fractionation of nitrogen by mycorrhizal fungi. New Phytol 151:503–511

Gaines GL III, Smith L, Neidle EL (1996) Novel nuclear magnetic resonance spectroscopy methods demonstrate preferential carbon source utilization by *Acetinobacter calcoaceticus*. J Bacteriol 178:6833–6841

Gebauer G, Dietrich P (1993) Nitrogen isotope ratios in different compartments of a mixed stand of spruce, larch and beech trees and of understorey vegetation including fungi. Isotopenpraxis 29:35–44

Gleixner G, Danier H-J, Werner RA, Schmidt H-L (1993) Correlations between the ^{13}C content of primary and secondary plant products in different cell compartments and that in decomposing basidiomycetes. Plant Physiol 102:1287–1290

Handley LL, Raven JA (1992) The use of natural abundance of nitrogen isotopes in plant physiology and ecology. Plant Cell Environ 15:965–985

Hayes JM (2002) Fractionation of the isotopes of carbon and hydrogen in biosynthetic processes. In: Valley JW, Cole DR (eds) Stable isotope geochemistry. Mineralogical Society of America and the Geochemical Society, Washington, DC, pp 225–277

Henn MR, Chapela IH (2000) Differential C isotope discrimination by fungi during decomposition of C_3- and C_4-derived sucrose. Appl Environ Microbiol 66:4180–4186

Henn MR, Chapela IH (2001) Ecophysiology of ^{13}C and ^{15}N isotopic fractionation in forest fungi and the roots of the saprotrophic-mycorrhizal divide. Oecologia 128:480–487

Hill SA, Waterhouse JS, Field EM, Switsur VR, ap Rees T (1995) Rapid recycling of triose phosphates in oak stem tissue. Plant Cell Environ 18:931–936

Hobbie EA, Colpaert JV (2003) Nitrogen availability and colonization by mycorrhizal fungi correlate with nitrogen isotope patterns in plants. New Phytol 157:115–126

Hobbie EA, Colpaert JV (2004) Nitrogen availability and mycorrhizal colonization influence water use efficiency and carbon isotope patterns in *Pinus sylvestris* L. New Phytol (in press)

Hobbie EA, Macko SA, Shugart HH (1999) Insights into nitrogen and carbon dynamics of ectomycorrhizal and saprotrophic fungi from isotopic evidence. Oecologia 118:353–360

Hobbie EA, Weber NS, Trappe JM (2001) Determining mycorrhizal or saprotrophic status of fungi from isotopic evidence: implications for element cycling and fungal evolution. New Phytol 150:601–610

Hobbie EA, Weber NS, Trappe JM, van Klinken GJ (2002) Using radiocarbon to determine mycorrhizal status in fungi. New Phytol 156:129–136

Hobbie EA, Watrud LS, Maggard S, Shiroyama S, Rygiewicz PT (2003) Carbohydrate use by litter and soil fungi assessed through stable isotopes and BIOLOG® assays. Soil Biol Biochem 35:303–311

Hobbie EA, Sanchez FS, Rygiewicz PT (2004) Carbon use, nitrogen use, and isotopic fractionation of ectomycorrhizal and saprotrophic fungi in natural abundance and ^{13}C-labelled cultures. Mycol Res 108:725–736

Hoefs J (1997) Stable isotope geochemistry, 4th edn. Springer, Berlin Heidelberg New York

Högberg P (1990) ^{15}N natural abundance as a possible marker of the ectomycorrhizal habit of trees in mixed African woodlands. New Phytol 115:483–486

Högberg P (1997) ^{15}N natural abundance in soil–plant systems. New Phytol 137:179–203

Högberg P, Plamboeck AH, Taylor AFS, Fransson PMA (1999a) Natural ^{13}C abundance reveals trophic status of fungi and host-origin of carbon in mycorrhizal fungi in mixed forests. Proc Natl Acad Sci USA 96:8534–8539

Högberg P, Högberg MN, Quist ME, Ekblad A, Näsholm T (1999b) Nitrogen isotope fractionation during nitrogen uptake by ectomycorrhizal and non-mycorrhizal *Pinus sylvestris*. New Phytol 142:569–576

Ingram LO, Calder JA, van Baalen C, Plucker FE, Parker PL (1973) Role of reduced exogenous organic compounds in the physiology of the blue-green bacteria (algae): photoheterotrophic growth of a "heterotrophic" blue-green bacterium. J Bacteriol 114:695–700

Jennings DH (1995) The physiology of fungal nutrition. Cambridge Univ Press, Cambridge

Johnsen AR, Winding A, Karlson U, Roslev P (2002) Linking of microorganisms to phenanthrene metabolism in soil by analysis of ^{13}C-labeled cell lipids. Appl Environ Microbiol 68:6106–6113

Jones TH, Bradford MA (2001) Assessing the functional implications of soil biodiversity in ecosystems. Ecol Res 16:845–858

Kohzu A, Yoshioka T, Ando T, Takahashi M, Koba K, Wada E (1999) Natural ^{13}C and ^{15}N abundance of field-collected fungi and their ecological implications. New Phytol 144:323–330

Kohzu A, Tateishi T, Yamada A, Koba K, Wada E (2000) Nitrogen isotope fractionation during nitrogen transport from ectomycorrhizal fungi, *Suillus granulatus*, to the host plant, *Pinus densiflora*. Soil Sci Plant Nutr 46:733–739

Lekkerkerk L, Lundkvist H, Ågren GI, Ekbohm G, Bosatta E (1990) Decomposition of heterogeneous substrates: an experimental investigation of a hypothesis on substrate and microbial properties. Soil Biol Biochem 22:161–167

Li C, Aber J, Stange F, Butterbach-Bahl K, Papen H (2000) A process-oriented model of N_2O and NO emissions from forest soils. 1. Model development. J Geophys Res 105:4369–4384

Lichtfouse E, Berthier G, Houot S, Barriuso E, Bergheaud V, Vallaeys T (1995) Stable carbon isotope evidence for the microbial origin of C_{14}-C_{18} n-alkanoic acids in soils. Org Geochem 23:849–852

Lilleskov EA, Hobbie EA, Fahey TJ (2002) Ectomycorrhizal fungal taxa differing in response to nitrogen deposition also differ in pure culture organic nitrogen use and natural abundance of nitrogen isotopes. New Phytol 154:219–231

Lundberg P, Ekblad A, Nilsson M (2001) C-13 NMR spectroscopy studies of forest soil microbial activity: glucose uptake and fatty acid biosynthesis. Soil Biol Biochem 33:621–632

Luo Y, Sternberg L (1994) Alterations in δ^{13}C values of seedling cellulose associated with respiration during germination. Phytochemistry 35:877–880

MacGregor BJ, Brüchert V, Fleischer S, Amann R (2002) Isolation of small-subunit rRNA for stable isotopic characterization. Environ Microbiol 4:451–464

Macko SA, Estep MLF (1984) Microbial alteration of stable nitrogen and carbon isotopic compositions of organic matter. Org Geochem 6:787–790

Monson KD, Hayes JM (1982) Carbon isotopic fractionation in the biosynthesis of bacterial fatty acids. Ozonolysis of unsaturated fatty acids as a means of determining the intramolecular distribution of carbon isotopes. Geochim Cosmochim Acta 46:139–149

Moore JC, de Ruiter PC, Hunt HW, Coleman DC, Freckman DW (1996) Microcosms and soil ecology: critical linkages between field studies and modelling food webs. Ecology 77:694–705

Nakano A, Takahashi K, Kimura M (1999) The carbon origin of arbuscular mycorrhizal fungi estimated from δ^{13}C values of individual spores. Mycorrhiza 9:141–147

Pelz O, Cifuentes LA, Hammer BT, Kelley CA, Coffin RB (1998) Tracing the assimilation of organic compounds using δ^{13}C analysis of unique amino acids in the bacterial peptidoglycan cell wall. FEMS Microbiol Ecol 25:229–240

Peterson B, Fry B, Deegan L, Hershey A (1993) The trophic significance of epilithic algal production in a fertilized tundra river ecosystem. Limnol Oceanogr 38:872–878

Pfeffer PE, Douds DD, Bécard G, Shachar-Hill Y (1999) Carbon uptake and the metabolism and transport of lipids in an arbuscular mycorrhiza. Plant Physiol 120:587–598

Pfeffer PE, Bago B, Shachar-Hill Y (2001) Exploring mycorrhizal function with NMR spectroscopy. New Phytol 150:534–553

Portais J-C, Delort A-M (2002) Carbohydrate cycling in micro-organisms: what can ^{13}C-NMR tell us? FEMS Microbiol Rev 26:375–402

Radajewski S, Ineson P, Parekh NR, Murrell JC (2000) Stable-isotope probing as a tool in microbial ecology. Nature 403:646–649

Rees GN, Rainey FA, Harfoot CG (1994) Characterization of a novel obligate anaerobe that ferments agar. Arch Microbiol 162:395–400

Ryan MC, Aravena R (1994) Combining ^{13}C natural abundance and fumigation-extraction methods to investigate soil microbial biomass turnover. Soil Biol Biochem 26:1583–1585

Šantrůčková H, Bird MI, Lloyd J (2000) Microbial processes and carbon-isotope fractionation in tropical and temperate grassland soils. Funct Ecol 14:108–114

Schmidt S, Stewart GR (1997) Waterlogging and fire impacts on nitrogen availability and utilization in a subtropical wet heathland (wallum). Plant Cell Environ 20:1231–1241

Silfer JA, Engel MH, Macko SA (1992) Kinetic fractionation of stable carbon and nitrogen isotopes during peptide bond hydrolysis: experimental evidence and geochemical implications. Chem Geol 101:211–221

Smith SE, Smith FA (1990) Structure and function of the interfaces in biotrophic symbioses as they relate to nutrient transport. New Phytol 114:1–38

Staddon PL, Robinson D, Graves JD, Fitter AH (1999) The $\delta^{13}C$ signature of the external phase of a *Glomus* mycorrhizal fungus: determination and implications. Soil Biol Biochem 31:1067–1070

Taylor AFS, Högbom L, Högberg M, Lyon AJE, Näsholm T, Högberg P (1997) Natural ^{15}N abundance in fruit bodies of ectomycorrhizal fungi from boreal forests. New Phytol 136:713–720

Werner RA, Schmidt H-L (2002) The in vivo nitrogen isotope discrimination among organic plant compounds. Phytochemistry 61:465–484

Whiticar MJ (1999) Carbon and hydrogen isotope systematics of bacterial formation and oxidation of methane. Chem Geol 161:291–314

Will OH, Tieszen LL, Kellen M, Gerlach T (1986) Stable carbon isotope discrimination in the smut fungus *Ustilago violacea*. Exp Mycol 10:83–88

Will OH, Tieszen LL, Gerlach T, Kellen M (1989) Alteration of carbon isotope ratios by eight *Ustilago* species on defined media. Bot Gaz 150:152–157

Zhang CL, Ye Q, Reysenbach A-L, Götz D, Peacock A, White DC, Horita J, Cole DR, Fong J, Pratt L, Fang J, Huang Y (2002) Carbon isotopic fractionations associated with thermophilic bacteria *Thermotoga maritima* and *Persephonella marina*. Environ Microbiol 4:58–64

Zyakun AM (1996) Stable carbon isotope discrimination by heterotrophic microorganisms. Appl Biochem Microbiol 32:153–159

Subject Index